Advances in Hydrogels

Advances in Hydrogels

Editors

Yang Liu
Kiat Hwa Chan

MDPI • Basel • Beijing • Wuhan • Barcelona • Belgrade • Manchester • Tokyo • Cluj • Tianjin

Editors
Yang Liu
School of Pharmacy
Hengyang Medical School
University of South China
Hengyang
China

Kiat Hwa Chan
Division of Science
Yale-NUS College
Singapore
Singapore

Editorial Office
MDPI
St. Alban-Anlage 66
4052 Basel, Switzerland

This is a reprint of articles from the Special Issue published online in the open access journal *Gels* (ISSN 2310-2861) (available at: www.mdpi.com/journal/gels/special_issues/advances_hydrogels).

For citation purposes, cite each article independently as indicated on the article page online and as indicated below:

LastName, A.A.; LastName, B.B.; LastName, C.C. Article Title. *Journal Name* **Year**, *Volume Number*, Page Range.

ISBN 978-3-0365-6124-0 (Hbk)
ISBN 978-3-0365-6123-3 (PDF)

© 2022 by the authors. Articles in this book are Open Access and distributed under the Creative Commons Attribution (CC BY) license, which allows users to download, copy and build upon published articles, as long as the author and publisher are properly credited, which ensures maximum dissemination and a wider impact of our publications.

The book as a whole is distributed by MDPI under the terms and conditions of the Creative Commons license CC BY-NC-ND.

Contents

About the Editors vii

Yang Liu
Editorial on Special Issue "Advances in Hydrogels"
Reprinted from: *Gels* 2022, 8, 787, doi:10.3390/gels8120787 1

Juan Liu, Chunyu Su, Yutong Chen, Shujing Tian, Chunxiu Lu and Wei Huang et al.
Current Understanding of the Applications of Photocrosslinked Hydrogels in Biomedical Engineering
Reprinted from: *Gels* 2022, 8, 216, doi:10.3390/gels8040216 5

José Luis Gradilla-Orozco, José Ángel Hernández-Jiménez, Oscar Robles-Vásquez, Jorge Alberto Cortes-Ortega, Maite Renteria-Urquiza and María Guadalupe Lomelí-Ramírez et al.
Physicomechanical and Morphological Characterization of Multi-Structured Potassium-Acrylate-Based Hydrogels
Reprinted from: *Gels* 2022, 8, 627, doi:10.3390/gels8100627 31

Tina Sabel-Grau, Arina Tyushina, Cigdem Babalik and Marga C. Lensen
UV-VIS Curable PEG Hydrogels for Biomedical Applications with Multifunctionality
Reprinted from: *Gels* 2022, 8, 164, doi:10.3390/gels8030164 45

Zhangkang Li, Cheng Yu, Hitendra Kumar, Xiao He, Qingye Lu and Huiyu Bai et al.
The Effect of Crosslinking Degree of Hydrogels on Hydrogel Adhesion
Reprinted from: *Gels* 2022, 8, 682, doi:10.3390/gels8100682 53

Richard Heger, Martin Kadlec, Monika Trudicova, Natalia Zinkovska, Jan Hajzler and Miloslav Pekar et al.
Novel Hydrogel Material with Tailored Internal Architecture Modified by "Bio" Amphiphilic Components—Design and Analysis by a Physico-Chemical Approach
Reprinted from: *Gels* 2022, 8, 115, doi:10.3390/gels8020115 63

Rubén D. Múnera-Tangarife, Efraín Solarte-Rodríguez, Carlos Vélez-Pasos and Claudia I. Ochoa-Martínez
Factors Affecting the Time and Process of CMC Drying Using Refractance Window or Conductive Hydro-Drying
Reprinted from: *Gels* 2021, 7, 257, doi:10.3390/gels7040257 87

Ion Călina, Maria Demeter, Anca Scărișoreanu and Marin Micutz
Development of Novel Superabsorbent Hybrid Hydrogels by E-Beam Crosslinking
Reprinted from: *Gels* 2021, 7, 189, doi:10.3390/gels7040189 103

Noam Y. Steinman and Abraham J. Domb
Instantaneous Degelling Thermoresponsive Hydrogel
Reprinted from: *Gels* 2021, 7, 169, doi:10.3390/gels7040169 121

Xiaoran Wang, Zizhen Yang, Zhaowei Meng and Shao-Kai Sun
Transforming Commercial Copper Sulfide into Injectable Hydrogels for Local Photothermal Therapy
Reprinted from: *Gels* 2022, 8, 319, doi:10.3390/gels8050319 133

Nutsarun Wangsawangrung, Chasuda Choipang, Sonthaya Chaiarwut, Pongpol Ekabutr, Orawan Suwantong and Piyachat Chuysinuan et al.
Quercetin/Hydroxypropyl--Cyclodextrin Inclusion Complex-Loaded Hydrogels for Accelerated Wound Healing
Reprinted from: *Gels* **2022**, *8*, 573, doi:10.3390/gels8090573 149

Yang Liu, Yu-Ning Luo, Pei Zhang, Wen-Fei Yang, Cai-Yao Zhang and Yu-Li Yin
The Preparation of Novel P(OEGMA-co-MEO$_2$MA) Microgels-Based Thermosensitive Hydrogel and Its Application in Three-Dimensional Cell Scaffold
Reprinted from: *Gels* **2022**, *8*, 313, doi:10.3390/gels8050313 167

Lihua Zou, Rong Ding, Xiaolei Li, Haohan Miao, Jingjing Xu and Guoqing Pan
Typical Fluorescent Sensors Exploiting Molecularly Imprinted Hydrogels for Environmentally and Medicinally Important Analytes Detection
Reprinted from: *Gels* **2021**, *7*, 67, doi:10.3390/gels7020067 179

Nidia Casas-Forero, Igor Trujillo-Mayol, Rommy N. Zúñiga, Guillermo Petzold and Patricio Orellana-Palma
Effects of Cryoconcentrated Blueberry Juice as Functional Ingredient for Preparation of Commercial Confectionary Hydrogels
Reprinted from: *Gels* **2022**, *8*, 217, doi:10.3390/gels8040217 191

Mohammad Zubair Ahmed, Anshul Gupta, Musarrat Husain Warsi, Ahmed M. Abdelhaleem Ali, Nazeer Hasan and Farhan J. Ahmad et al.
Nano Matrix Soft Confectionary for Oral Supplementation of Vitamin D: Stability and Sensory Analysis
Reprinted from: *Gels* **2022**, *8*, 250, doi:10.3390/gels8050250 211

Ayse Z. Sahan, Murat Baday and Chirag B. Patel
Biomimetic Hydrogels in the Study of Cancer Mechanobiology: Overview, Biomedical Applications, and Future Perspectives
Reprinted from: *Gels* **2022**, *8*, 496, doi:10.3390/gels8080496 227

Yosif Almoshari
Novel Hydrogels for Topical Applications: An Updated Comprehensive Review Based on Source
Reprinted from: *Gels* **2022**, *8*, 174, doi:10.3390/gels8030174 263

About the Editors

Yang Liu

Dr. Yang Liu received his B.E. degree in Chemistry from Zhejiang University in 2009 and received his Ph.D. degree in Polymer Chemistry and Physics from the Key Laboratory of Functional Polymer Materials, Nankai University, in 2014. Since then, he has been engaged in teaching and scientific research at the School of Pharmacy, Hengyang Medical School, University of South China. His research focuses on biomedical polymeric hydrogel materials.

Kiat Hwa Chan

Dr. Chan Kiat Hwa received his doctorate in Chemistry from Princeton University in 2010, under the auspices of an Agency for Science, Technology and Research (A*STAR) National Science Scholarship (PhD), after working on developing chemical tools to explore iron trafficking by Mycobacterium Tuberculosis in human macrophages. Thereafter, and prior to joining Yale-NUS College, he carried out postdoctoral research at the Institute of Bioengineering and Nanotechnology, A*STAR, where he worked on exploring the versatile properties of water-based peptide gels in biomaterial applications.

Editorial

Editorial on Special Issue "Advances in Hydrogels"

Yang Liu [1,2,3]

[1] School of Pharmacy, Hengyang Medical School, University of South China, Hengyang 421001, China; liuyanghxl@126.com; Tel.: +86-0734-8281296
[2] Hunan Provincial Key Laboratory of Tumor Microenvironment Responsive Drug Research, Hengyang 421001, China
[3] Hunan Province Cooperative Innovation Center for Molecular Target New Drug Study, Hengyang 421001, China

Hydrogels are a class of soft materials with crosslinked network structures. They show good biocompatibility, biodegradability, hydrophilicity, and mechanical properties similar to those of tissue, so they have a wide range of applications. In recent years, a variety of multifunctional hydrogels with excellent performance have been developed, greatly expanding the depth and breadth of their applications. This Special Issue focuses on the recent advances regarding hydrogels, aiming to provide reference for researchers in related fields. We have collected thirteen original research articles and three valuable reviews from thirteen different countries including Canada, China, Thailand, Mexico, India, Saudi Arabia, Chile, Germany, the Czech Republic, Colombia, Romania, Israel, and the USA.

Hydrogels can be prepared through different crosslinking methods. Photo-crosslinking has attracted much attention due to its advantages of mild preparation conditions, and convenient and simple operation. Liu and colleagues summarized the types of photo-crosslinked hydrogel monomers, the methods for the preparation of photo-crosslinked hydrogels with different morphologies, and their applications in biomedical engineering [1]. Gradilla-Orozco et al. prepared a multi-structured poly(potassium acrylate-co-acrylamide)-based hydrogel with a fractal-like structure via photo-crosslinking [2]. Sabel-Grau and co-workers developed a poly(ethylene glycol)-diacrylate (PEG-DA)-based photo-crosslinked hydrogel [3]. They used erythrosin B or eosin Y as the novel photoinitiator, which increased the multifunctionality of the prepared hydrogels. Different crosslinking methods and crosslinking degrees affect the properties of obtained hydrogels. To study the effect of the crosslinking degree on the adhesivity of hydrogels, Li et al. prepared polyvinyl alcohol (PVA), polyacrylamide (PAM) and polyvinyl alcohol-bearing styrylpyridinium group (PVA-SbQ) hydrogels through freeze–thaw cycles, thermal crosslinking and photo-crosslinking, respectively [4]. They found that the adhesion capability of these hydrogels decreased with an increase in the crosslinking degree. However, the adhesion could be improved by maintaining the adhesive functional groups and enhancing the flexibility of the polymer chains. Moreover, the synthesis of responsive hydrogels with excellent injectability and biodegradability is also attractive. Steinman and Domb synthesized a sensitive poly(ethylene glycol)-b-poly(lactic acid)-S-S-poly(lactic acid)-b-poly(ethylene glycol) (PEG-PLA-SS-PLA-PEG) copolymer by the grafting of PEG via urethane linkages, which could form a hydrogel above 32 °C, collapse immediately upon the reaction with $NaBH_4$, and be degraded slowly by hydrolytic degradation [5]. This hydrogel may be applied in drug-delivery vehicles with slow-release behavior and immediate-release behavior under the action of reducing agents.

According to the source, hydrogels can be divided into natural and synthetic hydrogels. Natural hydrogels generally have many potential sources, are low cost, and may show better biocompatibility, biodegradability, and other properties, which has attracted much attention. Heger and colleagues investigated the influence of a natural phospholipid lecithin (L-α-phosphatidylcholine) on three differently crosslinked hydrogels (physically

Citation: Liu, Y. Editorial on Special Issue "Advances in Hydrogels". *Gels* 2022, *8*, 787. https://doi.org/10.3390/gels8120787

Received: 26 November 2022
Accepted: 28 November 2022
Published: 30 November 2022

Publisher's Note: MDPI stays neutral with regard to jurisdictional claims in published maps and institutional affiliations.

Copyright: © 2022 by the author. Licensee MDPI, Basel, Switzerland. This article is an open access article distributed under the terms and conditions of the Creative Commons Attribution (CC BY) license (https://creativecommons.org/licenses/by/4.0/).

crosslinked agarose, ionically crosslinked alginate, and a chemically crosslinked mixture of PVA and chitosan) [6]. They found that lecithin could modify the internal architecture and mechanical properties of hydrogels. Múnera-Tangarife and co-workers prepared a natural carboxymethyl cellulose (CMC)-based transparent film with good barrier properties preventing the passage of oxygen and fats, and studied the factors affecting the drying process for CMC using refractance window-conductive hydro-drying (RW-CHD) [7]. Călina et al. also synthesized a series of superabsorbent hybrid hydrogel compositions constructed from natural xanthan gum (XG)/CMC/graphene oxide (GO) by e-beam radiation crosslinking [8].

With the deepening of research, various multifunctional hydrogels with excellent performance have played an increasingly important role in biomedical fields such as tumor treatment, cell scaffolds, wound repair, biological detection, the food industry and controlled drug release. Wang et al. developed a simple and powerful alginate–Ca^{2+} hydrogel (ACH) to load commercial copper sulfide (CuS) powders for local tumor NIR-II photothermal therapy (PTT) [9]. This hydrogel exhibited a good photothermal capacity and stability. Under 1064 nm NIR-II laser irradiation for 5 min, the temperature enhancement of hydrogels with different concentrations of CuS increased from 17.3 °C to 38.1 °C. These hydrogels had low cytotoxicity toward 4T1 cells, but the toxicity increased in a CuS-concentration- and laser-power-density-dependent manner under radiation. Animal experiments also showed that the large CuS particles in the hydrogel could very efficiently accumulate in tumor tissues and clearly suppress the tumors without causing evident inflammatory lesions or organ damage. Wangsawangrung and co-workers developed a quercetin/hydroxypropyl-β-cyclodextrin (HP-β-CD) inclusion complex-loaded polyvinyl alcohol (PVA) hydrogel that was physically crosslinked through multiple freeze–thaw cycles [10]. The quercetin/HP-β-CD inclusion complex was prepared via the solvent evaporation method, which showed significant antioxidant activity compared with free quercetin. An MTT assay also showed that the viability of mouse fibroblast NCTC 929 clone cells cultured with various concentrations of extracted media from these hydrogels was greater than 70%, indicating that these hydrogels showed low cytotoxicity. All the results illustrate that these hydrogels with the quercetin/HP-β-CD inclusion complex were attractive candidates for wound healing. We also constructed thermosensitive hydrogel scaffolds for cell culture [11]. These hydrogels were formed through the thermally induced gelation of thermosensitive microgels, which were prepared by the radical polymerization of 2-methyl-2-propenoic acid-2-(2-methoxyethoxy) ethyl ester (MEO_2MA) and oligoethylene glycol methyl ether methacrylate (OEGMA). The prepared thermosensitive hydrogels could be used for the in situ embedding and three-dimensional (3D) culture of MCF-7 breast cancer cells. The cells grew rapidly in the 3D scaffold and maintained a high proliferative capacity.

Molecularly imprinted polymeric hydrogel (MIPG)-based fluorescent sensors for commercial applications were developed by Zou and colleagues [12]. The MIPG for the detection of zearalenone (ZON) (MIPG_ZON) was first prepared using 4-vinylpyridine (4-VPY) as the functional monomer and ethylene glycol dimethacrylate (EGDMA) as the crosslinker for the ZON. This MIPG_ZON optical sensor showed excellent stability and reproducibility, and could detect ZON in commercial corn juice in a linear fashion across the concentration range 0–10 µM, with a limit of detection (LOD) of 1.6 µM. In parallel, a MIP-based fluorescent probe for the detection of glucuronic acid was also fabricated for cell imaging. Casas-Forero et al. prepared four types of commercial confectionary hydrogels through the introduction of a cryoconcentrated blueberry juice (CBJ) rich in polyphenols, anthocyanins, and flavonoids into a gelatin gel (GG), aerated gelatin gel (AGG), gummy (GM), and aerated gummy (AGM), respectively [13]. The structures of the hydrogels could protect the CBJ during in vitro digestion, which enhanced the bioaccessibility of the CBJ. Ahmed and colleagues also developed a type of nanogel matrix soft confectionary for the oral supplementation of vitamin D, which resulted in the enhancement of bioavailability, stability, and patient compliance [14]. Moreover, Sahan et al. highlighted the biomedical applications and future perspectives of biomimetic hydrogels in the study of cancer mechanobiology [15]. Additionally, novel hydrogels with new formulations for the topical administration of therapeutic agents were also reviewed by Almoshari [16].

The articles and reviews presented in this Special Issue offer a real insight into the advances regarding hydrogels. It is very clear that many novel multifunctional hydrogels will continue to emerge with bright application prospects from the efforts of researchers.

Funding: This research received no external funding.

Acknowledgments: The Guest Editors would like to thank to all the contributors to this Special Issue. Special thanks go to all the reviewers for helping us to ensure the quality of each published article; special thanks also go to the Editor in Chief and assistant editorial team of *Gels* for helping us to complete this work.

Conflicts of Interest: The authors declare no conflict of interest.

References

1. Liu, J.; Su, C.; Chen, Y.; Tian, S.; Lu, C.; Huang, W.; Lv, Q. Current Understanding of the Applications of Photocrosslinked Hydrogels in Biomedical Engineering. *Gels* **2022**, *8*, 216. [CrossRef] [PubMed]
2. Gradilla-Orozco, J.L.; Hernández-Jiménez, J.; Robles-Vásquez, O.; Cortes-Ortega, J.A.; Renteria-Urquiza, M.; Lomelí-Ramírez, M.G.; Rendón, J.G.T.; Jiménez-Amezcua, R.M.; García-Enriquez, S. Physicomechanical and Morphological Characterization of Multi-Structured Potassium-Acrylate-Based Hydrogels. *Gels* **2022**, *8*, 627. [CrossRef] [PubMed]
3. Sabel-Grau, T.; Tyushina, A.; Babalik, C.; Lensen, M.C. UV-VIS Curable PEG Hydrogels for Biomedical Applications with Multifunctionality. *Gels* **2022**, *8*, 164. [CrossRef] [PubMed]
4. Li, Z.; Yu, C.; Kumar, H.; He, X.; Lu, Q.; Bai, K.; Kim, K.; Hu, J. The Effect of Crosslinking Degree of Hydrogels on Hydrogel Adhesion. *Gels* **2022**, *8*, 682. [CrossRef] [PubMed]
5. Heger, R.; Kadlec, M.; Trudicova, M.; Zinkovska, N.; Hajzler, J.; Pekar, M.; Smilek, J. Novel Hydrogel Material with Tailored Internal Architecture Modified by "Bio" Amphiphilic Components—Design and Analysis by a Physico-Chemical Approach. *Gels* **2022**, *8*, 115. [CrossRef] [PubMed]
6. Múnera-Tangarife, R.D.; Solarte-Rodríguez, E.; Vélez-Pasos, C.; Ochoa-Martínez, C.I. Factors Affecting the Time and Process of CMC Drying Using Refractance Window or Conductive Hydro-Drying. *Gels* **2021**, *7*, 257. [CrossRef] [PubMed]
7. Călina, I.; Demeter, M.; Scărișoreanu, A.; Micutz, M. Development of Novel Superabsorbent Hybrid Hydrogels by E-Beam Crosslinking. *Gels* **2021**, *7*, 189. [CrossRef] [PubMed]
8. Steinman, N.Y.; Domb, A.J. Instantaneous Degelling Thermoresponsive Hydrogel. *Gels* **2021**, *7*, 169. [CrossRef] [PubMed]
9. Wang, X.; Yang, Z.; Meng, Z.; Sun, S.-K. Transforming Commercial Copper Sulfide into Injectable Hydrogels for Local Photothermal Therapy. *Gels* **2022**, *8*, 319. [CrossRef] [PubMed]
10. Wangsawangrung, N.; Choipang, C.; Chaiarwut, S.; Ekabutr, P.; Suwantong, O.; Chuysinuan, P.; Techasakul, S.; Supaphol, P. Quercetin/Hydroxypropyl-β-Cyclodextrin Inclusion Complex-Loaded Hydrogels for Accelerated Wound Healing. *Gels* **2022**, *8*, 573. [CrossRef] [PubMed]
11. Liu, Y.; Luo, Y.-N.; Zhang, P.; Yang, W.-F.; Zhang, C.-Y.; Yin, Y.-L. The Preparation of Novel P(OEGMA-co-MEO$_2$MA) Microgels-Based Thermosensitive Hydrogel and Its Application in Three-Dimensional Cell Scaffold. *Gels* **2022**, *8*, 313. [CrossRef] [PubMed]
12. Zou, L.; Ding, R.; Li, X.; Miao, H.; Xu, J.; Pan, G. Typical Fluorescent Sensors Exploiting Molecularly Imprinted Hydrogels for Environmentally and Medicinally Important Analytes Detection. *Gels* **2021**, *7*, 67. [CrossRef] [PubMed]
13. Casas-Forero, N.; Trujillo-Mayol, I.; Zúñiga, R.N.; Petzold, G.; Orellana-Palma, P. Effects of Cryoconcentrated Blueberry Juice as Functional Ingredient for Preparation of Commercial Confectionary Hydrogels. *Gels* **2022**, *8*, 217. [CrossRef] [PubMed]
14. Ahmed, M.Z.; Gupta, A.; Warsi, M.H.; Ali, A.M.A.; Hasan, N.; Ahmad, F.J.; Zafar, A.; Jain, G.K. Nano Matrix Soft Confectionary for Oral Supplementation of Vitamin D: Stability and Sensory Analysis. *Gels* **2022**, *8*, 250. [CrossRef] [PubMed]
15. Sahan, A.Z.; Baday, M.; Patel, C.B. Biomimetic Hydrogels in the Study of Cancer Mechanobiology: Overview, Biomedical Applications, and Future Perspectives. *Gels* **2022**, *8*, 496. [CrossRef] [PubMed]
16. Almoshari, Y. Novel Hydrogels for Topical Applications: An Updated Comprehensive Review Based on Source. *Gels* **2022**, *8*, 174. [CrossRef] [PubMed]

Review

Current Understanding of the Applications of Photocrosslinked Hydrogels in Biomedical Engineering

Juan Liu [1], Chunyu Su [1], Yutong Chen [1], Shujing Tian [1], Chunxiu Lu [1], Wei Huang [1,*] and Qizhuang Lv [1,2,*]

[1] College of Biology & Pharmacy, Yulin Normal University, Yulin 537000, China; liujuan5269@163.com (J.L.); suchunyu10@163.com (C.S.); c2922095838@163.com (Y.C.); tian011028@163.com (S.T.); lu18934908250@163.com (C.L.)
[2] Guangxi Key Laboratory of Agricultural Resources Chemistry and Biotechnology, Yulin 537000, China
* Correspondence: huangwei@ylu.edu.cn (W.H.); lvqizhuang062@163.com (Q.L.)

Abstract: Hydrogel materials have great application value in biomedical engineering. Among them, photocrosslinked hydrogels have attracted much attention due to their variety and simple convenient preparation methods. Here, we provide a systematic review of the biomedical-engineering applications of photocrosslinked hydrogels. First, we introduce the types of photocrosslinked hydrogel monomers, and the methods for preparation of photocrosslinked hydrogels with different morphologies are summarized. Subsequently, various biomedical applications of photocrosslinked hydrogels are reviewed. Finally, some shortcomings and development directions for photocrosslinked hydrogels are considered and proposed. This paper is designed to give researchers in related fields a systematic understanding of photocrosslinked hydrogels and provide inspiration to seek new development directions for studies of photocrosslinked hydrogels or related materials.

Keywords: water gel; photocrosslinking; synthetic polymer; natural polymer modification; biomedical-engineering applications

1. Introduction

Hydrogels are crosslinked networks of polymers. They have excellent hydrophilicity and can absorb large amounts of water or tissue fluid. At the same time, their volumes can expand to thousands of times that of the anhydrous state [1]. Water absorption and water retention of hydrogels are closely related to the molecular structure of crosslinking network, hydrophilicity, and crosslinking degree of monomers. If the crosslinking network structure is too tight, its water absorption will be reduced. The higher the hydrophilicity of the monomer structure, the better the hydroscopicity. Too high a crosslinking degree will lead to a dense structure of hydrogel, thus reducing water absorption. When the water content is within a certain range, hydrogels have softness and a rubbery consistency similar to that of living tissue, which demonstrates their excellent biocompatibility for cells and tissues [2]. Therefore, hydrogels have great application value in biomedical engineering.

In hydrogels, crosslinking is the key to avoiding dissolution of hydrophilic-polymer chains or segments. Hydrogels can be divided into physical and chemical hydrogels according to differences in crosslinking modes between polymers. In physically crosslinked hydrogels, polymers usually form three-dimensional network structures through hydrogen bonding, ionic bonding, hydrophobic bonding, chain entanglement, microcrystal formations, electrostatic interactions, etc. [3]. Such crosslinking is relatively simple and convenient, but the resulting network structure is not uniform, the mechanical strength is poor, and the crosslinking is usually reversible. Physical gels can subsequently be degraded by changes in temperature, pH, or ionic strength in the environment. This limits their use in complex internal environments. In contrast, chemical crosslinking that usually occurs by covalent bonds is usually irreversible and thus stable under changing conditions. In addition, hydrogels obtained by chemical crosslinking generally have better mechanical stability [4]. Therefore, chemically crosslinked hydrogels are more common in biomedical applications.

There are many methods for chemically crosslinking hydrogels, such as crosslinking polymerization of complementary functional groups, enzyme-induced polymerization, photo- or heat-induced free-radical polymerization, and high-energy irradiation-crosslinking polymerization [5]. Among them, the conditions of crosslinking polymerization with complementary functional groups and enzyme-induced polymerization are mild, and unnecessary functional molecules are not introduced, so they are the two preferred crosslinking methods [6,7]. However, in practical applications, the small molecules and macromolecules suitable for these two crosslinking methods are relatively limited, so there are few hydrogels prepared by using these two polymerization methods [8]. In contrast, light- or heat-induced radical polymerization is a common method in chemical crosslinking. In these methods, initiators must be added to a solution of crosslinked molecules to induce cracking into free radicals with light or heat, attacking the polymer chain withthe crosslinking molecules and initiating a chain reaction to complete polymerization [9]. Therefore, photo- or heat-induced free-radical polymerization is especially widely used because of its simplicity and speed.

With the development of cross-disciplines and deepening of scientific research, an increasing number of photocrosslinked hydrogels have been developed and applied in biomedical engineering. It is therefore necessary to give a summary and a review of the subject, as well as a glimpse into possible future development prospects. First, we introduce the compositions of photocrosslinked hydrogels, including synthetic monomers and modified monomers based on natural materials. Subsequently, we summarize hydrogels with different morphologies, such as films, fibers, microspheres, microneedles, and amorphous (injectable) hydrogels based on these monomers. Then, we summarize recent applications of these hydrogels in biomedical engineering. Finally, based on current development status, we put forward views on the development prospects forphotocrosslinked hydrogels in biomedical engineering. Overall, we hope that this review will give researchers a better understanding of this field and promote further development.

2. Chemically Synthesized Molecules for Preparation of Photocrosslinked Hydrogels

To prepare photocrosslinked hydrogels exhibitinggood biocompatibility, many polymerizable small molecules have been synthesized. These small molecules usually have carbon–carbon unsaturated bonds, and the polymer network is formed by bond breaking and addition reaction under initiator and light. Depending on the type of polymerizable group, these molecules can be roughly divided into four classes: ethylene, acrylic, acrylamide, and acrylate (Table 1).

Table 1. Four different photocrosslinked molecules and their molecular structures.

Functional Group Category	Single Structural Formula	Polymer Structural Formula
Vinyl	$H_2C=C{<}^{R_1}_{R_2}$	$(-H_2C-\underset{R_2}{\overset{R_1}{C}}-)_n$
Acrylic class	$R-HC=CH-\overset{O}{\underset{}{C}}-OH$	$(-HC-\underset{H}{\overset{R}{\underset{}{C}}}-)_n$ with $C=O$, OH

Table 1. Cont.

Functional Group Category	Single Structural Formula	Polymer Structural Formula
Acrylamide	R—HC=CH—C(=O)—NH$_2$	$\left(-HC(R)-C(H)(C(=O)NH_2)-\right)_n$
Acrylates	R$_1$—HC=CH—C(=O)—OR$_2$	$\left(-HC(R_1)-C(H)(C(=O)OR_2)-\right)_n$

2.1. Ethylenes

Structurally, ethylene is the simplest functional group among polymerizable molecules. Photopolymerization of ethylene molecules was also the earliest method discovered and applied. In early studies, styrene and other monomers were mainly used to synthesize resins due to their poor water solubility, and they were usually polymerized by high-energy radiation or heat [10]. At present, the vinyl molecules used in hydrogel preparations are mainly N-vinyl pyrrolidone (NVP). NVP itself has low viscosity, high reactivity, and limited skin irritation [11], so it is widely used as a typical ethylene molecule [12]. Kao et al. first reported NVP photocrosslinking with four different comonomers to prepare a series of UV-curable bioadhesives with a high water uptake ranging from 25 to 350 wt% [13]. Subsequently, Fechine et al. studied the effects of crosslinking NVP with other polymerizable molecules such as hydrogen peroxide [14,15]. Lee and Devine et al. provided a detailed discussion on the network structure of copolymerized NVP and polyethylene glycol diacrylate (PEGDA) hydrogels and found that the molecular weight of the main chain was mainly related to the NVP content and was not affected by polymerization time [16,17]. In addition to NVP, a variety of other ethylene molecules have been reported for photocrosslinking polymerizations. For example, Sahiner et al. successfully prepared photocrosslinked bulk polyethylene phosphonic acid (PVPA) by mixing and crosslinking PEGDAs with different molecular weights [18,19]. Ren et al. synthesized a hydrogen-bonded calcium-crosslinked PVDT-PAA hydrogel from 2-vinyl-4,6-diamino-1,3,5-triazine (VDT), acrylic acid (AA), and PEGDA through one-step photopolymerization [20]. However, due to the restriction of water solubility and reactivity, vinyl monomers are still rarely used in the preparation of photocrosslinked hydrogels [21,22].

2.2. Acrylic Acid

Acrylic-acid (AA) molecules have photoactive groups and good water solubility, and due to the free carboxyl structure remaining after polymerization, they can swell or shrink with environmental pH changes, electric fields, enzyme reactions, and temperature, so that the hydrogel exhibits a tunable response. In addition, these free carboxyl groups can interact with other groups under certain conditions, so the hydrogel network can be functionalized by reactions with carboxyl groups or interactions with solutes [23–26]. Therefore, photocrosslinked hydrogels based on AA molecules are very common and are often used for biological adsorption and environmental purification. For example, Liu et al. successfully synthesized a b-cyclodextrin or polyacrylic-acid nanocomposite (b-CD/PAA/GO) grafted with graphene oxide (GO) based on polyacrylic acid (PAA) through an esterification reaction, and prepared a composite PAA hydrogel, with an adsorption capacity of up to 248 mg/g for dye molecules in wastewater [27]. Hu et al. prepared a new PAA hydrogel by improving and optimizing the crosslinking agent used in the AA polymerization process, and the adsorption capacity for dye molecules reached a record-breaking 2100 mg/g under neutral conditions [28]. Ma and Kong et al. combined PAA hydrogels with organic montmorillonite or GO, and the resulting composite hydrogel showed good adsorption capacities for lead ions (Pb^{2+}) with an adsorption capacity of 223.84 mg/g [29] and cadmium ions (Cd^{2+})

with maximum adsorption capacity up to 316.4 mg/g [30]. In addition, PAA hydrogels can be used in the construction of flexible devices due to their good biocompatibility and flexibility. Based on this, Lu et al. combined PAA with nanocellulose to prepare a flexible hydrogel that can be used for skin sensing [31]. Clearly photocrosslinked hydrogels based on AA derivatives show a wide range of applications.

2.3. Acrylamide

Unlike liquid AA, acrylamide (AAm) molecules are usually solid and exhibit good water solubility. The AAm molecule does not have a free carboxyl group, but instead has a neutral amide bond, so polyacrylamide (PAAm) hydrogels do not exhibit pH responsiveness but show an equilibrium water content in the range of 94.73–96.26 wt%. Thanks to this, PAAm hydrogels maintain stability in environments with variable solutes [32]. However, the amide bond can also be partially hydrolyzed into a carboxyl group under certain conditions, and the PAAm hydrogel can function as a partial PAA hydrogel. For example, Wang et al. used this principle to prepare enzyme-functionalized microspheres for detection and cleaning of objects [33].

If a hydrophobic isopropyl group is introduced to the other end of the acrylamide molecule, isopropyl acrylamide (NIPAm) is obtained. After photocrosslinking, the resulting polyisopropyl acrylamide (PNIPAm) has a critical phase-transition temperature. When the temperature of the PNIPAm hydrogel is increased above the critical phase-transition temperature, the volume of the hydrogel shrinks significantly, and vice versa. It is worth mentioning that the critical phase-transition temperature of PNIPAm hydrogels is 32 °C and its low critical-dissolution temperature is close to the physiological temperature, it has important application value in biomedical engineering [34]. Zhang et al. prepared PNIPAm-hydrogel microspheres loaded with drugs, which shrank at physiological temperatures and released the loaded drugs for wound repair and disease treatment [35]. When PNIPAm hydrogel is combined with a substance susceptible to photothermal conversions, the new material exhibits photoresponsiveness, which can be used to deter counterfeiting [36,37].

2.4. Acrylates

The reaction of an acrylic derivative (acrylic or methacrylic acid) with the terminal hydroxyl group of another molecule yields an acrylate that can be used for photocrosslinking. For example, polyethylene glycol acrylate (PEGMA), polyethylene glycol diacrylate (PEGDA), polyethylene glycol dimethacrylate (PEGDMA) and other molecules that were obtained after modification at the end of polyethylene glycol (PEG) can be photocrosslinked to form the corresponding polymer network. Compared with acrylic acid and acrylamide hydrogels, acrylate hydrogels are not sensitive to changes in environmental temperature, pH, or other conditions, and are not convenient for functional-group modification; when the surrounding environment changes, this kind of hydrogel maintains good stability. In addition, some acrylate hydrogels, such as PEGDA hydrogels, show good water absorption (about 60 wt%), good biocompatibility, and good adhesion resistance [38]. Therefore, this kind of hydrogel can be used in biomedicine to construct a stable hydrogel skeleton for direct contact with cells or human tissues, biological analyses, and other applications [39]. Hou et al. prepared photocrosslinked fibers composed of PEGDA with different molecular weights and studied the conversion of acrylate bonds in the hydrogel and the mechanical properties of the fibers in detail. Hou et al. prepared a hydrogel microsphere based on a PEGDA hydrogel and realized the detection of glycoprotein molecules in the solution [40].

3. Chemically Modified Natural Materials

Compared with synthetic materials, natural materials exist widely in nature and have incomparable advantages in sourcing and storage. In addition, natural materials usually have good biocompatibility; they are safer than synthetic materials for use in the biomedical field. At present, hydrogels prepared from natural materials play important roles in cell culture, tissue engineering, and other fields. However, natural materials usually do not have photocrosslinkable groups. In addition to a few complementary functional-group reactions, crosslinking based on natural materials usually relies on physical crosslinking. In practice, the mechanical strength and stability are poor. Therefore, researchers have proposed various strategies for modifying natural materials. The introduction of photocrosslinking groups makes the preparation of natural material hydrogels simpler and more convenient, solves the problems of mechanical strength and poor stability, and further expands their applications. Natural materials used for modification and preparation of crosslinked hydrogels are mainly divided into two categories depending on their composition: polysaccharides and proteins or peptides.

3.1. Polysaccharides

Polysaccharides are carbohydrates with complex and large molecular structures formed by dehydration and condensation of multiple monosaccharide molecules. They are widely distributed in nature and play important roles [41]. For example, peptidoglycan and cellulose are components of the cytoskeletal structure of plants and animals, and starch and glycogen are important energy-storage materials used by animals and plants. The monosaccharides that make up various polysaccharides often have free functional groups; there are many active groups on the polysaccharide chain that can be modified to allow photocrosslinking. Common modified crosslinked hydrogels include alginate, hyaluronic acid, chitosan, heparin, chondroitin sulfate, gellan gum, cyclodextrin, and dextran, among others.

Alginate is a natural polysaccharide that can form a physical hydrogel by chelating its independent hydroxyl group with divalent and trivalent metal and heavy-metal ions. It has wide application value for drug delivery, wound repair, and tissue engineering [42,43]. However, physically crosslinked alginate hydrogels have limitations in practical applications due to their uncontrollable gluing speed and instability after gelation. Therefore, many researchers have modified alginate to graft photocrosslinking groups, and realize chemical crosslinking [44]. For example, Xu et al. chemically grafted aminopropyl vinyl ether after activating the carboxyl groups of sodium alginate and obtained alginic acid functionalized with vinyl ethers (Figure 1a) [45]. Photoinitiators can affect the crosslinking of hydrogels under UV irradiation. Bukhair et al. obtained cinnamoyl-modified photocrosslinkable alginic acid via multistep modification and demonstrated that these crosslinked alginate hydrogels have better mechanical properties and stabilities, as well as biocompatibility with physically crosslinked alginate hydrogels due to the formation of cyclobutane bridges connecting the alginate polysaccharide chains through the $(2\pi + 2\pi)$ cycloaddition reaction of the inserted cinnamoyl moieties [46]. They have great potential in biomedical applications.

Hyaluronic acid is also a linear macromolecular mucopolysaccharide. It has unique viscoelasticity, excellent water retention, biocompatibility, and nonimmunogenicity. It also has important physiological and biological functions and is widely used in clinical procedures [47,48]. However, natural hyaluronic acid has poor stability, is sensitive to hyaluronidase, and lacks mechanical strength. Therefore, chemical modifications are needed to prevent degradation and improve its mechanical strength, and the modified groups mainly include carboxyl, hydroxyl, and the amino group exposed by deacetylation. Lee et al. used Pluronic F127 to modify hyaluronic acid. The resulting polymers exhibited thermosensitive sol–gel transition behaviors over the temperature range of 20–40 °C. After modifying the functional groups of N-(3-dimethylamino propyl)methacrylamide, the HA-F127 polymer was polymerized to form a stable photocrosslinked hydrogel (Figure 1b) [49]. Jenjob et al. prepared microspheres by photocrosslinking bisphosphonates (alendronate) with methyl methacrylate-modified hyaluronic acid. The adsorption efficiency of the microspheres for cationic bone morphogenetic protein 2 (BMP2) reached 91.0% [50].

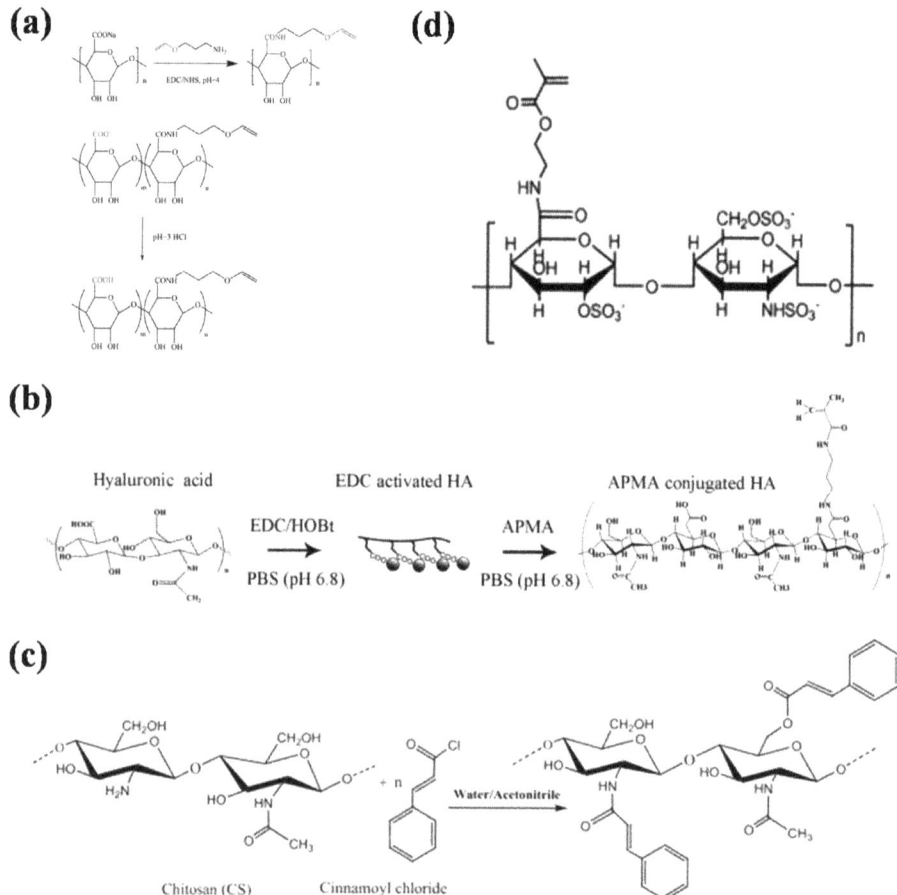

Figure 1. Synthetic route map or structural formula of the photocrosslinked hydrogel. (**a**) Synthetic route for aminopropyl vinyl ether-modified alginic acid [45]. (**b**) Synthetic route to N-(3-aminopropyl) methylacrylamide-modified hyaluronic acid [49]. (**c**) Synthetic roadmap for cinnamoyl chloride chloride-modified [51]. (**d**) Structural formula of heparin modified by methyl acrylate [52].

Chitosan is formed by removing some acetyl groups from the natural polysaccharide chitin. Its chemical structure is similar to that of hyaluronic acid. It is the only cationic polymer in nature [53]. It shows good biodegradability, biocompatibility, bacteriostasis, and other functions. Chitosan is usually insoluble in water and alkali solutions and needs to be dissolved with acid, so chitosan solutions are usually acidic. In addition, the commonly used gelation method for chitosan involves use of aldehyde-containing molecules (such as glutaraldehyde) as crosslinking agents to react with the amino groups of chitosan to form a Schiff base, leading to gelation [34]. In biomedical applications, acidic materials cause irritation to skin or wounds, while residual crosslinking agents have strong biological toxicity. To overcome these shortcomings, Zhou et al. modified chitosan with ethylene groups; the material was dissolved in water and mixed with polyvinyl alcohol modified by methacrylic acid. Photocrosslinked hydrogels were formed after a photoinitiator was added [54]. Monier et al. modified chitosan molecules with cinnamyl chloride. The crosslinked chitosan hydrogel showed good stability (Figure 1c) [51].

Heparin is a mucopolysaccharide sulfate composed of glucosamine, L-ieduraldehyde glycoside, N-acetylglucosamine, and D-glucuronic acid. It is strongly acidic. It is a natural anticoagulant in animals. Heparin specifically binds to growth factors, so it has important value in growth-factor-delivery systems [55,56]. Based on this, Yoon et al. modified heparin with N-methylacrylamide hydrochloride and then crosslinked it with cryloyl-modified Pluronic F127 to obtain composite hydrogels for the controlled release of growth factors [57]. Jeong et al. successfully prepared electrospun fiber scaffolds by mixing methyl acrylate-modified alginate and methyl acrylate-modified heparin with polyoxyethylene (PEO) through electrospinning and photocrosslinking, and found that they were effective inregulating cell behavior (Figure 1d) [52].

Chondroitin sulfate is a glycosaminoglycan covalently linked to proteins to form proteoglycans. It is widely distributed in the extracellular matrix and cell surfaces of animal tissues. The sugar chain is polymerized by alternating glucuronic acid and N-acetylgalactosamine disaccharides. It relieves pain and promotes cartilage regeneration [58]. After chemical modification, a chondroitin sulfate hydrogel can be obtained through photocrosslinking. The hydrogel has good biocompatibility and biodegradability, and has broad application prospects in drug delivery and bone repair [59,60]. For example, Kim and others prepared methacrylated PEGDA or chondroitin sulfate hydrogel and used it as a biomineralized three-dimensional scaffold for binding and deposition of charged ions [61]. Ornell et al. used the photocrosslinked methacryl group to covalently modify chondroitin sulfate and form injectable hydrogels, which can be used for sustained release of drugs over a certain period of time (Figure 2a) [62].

Gellan gum is a linear polysaccharide composed of glucose, glucuronic acid, and rhamnose. It is heat-resistant, acid-resistant, and enzyme-resistant and has good chemical stability. Gellan gum is insoluble in nonpolar organic solvents and cold water, but can be dissolved in hot water to form a transparent solution. After cooling, it becomes a transparent and solid gel [63–65]. Because of its good biocompatibility and tunability, researchers have tried to use gellan glue for tissue engineering. However, the gel formed by cations is hard and brittle, which limits its application. For this reason, a variety of photocrosslinking groups were used to modify gellan gum and obtain hydrogels with good biocompatibility and mechanical properties. For example, Oliveiraet al. modified gellan gum with trans-4-aminophenyl pyridine, and the product can be used for catalase immobilization after photocrosslinking (Figure 2b) [66]. Mano et al. modified gellan gum with methacrylic acid to prepare injectable gellan gum, which can be used as a self-generated osteogenic material [67].

Cyclodextrins are a series of cyclic oligosaccharides produced with amylose under the action of cyclodextrin glucosyltransferase. The three common cyclodextrins contain 6, 7, and 8 glucose units and arecalled α-, β-, and γ-cyclodextrins, respectively. These molecules are particularly attractive because they can form inclusion complexes with hydrophobic guests exhibiting appropriate molecular sizes and have high versatility. In addition, they can be chemically modified by hydroxyl substitution [68,69]. For example, Cosola et al. prepared a cyclodextrin modified with polyacrylate, which can prepare photocrosslinked materials with different morphologies via 3D technology [70]. Yamasaki et al. modified cyclodextrin with isophorone diisocyanate and 2-hydroxyethyl acrylate to obtain photocrosslinked cyclodextrin [71]. Microspheres of the prepared photocrosslinked polymer showed good separation efficiency for phenol (Figure 2c) [72].

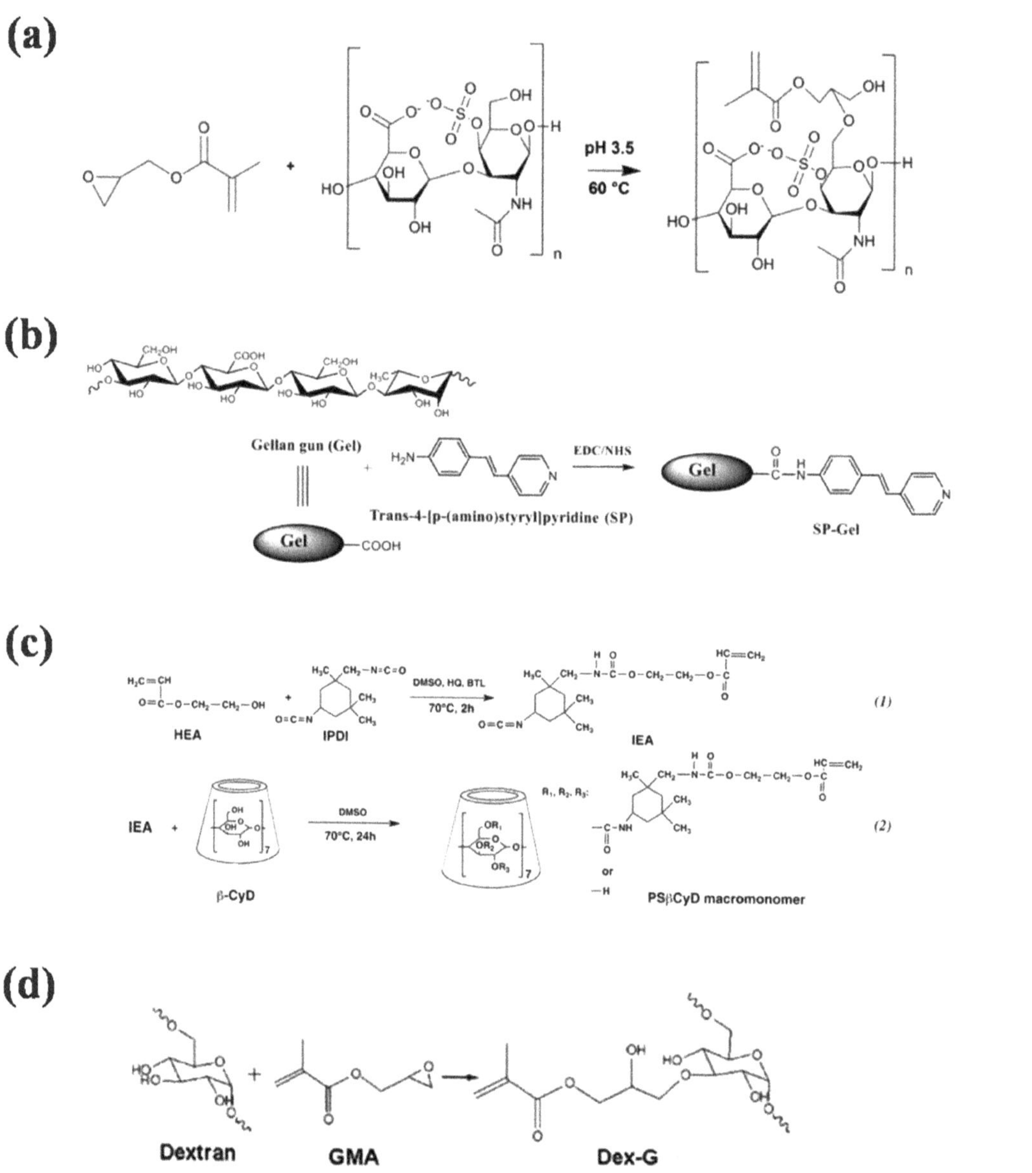

Figure 2. Synthetic routes for some crosslinked hydrogels. (**a**) Route map for the synthesis of chondroitin methacryloyl sulfate [62]. (**b**) Synthetic route for gellan gum modified by trans-4-aminophenyl pyridine [66]. (**c**) Synthetic route to cyclodextrin modified by isophorone diisocyanate and 2-hydroxyethyl acrylate [72]. (**d**) Synthetic route to glycidyl methacrylate modified dextran [73].

Glucan is a homopolysaccharide formed with glycosidic linkages between glucose. Based on the types of glycosidic bonds, it can be divided into α-glucan and β-glucan. The common glucan is dextran, a common α-glucan. It has good biocompatibility and is widely used in the biomedical field [74]. Dextran is rich in active hydroxyl functional groups, so it can be chemically modified. Casadei et al. modified it with methacrylate to obtain photocrosslinked dextran, which enables the controllable release of polymers [75].

Yin et al. used glycidyl methacrylate to modify dextran and crosslinked it with methacrylic acid ethylene glycol-modified concanavalin A and polyethylene glycol dimethacrylateto obtain hydrogels (Figure 2d) [73]. The hydrogelsshowed good biocompatibility and glucose responsiveness and are expected to be used in glucose biosensors and intelligent insulin delivery.

3.2. Proteins/Peptides

Polypeptides are compounds composed of α-amino acids connected by peptide bonds. They are also intermediate products in protein hydrolysis. Amino acids form polypeptides via dehydration condensation. After winding and folding, polypeptides can form macromolecules with certain spatial structures, namely proteins. Proteins and peptides play important roles in the growth and development of life. Because of their widespread sources, nontoxic degradation products, and even benefits to the human body, proteins and polypeptide products are widely used in the biomedical field. Because the amino acids that make up proteins and peptides have free side-chain groups, they can be endowed with new functions through chemical modification, such as photocrosslinking groups. Common proteins used for modification with photocrosslinking groups include collagen, gelatin, keratin, silk fibroin, and albumin. Their application scope is greatly expanded through chemical modification, which gives them wide influence in the biomedical field.

Collagen is a protein formed by three peptide super-chain helices; it accounts for 25–30% of all proteins in mammals and is the most abundant protein [76,77]. Common collagens include type I, type II, type III, type V, and type XI. Collagen is widely used in foods, medicines, tissue engineering, cosmetics, and other fields because of its good biocompatibility, biodegradability, and biological activity [78]. Photocrosslinking of collagen hydrogel can be realized with chemical modifications and photocrosslinking. The degradation rate after the collapse of collagen hydrogels is slow, and they can be maintained for a long time to achieve specific functions. For example, Yang et al. photocrosslinked methacrylated type II collagen and used it for encapsulation of bone-marrow mesenchymal stem cells (BMSCs) (Figure 3a). The crosslinked collagen maintained its triple-helix structure, which provides a good microenvironment for proliferation and differentiationof BMSCs [79].

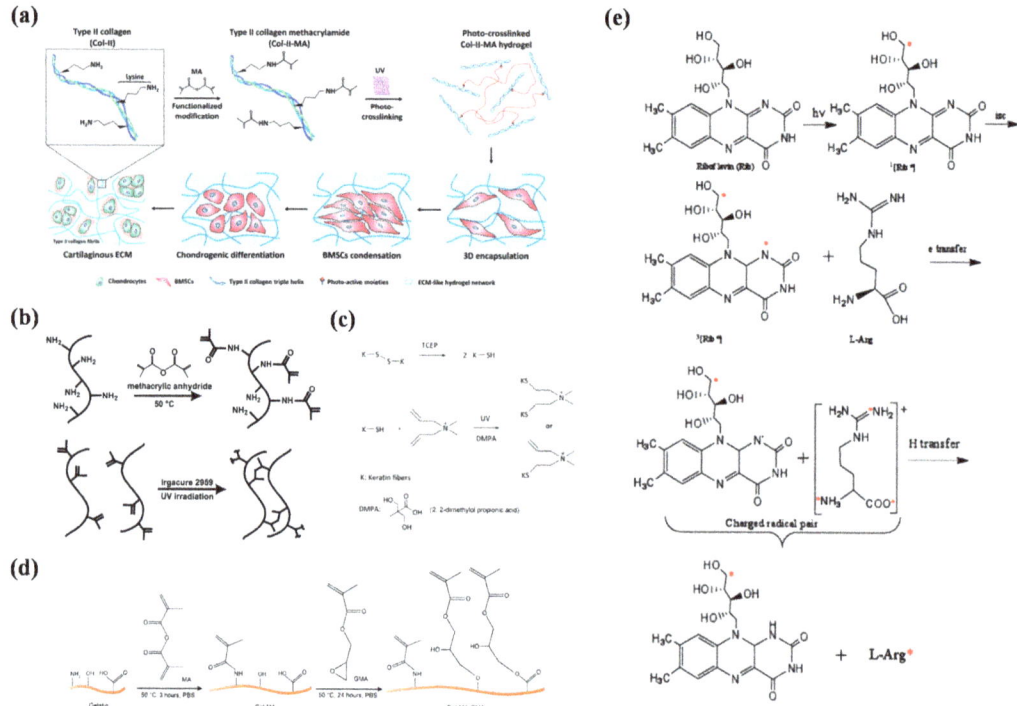

Figure 3. Part of the synthetic route maps or related schematic diagrams for crosslinked hydrogels. (**a**) Chemical modification of type II collagen and photocrosslinking hydrogel for bone-marrow mesenchymal stem-cell culture [79]. (**b**) Synthetic route of methacryloylated gelatin and its gelling diagram [80]. (**c**) Synthesis roadmap of two-step modified silk fibroin [81]. (**d**) Schematic diagram for crosslinking between keratin and 2,2-dimethoxy-2-phenylacetophenone [82]. (**e**) Schematic diagram of free-radical formation in the rib/L-Arg system [83].

Gelatineis formed by partial hydrolysis of collagen, and it can be dissolved in hot water and forms a gel after cooling. This excellent feature provides flexible application scenarios. However, in most cases, a more stable hydrogel structure is desired, and the temperature range for gelatine gels is relatively narrow, which makes it difficult to meet the demand. Therefore, researchers have modified gelatinewith methacryloyl to enable photocrosslinking. Methacryloylated gelatine (GelMA) has good biocompatibility and structural stability after photocrosslinking. Therefore, methacryloylated gelatine shows great application value in the biomedical field [84]. The Khademhosseini team used gelatineas a raw material to synthesize GelMA, and used it in cell culture and microchips to verify its application value in cell-responsive microengineered hydrogels (Figure 3b) [80]. Fu et al. prepared structural colored microspheres with GelMA and used them for cell culture. They found that they had good biocompatibility and are expected to be used to construct liver chips [85].

Keratin is the main protein constituting hair, horn, claws, and the outer layers of animal skin. Because keratin contains cystine, it has a high proportion of disulfide bonds and plays acrosslinking role in protein peptide chains. Therefore, keratin has particularly stable chemical properties and high mechanical strength [86,87]. Studies have shown that keratin extracted from human hair fibers contains a leucine–aspartate–valine (LDV) cell-adhesion motif. Therefore, keratin has great application potential for cell culture and tissue engineering. To provide more application scenarios for keratin, it can be modified.

For example, Yu and Hu et al. successfully realized photocrosslinking of keratin by using click chemistry and introducing 2,2-dimethoxy-2-phenylacetophenone (Figure 3c) [81,88].

Silk fibroin is a natural high-molecular fibrin extracted from silk. It has good mechanical and physicochemical properties and a long application history. To improve plasticity, many researchers have tried a variety of methods to modify it chemically (Figure 3d) [82]. Qi and others used silk fibroin modified by methacryl groups to obtain an injectable silk-fibroin hydrogel after photocrosslinking [89]. However, the lower side chains of silk fibroin decreased the content of methacryl groups, so the photocrosslinked silk-fibroin hydrogel had lower mechanical strength. Therefore, it was necessary to modify other active groups on silk fibroin to improve the mechanical strength. For example, Ju et al. obtained silk-fibroin hydrogels with good biocompatibility and mechanical properties by modifying the silk-fibroin hydroxyl with methacrylic acid and then performing crosslinking [90].

Albumin is a protein in plasma that maintains body nutrition and osmotic pressure, and it accounts for approximately 50% of all plasma proteins. It has good biocompatibility and solubility. Therefore, when preparing biomedical materials with albumin as a raw material, the inherent defects of synthetic materials can be avoided. For example, Chiriac and collaborators used riboflavin and arginine as natural initiators to crosslink bovine serum albumin (BSA) and obtain BSA hydrogels (Figure 3e) [83]. In vitro and in vivo experiments showed that the hydrogel had good biocompatibility.

4. Preparation of Photocrosslinked Hydrogels with Different Morphologies

In biomedical applications, different applications have different morphological requirements for common hydrogels. Therefore, it is crucial to prepare a hydrogel with a specific morphology. Common photocrosslinked hydrogel morphologies mainly include fibers, microspheres, thin films, microneedles, amorphous shapes, and so on. At present, many researchers have effectively explored processing methods for hydrogels with different morphologies, which has greatly expanded the application potential of hydrogels in biomedicine.

4.1. Fiber

Fibrous products such as gauze are widely used in the life science and biomedical fields. Inspired by this, many researchers have developed new hydrogel-fiber products. Due to their high surface-to-volume ratios and high porosities, they exhibit good water absorbance and air permeability and can replace some functions, such as those of traditional gauze in the biomedical field [91,92]. Usually, a hydrogel precursor solution is converted into fibers, photocrosslinked, and finally accumulated in woven-fiber products. At present, the main method used for fiber preparation is electrospinning [93–95]. When the solution is squeezed out under a high-pressure electric field, the liquid becomes filamentous due to the electric field. After the solvent evaporates, the polymer, polymer mixture, composite materials, and other molecules form fiber shapes. Electrospun fibers have many remarkable properties, such as high surface-to-volume ratios and functional tunability; they can be applicable in many areas including drug delivery, wound dressings, tissue engineering, membranes or filters, electronics, sensors, and energy [92]. Furthermore, when photocrosslinking technology is applied, the mechanical performance and stability of the fiber can be improved further [96,97]. For example, Tang et al. developed a novel high-throughput nanofiber-composite ultrafiltration membrane (Figure 4a) [98]. They first prepared chemically crosslinked polyvinyl-alcohol (PVA) nanofiber scaffolds on a nonwoven substrate, and then prepared a polyvinyl-alcohol (UV-PVA) barrier layer via UV crosslinking. The results showed that the 5 wt% UV-PVA solution coating provided an ultrafiltration membrane with high throughput and high retention rate after UV curing for 20 s. It had good pollution resistance and could be used for the separation of oil and water emulsions.

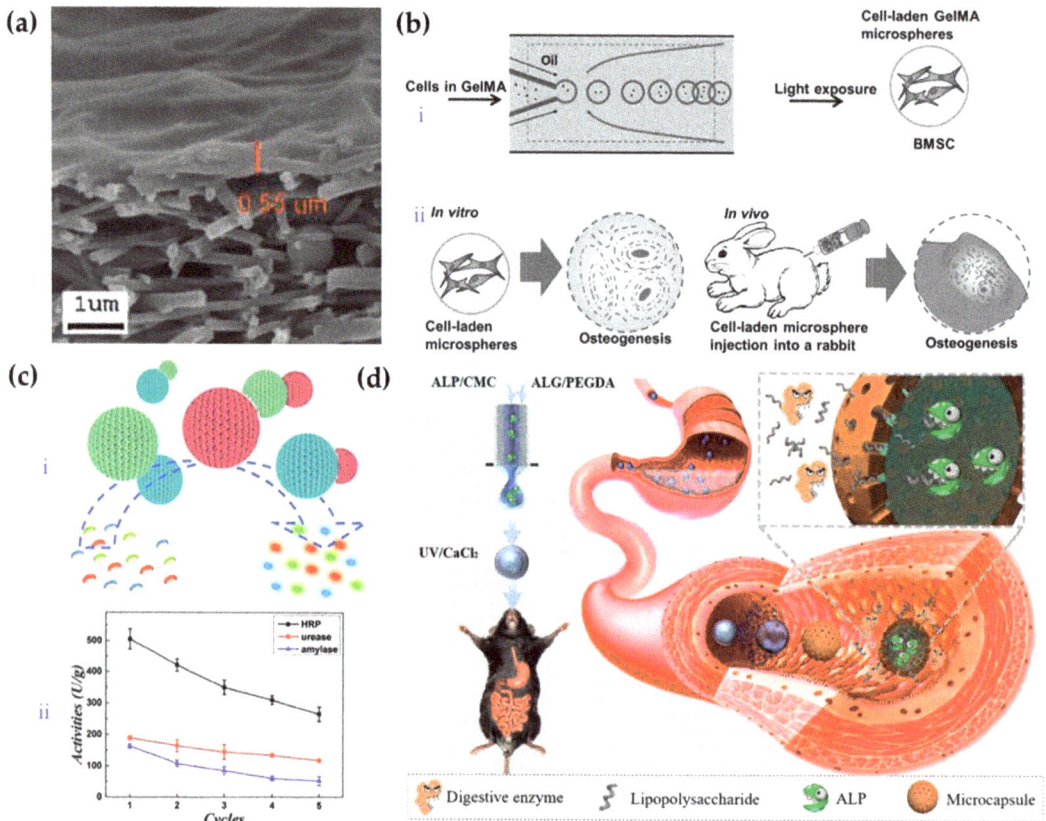

Figure 4. Schematic diagram of the preparation and application of hydrogels with fiber and microspherical morphologies. (**a**) Electron micrograph of ahigh-flux nanofiber-composite ultrafiltration membrane [98]. (**b**) Schematic diagram of the preparation and application of photocrosslinked GelMA-hydrogel microsphere [99]. (**c**) Enzyme-functionalized antiproteolytic-hydrogel microspheres were used for biocatalysis [100]. (**d**) Schematic diagram for preparation and application of ALP microcapsules [101].

Microspherical hydrogels have excellent mobility and mass delivery, and thus have received considerable attention in fields such as biomedical detection and drug delivery. Specific structural differences can be used to subdivide the micropellets into homogeneous micropellets, antiopal micropellets [32], and nuclear-shell micropellets (microcapsules) [101].

4.2. Microballoons

4.2.1. Homogeneous Micropellets

Homogenous hydrogel microspheres are usually formed by photocrosslinking after the hydrogel precursor solution is dispersed into liquid droplets. There are many ways to disperse solutions into droplets, but the principle is basically consistent; all methods use emulsification to generate droplets, such as with stirring, microfluidics, and electrospraying. Among these techniques, microfluidic technology has obvious advantages in preparing microspheres due to its accurate fluid control [99,102]. For example, Zhao et al. obtained hydrogel microspheres loaded with cells and growth factors by mixing cells and growth factors with GelMA solution, using microfluidics to cut them into droplets, and then photocrosslinking them. BMSCs in microspheres showed significant osteogenic effects both in vitro and in vivo, and significantly increased mineralization, thus promoting bone regeneration (Figure 4b) [99]. Combining microfluidic technology with electrojet technology allowed the application of an electric field as an additional shear force to simplify the microfluidic device. For example, Zhang et al. successfully used microfluidic electroinjection technology to prepare photocrosslinked chondroitin-sulfate microspheres, which showed good value for drug loading and wound repair [103].

4.2.2. Antiopal Micropellets

Opal has a periodic, ordered structure formed by nanoparticles. When the nanoparticles have a specific size, the opal reflects light of a specific wavelength and thus shows a structural color. Negative replication using opal as a template yields an antiopal material. This material has the same structural color characteristics and a continuous porous structure, which provide good prospects for applications in mass transfer (especially in the fields of drug loading and sensing) [100]. For example, Wang and others prepared inverse-opal hydrogel microspheres by using colloidal-crystal microspheres assembled with silicon-dioxide nanoparticles as templates and AAm and N,N'-methylenebis-(acrylamide) (Bis) as photocrosslinked hydrogel skeletons to copy the template microspheres [32]. Afterhydrolysis and enzymatic immobilization of the microspheres, the resulting material had biocatalytic functionality and is expected to be used for treatment of complex water bodies (Figure 4c).

4.2.3. Nuclear-Shell Microspheres (Microcapsules)

Antiopal hydrogel microspheres were obtained by using corrosion for removal of nanoparticle templates. If the corrosion was incomplete, the prepared microspheres had a nuclear-shell structure consisting of a hydrogel or nanoparticle-composite core and an antiopal hydrogel shell. These nuclear-shell microspheres exhibit responsiveness similar to that of antiopal hydrogel microspheres, and they also have high stability and can be used for encoding [104]. Based on this, Xu et al. developed photocrosslinked nuclear-shell hydrogel microspheres that can be used for miRNA detection by modifying the corresponding aptamer [105].

Homogeneous hydrogel microspheres are generally prepared by single-emulsion devices. Core-shell hydrogel microcapsules can be prepared if a double emulsion device is used and the inner solute is not crosslinked. For example, Zhao et al. developed microcapsules containing an alkaline phosphatase (ALP) solution by applying microfluidic electrojet technology. The shells of the microcapsule were formed by calcium alginate and photocrosslinked PEGDA, which can preserve the activity of ALP in the digestive tract and thus be used for intestinal endotoxin cleaning (Figure 4d) [101].

4.3. Thin Film

Among the many morphologies of photocrosslinked hydrogels, hydrogels with thin-film morphologies are most widely used because of their excellent size ranges and mechanical strength. Thin films are usually made of homogeneous structures, and by adding functional materials, hydrogel functionality can be enhanced in various ways. For example, incorporation of mesoporous active nanoparticles or a vascular endothelial growth factor can allow the hydrogel membrane to play an obvious promoting role in bone and wound healing [106,107]. Using the opal structure as the template, a hydrogel thin film with the antiopal structure was prepared. The nanoporosityand structure color of the antiopal structure enabled enhanced drug loading, sensing, and driving functions for hydrogels, withgood application prospects [37,108]. For example, Zhang et al. constructed photoresponsive antiopal hydrogel films that can be driven to capture and release objects under the action of light (Figure 5a) [109]. In some cases, two sides of the hydrogel film can be made to have different properties and functions by using different functionalization treatments to obtain a Janus structure thin film. Such films show good results when applied in complex environments. Zhang et al. prepared a Janus sponge dressing that was hydrophilic on one side and hydrophobic on the other. When the dressing was applied to the wound, the exudate was discharged from the wound to maintain the microenvironment of the wound, thus accelerating wound repair (Figure 5b) [110].

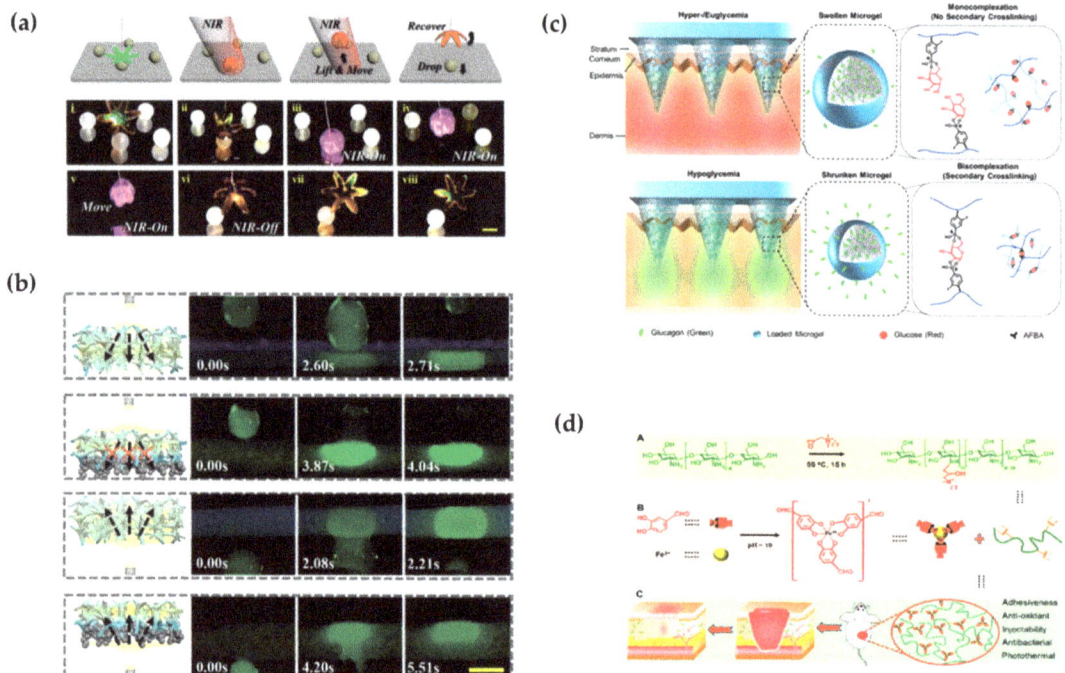

Figure 5. Schematic diagram for application of thin films, microneedles, and injectable morphologies of hydrogels. (**a**) A light-responsive antiopal hydrogel thin film captures and releases objects under light [109]. (**b**) One-way liquid discharge capacity of Janus sponge dressings [110]. (**c**) Schematic representation of the intelligent release of glucagon drugs contained by microneedles in vivo [111]. (**d**) Preparation principle for an injectable hydrogel with a double dynamic covalent bond and schematic diagram of its application in wound healing [112].

4.4. Microacupuncture Needle

Microneedles are usually arrays of microneedles, typically less than 1mm in length. Due to their size, they only penetrate the surface of the skin without causing bleeding; they cause much less pain than a traditional needle, and the trauma can heal in hours. Therefore, microneedles have obvious advantages in transdermal administration [113–115]. Compared with traditional metal- and silicon-based microneedles, hydrogel microneedles have good biocompatibility, large drug loads, and controllable drug release, and they have great development prospects. For example, Ghavaminejad et al. prepared microgel-loaded glucagon using phenylboric acid as a functional group, and then mixed the microgel with methacrylated hyaluronic acid (MeHA) and added it to the template [111]. After photocrosslinking, a photocrosslinked hydrogel microneedle was obtained. This microneedle can release glucagon to raise blood glucose under hypoglycemic conditions, thus preventing hypoglycemia after insulin injections for patients with diabetes (Figure 5c). The mechanical properties of the microneedles can also be controlled by adjusting the morphology. For example, Yu et al. designedan amifostine-loaded armored microneedle AAMN to havestronger mechanical properties than traditional conical microneedles, much higher mechanical strength than conical structuresand high skin permeability, and they can be better used for transdermal administration of amifostine [115].

4.5. Amorphous (Injectable) Hydrogels

Some hydrogels have poor mechanical properties due to limited high-molecular content or minimal crosslinking between large molecules. However, although these hydrogels do not maintain their specific fine structures and appearance, they can be used to cover irregular wounds and can be injected into irregularly shaped defect sites, which provides unique advantages in medical trauma repair. Therefore, such hydrogels have great application value in biomedical engineering. Liang et al. modified chitosan with quaternary ammonium moieties and formed an injectable hydrogel with trivalent iron, a protocatechuic aldehyde containing catechol and aldehyde groups. It can be used for wound healing and healing ofinfectionswith methicillin-resistant Staphylococcus aureus (Figure 5d) [112]. In addition, other functional materials can be added to injectable hydrogels to meet specific needs. For example, Zhao et al. combined drug-loaded hydrogel microspheres with injectable hydrogels to successfully prepare injectable hydrogels that were used for diabetes treatment [116].

5. Biomedical Applications

Photocrosslinked hydrogels are widely used in biomedicine, such as for biosensors, flexible wearable devices, medicine or tissue engineering, cell microcarriers, organ chips, and so on. It should be noted that when constructing hydrogels, appropriate components should be selected according to the application scenarios. For example, when a hydrogel is used for sensing, it should exhibit antiadhesion properties. When applied directly to humans, the hydrogel components should be biocompatible and preferably adaptable.

5.1. Biomedical Sensor

The components of hydrogels usually contain various functional groups, which can combine with molecules or ions in the surrounding environment and cause changes in the physical and chemical properties of hydrogels. By detecting these changes, the corresponding molecules or ions can be analyzed and sensed. For example, Qin et al. prepared hydrogel microspheres with a Janus structure, which changed its volume when the pH of the environment was changed; this led to changes in the intensitiesof surface-enhanced Raman scattering (SERS) signals and fluorescence signal. By collecting and analyzing these two signals, the pH of the solution could be determined (Figure 6a) [117].

When a hydrogel has an ordered micro/nanostructure, it can show structural color, which will change with volume changes of the hydrogel. Therefore, structural color can be used to detect and analyzethe target molecule. The Sun research group has performed a series of studies in this area. They used microspheres or membranes self-assembled with colloidal nanoparticles of silicon wafers as templates, used hydrogels for repeated preparation, and prepared a series of structurally colored hydrogel microspheres and films with different functions. These microspheres showed good sensing ability for detection of tumor markers, glycoproteins, DNA, and heavy-metal ions (Figure 6b) [118].

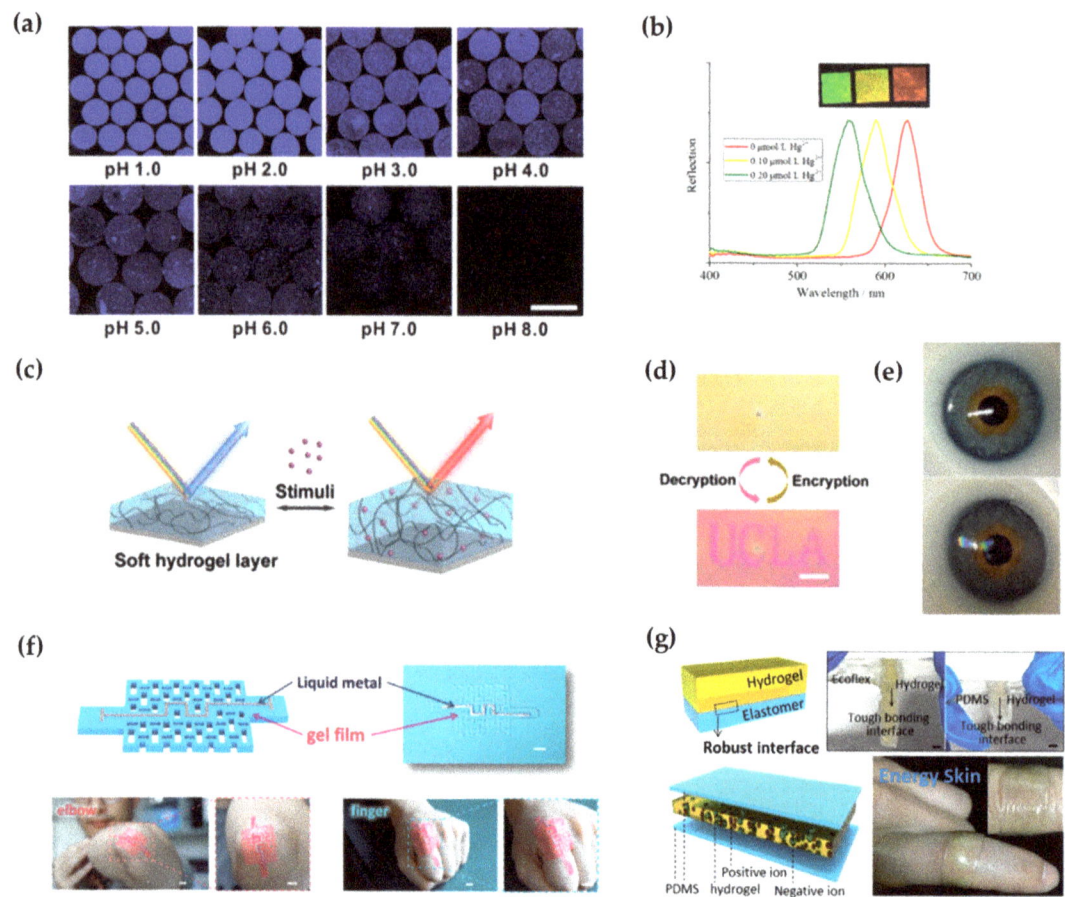

Figure 6. Applications of photocrosslinked hydrogels in biomedical sensors and flexible wearable devices. (**a**) The volume and fluorescence intensity of pH-responsive hydrogel microspheres changed under different pH conditions [117]. (**b**) The color of a mercury-ion-responsive hydrogels with structural color in mercury-ion solutions with different concentrations and reflection spectra of responses [118]. (**c**,**d**) Schematic diagram of hydrogel interferometer response to the target and demonstration of information encryption [119]. (**e**) Commercial contact lenses (top) and biosensing contact lenses (bottom) [120]. (**f**) Origami hydrogel for motion-signal sensing [121]. (**g**) Self-powered triboelectric nanogenerator that collects motion energy [122].

In addition to construction with micro/nanostructures, optical hydrogels can be directly prepared by using the principle of thin-film interference and the swelling characteristics of hydrogels. For example, Qin's team proposed a hydrogel interferometercolor-change system that was simple to prepare, quick to respond, and easily patterned, and demonstrated various applications in detection and information encryption [119]. This kind of hydrogel does not need a fine micro/nanoconstruction unit, but is used to fabricate a film by rotating the coating on a highly reflective substrate with grafted functional groups. According to the principle of film interference, the hydrogel-film thickness is controlled to cause instant discoloration (Figure 6c). To achieve a portable detection device, researchers have also developed the necessary mobile phone software [119]. When an unknown sample must be analyzed, one takes a picture with the mobile phone, and the software automatically analyzes the subject and provides the test results. Reversible encryption and decryption can be realized according to whether the film is damp or not (Figure 6d) [119].

5.2. Flexible Wearable Devices

The concept of hydrogels was first proposed by Wichterle et al. in 1960. They constructed a three-dimensional hydrogel network comprising hydroxyethyl methacrylate (HEMA) containing a small amount of the crosslinking agent EGDMA to overcome the poor biocompatibility and stability of plastic products at that time. This hydrogel has a soft texture, good light permeability, adjustable mechanical properties and water content, and a certain degree of biological inertia. Based on these excellent properties, Wichterle used it for the preparation of contact lenses [123]. Subsequently, many researchers made improvements on this material and developed cosmetic pupils with their own structural colors, which can be used for detection of substances in tear drops (Figure 6e) [120,124–126].

In addition to contact lenses, many flexible hydrogels are used in combination with sensing elements to produce flexible electronic devices that convert and conduct signals [127,128]. For example, Yu et al. developed a photocrosslinked acrylic hydrogel film with controllable thickness and excellent mechanical properties. During polymerization of the hydrogels, Zr^{4+} added to the solution coordinated with some of the carboxyl groups in polyacrylic acid, and the resulting hydrogels exhibited high stability. Akirigami structure for the hydrogel was obtained by photolithographic polymerization, so the hydrogel exhibited elevated ductility and flexibility to wrap the curved surface. After combining this origami hydrogel with a liquid metal, the resulting hydrogel sensor was used to sense arm or finger-bending motions (Figure 6f) [121]. To solve the need for a power supply for flexible electronic devices, Wang's research group developed a stretchable triboelectricnanogenerator (TENG) based on elastomer hydrogels. The tensioning, transparent, ultrathin single-electrode TENG with a double-layer structure fit firmly to human skin and deformed as the human body moved. TENGs can also capture energy during deformation processes (pressing, stretching, bending, and twisting) to drive electronic devices (Figure 6g) [122].

5.3. Drug Delivery and Tissue Engineering

The traditional methods of drug administration such as injection and oral administration need to be given frequently, and will cause the problem of high drug concentration in a short time, resulting in side effects. Therefore, sustainable and low-dose drug delivery has important application prospects. The hydrogel material has a three-dimensional network structure that "locks" the drug in a grid, allowing it to be released slowly. As a result, hydrogels could be used for drug delivery, which in turn could play a role in tissue engineering. For example, Lei et al. prepared a photocrosslinked methacrylated hyaluronic acid (HAMA) microsphere loaded with vascular endothelial growth factor (VEGF) by a microfluidic electrospray technique and used in the treatment of thin endometrium. The combination of VEGF and HAMA can promote endometrial regeneration and embryo implantation, while hyaluronic-acid hydrogel scaffolds can work with VEGF to promote endometrial hyperplasia after degradation. Therefore, this drug-loaded hydrogel has a good application prospect (Figure 7a) [129–131].

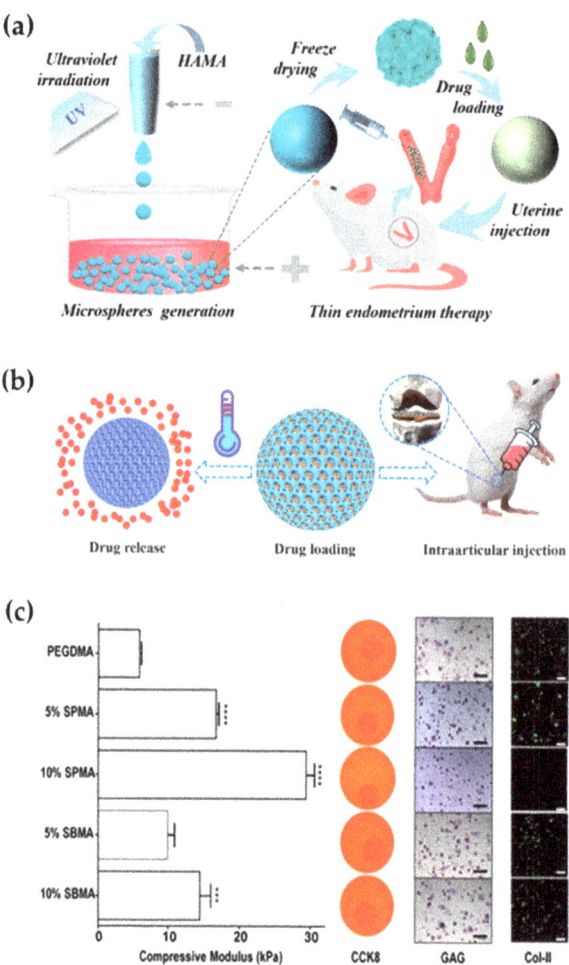

Figure 7. Application of photocrosslinked hydrogels in drug delivery, tissue engineering, and cell microcarriers. (**a**) Schematic diagram for the preparation of hydrogel microspheres for thin endometrium therapy [129]. (**b**) Application and principles of structural colored microspheres for treatment of osteoarthritis [132]. (**c**) Effects of different hydrogels on regulation of cell metabolism [133].

By selecting suitable responsive hydrogels or changing the structure of nanopores, hydrogels can be designed to have tunable drug-delivery effects. The stimuli used to trigger the hydrogel's response are usually light, heat, molecules, or ions. For example, Zhao et al. developed a glucose-responsive injectable hydrogel. Injected into the body, this hydrogel can maintain quasi-homeostasis in the normal range of blood glucose levels for approximately two weeks [116]. Salahuddin et al. prepared a light-controlled hydrogel. When exposed to light, the mesh of the hydrogel changes, altering the diffusion rate of the molecules being convenient for the controlled release of the drug [134]. Yang et al. prepared an antiopal microsphere and used its nanoporous structure to load drugs injected into the joint cavity. Elevated local temperatures during exercise or arthritis can promote the release of drugs, and vice versa. Therefore, this ingenious drug-delivery system could play an important role in the treatment of osteoarthritis (Figure 7b) [132].

5.4. Cellular Microcarrier

Photocrosslinked hydrogels can be used to load cells as microcarriers for cell culture, and biocompatibility and bioadhesion of hydrogels can be evaluated according to the growth status of the cells. For example, Liu et al. used microcarriers constructed from different hydrogels for cell culture, and found that the growth status of the cells with different microcarriers was inconsistent. This indicated that hydrogel-cell microcarriers can be used to study biological adhesion of different hydrogels [135].

When photocrosslinked hydrogels are used in cell culture, they can also be used as a platform to study cell growth, proliferation, and interaction between cells, which is also the preliminary basis for the application of hydrogels in tissue engineering in vivo. In earlier studies, Burdick's research group prepared hydrogels with gradient-crosslinking density using microfluidic technology, which were used to study the migration and other behaviors of cells in vitro [136]. Dadsetan et al. obtained hydrogels with different crosslinking degrees and mechanical properties by changing the ratio of crosslinking agent and polymerizer in photocrosslinking hydrogels. When these hydrogels were used for chondrocyte culture, cells showed different adhesion effects and morphologies on hydrogels with different crosslinking densities [137]. Huang et al. found that by adding different ionic residues to the hydrogel, it can improve its mechanical properties and regulate the metabolic activity and collagen secretion of the loaded chondrocytes without changing the polymer content and swelling behavior (Figure 7c) [133]. These results indicated that hydrogels, when used as cell microcarriers, can simulate tissues in vitro and preliminarily monitor and regulate cell behavior [138].

5.5. Bionic Organ

As mentioned earlier, photocrosslinked hydrogels can be used in cell culture. If microfluidic and other technologies are integrated into the cell-culture process, multiple cell cocultures can be realized, and the structures and functions of some organs can be simulated and realized through interactions between cells. The system is called an organ chip or organoid. These biomimetic organs can have important application value in the biomedical field. One of the most important applications is for drug evaluation. Today's drug development process often requires multiple rounds of animal or human trials to verify the drug's efficacy. Due to the differences between species, animal experiments often do not truly reflect the effects of drugs on humans, and both animal experiments and human experiments are faced with ethical problems. In addition, animal trials and human trials often take a long time, which greatly increases the cost of new drug development. Comparatively speaking, organ chips or organoids can prevent ethical problems and reduce time cycles, which is a good way to solve these problems. For example, Zhang et al. used a 3D microfluidic cell-culture system to construct microchips for liver, lung, kidney, and other organs for drug screening for related organ diseases [139]. Huh's team constructed a biomimetic lung-chip microdevice, that simulated lung respiration by applying mechanical force, and this can be used to evaluate the biotoxicities of nanoparticles [140]. Furthermore, they conducted an evaluation study on the toxicities of drugs with lung microchips [141]. Toh et al. cocultured a variety of cells to construct a three-dimensional liver-organ chip. Studies have shown that this chip can simulate the function of the liver and be used for toxicity testing [142]. Zhao et al. combined structurally colored hydrogels with cardiomyocytes to construct a novel heart chip [143] and used optical signals to monitor cardiomyocyte activity and drug evaluation (Figure 8a) [144].

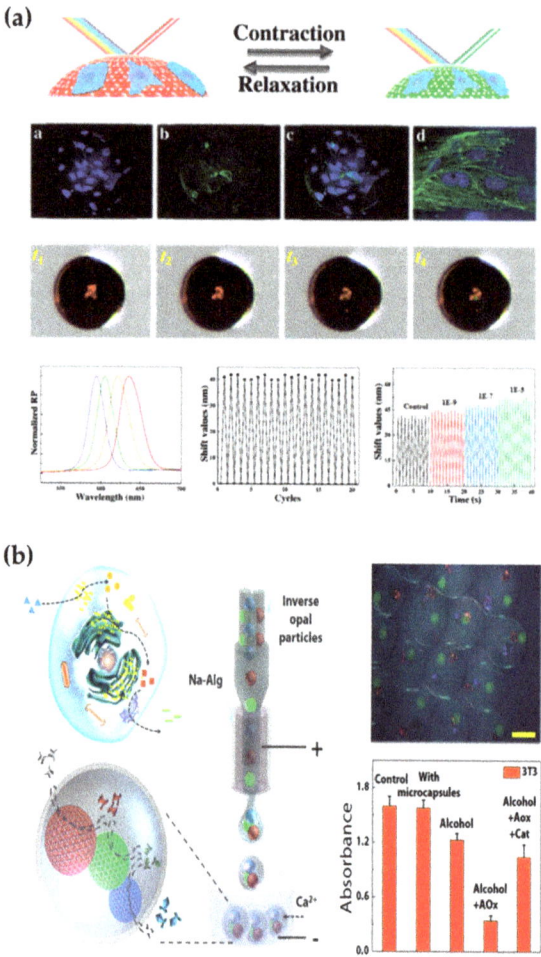

Figure 8. Application of photocrosslinked hydrogels in bionic organoids. (**a**) Bionic heart chip for cardiac drug evaluation [144]. (**b**) Bionic liver chip for the study of alcohol metabolism [145].

In addition to using bionic organs for drug evaluation, researchers have built organ chips for direct treatment of diseases. For example, Fu's team used 3D printing technology to construct a bionic liver with microchannels and nanoparticles for metabolism and adsorption of toxic substances [85]. Wang et al. constructed a bionic enzyme-cascade microcapsule by using microfluidic electrojet technology and a structurally e-colored microsphere enzyme carrier. In such microcapsules, the cascade metabolism of alcohol in the liver can be simulated with an enzyme-cascade reaction (Figure 8b) [145].

6. Conclusions and Prospects

In general, synthetic photocrosslinked hydrogels are usually constructed by introducing molecules containing double bonds that can be polymerized by light at the end of the molecule, such as ethylene, acrylic acid, acrylamide, acrylic ester, etc. Natural-polymer photocrosslinked hydrogels are usually modified by acrylates or acrylamide derivatives, which are prepared by acylation of natural polymers containing reactive groups. In the presence of photoinitiator, photocrosslinked hydrogels can be carried out under mild conditions. Compared with other chemical crosslinking, photocrosslinked hydrogels can achieve in situ polymerization crosslinking and have the characteristics of fast reaction rate, mild reaction conditions, easy control of geometric shape, and low reaction heat release.

Photocrosslinked hydrogels are biomedical materials, and they have good development prospects. However, they still have some limitations. The synthetic monomers used to prepare photocrosslinked hydrogels are usually toxic. Removing these harmful substances and free-radical residues from photocrosslinked hydrogels is a challenging problem. Although chemically modified natural materials can vent monomer toxicity, their types are still limited, and the resulting mechanical properties are not as easy to control as those of hydrogels prepared from synthetic monomers. In addition, their high biocompatibility allows easy contamination with many bacteria, which limits their application in other fields. Therefore, in future research, more natural materials with chemical modifications designed to meet actual demand should be developed. In terms of preparation and application, although a variety of techniques have been used to prepare photocrosslinked hydrogels with different morphologies, it is still a great challenge to prepare photocrosslinked hydrogels that effectively simulate tissues or organs in vivo. In future research, the intersection of advanced processing technology, new materials and human-tissue mechanics will be important in solving this problem. We hope that with further combinations of biology, chemistry, engineering, and other disciplines, applications of photocrosslinked hydrogels in biomedical fields will be expanded and more remarkable achievements will be realized.

Author Contributions: Conceptualization, Q.L.; methodology, Q.L. and W.H.; investigation, J.L.; C.S.; Y.C.; S.T. and C.L.; resources, W.H.; data curation, S.T. and C.L.; writing—original draft preparation, J.L. and Q.L.; writing—review and editing, C.L. and W.H.; visualization, Y.C., S.T. and C.L.; supervision, Q.L.; funding acquisition, W.H. and Q.L. All authors have read and agreed to the published version of the manuscript.

Funding: National natural science foundation of China (No. 31860708), Improvement project of the basic scientific research capacity of young and middle-aged teachers in Guangxi universities (No. 2022KY0565) and scientific research start-up fund for high-level talents of Yulin Normal University (No. G2021ZK12, G2022ZK02).

Institutional Review Board Statement: Not applicable.

Informed Consent Statement: Not applicable.

Data Availability Statement: Data are contained within the article.

Acknowledgments: The authors thank Tao Wang (College of Veterinary Medicine, Northwest A&F University) for helping to polish the English language.

Conflicts of Interest: The authors declare no conflict of interest.

References

1. Naseri, N.; Deepa, B.; Mathew, A.P.; Oksman, K.; Girandon, L. Nanocellulose-based interpenetrating polymer network (IPN) hydrogels for cartilage applications. *Biomacromolecules* **2011**, *3*, 1877–1888. [CrossRef] [PubMed]
2. Dragan, E.S. Design and applications of interpenetrating polymer network hydrogels. A review. *Chem. Eng. J.* **2014**, *243*, 572–590. [CrossRef]
3. Norioka, C.; Inamoto, Y.; Hajime, C.; Kawamura, A.; Miyata, T. A universal method to easily design tough and stretchable hydrogels. *NPG Asia Mater.* **2021**, *13*, 34. [CrossRef]
4. Hennink, W.E.; van Nostrum, C.F. Novel crosslinking methods to design hydrogels. *Adv. Drug Deliv. Rev.* **2012**, *64*, 223–236. [CrossRef]
5. Zhao, G.; Li, T.; Brochet, D.X.; Rosenberg, P.B.; Lederer, W.J. STIM1 enhances SR Ca^{2+} content through binding phospholamban in rat ventricular myocytes. *Proc. Natl. Acad. Sci. USA* **2015**, *112*, 4792–4801. [CrossRef]
6. Sperinde, J.J.; Griffith, L.G. Synthesis and characterization of enzymatically-cross-linked poly(ethylene glycol) hydrogels. *Macromolecules* **1997**, *30*, 5255–5264. [CrossRef]
7. Sperinde, J.J.; Griffith, L.G. Control and prediction of gelation kinetics in enzymatically cross-linked poly(ethylene glycol) hydrogels. *Macromolecules* **2000**, *33*, 5476–5480. [CrossRef]
8. Westhaus, E.; Messersmith, P.B. Triggered release of calcium from lipid vesicles: A bioinspired strategy for rapid gelation of polysaccharide and protein hydrogels. *Biomaterials* **2001**, *22*, 453–462. [CrossRef]

9. Gottlieb, R.; Schmidt, T.; Arndt, K.F. Synthesis of temperature-sensitive hydrogel blends by high-energy irradiation. *Nucl. Instrum. Methods Phys. Res.* **2005**, *236*, 371–376. [CrossRef]
10. Oezdemir, S.; Oezdemir, E.; Tunca, R.; Haziroglu, R.; Sen, M.; Kantoglu, O.E.; Gueven, O. In vivo biocompatibility studies of poly hydrogels synthesized by γ-rays. *Nucl. Instrum. Methods Phys. Res. Sect. B Beam Interact. Mater. Atoms* **2003**, *208*, 395–399. [CrossRef]
11. Karakecili, A.G.; Satriano, C.; Gumusderelioglu, M.; Marletta, G. Thermoresponsive and bioactive poly(vinyl ether)-based hydrogels synthesized by radiation copolymerization and photochemical immobilization. *Radiat. Phys. Chem.* **2008**, *77*, 154–161. [CrossRef]
12. Chen, M.; Shen, Y.; Xu, L.; Xiang, G.; Ni, Z. Highly efficient and rapid adsorption of methylene blue dye onto vinyl hybrid silica nano-cross-linked nanocomposite hydrogel. *Colloids Surf. A Physicochem. Eng. Asp.* **2021**, *613*, 126050. [CrossRef]
13. Kao, F.J.; Manivannan, G.; Sawan, S.P. UV curable bioadhesives: Copolymers of N-vinyl pyrrolidone. *J. Biomed. Mater. Res.* **1997**, *38*, 191–196. [CrossRef]
14. Lopergolo, L.C.; Lugao, A.B.; Catalani, L.H. Direct UV photocrosslinking of poly(N-vinyl-2-pyrrolidone) (PVP) to produce hydrogels. *Polymer* **2003**, *44*, 6217–6222. [CrossRef]
15. Fechine, G.; Barros, J.; Catalani, L.H. Poly(N-vinyl-2-pyrrolidone) hydrogel production by ultraviolet radiation: New methodologies to accelerate crosslinking. *Polymer* **2004**, *45*, 4705–4709. [CrossRef]
16. Lee, C.Y.; Teymour, F.; Camastral, H.; Tirelli, N.; Hubbell, J.A.; Elbert, D.L.; Papavasiliou, G. Characterization of the network structure of PEG diacrylatehy drogels formed in the presence of N-Vinyl pyrrolidone. *Macromol. React. Eng.* **2014**, *8*, 314–328. [CrossRef]
17. Devine, D.M.; Higginbotham, C.L. Synthesis and characterisation of chemically crosslinked N-vinyl pyrrolidinone (NVP) based hydrogels. *Eur. Polym. J.* **2005**, *41*, 1272–1279. [CrossRef]
18. Sahiner, N.; Sagbas, S. The preparation of poly(vinyl phosphonic acid) hydrogels as new functional materials for in situ metal nanoparticle preparation. *Colloids Surf. A Physicochem. Eng. Asp.* **2013**, *418*, 76–83. [CrossRef]
19. Sagbas, S.; Sahiner, N. A novel p (AAm-co-VPA) hydrogel for the Co and Ni nanoparticle preparation and their use in hydrogel generation from NaBH4. *Fuel Process. Technol.* **2012**, *104*, 31–36. [CrossRef]
20. Ren, Z.Q.; Zhang, Y.Y.; Li, Y.M.; Xu, B.; Liu, W.G. Hydrogen bonded and ionically crosslinked high strength hydrogels exhibiting Ca^{2+}-triggered shape memory properties and volume shrinkage for cell detachment. *J. Mater. Chem. B Mater. Biol. Med.* **2015**, *3*, 6347–6354. [CrossRef]
21. Hou, N.; Wang, R.; Wang, F.; Bai, J.; Zhou, J.; Hu, J.; Liu, S.; Jiao, T. Fabrication of hydrogels via host–guest polymers as highly efficient organic dye adsorbents for wastewater treatment. *ACS Omega* **2020**, *5*, 5470–5479. [CrossRef] [PubMed]
22. Zhang, R.; Peng, H.; Zhou, T.; Li, M.; Guo, X.; Yao, Y. Selective adsorption and separation oforganic dyes by poly(acrylic acid) hydrogels formed withspherical polymer brushes and chitosan. *Aust. J. Chem.* **2018**, *71*, 846–854. [CrossRef]
23. Jana, S.; Ray, J.; Jana, D.; Mondal, B.; Bhanja, S.K.; Tripathy, T. Removal of vanadium (IV) from water solution by sulfated Katira gum-cl-poly(acrylic acid) hydrogel. *Colloids Surf. A Physicochem. Eng. Asp.* **2019**, *566*, 70–83. [CrossRef]
24. Anah, L.; Astrini, N. Isotherm adsorption studies of Ni (II) ion removal from aqueous solutions by modified carboxymethyl cellulose hydrogel. *Conf. Ser. Earth Environ. Sci. IOP Publ.* **2018**, *160*, 012017. [CrossRef]
25. Makhado, E.; Pandey, S.; Nomngongo, P.N.; Ramontja, J. Preparation and characterization of xanthan gum-cl-poly(acrylic acid)/o-MWCNTs hydrogel nanocomposite as highly effective re-usable adsorbent for removal of methylene blue from aqueous solutions. *J. Colloid Interface Sci.* **2018**, *513*, 700–714. [CrossRef]
26. Gwon, S.; Park, S. Preparation of uniformly sized interpenetrating polymer network polyelectrolyte hydrogel droplets from a solid-state liquid crystal shell. *J. Ind. Eng. Chem.* **2021**, *99*, 235–245. [CrossRef]
27. Liu, J.; Liu, G.; Liu, W. Preparation of water-soluble β-cyclodextrin/poly(acrylic acid)/graphene oxide nanocomposites as new adsorbents to remove cationic dyes from aqueous solutions. *Chem. Eng. J.* **2014**, *257*, 299–308. [CrossRef]
28. Hu, X.S.; Liang, R.; Sun, G. Super-adsorbent hydrogel for removal of methylene blue dye from aqueous solution. *J. Mater. Chem. A* **2018**, *6*, 17612–17624. [CrossRef]
29. Ma, Y.; Lyu, L.; Guo, Y.; Fu, Y.; Wu, T.; Guo, S.; Sun, K.; Wujcik, E.K.; Guo, Z. Porous lignin based poly(acrylic acid)/organo-montmorillonite nanocomposites: Swelling behaviors and rapid removal of Pb (II) ions. *Polymer* **2017**, *128*, 12–23. [CrossRef]
30. Kong, W.; Yue, Q.; Li, Q.; Gao, B. Adsorption of Cd^{2+} on GO/PAA hydrogel and preliminary recycle to GO/PAA-CdS as efficient photocatalyst. *Sci. Total Environ.* **2019**, *668*, 1165–1174. [CrossRef]
31. Lu, F.; Wang, Y.; Wang, C.; Kuga, S.; Huang, Y.; Wu, M. Two-dimensional nanocellulose-enhanced high-strength, self-adhesive, and strain-sensitive poly(acrylic acid) hydrogels fabricated by a radical-induced strategy for a skin sensor. *ACS Sustain. Chem. Eng.* **2020**, *8*, 3427–3436. [CrossRef]
32. Wang, H.; Gu, H.; Chen, Z.; Shang, L.; Zhao, Z.; Gu, Z.; Zhao, Y. Enzymatic Inverse Opal Hydrogel Particles for Biocatalyst. *ACS Appl. Mater. Interfaces* **2017**, *9*, 12914–12918. [CrossRef] [PubMed]
33. Wang, H.; Zhang, H.; Zhang, D.; Wang, J.; Tan, H.; Kong, T. Enzyme-functionalized structural color hydrogel particles for urea detection and elimination. *J. Clean. Prod.* **2021**, *315*, 128149. [CrossRef]
34. Chen, C.; Liu, Y.; Wang, H.; Chen, G.; Ren, J.; Zhang, H. Multifunctional chitosan inverse opal particles for wound healing. *ACS Nano* **2018**, *12*, 10493–10500. [CrossRef] [PubMed]

35. Zhang, B.; Cheng, Y.; Wang, H.; Ye, B.; Shang, L.; Zhao, Y.; Gu, Z. Multifunctional inverse opal particles for drug delivery and monitoring. *Nanoscale* **2015**, *7*, 10590. [CrossRef]
36. Zhao, Z.; Wang, H.; Shang, L.; Yu, Y.; Fu, F.; Zhao, Y.; Gu, Z. Bioinspired heterogeneous structural color stripes from capillaries. *Adv. Mater.* **2017**, *29*, 1704569. [CrossRef]
37. Fu, F.; Chen, Z.; Wang, H.; Liu, C.; Liu, Y.; Zhao, Y. Graphene hybrid colloidal crystal arrays with photo-controllable structural colors. *Nanoscale* **2019**, *11*, 10846–10851. [CrossRef]
38. Li, L.; Chen, Z.; Shao, C.; Sun, L.; Zhao, Y. Graphene hybrid anisotropic structural color film for cardiomyocytes'monitoring. *Adv. Funct. Mater.* **2020**, *30*, 1906353. [CrossRef]
39. Manz, A.; Graber, N.; Widmer, H.M. Miniaturized total chemical analysis systems: A novel concept for chemical sensing. *Sens. Actuators B Chem.* **1990**, *1*, 244–248. [CrossRef]
40. Hou, K.; Hu, Z.; Mugaanire, I.T.; Li, C.; Chen, G.; Zhu, M. Fiber forming mechanism and reaction kinetics of novel dynamic-crosslinking-spinning for poly(ethylene glycol) diacrylate fiber fabrication. *Polymer* **2019**, *183*, 121903. [CrossRef]
41. Yang, J.; Zhang, Y.S.; Yue, K.; Khademhosseini, A. Cell-laden hydrogels for osteochondral and cartilage tissue engineering. *Acta Biomater.* **2017**, *57*, 1–25. [CrossRef] [PubMed]
42. Colinet, I.; Dulong, V.; Mocanu, G.; Picton, L.; Cerf, D.L. New amphiphilic and pH-sensitive hydrogel for controlled release of a model poorly water-soluble drug. *Eur. J. Pharm. Biopharm.* **2009**, *73*, 345–350. [CrossRef] [PubMed]
43. Gupta, A.; Kowalczuk, M.; Heaselgrave, W.; Britland, S.T.; Martin, C.; Radecka, I. The production and application of hydrogels for wound management: A review. *Eur. Polym. J.* **2019**, *111*, 134–151. [CrossRef]
44. Goh, C.H.; Heng, P.W.S.; Chan, L.W. Alginates as a useful natural polymer for microencapsulation and therapeutic applications. *Carbohydr. Polym.* **2012**, *88*, 1–12. [CrossRef]
45. Xu, S.; Liang, W.; Xu, G.; Huang, C.; Zhang, J.; Lang, M. A fast and dual crosslinking hydrogel based on vinyl ether sodium alginate. *Appl. Surf. Sci.* **2020**, *515*, 145811. [CrossRef]
46. Bukhari, A.A.H.; Elsayed, N.H.; Monier, M. Development and characterization of photo-responsive cinnamoly modified alginate. *Carbohydr. Polym.* **2021**, *260*, 117771. [CrossRef]
47. Huang, H.; Feng, J.; Wismeijer, D.; Wu, G.; Hunziker, E.B. Hyaluronic acid promotes the osteogenesis of BMP-2 in an absorbable collagen sponge. *Polymers* **2017**, *9*, 339. [CrossRef]
48. Nielsen, J.J.; Low, S.A. Bone-targeting systems to systemically deliver therapeutics to bone fractures for accelerated healing. *Curr. Osteoporos. Rep.* **2020**, *18*, 449–459. [CrossRef]
49. Lee, H.; Park, T.G. Photocrosslinkable, biomimetic, and thermo-sensitive pluronic grafted hyaluronic acid copolymers for injectable delivery of chondrocytes. *J. Biomed. Mater. Res. Part A* **2009**, *88*, 797–806. [CrossRef]
50. Jenjob, R.; Nguyen, H.P.; Kim, M.K.; Jiang, Y.; Kim, J.J.; Yang, S.G. Bisphosphonate-conjugated photo-crosslinking polyanionic hyaluronic acid microbeads for controlled BMP2 delivery and enhanced bone formation efficacy. *Biomacromolecules* **2021**, *22*, 4138–4145. [CrossRef]
51. Monier, M.; Youssef, I.; Abdel-Latif, D.A. Synthesis of photo-responsive chitosan-cinnamate for efficient entrapment of β-galactosidase enzyme. *React. Funct. Polym.* **2018**, *124*, 129–138. [CrossRef]
52. Jeong, S.I.; Jeon, O.; Krebs, M.D.; Hill, M.C.; Alsberg, E. Biodegradable photo-crosslinked alginate nanofibre scaffolds with tuneable physical properties, cell adhesivity and growth factor release. *Eur. Cells Mater.* **2012**, *24*, 331–343. [CrossRef] [PubMed]
53. Younes, I.; Rinaudo, M. Chitin and chitosan preparetion from marine sources. Structure, properties and applications. *Mar. Drugs* **2015**, *13*, 1133–1174. [CrossRef] [PubMed]
54. Zhou, Y.; Dong, Q.; Yang, H.; Liu, X.; Yin, X.; Tao, Y.; Bai, Z. Photocrosslinked maleilated chitosan/methacrylated poly (vinyl alcohol) bicomponent nanofibrous scaffolds for use as potential wound dressings. *Carbohydr. Polym.* **2017**, *168*, 220–226. [CrossRef]
55. Gospodarowicz, D.; Cheng, J. Heparin protects basic and acidic FGF from inactivation. *J. Cell. Physiol.* **1986**, *128*, 475–484. [CrossRef]
56. Wissink, M.J.B.; Beernink, R.; Pieper, J.S.; Poot, A.A.; Engbers, G.H.M.; Beugeling, T.; Aken, W.G.V.; Feijen, J. Binding and release of basic fibroblast growth factor from heparinized collagen matrices. *Biomaterials* **2001**, *22*, 2291–2299. [CrossRef]
57. Yoon, J.J.; Chung, H.J.; Park, T.G. Photocrosslinkable and biodegradable pluronic/heparin hydrogels for local and sustained delivery of angiogenic growth factor. *J. Biomed. Mater. Res. Part A* **2007**, *83*, 597–605. [CrossRef]
58. Ma, L.; Li, X.; Guo, X.; Jiang, Y.; Li, X.M.; Guo, H.; Zhang, B.; Xu, Y.; Wang, X.; Li, Q. Promotion of endothelial cell adhesion and antithrombogenicity of polytetrafluoroethylene by chemical grafting of chondroitin sulfate. *ACS Appl. Bio. Mater.* **2019**, *3*, 891–901. [CrossRef]
59. Bai, X.; Lü, S.; Cao, Z.; Ni, B.; Wang, X.; Ning, P.; Ma, D.; Wei, H.; Liu, M. Dual crosslinked chondroitin sulfate injectable hydrogel formed via continuous Diels-Alder (DA) click chemistry for bone repair. *Carbohydr. Polym.* **2017**, *166*, 123–130. [CrossRef]
60. Li, S.; Ma, F.; Pang, X.; Tang, B.; Lin, L. Synthesis of chondroitin sulfate magnesium for osteoarthritis treatment. *Carbohydr. Polym.* **2019**, *212*, 387–394. [CrossRef]
61. Kim, H.D.; Lee, E.A.; An, Y.H.; Kim, S.L.; Lee, S.S.; Yu, S.J.; Jang, H.L.; Nam, K.T.; Lm, S.G.; Hwang, N.S.Y. Chondroitin sulfate-based biomineralizing surface hydrogels for bone tissue engineering. *ACS Appl. Mater. Interfaces* **2017**, *9*, 21639–21650. [CrossRef] [PubMed]
62. Ornell, K.J.; Lozada, D.; Phan, N.V.; Coburn, J.M. Controlling methacryloyl substitution of chondroitin sulfate: Injectable hydrogels with tunable long-term drug release profiles. *J. Mater. Chem. B* **2019**, *7*, 2151–2161. [CrossRef] [PubMed]

63. Cerqueira, M.T.; Silva, L.P.D.; Santos, T.C.; Pirraco, R.P.; Correlo, V.M.; Reis, R.L.; Marques, A.P. Gellan gum-hyaluronic acid spongy-like hydrogels and cells from adipose tissue synergize promoting neoskin vascularization. *ACS Appl. Mater. Interfaces* **2014**, *6*, 19668–19679. [CrossRef] [PubMed]
64. Cerqueira, M.T.; Silva, L.P.D.; Santos, T.C.; Pirraco, R.P.; Correlo, V.M.; Marques, A.P.; Reis, R.L. Human skin cell fractions fail to self-organize within a gellan gum/hyaluronic acid matrix but positively influence early wound healing. *Tissue Eng. Part A* **2014**, *20*, 1369–1378. [CrossRef] [PubMed]
65. Ichibouji, T.; Miyazaki, T.; Ishida, E.; Sugino, A.; Ohtsuki, C. Apatite mineralization abilities and mechanical properties of covalently crosslinked pectin hydrogels. *Mater. Sci. Eng. C* **2009**, *29*, 1765–1769. [CrossRef]
66. Monier, M.; Shafik, A.L.; El-Mekabaty, A. Designing and investigation of photo-active gellan gum for the efficient immobilization of catalase by entrapment. *Int. J. Biol. Macromol.* **2020**, *161*, 539–549. [CrossRef]
67. Oliveira, M.B.; Custódio, C.A.; Gasperini, L.; Reis, R.L.; Mano, J.F. Autonomous osteogenic differentiation of hASCs encapsulated in methacrylated gellan-gum hydrogels. *Acta Biomater.* **2016**, *41*, 119–132. [CrossRef]
68. Serafini, M.R.; Menezes, P.P.; Costa, L.P. Interaction of p-cymene with β-cyclodextrin. *J. Therm. Anal. Calorim.* **2010**, *109*, 951–955. [CrossRef]
69. Kurkov, S.V.; Loftsson, T. Cyclodextrins. *Int. J. Pharm.* **2013**, *453*, 167–180. [CrossRef]
70. Cosola, A.; Conti, R.; Grützmacher, H.; Sangermano, M.; Roppolo, L.; Pirri, C.F.; Chiappone, A. Multiacrylated Cyclodextrin: A Bio-Derived Photocurable Macromer for VAT 3D Printing. *Macromol. Mater. Eng.* **2020**, *305*, 2000350. [CrossRef]
71. Rezanka, M. Synthesis of substituted cyclodextrins. *Environ. Chem. Lett.* **2019**, *17*, 49–63. [CrossRef]
72. Yamasaki, H.; Odamura, A.; Makihata, Y.; Fukunaga, K. Preparation of new photo-crosslinked β-cyclodextrin polymer beads. *Polym. J.* **2017**, *49*, 377–383. [CrossRef]
73. Yin, R.; Wang, K.; Han, J.; Nie, J. Photo-crosslinked glucose-sensitive hydrogels based on methacrylate modified dextran–concanavalin A and PEG dimethacrylate. *Carbohydr. Polym.* **2010**, *82*, 412–418. [CrossRef]
74. Lee, M.H.; Boettiger, D.; Composto, R.J. Biomimetic carbohydrate substrates of tunable properties using immobilized dextran hydrogels. *Biomacromolecules* **2008**, *9*, 2315–2321. [CrossRef]
75. Pacelli, S.; Paolicelli, P.; Casadei, M.A. New biodegradable dextran-based hydrogels for protein delivery: Synthesis and characterization. *Carbohydr. Polym.* **2015**, *126*, 208–214. [CrossRef] [PubMed]
76. Yang, K.; Sun, J.; Guo, Z.; Yang, J.; Wei, D.; Tan, Y.; Guo, L.; Luo, H.; Fan, H.; Zhang, X. Methacrylamide-modified collagen hydrogel with improved anti-actin-mediated matrix contraction behavior. *J. Mater. Chem. B* **2018**, *6*, 7543–7555. [CrossRef]
77. Tytgat, L.; Markovic, M.; Qazi, T.H.; Vagenende, M.; Bray, F.; Martins, J.C.; Rolando, C.; Thienpont, H.; Ottevaere, H.; Ovsianikov, A.; et al. Photo-crosslinkable recombinant collagen mimics for tissue engineering applications. *J. Mater. Chem. B* **2019**, *7*, 3100–3108. [CrossRef]
78. Song, X.; Dong, P.; Gravesande, J.; Cheng, B.; Xing, J. UV-mediated solid-state crosslinking of electrospinning nanofibers of modified collagen. *Int. J. Biol. Macromol.* **2018**, *120*, 2086–2093. [CrossRef]
79. Yang, K.; Sun, J.; Wei, D.; Yuan, L.; Yang, J.; Guo, L.; Fan, H.; Zhang, X. Photo-crosslinked mono-component type II collagen hydrogel as a matrix to induce chondrogenic differentiation of bone marrow mesenchymal stem cells. *J. Mater. Chem. B* **2017**, *5*, 8707–8718. [CrossRef]
80. Nichol, J.W.; Koshy, S.T.; Bae, H.; Hwang, C.M.; Yamanlar, S.; Khademhosseini, A. Cell-laden microengineered gelatin methacrylate hydrogels. *Biomaterials* **2010**, *31*, 5536–5544. [CrossRef]
81. Hu, Y.; Wang, W.; Yu, D. Preparation of antibacterial keratin fabrics via UV curing and click chemistry. *RSC Adv.* **2016**, *6*, 81731–81735. [CrossRef]
82. Xuan, M.; Sahoo, J.K.; Cebe, P.; Kaplan, D.L. Photo-crosslinked silk fibroin for 3D printing. *Polymers* **2020**, *12*, 2936. [CrossRef]
83. Rusu, A.G.; Chiriac, A.P.; Nita, L.E.; Mititelu-Tartau, L.; Tudorachi, N.; Ghilan, A.; Rusu, D. Multifunctional BSA scaffolds prepared with a novel combination of UV-crosslinking systems. *Macromol. Chem. Phys.* **2019**, *220*, 1900378. [CrossRef]
84. Chiriac, A.P.; Ghilan, A. Advancement in the biomedical applications of the nanogel structures based on particular polysaccharides. *Macromol. Biosci.* **2019**, *15*, 1903104.
85. Fu, F.; Shang, L.; Zheng, F.; Chen, Z.; Wang, H.; Wang, J.; Gu, Z. Cells cultured on core–shell photonic crystal barcodes for drug screening. *ACS Appl. Mater. Interfaces* **2016**, *8*, 13840–13848. [CrossRef]
86. Chevallay, B.; Abdul-Malak, N.; Herbage, D. Mouse fibroblasts in long-term culture within collagen three-dimensional scaffolds: Influence of crosslinking with diphenylphosphorylazide on matrix reorganization, growth, and biosynthetic and proteolytic activities. *J. Biomed. Mater. Res.* **2000**, *49*, 448–459. [CrossRef]
87. Wu, Y.L.; Lin, C.W.; Cheng, N.C.; Yang, K.C.; Yu, J. Modulation of keratin in adhesion, proliferation, adipogenic, and osteogenic differentiation of porcine adipose-derived stem cells. *J. Biomed. Mater. Res. Part B Appl. Biomater.* **2017**, *105*, 180–192. [CrossRef]
88. Yu, D.; Cai, J.Y.; Church, J.S.; Wang, L. Modifying surface resistivity and liquid moisture management property of keratin fibers through thiol-Ene click reactions. *ACS Appl. Mater. Interfaces* **2014**, *6*, 1236–1242. [CrossRef]
89. Qi, C.; Liu, J.; Jin, Y.; Xu, L.; Wang, G.; Wang, Z.; Wang, L. Photo-crosslinkable, injectable sericin hydrogel as 3D biomimetic extracellular matrix for minimally invasive repairing cartilage. *Biomaterials* **2018**, *163*, 89–104. [CrossRef]
90. Ju, J.; Hu, N.; Cairns, D.M.; Liu, H.; Timko, B.P. Photo-cross-linkable, insulating silk fibroin for bioelectronics with enhanced cell affinity. *Proc. Natl. Acad. Sci. USA* **2020**, *117*, 15482–15489. [CrossRef]

91. Aduba, D.C.; Hammer, J.A.; Yuan, Q.; Andrew Yeudall, W.; Bowlin, G.L.; Yang, H. Semi-interpenetrating network (sIPN) gelatin nanofiber scaffolds for oral mucosal drug delivery. *Acta Biomater.* **2013**, *9*, 6576–6584. [CrossRef] [PubMed]
92. Rong, L.; Dou, J.; Jiang, Q.; Jing, L.; Ren, X. Preparation and antimicrobial activity of β-cyclodextrin derivative copolymers/cellulose acetate nanofibers. *Chem. Eng. J.* **2014**, *248*, 264–272.
93. Maciejewska, B.M.; Wychowaniec, J.K.; Woniak-Budych, M.; Popenda, L.; Jurga, S. UV cross-linked polyvinylpyrrolidone electrospun fibres as antibacterial surfaces. *Sci. Technol. Adv. Mater.* **2019**, *20*, 979–991. [CrossRef]
94. Rosa, R.M.; Silva, J.C.; Sanches, I.S.; Henriques, C. Simultaneous photo-induced crosslinking and silver nanoparticle formation in a PVP electrospun wound dressing. *Mater. Lett.* **2017**, *207*, 145–148. [CrossRef]
95. Son, W.K.; Ji, H.Y.; Park, W.H. Antimicrobial cellulose acetate nanofibers containing silver nanoparticles. *Carbohydr. Polym.* **2006**, *65*, 430–434. [CrossRef]
96. Wang, Y.S.; Cheng, C.C.; Ye, Y.S.; Yen, Y.C.; Chang, F.C. Bioinspired photo-cross-linked nanofibers from uracil-functionalized polymers. *ACS Macro Lett.* **2012**, *1*, 159–162. [CrossRef]
97. Singh, U.; Mohan, S.; Davis, F.; Mitchell, G. Modifying the thermomechanical properties of electrospun fibres of poly-vinyl cinnamate by photo-cross-linking. *SN Appl. Sci.* **2019**, *1*, 31. [CrossRef]
98. Tang, Z.; Jie, W.; Yung, L.; Ji, B.; Ma, H.; Qiu, C.; Yoon, K.; Wan, F.; Fang, D.; Hsiao, B.S. UV-cured poly(vinyl alcohol) ultrafiltration nanofibrous membrane based on electrospun nanofiber scaffolds. *J. Membr. Sci.* **2009**, *328*, 1–5. [CrossRef]
99. Zhao, X.; Liu, S.; Yildirimer, L.; Zhao, H.; Ding, R.H.; Wang, H.N.; Cui, W.G.; Weitz, D. Injectable stem cell-laden photocrosslinkable microspheres fabricated using microfluidics for rapid generation of osteogenic tissue constructs. *Adv. Funct. Mater.* **2016**, *26*, 2809–2819. [CrossRef]
100. Lee, G.H.; Jeon, T.Y.; Kim, J.B.; Lee, B.; Lee, C.S.; Lee, S.Y.; Kim, S.H. Multicompartment photonic microcylinders toward structural color inks. *Chem. Mater.* **2018**, *30*, 3789–3797. [CrossRef]
101. Zhao, C.; Chen, G.; Wang, H.; Zhao, Y.; Chai, R. Bio-inspired intestinal scavenger from microfluidic electrospray for detoxifying lipopolysaccharide. *Bioact. Mater.* **2021**, *6*, 1653–1662. [CrossRef]
102. Schonberg, J.N.; Zinggeler, M.; Fosso, P.L.; Brandstetter, T.; Ruhe, J. One-step photochemical generation of biofunctionalized hydrogel particles via two-phaseflow. *ACS Appl. Mater. Interfaces* **2018**, *10*, 39411–39416.
103. Zhang, H.Z.; Xu, R.H.; Yin, Z.W.; Yu, J.; Liang, N.; Geng, Q. Drug-loaded chondroitin sulfate microspheres generated from microfluidic electrospray for wound healing. *Macromol. Res.* **2022**, *30*, 36–42. [CrossRef]
104. Ye, B.; Ding, H.; Cheng, Y.; Gu, H.C.; Gu, Z.Z. Photonic crystal microcapsules for label-free multiplex detection. *Adv. Mater.* **2014**, *26*, 3270–3274. [CrossRef] [PubMed]
105. Xu, Y.S.; Wang, H.; Luan, C.X.; Fu, F.F.; Chen, B.A.; Liu, H.; Zhao, Y.J. Porous hydrogel encapsulated photonic barcodes for multiplex microRNA quantification. *Adv. Funct. Mater.* **2018**, *28*, 1704458. [CrossRef]
106. Xin, T.; Gu, Y.; Cheng, R.; Tang, J.; Sun, Z.; Cui, W.; Chen, L. Inorganic strengthened hydrogel membrane as regenerative periosteum. *ACS Appl. Mater. Interfaces* **2017**, *9*, 41168–41180. [CrossRef] [PubMed]
107. Luo, Z.; Che, J.; Sun, L.; Yang, L.; Zu, Y.; Wang, H.; Zhao, Y. Microfluidic electrospray photo-crosslinkable κ-Carrageenan microparticles for wound healing. *Eng. Regen.* **2021**, *2*, 257–262. [CrossRef]
108. Zhao, Z.; Wang, J.; Lu, J.; Yu, Y.R.; Fu, F.F.; Wang, H.; Liu, Y.X.; Zhao, Y.J.; Gu, Z.Z. Tubular inverse opal scaffolds for biomimetic vessels. *Nanoscale* **2016**, *8*, 13574–13580. [CrossRef]
109. Zhang, Z.H.; Chen, Z.Y.; Wang, Y.; Chi, J.J.; Wang, Y.T.; Zhao, Y.J. Bioinspired bilayer structural color hydrogel actuator with multienvironment responsiveness and survivability. *Small Methods* **2019**, *3*, 1900519. [CrossRef]
110. Zhang, H.; Chen, C.W.; Zhang, H.; Chen, G.P.; Wang, Y.T.; Zhao, Y.J. Janus medical sponge dressings with anisotropic wettability for wound healing. *Appl. Mater. Today* **2021**, *23*, 101068. [CrossRef]
111. GhavamiNejad, A.; Li, J.; Lu, B.; Zhou, L.; Lam, L.; Giacca, A.; Wu, X.Y. Glucose-responsive composite microneedle patch for hypoglycemia-triggered delivery of native glucagon. *Adv. Mater.* **2019**, *31*, 1901051. [CrossRef]
112. Liang, Y.; Li, Z.; Huang, Y.; Yu, R.; Guo, B. Dual-dynamic-bond cross-linked antibacterial adhesive hydrogel sealants with on-demand removability for post-wound-closure and infected wound healing. *ACS Nano* **2021**, *15*, 7078–7093. [CrossRef] [PubMed]
113. Liu, S.; Yeo, D.C.; Wiraja, C.; Tey, H.L.; Mrsksich, M.; Xu, C. Peptide delivery with poly(ethylene glycol) diacrylate microneedles through swelling effect. *Bioeng. Transl. Med.* **2017**, *2*, 258–267. [CrossRef] [PubMed]
114. Zhang, T.; Sun, B.; Guo, J.; Wang, M.; Yan, F. Active pharmaceutical ingredient poly(ionic liquid)-based microneedles for the treatment of skin acne infection. *Acta Biomater.* **2020**, *115*, 136–147. [CrossRef]
115. Yu, X.; Li, M.; Zhu, L.; Li, J.; Jin, Y. Amifostine-loaded armored dissolving microneedles for long-term prevention of ionizing radiation-induced injury. *Acta Biomater.* **2020**, *112*, 87–100. [CrossRef] [PubMed]
116. Zhao, F.; Wu, D.; Yao, D.; Guo, R.; Wang, W.; Dong, A.; Kong, D.; Zhang, J. An injectable particle-hydrogel hybrid system for glucose-regulatory insulin delivery. *Acta Biomater.* **2017**, *64*, 334–345. [CrossRef]
117. Yue, S.; Sun, X.T.; Wang, N.; Wang, Y.N.; Wang, Y.; Xu, Z.R.; Chen, M.L.; Wang, J.H. SERS–fluorescence dual-mode pH-sensing method based on Janus microparticles. *ACS Appl. Mater. Interfaces* **2017**, *9*, 39699–39707. [CrossRef]
118. Ye, B.F.; Wang, H.; Ding, H.B.; Zhao, Y.J.; Pu, Y.P.; Gu, Z.Z. Colorimetric logic response based on aptamer functionalized colloidal crystal hydrogels. *Nanoscale* **2015**, *7*, 7565–7568. [CrossRef]

119. Qin, M.; Sun, M.; Bai, R.B.; Mao, Y.Q.; Qin, X.S.; Sikka, D.; Zhao, Y.; Qi, H.J.; Suo, Z.G.; He, X.M. Bioinspired hydrogel interferometer for adaptive coloration and chemical sensing. *Adv. Mater.* **2018**, *30*, 1800468. [CrossRef]
120. Elsherif, M.; Hassan, M.U.; Yetisen, A.K.; Butt, H. Wearable contact lens biosensors for continuous glucose monitoring using smartphones. *ACS Nano* **2018**, *2*, 5452–5462. [CrossRef]
121. Yu, H.C.; Hao, X.P.; Zhang, C.W.; Zheng, S.Y.; Du, M.; Liang, S.M.; Wu, Z.L.; Zheng, Q. Engineering tough metallosupramolecular hydrogel films with kirigami structures for compliant soft electronics. *Small* **2021**, *17*, 2103836. [CrossRef] [PubMed]
122. Liu, T.; Liu, M.M.; Dou, S.; Sun, J.M.; Cong, Z.F.; Jiang, C.Y.; Du, C.H.; Pu, X.; Hu, W.G.; Wang, Z.L. Triboelectric-nanogenerator-based soft energy-harvesting skin enabled by toughly bonded elastomer/hydrogel hybrids. *ACS Nano* **2018**, *12*, 2818–2826. [CrossRef] [PubMed]
123. Wichterle, O.; Lim, D. Hydrophilic gels for biological use. *Nature* **1960**, *185*, 117–118. [CrossRef]
124. Alexeev, V.L.; Das, S.; Finegold, D.N.; Asher, S.A. Photonic crystal glucose-sensing material for noninvasive monitoring of glucose in tear fluid. *Clin. Chem.* **2004**, *50*, 2353–2360. [CrossRef]
125. Xie, Z.Y.; Li, L.L.; Liu, P.M.; Zheng, F.Y.; Guo, L.Y.; Zhao, Y.J.; Jin, L.; Li, T.T.; Gu, Z.Z. Self-assembled coffee-ring colloidal crystals for structurally colored contact lenses. *Small* **2015**, *11*, 926–930. [CrossRef] [PubMed]
126. Deng, J.Z.; Chen, S.; Chen, J.L.; Ding, H.L.; Deng, D.W.; Xie, Z.Y. Self-reporting colorimetric analysis of drug release by molecular imprinted structural color contact lens. *ACS Appl. Mater. Interfaces* **2018**, *10*, 34611–34617. [CrossRef]
127. Lin, S.T.; Yuk, H.; Zhang, T.; Parada, G.A.; Koo, H.; Yu, C.J.; Zhao, X.H. Stretchable hydrogel electronics and devices. *Adv. Mater.* **2016**, *28*, 4497–4505. [CrossRef]
128. Wang, M.X.; Chen, Y.M.; Gao, Y.; Hu, C.; Hu, J.; Tan, L.; Yang, Z.M. Rapid self-recoverable hydrogels with high toughness and excellent conductivity. *ACS Appl. Mater. Interfaces* **2018**, *10*, 26610–26617. [CrossRef]
129. Lei, L.J.; Lv, Q.Z.; Jin, Y.; An, H.; Shi, Z.; Hu, G.; Yang, Y.Z.; Wang, X.G.; Yang, L. Angiogenic microspheres for the treatment of a thin endometrium. *ACS Biomater. Sci. Eng.* **2021**, *7*, 4914–4920. [CrossRef]
130. Lei, L.J.; Wang, X.G.; Zhu, Y.L.; Su, W.T.; Lv, Q.Z.; Li, D. Antimicrobial hydrogel microspheres for protein capture and wound healing. *Mater. Des.* **2022**, *215*, 110478. [CrossRef]
131. Lei, L.J.; Zhu, Y.L.; Qin, X.Y.; Chai, S.L.; Liu, G.X.; Su, W.T.; Lv, Q.Z.; Li, D. Magnetic biohybrid microspheres for protein purification and chronic wound healing in diabetic mice. *Chem. Eng. J.* **2021**, *425*, 130671. [CrossRef]
132. Yang, L.; Liu, Y.X.; Shou, X.; Ni, D.; Kong, T.T.; Zhao, Y.J. Bio-inspired lubricant drug delivery particles for the treatment of osteoarthritis. *Nanoscale* **2020**, *12*, 17093–17102. [CrossRef] [PubMed]
133. Huang, H.; Tan, Y.; Ayers, D.C.; Song, J. Anionic and zwitterionic residues modulate stiffness of photocrosslinked hydrogels and cellular behavior of encapsulated chondrocytes. *ACS Biomater. Sci. Eng.* **2018**, *4*, 1843–1851.
134. Salahuddin, B.; Wang, S.; Sangian, D.; Aziz, S.; Gu, Q. Hybrid gelatinhydrogels in nanomedicine applications. *ACS Appl. Bio Mater.* **2021**, *4*, 2886–2906. [CrossRef] [PubMed]
135. Liu, W.; Shang, L.R.; Zheng, F.Y.; Lu, J.; Qian, J.L.; Zhao, Y.J.; Gu, Z.Z. Photonic crystal encoded microcarriers for biomaterial evaluation. *Small* **2014**, *10*, 88–93. [CrossRef]
136. Burdick, J.A.; Khademhosseini, A.; Langer, R. Fabrication of gradient hydrogels using a microfluidics/photopolymerization process. *Langmuir* **2004**, *20*, 5153–5156. [CrossRef]
137. Dadsetan, M.; Szatkowski, J.P.; Yasemski, M.J.; Lu, L.C. Characterization of photocrosslinked oligo [poly(ethylene glycol) fumarate] hydrogels for cartilage tissue engineering. *Biomacromolecules* **2007**, *8*, 1702–1709. [CrossRef]
138. Sontheimer-Phelps, A.; Hassell, B.A.; Ingber, D.E. Modelling cancer in microfluidic human organs-on-chips. *Nat. Rev. Cancer* **2019**, *19*, 65–81. [CrossRef]
139. Zhang, C.; Zhao, Z.Q.; Rahim, N.A.A.; Noort, D.V.; Yu, H. Towards a human-on-chip: Culturing multiple cell types on a chip with compartmentalized microenvironments. *Lab Chip* **2009**, *9*, 3185–3192. [CrossRef]
140. Huh, D.; Matthews, B.D.; Mammoto, A.; Montoya-Zavala, M.; Hsin, H.Y.; Ingber, D.E. Reconstituting organ-level lung functions on a chip. *Science* **2010**, *328*, 1662–1668. [CrossRef]
141. Huh, D.; Leslie, D.C.; Matthews, B.D.; Fraser, J.P.; Jurek, S.; Hamiton, G.A.; Throneloe, K.S.; Mcalexander, M.A.; Ingber, D.E. A human disease model of drug toxicity–induced pulmonary edema in a lung-on-a-chip microdevice. *Sci. Transl. Med.* **2012**, *4*, 147. [CrossRef] [PubMed]
142. Toh, Y.C.; Lim, T.C.; Tai, D.; Xiao, G.F.; Noort, D.V.; Yu, H. A microfluidic 3D hepatocyte chip for drug toxicity testing. *Lab Chip* **2009**, *9*, 2026–2035. [CrossRef] [PubMed]
143. Fu, F.F.; Shang, L.R.; Chen, Z.Y.; Yu, Y.R.; Zhao, Y.J. Bioinspired living structural color hydrogels. *Sci. Robot.* **2018**, *3*, 8580. [CrossRef] [PubMed]
144. Wang, H.; Liu, Y.X.; Chen, Z.Y.; Sun, L.Y.; Zhao, Y.J. Anisotropic structural color particles from colloidal phase separation. *Sci. Adv.* **2020**, *6*, 1438. [CrossRef] [PubMed]
145. Wang, H.; Zhao, Z.; Liu, Y.X.; Shao, C.M.; Boan, F.K.; Zhao, Y.J. Biomimetic enzyme cascade reaction system in microfluidic electrospray microcapsules. *Sci. Adv.* **2018**, *4*, 2816. [CrossRef] [PubMed]

Article

Physicomechanical and Morphological Characterization of Multi-Structured Potassium-Acrylate-Based Hydrogels

José Luis Gradilla-Orozco [1], José Ángel Hernández-Jiménez [2], Oscar Robles-Vásquez [3], Jorge Alberto Cortes-Ortega [4], Maite Renteria-Urquiza [4], María Guadalupe Lomelí-Ramírez [2], José Guillermo Torres Rendón [2], Rosa María Jiménez-Amezcua [3,*] and Salvador García-Enriquez [2,*]

1. Department of Engineering, Centre for Industrial Technical Education, Guadalajara 44630, Mexico
2. Department of Wood Cellulose and Paper, University of Guadalajara, Guadalajara 44430, Mexico
3. Department of Chemical Engineering, University of Guadalajara, Guadalajara 44430, Mexico
4. Department of Chemistry, University of Guadalajara, Guadalajara 44430, Mexico
* Correspondence: rosa.jamezcua@academicos.udg.mx (R.M.J.-A.); salvador.genriquez@academicos.udg.mx (S.G.-E.)

Abstract: In this work, a photo-polymerization route was used to obtain potassium acrylate-co-acrylamide hydrogels with enhanced mechanical properties, well-defined microstructures in the dry state, and unique meso- and macrostructures in the hydrated state. The properties of the hydrogels depended on the concentration of the crosslinking agent. Mechanical properties, swelling capacity, and morphology were analyzed, showing a well-defined transition at a critical concentration of the crosslinker. In terms of morphology, shape-evolving surface patterns appeared at different scales during swelling. These surface structures had a noticeable influence on the mechanical properties. Hydrogels with structures exhibited better mechanical properties compared to unstructured hydrogels. The critical crosslinking concentration reported in this work (using glycerol diacrylate) is a reference point for the future preparation of multistructured acrylic hydrogel with enhanced properties.

Keywords: hydrogels; photo-polymerization; potassium acrylate-co-acrylamide; swelling capacity; multistructured

1. Introduction

Hydrogels are polymeric tridimensional networks that can absorb large amounts of water and other fluids without compromising their structure [1–3]. Their swelling depends on the presence of certain functional groups, crosslinking degree, chain flexibility, tacticity, crystallinity of components [4], and thermal history [5]. These unique materials are being used in several commercial applications, for example, as ophthalmic devices, biosensors, biological membranes, and drug carriers [1,2]. Potential applications of hydrogels as soil conditioners and as removal agents for heavy metal ions have also been mentioned in the literature [6–8]. The properties of hydrogels strongly depend on the synthesis, concentration, and nature of components and the polymerization process. In this regard, several studies dealing with methods of synthesis, effects of swelling, and crosslinkers on properties have been reported [9–22].

UV curing (photopolymerization) is an alternative process for synthesis of hydrogels. It allows for a better control over the reaction kinetics and it is not affected by the presence of oxygen in the system [23–25]. Additionally, photopolymerization can be utilized with most monomers, and only one additive (photoinitiator) is needed [26,27]. Other advantages of photopolymerization are short times of synthesis and minimum generation of heat [28]. It is also important to mention that photopolymerization is widely used to prepare hydrogels with applicability in medical and biological fields. For example, successful cell encapsulation and high cell viability in photopolymerized hydrogels [23] and the development of advanced photoinitiators suitable for several medical conditions [29] have been reported. Moreover, excellent resistance against bacteria in photopolymerizable hydrogels have also been reported. Lin et al. (2011) prepared a biocompatible silicone based on carboxybetaine and a macromer (bis-α, ω-(methacryloxypropyl) poly-dimethylsiloxane) via photopolymerization using Darocur® TPO as the photo-initiator. These hydrogels showed excellent resistance against bacterial adhesion and protein adsorption [30].

In the case of acrylic hydrogels prepared via photopolymerization, there are also plenty of published studies, from copolymer systems [31] to reinforced hydrogels [22,32,33], among others. In the case of photopolymerized hydrogels based on potassium polyacrylates, there are only a few studies reported. Ruan et al. (2004) studied polyacrylate potassium and polyacrylate sodium hydrogels using different photoinitiators in the synthesis. They found that the highest water absorptions were exhibited by hydrogels synthesized with Irgacure 1700 and Irgacure 1800 [34]. In this work, we utilized Irgacure 1700.

The superficial instabilities that appear during swelling in gels are a phenomenon known since the XIX century [35]. Tanaka et al. (1987) observed the appearance of patterns in polyacrylamide-based gels during phase transition. This phenomenon affected the understanding of the kinetic process in the gels [36]. In another study by Tanaka et al. (1992), the morphologic evolution and kinetics of superficial patterns in acrylic gels during swelling was reported. A dynamic ordering of patterns was observed [37]. Li et al. (1994) reported the presence of hexagonal-, grain-, and bubble-like shape patterns in ionic N-isopropylacrylamide (N-IPA)-based gels. These patterns were observed below, near, and above the transition phase temperature, and their behavior depended on the temperature, time, external constraint, and thermal history [38]. In a study dealing with photopolymerizable polyhydroxythylmethacrylate (PHEMA) hydrogel films with a crosslinking gradient, Guvendiren et al. (2009) reported a method that allowed them to form various osmotically driven surface patterns without organic solvents for swelling. They observed and captured the shape evolution of such patterns, being first hexagonal structures, then peanut shapes, and then lamellar and finally worm-like patterns [39]. The same research group later reported creasing formation in the gradient PHEMA hydrogels using various solvents. They found that the morphology of patterns depended on the equilibrium linear expansion, which was as a function of the solvent–polymer interaction and the concentration of the crosslinker [40]. Recently, Chuang et al. (2021) demonstrated that the UV irradiation dose and the immersion conditions in DI water determined the characteristics of surface patterns in pHEMA-based hydrogels [41]. The maximum characteristic wavelength of the formed wrinkles depended on the initial immersion time. This dependency had a relationship that followed the power law. Furthermore, it is important to mention that surface structures seem to have an important effect on cells attached on hydrogels. In this regard, Saha et al. (2010) reported that wrinkled patterns on the surface of soft hydrogels made of polyacrylamide greatly influenced cell attachment and cell behavior [42]. Figure 1 displays some of the surface patterns that can be generated in acrylic hydrogels.

It is clear that there is a great interest in polymeric systems that display spontaneous formation of patterns that can be controlled in terms of size, order, morphology, and complexity. Such materials could be useful in many applications such as coatings, optical filters, batteries, actuators, valves, microfluidic devices, and flexible electronics [43–47]. In this work, we prepared multi-structured hydrogels of poly(potassium acrylate-co-acrylamide) via photopolymerization. Photo initiation allowed us to control the temperature of the synthesis, which led a higher structural stability of hydrogels during swelling. We used the term multi-structured due to the ability of these hydrogels to display patterns/structures at different scales (micro-, meso-, and macroscales). The effect of the crosslinking agent (glycerol diacrylate or DAG) on the swelling capacity, morphology, and mechanical properties was evaluated.

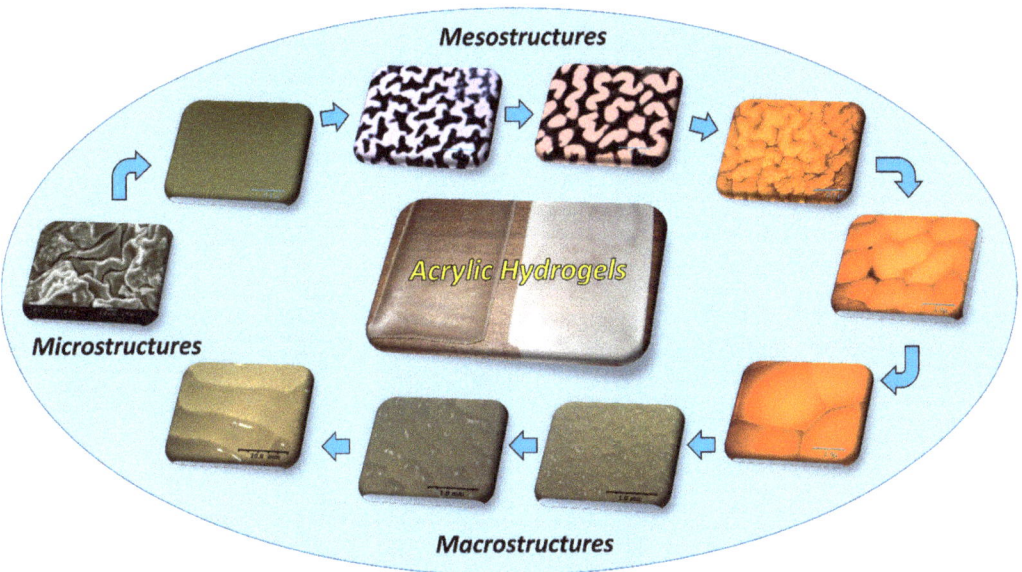

Figure 1. Surface patterns of acrylic hydrogels.

2. Results and Discussion

Figure 2 shows the probable reaction between the acrylic monomers and the crosslinking agent to form the polymeric network. Undesired residues such as unreacted molecules from the pho-initiator and the monomers, as well as free oligomers, were removed in the cleaning step [48].

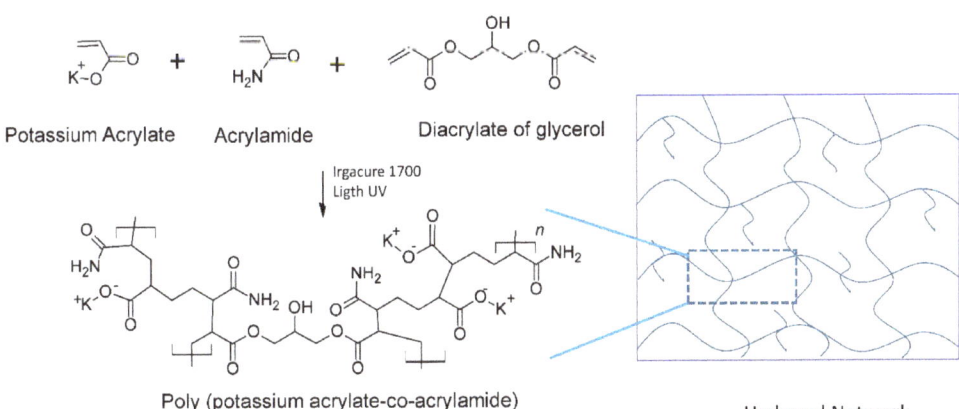

Figure 2. Proposed formation of the polymeric network.

During preparation of hydrogels, it was possible to control the temperature of the reaction by means of photo-polymerization. By controlling the entropy of the reaction solution at low temperatures, an ordered polymeric network was formed, allowing for the preparation of transparent and homogeneous hydrogels. These hydrogels presented a fractal-like structure in three different scales: micro (10^{-7} to 10^{-8} m), meso (10^{-5} to 10^{-7} m), and macro (10^{-3} to 10^{-5} m). Such properties were consistent with those from hydrogels synthetized via redox initiation [20]. The polymer gel fraction (G_F) was calculated for all hydrogels (shown in Table 1), as mentioned in the experimental section. All values were higher than 91%, which is consistent with values reported by Rodgers et al., who determined G_F's of 89% and 90% for acrylamide and 93% for acrylic acid [29]. Lara-Valencia et al. [10] reported values lower than 90% for acrylic hydrogels obtained by redox initiation and higher than 90% for photoinitiated ones. Wen et al. [31] reported values lower than ours for poly acrylic acid/cellulose nanofibers hydrogels. This could be attributed to the nanofibers, which could have had decreased the UV light transmittance.

Table 1. Maximum swelling and mechanical tests results for hydrogel samples.

Concentration of DAG (wt %)	Polymer Gel Fraction, G_F(%)	Maximum Swelling	$K \times 10^8$	Characteristic Length, $\lambda \times 10^6$ m	Tensile Strength (Pa)	Young's Modulus (MPa)	Elongation at Break(%)
0.5	91.2 ± 0.8	216.2 ± 7.5	4.28 ± 0.15	1.46 ± 0.038	1979 ± 19	2.64 ± 0.09	1647 ± 52.4
1.0	92.5 ± 1.1	155.1 ± 5.9	4.72 ± 0.18	1.41 ± 0.036	2087 ± 17	3.66 ± 0.10	1036 ± 43.1
2.0	92.4 ± 0.9	136.8 ± 4.0	5.62 ± 0.16	1.34 ± 0.039	2121 ± 32	3.73 ± 0.18	871 ± 36.0
3.0	93.6 ± 1.4	120.2 ± 4.3	6.03 ± 0.21	1.18 ± 0.025	2206 ± 16	6.48 ± 0.19	660 ± 34.3
4.0	92.1 ± 0.9	99.7 ± 3.9	5.54 ± 0.22	1.08 ± 0.031	2262 ± 34	7.80 ± 0.22	517 ± 32.0
5.0	93.4 ± 1.2	93.7 ± 3.4	5.93 ± 0.21	1.04 ± 0.016	2261 ± 28	8.84 ± 0.43	391 ± 15.2
6.0	93.1 ± 1.3	93.0 ± 3.5	6.86 ± 0.26	0.99 ± 0.007	2343 ± 24	13.43 ± 0.38	349 ± 23.7
7.0	92.9 ± 1.4	79.2 ± 2.3	8.08 ± 0.23	0.97 ± 0.008	2427 ± 20	14.38 ± 0.86	279 ± 21.0
8.0	91.6 ± 0.9	75.0 ± 2.7	7.38 ± 0.26	0.95 ± 0.005	2435 ± 18	15.25 ± 0.44	251 ± 15.5
9.0	92.0 ± 0.9	70.7 ± 2.8	8.34 ± 0.33	0.92 ± 0.008	2482 ± 26	22.34 ± 1.45	210 ± 19.2
10.0	92.7 ± 1.4	69.4 ± 3.3	10.21 ± 0.49	0.91 ± 0.006	2515 ± 33	28.47 ± 0.81	128 ± 12.5

± standard deviation.

2.1. Infrared Spectroscopy FTIR

Figure 3A shows the FTIR spectrum corresponding to acrylic acid. Here, the bands at 3260 cm^{-1} were attributed to the OH groups, while the bands corresponding to the asymmetric and symmetric tension vibrations of CH$_2$ were detected at 2968 cm^{-1} and 1460 cm^{-1}, respectively. The band at 1733 cm^{-1} corresponded to the carboxylic carbonyl group, while the one at 1635 cm^{-1} was attributed to the vibration of the C=C double bond. The out-of-plane deformation (=C-H) was represented by the band at 812 cm^{-1}. Figure 3B corresponds to the FTIR spectrum of the glycerol diacrylate (DAG), in which the same characteristic bands were observed as in the spectrum shown in Figure 3A. This was because both compounds possess the same functional groups. However, here, we had the presence of a secondary alcohol (-CHOH) that generated bands at 1059 cm^{-1}, 1300 cm^{-1}, and 1413 cm^{-1}. Figure 3C shows the FTIR spectrum of acrylamide. Here, we had the combination of N-H strain and C-N bending in the bands at 1356 cm^{-1} and 675 cm^{-1}, respectively, while the combination of N-H strain and C-N strain occurred at 1610 cm^{-1}. Bands corresponding to the double bond C=O ($\approx$1735 cm^{-1}) and the N-H strain (from 3342 to 3200 cm^{-1}) were also observed. It can be noted that the band for the carbonyl group (C=O) appeared at 1670 cm^{-1}, which could be attributed to the presence of the amino group [10]. In Figure 3D, the spectrum obtained for the acrylamide/potassium acrylate copolymer, with 4 wt % glycerol diacrylate (DAG), is shown. Here, the absence of the C=C double bond signal was evident, and the presence of the characteristic signals for both monomers can be noted. Bands located at 1677 and 1572 cm^{-1} can be attributed to the carbonyl and amino groups, respectively.

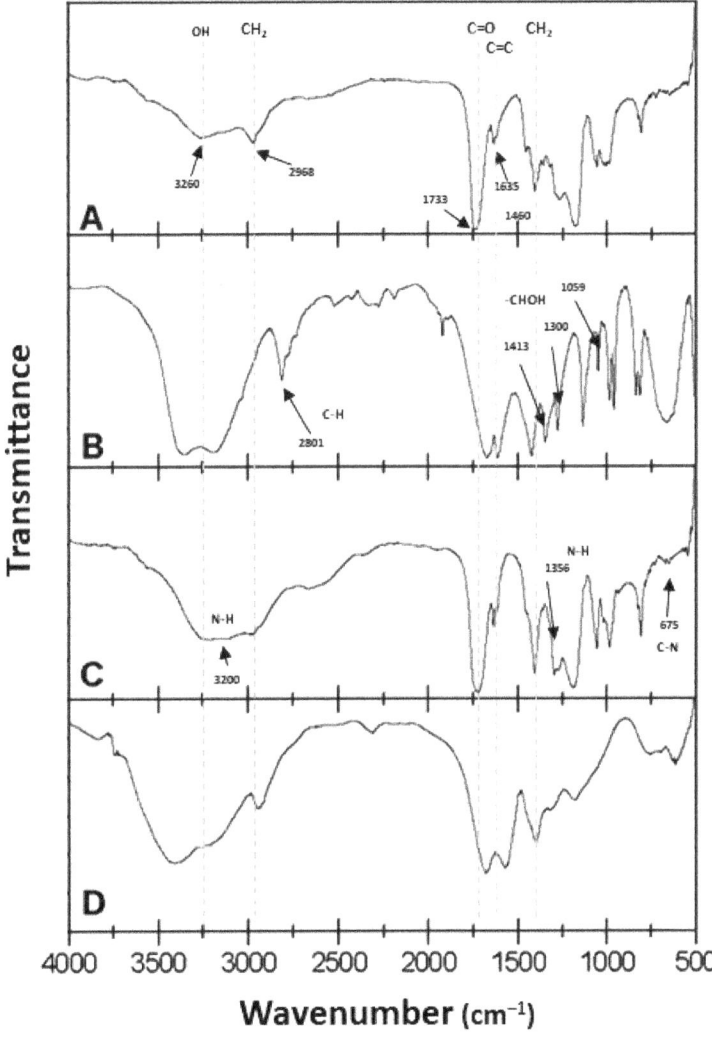

Figure 3. FTIR spectra for (**A**) acrylic acid, (**B**) DAG, (**C**) acrylamide, and (**D**) hydrogel with 4 wt % DAG.

2.2. Swelling Kinetics

Figure 4a shows the swelling kinetics of the obtained hydrogels. The swelling of hydrogels with 0.5 wt % DAG was significantly higher than the rest. On the other hand, hydrogels with DAG concentrations from 1 wt % to 10 wt % showed a very similar behavior, having their swelling capacities reduced with increasing crosslinking. It can also be observed that groups of samples followed practically the same behavior (groups of 1, 2, and 3 wt %; groups of 4, 5, and 6 wt %; and groups of 7, 8, and 9 wt %). The maximum swelling ranged from 69 to 217 times the weight of the xerogel. This behavior depended on the amount of DAG (Figure 4). Table 1 shows the values of the maximum swelling for all formulations. Magalhães et al. [49] reported swelling degree values from 144 to 189 for hydrogels where they varied the amount of sodium acrylate in the formulation of poly(sodium acrylate-co-acrylamide) hydrogels. Leitão et al. [50] reported swelling degrees from 620 to 1100 times in acrylamide/potassium acrylate hydrogels crosslinked at 0.05, 0.1, and 0.2% mol with respect to the total mass of monomers.

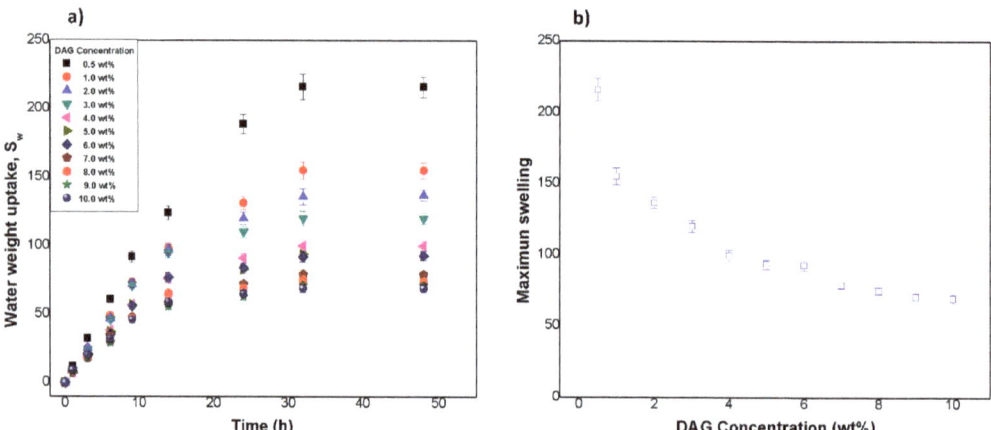

Figure 4. (**a**) Water weight uptakes of hydrogels as a function of time. (**b**) Maximum swelling as a function of the crosslinking agent (DAG).

Figure 4b shows the maximum swelling as a function of the crosslinking agent (DAG). A change in slope was observed at 4 wt % of DAG, similar to the characteristic length (λ) of the mesostructures and macrostructures at different times (as can be seen in the figure of the characteristic lengths). This critical value of 4 wt % of DAG is consistent with a critical crosslinking concentration reported in a study from Lopez-Ureta et al. (2008), in which the authors synthetized acrylic acid/acrylamide hydrogels via redox initiation. They also used DAG as a crosslinker [20]. We believe that, starting from this critical concentration (4 wt %), stronger networks were formed, which led to less swelling in the hydrogels. At lower DAG concentrations, the hydrogels physically collapsed near the equilibrium, which can be attributed to weaker networks.

The Schott's model was used in order to analyze the adsorption behavior of hydrogels in water. Figure 5a shows the linear behavior obtained from Equation (5) (see below in the Experimental section), which showed linearity values close to 1, demonstrating that it followed the model adequately.

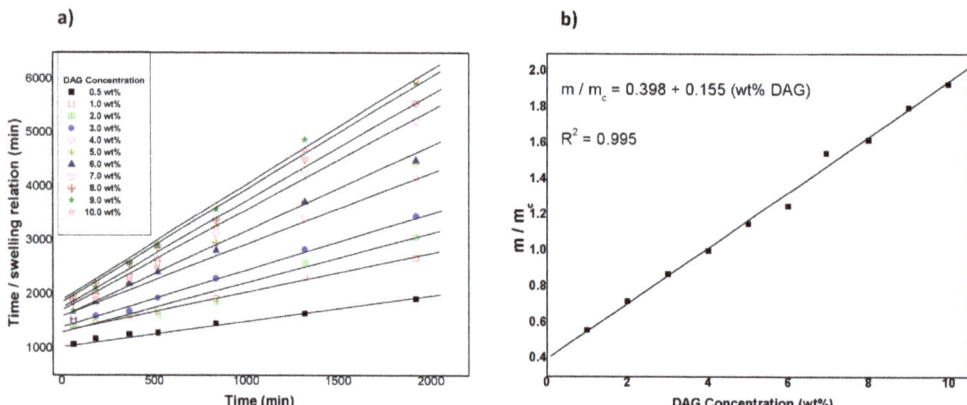

Figure 5. (**a**) Linearization of the Schott model for all samples. (**b**) Normalized slopes from the linearization of the Schott model (m/m_c) as a function of DAG concentration.

In this context, Figure 5b shows the normalized slopes for all samples ($m/m_{critical}$). Here, a linear dependence can be observed. Table 1 shows the values of the constant K for all the samples evaluated.

2.3. Surface Analysis

Figure 6a shows images of the surface of the xerogels where it can be observed that with a higher the amount of DAG, the structuring of the hydrogel was more noticeable. A critical crosslinking concentration (CCLC) of around 4 wt % was observed. For DAG concentrations below 4 wt %, an irregular surface arrangement was noted, while for DAG concentrations higher than 4 wt %, a compact surface, aligned preferentially in one direction, and with a more regular surface texture, was observed. Similar results were reported by López-Ureta et al. [20]. The characteristic length of the structure was in the microscale in xerogels with 10 wt % of DAG but increased when the concentration of DAG decreased. The mesostructure of hydrogels was observed by optical microscopy. When the water penetrated the hydrogels, four distinctive wrinkle patterns were spontaneously formed and transited to random worms, lamellae, peanuts, and ordered hexagonal patterns during swelling, until the maximum swelling was reached. Figure 6b shows images of hydrogels with 1 wt %, 2 wt %, 4 wt %, 6 wt %, and 10 wt % of DAG at short hydration times (10, 40, and 180 s). The order of magnitude of the structures was in the range of 10^{-6} m. The size of structures increased with hydration time, becoming more defined as the amount of DAG increased. There was evidence in this research that the crosslinking gradient plays a critical role in the evolution of surface patterns and their ordering along with the lateral confinement of the hydrogel, promoting anisotropic osmotic pressure along with the thickness. Similar systems were described [35,36,39,40].

Figure 5b shows images of hydrogels exhibiting meso- and macrostructures at different swelling times as a function of DGA. The macrostructure was defined from the first minutes of contact of the xerogel with water. After 24 h, all macrostructures disappeared. Equilibrium was reached approximately 32 h after the start of the swelling process. Depending on the degree of crosslinking of the hydrogel, the hydrogel can undergo large volume changes during swelling. During the water absorption process, the resulting compressive stresses can be quite large, even exceeding the elastic modulus of the gel. When this compressive stress becomes large enough (and the material cannot delaminate and buckle macroscopically), an elastic instability arises in which the free surface folds in on itself to locally relieve the compressive stress.

Table 1 shows the average characteristic length values for the mesostructure, measured at 60 s of hydration. Initially, the gel was in an almost stress-free state; however, immersion in a solvent led to swelling of the network until the osmotic stress, due to the mixing of solvent with polymer chains and counterions, was balanced by the elastic strain due to chain stretching. Because of the mechanical constraint provided by the substrate, a gel attached to the surface that was much thinner than its lateral dimensions can only expand in the direction normal to the surface. The result of this uniaxial expansion is that the gel undergoes a state of equibiaxial compressive stress [35].

Figure 7a shows the characteristic length (λ) of the mesostructures at 60 s and 180 s as a function of DAG concentration. The order of magnitude (10^{-6} m) increased with the water absorption. Similar to Figures 3B and 7b, when the DAG concentration was 4 wt %, there was a clear change in the behavior of the data. This was probable due to the reordering of the polymeric network. Figure 7b displays the relationship between λ and the DAG concentration, but this time at 60 and 360 min. The order of magnitude was 10^{-3} m. The mesoestructures disappeared when the hydrogels reached their swelling equilibrium.

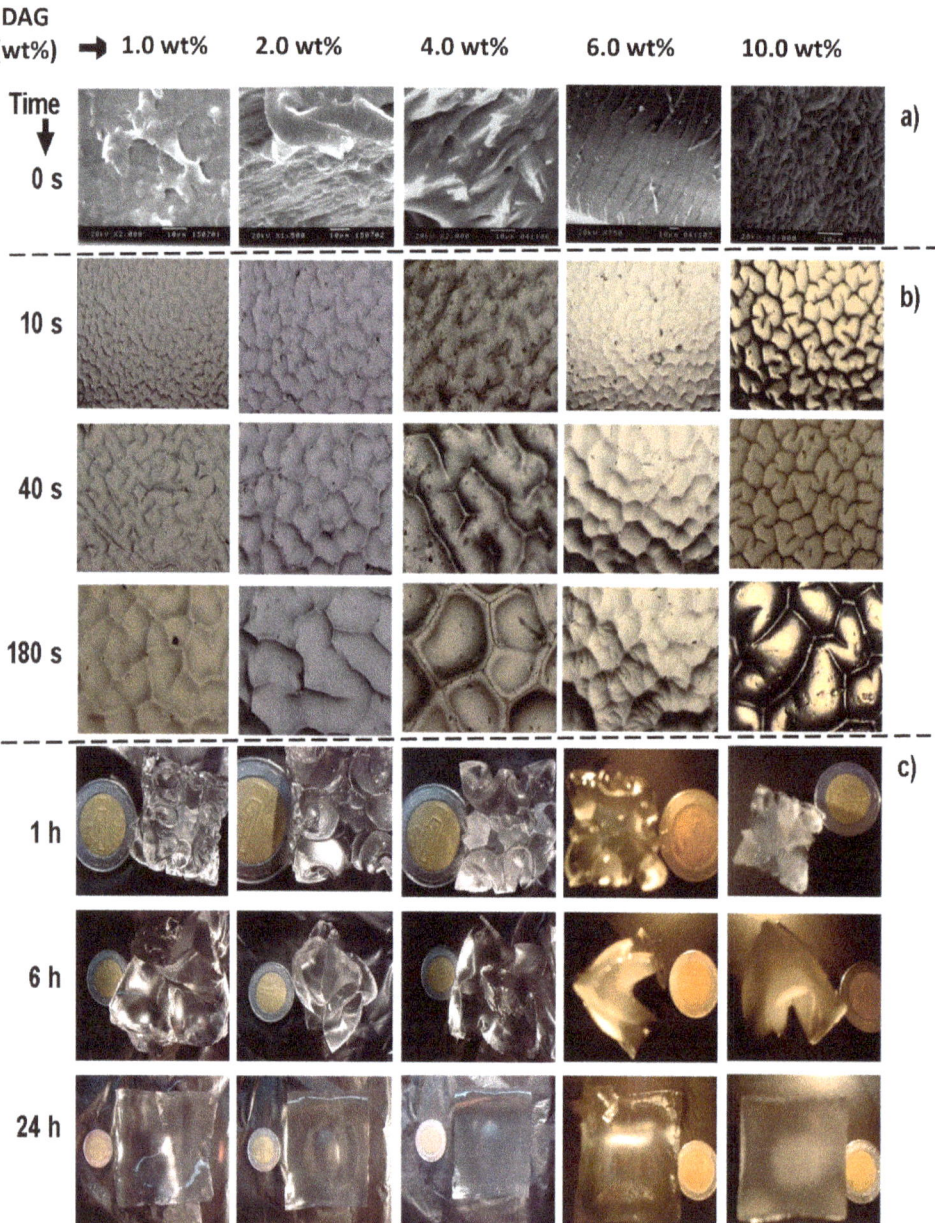

Figure 6. Images for hydrogels with 1 wt %, 2 wt %, 4 wt %, 6 wt %, and 10 wt % of DAG. (**a**) Micrograph (SEM) of xerogels; (**b**) polarized light micrographs of swollen hydrogels; (**c**) macrostructures of swollen hydrogels; all as functions of swelling time.

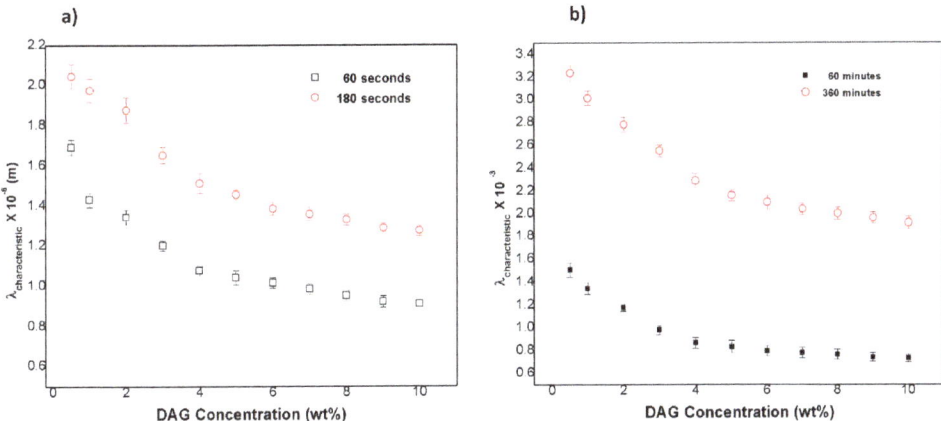

Figure 7. Characteristic length (λ) of structures at (**a**) 60 s and 180 s and at (**b**) 60 min and 360 min as a function of the DAG concentration.

The size of the meso- and macrostructures slightly decreased when the DAG concentration also decreased, generating regular "packed" shapes and thus more compacted hydrogels. This type of superficial morphology was observed in all sides of the samples (cubes), demonstrating the tridimensional nature of the patterns. The size of these well-defined structures augmented with the time after immersion. However, the number of these structures remained constant.

2.4. Mechanical Tests

Table 2 shows the values of tensile strength, elongation at break, and Young's modulus. Figure 8a displays tensile test curves for all compositions evaluated, and Figure 8b shows the Young's modulus as a function of the concentration of DAG. It can be observed that the elastic modulus increased when the concentration of DAG increased. In the case of the elongation at break, it was observed that it decreased when DAG concentration increased. For the tensile strength, it generally increased when the concentration of DAG increased as well. These results were expected due to the formation of stronger and stiffer polymeric networks as the DAG concentration increased.

Table 2. Formulations of hydrogels obtained.

Substance	Amount (g)									
Acrylamide	49.0									
Acrylic acid	51.0									
DAG	0.5	1.0	2.0	3.0	4.0	5.0	6.0	7.0	8.0	9.0	10.0
Irgacure 1700	0.03									
Water	100.0									
KOH	neutralizer and generating potassium acrylate									

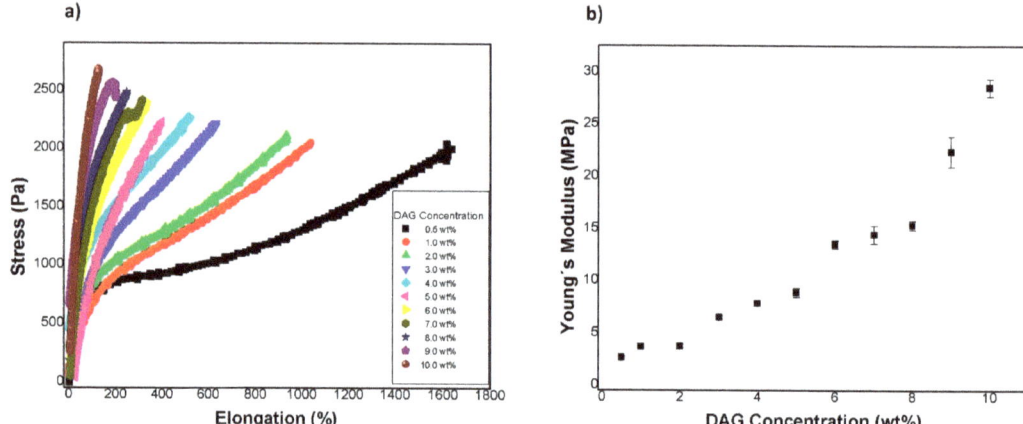

Figure 8. (**a**) Tensile test curves for all compositions evaluated. (**b**) Average Young's modulus as a function of the concentration of the crosslinking agent (DAG).

3. Conclusions

By means of the photo-polymerization of potassium acrylate and acrylamide, the temperature of the synthesis was controlled. This allowed for the preparation of transparent and homogeneous hydrogels based on potassium acrylate-co-acrylamide that were obtained via photo-polymerization. The resulting hydrogels presented a fractal-like structure in the micro-, meso-, and macroscales. The maximum swelling ranged from 69 to 217 times the weight of the xerogel. This swelling behavior depended on the amount of the crosslinking agent (DAG). The critical concentration of DAG was 4 wt %, which could be a reference point to produce a hydrogel with better mechanical properties and structure characteristics. Upon increasing the amount of DAG, the mechanical resistance increased, and simultaneously, values of elongation at break and swelling capacity decreased. The morphology, swelling capacity, and Young's modulus showed a transition between 4 and 5 wt % of DAG. This work is a starting point for the future preparation of advanced multi-structured hydrogel materials that could have a wide range of applications, such as coatings, batteries, flexible electronics, actuators, and optical filters.

4. Materials and Methods

4.1. Materials

The monomers used for hydrogel preparation were acrylamide and acrylic acid from Aldrich with purities of 98.5% and 99.3%, respectively. Potassium hydroxide (KOH), 99%, was also purchased from Aldrich. As a photoinitiator, a mixture of 25% bis(2,6-dimethoxybenzoyl)-2,4,4-trimethyl pentylphosphineoxide and 75% 2-hydroxy-2-methyl-1-phenyl-propane-1-one (Irgacure 1700) from Ciba Speciality Chemicals Inc. was used. Glycerol diacrylate (DAG), 97% purity, was obtained from Industria Azteca Integral.

4.2. Synthesis of Hydrogels

Figure 9 shows a scheme of the complete methodology of this work. The acrylic acid was dissolved in bidistilled water according to the formulation shown in Table 2. The solution was taken to a pH 7 with the addition of a KOH solution at 47 wt %. The solution was kept below 25 °C to avoid thermal polymerization. Subsequently, the acrylamide and DAG were added, the mixture was stirred until the system was homogeneous, and the temperature was lowered to 2 °C. Finally, 1 mL of photoinitiator solution (3 wt % in methanol) was added. The reaction solution was poured into a 0.450 L glass semi-infinite plate reactor, and nitrogen was bubbled for 15 min. It was placed 15 cm away from the lamp (Tecno F15T8-BLB 20 W, 127 v) rich in 366 nm wavelength radiation, inside an isothermal bath at 2 °C for 1 h (to achieve higher conversion). After the reaction time, the hydrogels were removed, and 0.5 × 0.5 × 0.3 mm³ samples were cut and allowed to dry until constant weight, and then were introduced into double-distilled water at 25 °C for three days. The water was changed every 24 h, and they were submerged in the water again for three days at 45 °C. Once the cleaning process was completed, the hydrogels were left to dry at room temperature for 5 days and then put in a vacuum oven at 40 °C until constant weight. The polymer gel fraction (G_F) was calculated as follows:

$$G_F(crosslinked\ polymer\ \%) = \left(\frac{W_d}{W_0}\right) * 100 \tag{1}$$

where W_d is the weight of the dry insoluble part of the hydrogel after extraction with water, and W_0 is the initial weight of the xerogel [10].

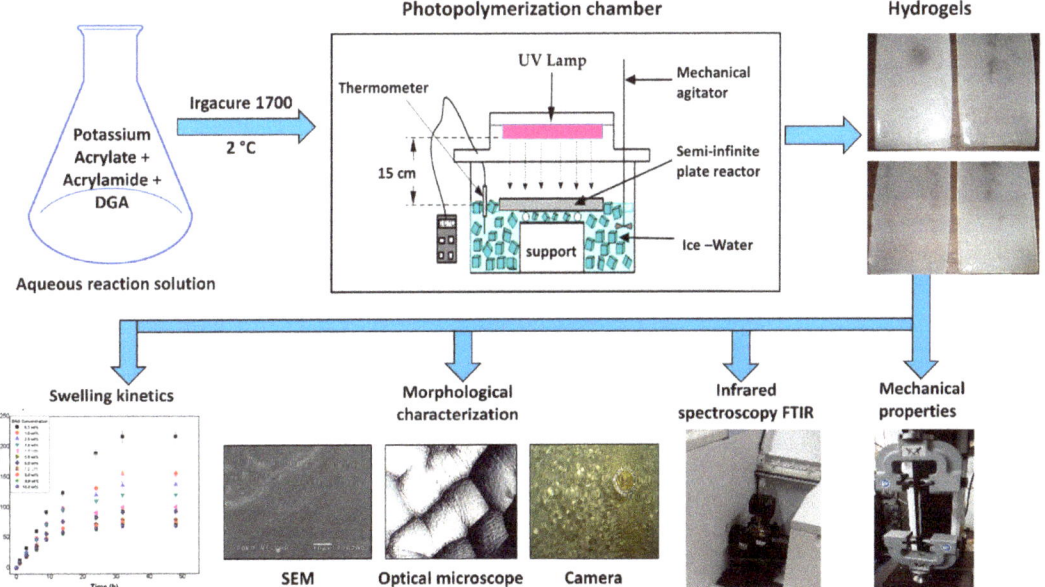

Figure 9. Schematic view of the preparation and characterization of hydrogels.

4.3. Swelling Kinetics

Dried samples were weighed and placed in double-distilled water at 25 °C. The samples were taken out and weighed at different times. Absorbent paper was used to remove the excess of water on the surface of samples. Water absorption was calculated by the difference in weight between the weight of the dry sample and the weight of the swollen sample, using the following equation:

$$S_w \left(\frac{W_t - W_0}{W_0}\right) \tag{2}$$

where W_t and W_0 are the weight of the hydrogel at time t and the weight of the xerogel, respectively. To model the swelling kinetics, the second-order model is commonly used, as proposed by Schott [51], which has been used to predict the swelling in acrylic hydrogels [10,20,52]:

$$\frac{dS_w}{dt} = K(S_{w\infty} - S_w)^2 \tag{3}$$

where S_w and $S_{w\infty}$ are the swellings at time t and at equilibrium swelling, respectively, and K is a constant of the system.

Integrating Equation (3), in the limits of t, S_w, and 0, this gives

$$S_w = \frac{K\, S_w^2\, t}{1 + K\, S_w\, t} \qquad (4)$$

Rewriting Equation (4):

$$\frac{t}{S_w} = \frac{1}{K\, S_{w\infty}^2} + \frac{1}{S_{w\infty}} t \qquad (5)$$

Equation (5) represents second-order kinetics. In this case, the rate of swelling at any time is directly proportional to the square of the still available swelling capacity, that is, to the solvent uptake that has not yet occurred before reaching the maximum or equilibrium uptake [52].

4.4. Morphological Characterization

For morphological characterization, a JEOL JSM 5400 LV scanning electron microscope was used. SEM micrographs were obtained for xerogels, which were coated with gold. The mesostructure of hydrated samples was visualized at different times (10, 20, 30, 40, 40, 60, 120, and 180 s), using an OLIMPUS BX4OF optical microscope with a 40× objective. Images were obtained with a SCC-131A SAMSUNG digital video camera adapted to the microscope. The macrostructure was observed with a Hitachi CCD camera. A coin (21.0 mm of diameter) was used for size comparison.

4.5. Mechanical Characterization

Mechanical measurements of the hydrogels were carried out by an SFM-10 Universal Testing Machine. The strain rate was set at 50 mm per minute at room temperature. The experimental data were obtained using an ASTM D-638-14 Standard Test Method [53] as a reference for tensile properties of plastics with eight specimens. Young's modulus was calculated from the initial slope of the tensile curve.

Author Contributions: Conceptualization, R.M.J.-A. and S.G.-E.; methodology, J.L.G.-O., J.Á.H.-J., M.R.-U. and J.A.C.-O.; formal analysis, O.R.-V. and R.M.J.-A.; analysis of data, J.G.T.R., M.G.L.-R. and S.G.-E.; investigation, R.M.J.-A., J.G.T.R. and S.G.-E.; resources, R.M.J.-A. and S.G.-E.; writing—original draft preparation, R.M.J.-A., J.G.T.R. and S.G.-E. writing—review and editing, M.R.-U., R.M.J.-A., J.G.T.R. and S.G.-E.; visualization, R.M.J.-A. and S.G.-E.; supervision, R.M.J.-A. and S.G.-E.; project administration, R.M.J.-A. and S.G.-E.; funding acquisition, R.M.J.-A., J.G.T.R. and S.G.-E. All authors have read and agreed to the published version of the manuscript.

Funding: The authors would like to thank the Mexican Secretariat of Public Education (SEP) for financially supporting the project "Development and innovation in nanomaterials and nanocomposites, from the International Thematic Network", call 2015.

Institutional Review Board Statement: Not applicable.

Informed Consent Statement: Not applicable.

Data Availability Statement: Data Availability Statement: The data that support the findings of this study are available from the corresponding authors, R.M.J.-A., and S.G.-E., upon reasonable request.

Conflicts of Interest: The authors declare no conflict of interest.

References

1. Amashta, I.A.K.; Trabanca, O.K.; Trabanca, D.K. Materiales inteligentes: Hidrogeles macromoleculares. Algunas aplicaciones biomédicas. In *Anales de la Real Sociedad Española de Química*; Real Sociedad Española de Química: Madrid, Spain, 2005; Volume 4, pp. 35–50.
2. Hoffman, A.S. Hydrogels for biomedical applications. *Adv. Drug Deliv. Rev.* **2012**, *64*, 18–23. [CrossRef]
3. Kiatkamjornwong, S. Superabsorbent polymers and superabsorbent polymer composites. *ScienceAsia* **2007**, *33*, 39–43. [CrossRef]
4. Khare, A.R.; Peppas, N.A. Swelling/deswelling of anionic copolymer gels. *Biomaterials* **1995**, *16*, 559–567. [CrossRef]

5. Bautista, F.; Garcia, S.; Jimenez, R.; Lopez, L.; Orozco, E.; Prado, M.; Reyes, I. Influence of the Synthesis Thermal History on the Structure of Acrylate Based Hydrogels; ANTEC-CONFERENCE PROCEEDINGS: Boston, MA, USA, 2005; Volume 5, p. 373.
6. Costa, M.C.G.; Freire, A.G.; Lourenço, D.V.; Sousa, R.R.D.; Feitosa, J.P.D.A.; Mota, J.C.A. Hydrogel composed of potassium acrylate, acrylamide, and mineral as soil conditioner under saline conditions. *Sci. Agric.* **2021**, *79*, e20200235. [CrossRef]
7. Jasim, L.S.; Aljeboree, A.M. Removal of Heavy Metals by Using Chitosan/Poly (Acryl Amide-Acrylic Acid) Hydrogels: Characterization and Kinetic Study. *NeuroQuantology* **2021**, *19*, 31–38. [CrossRef]
8. Zdravković, A.; Nikolić, L.; Ilić-Stojanović, S.; Nikolić, V.; Najman, S.; Mitić, Ž.; Ćirić, A.; Petrović, S. The removal of heavy metal ions from aqueous solutions by hydrogels based on N-isopropylacrylamide and acrylic acid. *Polym. Bull.* **2018**, *75*, 4797–4821. [CrossRef]
9. Washington, R.P.; Steinbock, O. Frontal Polymerization Synthesis of Temperature-Sensitive Hydrogels. *J. Am. Chem. Soc.* **2001**, *123*, 7933–7934. [CrossRef]
10. Lara-Valencia, V.A.; Dávila-Soto, H.; Moscoso-Sánchez, F.J.; Figueroa-Ochoa, E.B.; Carvajal-Ramos, F.; Fernández-Escamilla, V.V.A.; González-Álvarez, A.; Soltero-Martínez, J.F.A.; Macías-Balleza, E.R.; Enríquez, S.G. The use of polysaccharides extracted from seed of Persea americana var. Hass on the synthesis of acrylic hydrogels. *Quim. Nova* **2018**, *41*, 140–150. [CrossRef]
11. Isik, B.; Kis, M. Preparation and determination of swelling behavior of poly (acrylamide-co-acrylic acid) hydrogels in water. *J. Appl. Polym. Sci.* **2004**, *94*, 1526–1531. [CrossRef]
12. Thakur, A.; Wanchoo, R.K.; Singh, P. Structural parameters and swelling behavior of pH sensitive poly (acrylamide-co-acrylic acid) hydrogels. *Chem. Biochem. Eng. Q.* **2011**, *25*, 181–194.
13. Cheng, W.M.; Hu, X.M.; Zhao, Y.Y.; Wu, M.Y.; Hu, Z.X.; Yu, X.T. Preparation and swelling properties of poly (acrylic acid-co-acrylamide) composite hydrogels. *e-Polymers* **2017**, *17*, 95–106. [CrossRef]
14. Sennakesavan, G.; Mostakhdemin, M.; Dkhar, L.K.; Seyfoddin, A.; Fatihhi, S.J. Acrylic acid/acrylamide based hydrogels and its properties—A review. *Polym. Degrad. Stab.* **2020**, *180*, 109308. [CrossRef]
15. Argade, A.B.; Peppas, N.A. Poly (acrylic acid)-poly (vinyl alcohol) copolymers with superabsorbent properties. *J. Appl. Polym. Sci.* **1998**, *70*, 817–829. [CrossRef]
16. Kiatkamjornwong, S.; Wongwatthanasatien, R. Superabsorbent polymer of poly [acrylamide-co-(acrylic acid)] by foamed polymerization. I. synthesis and water swelling properties. *Macromol. Symp.* **2004**, *207*, 229–240. [CrossRef]
17. Pourjavadi, A.; Kurdtabar, M. Collagen-based highly porous hydrogel without any porogen: Synthesis and characteristics. *Eur. Polym. J.* **2007**, *43*, 877–889. [CrossRef]
18. Tomar, R.S.; Gupta, I.; Singhal, R.; Nagpal, A.K. Synthesis of Poly (Acrylamide-co-Acrylic Acid) based Superabsorbent Hydrogels: Study of Network Parameters and Swelling Behaviour. *Polym. -Plast. Technol. Eng.* **2007**, *46*, 481–488. [CrossRef]
19. Xie, J.; Liu, X.; Liang, J. Absorbency and adsorption of poly (acrylic acid-co-acrylamide) hydrogel. *J. Appl. Polym. Sci.* **2007**, *106*, 1606–1613. [CrossRef]
20. Lopez-Ureta, L.C.; Orozco-Guareño, E.; Cruz-Barba, L.E.; Gonzalez-Alvarez, A.; Bautista-Rico, F. Synthesis and characterization of acrylamide/acrylic acid hydrogels crosslinked using a novel diacrylate of glycerol to produce multistructured materials. *J. Polym. Sci. Part A Polym. Chem.* **2008**, *46*, 2667–2679. [CrossRef]
21. Nesrinne, S.; Djamel, A. Synthesis, characterization and rheological behavior of pH sensitive poly (acrylamide-co-acrylic acid) hydrogels. *Arab. J. Chem.* **2017**, *10*, 539–547. [CrossRef]
22. Jiménez-Amezcua, R.M.; Villanueva-Silva, R.J.; Muñoz-García, R.O.; Macias-Balleza, E.R.; Flores-Sahagun, T.H.S.; Lomelí-Ramírez, M.G.; Torres-Rendon, J.G.; Garcia-Enriquez, S. Preparation of Agave tequilana Weber Nanocrystalline Cellulose and its Use as Reinforcement for Acrylic Hydrogels. *BioResources* **2021**, *16*, 2731–2746. [CrossRef]
23. Mironi-Harpaz, I.; Wang, D.Y.; Venkatraman, S.; Seliktar, D. Photopolymerization of cell-encapsulating hydrogels: Crosslinking efficiency versus cytotoxicity. *Acta Biomater.* **2012**, *8*, 1838–1848. [CrossRef] [PubMed]
24. O'Brien, A.K.; Bowman, C.N. Impact of Oxygen on Photopolymerization Kinetics and Polymer Structure. *Macromolecules* **2006**, *39*, 2501–2506. [CrossRef]
25. O'Brien, A.K.; Bowman, C.N. Modeling the Effect of Oxygen on Photopolymerization Kinetics. *Macromol. Theory Simul.* **2006**, *15*, 176–182. [CrossRef]
26. Fouassier, J.P.; Allonas, X.; Burget, D. Photopolymerization reactions under visible lights: Principle, mechanisms and examples of applications. *Prog. Org. Coat.* **2003**, *47*, 16–36. [CrossRef]
27. Nguyen, K.T.; West, J.L. Photopolymerizable hydrogels for tissue engineering applications. *Biomaterials* **2002**, *23*, 4307–4314. [CrossRef]
28. Decker, C.; Moussa, K. Photopolymerization of multifunctional monomers in condensed phase. *J. Appl. Polym. Sci.* **1987**, *34*, 1603–1618. [CrossRef]
29. Rodgers, Z.L.; Hughes, R.M.; Doherty, L.M.; Shell, J.R.; Molesky, B.P.; Brugh, A.M.; Forbes, M.D.E.; Moran, A.M.; Lawrence, D.S. B_{12}-Mediated, Long Wavelength Photopolymerization of Hydrogels. *J. Am. Chem. Soc.* **2015**, *137*, 3372–3378. [CrossRef]
30. Lin, W.; Zhang, J.; Wang, Z.; Chen, S. Development of robust biocompatible silicone with high resistance to protein adsorption and bacterial adhesion. *Acta Biomater.* **2011**, *7*, 2053–2059. [CrossRef]
31. Weiqing, R.; Jinliang, Q.; Yuli, H.; Aijie, N. Superabsorbent resin of acrylic acid/ammonium acrylate copolymers synthesized by ultraviolet photopolymerization. *J. Appl. Polym. Sci.* **2004**, *95*, 546–555. [CrossRef]

32. Wen, Y.; Zhu, X.; Gauthier, D.E.; An, X.; Cheng, D.; Ni, Y.; Yin, L. Development of poly (acrylic acid)/nanofibrillated cellulose superabsorbent composites by ultraviolet light induced polymerization. *Cellulose* **2015**, *22*, 2499–2506. [CrossRef]
33. Martínez-Salcedo, S.L.; Torres-Rendón, J.G.; García-Enriquez, S.; Anzaldo-Hernández, J.; Silva-Guzmán, J.A.; de Muniz, G.I.B.; Lomelí-Ramírez, M.G. Physicomechanical Characterization of Poly (acrylic acid-co-acrylamide) Hydrogels Reinforced with TEMPO-oxidized Blue Agave Cellulose Nanofibers. *Fibers Polym.* **2022**, *23*, 1161–1170. [CrossRef]
34. Ruan, W.; Qiao, J.; Huang, Y.; Niu, A. Synthesis of superabsorbent resin by ultraviolet photopolymerization. *J. Appl. Polym. Sci.* **2004**, *92*, 1618–1624. [CrossRef]
35. Trujillo, V.; Kim, J.; Hayward, R.C. Creasing instability of surface-attached hydrogels. *Soft Matter* **2008**, *4*, 564–569. [CrossRef]
36. Tanaka, T.; Sun, S.T.; Hirokawa, Y.; Katayama, S.; Kucera, J.; Hirose, Y.; Amiya, T. Mechanical instability of gels at the phase transition. *Nature* **1987**, *325*, 796–798. [CrossRef]
37. Tanaka, H.; Tomita, H.; Takasu, A.; Hayashi, T.; Nishi, T. Morphological and kinetic evolution of surface patterns in gels during the swelling process: Evidence of dynamic pattern ordering. *Phys. Rev. Lett.* **1992**, *68*, 2794–2797. [CrossRef]
38. Li, C.; Hu, Z.; Li, Y. Temperature and time dependencies of surface patterns in constrained ionic N-isopropylacrylamide gels. *J. Chem. Phys.* **1994**, *100*, 4645–4652. [CrossRef]
39. Guvendiren, M.; Yang, S.; Burdick, J.A. Swelling-induced surface patterns in hydrogels with gradient crosslinking density. *Adv. Funct. Mater.* **2009**, *19*, 3038–3045. [CrossRef]
40. Guvendiren, M.; Burdick, J.A.; Yang, S. Solvent induced transition from wrinkles to creases in thin film gels with depth-wise crosslinking gradients. *Soft Matter* **2010**, *6*, 5795–5801. [CrossRef]
41. Chuang, Y.F.; Wei, M.K.; Yang, F.; Lee, S. Water-driven surface wrinkling of poly (2-hydroxyethyl methacrylate) after ultraviolet irradiation. *J. Polym. Res.* **2021**, *28*, 1–11. [CrossRef]
42. Saha, K.; Kim, J.; Irwin, E.; Yoon, J.; Momin, F.; Trujillo, V.; Schaffeer, D.V.; Healy, K.E.; Hayward, R.C. Surface creasing instability of soft polyacrylamide cell culture substrates. *Biophys. J.* **2010**, *99*, L94–L96. [CrossRef]
43. Stafford, C.M.; Harrison, C.; Beers, K.L.; Karim, A.; Amis, E.J.; VanLandingham, M.R.; Kim, H.C.; Volksen, W.; Miller, R.D.; Simonyi, E.E. A buckling-based metrology for measuring the elastic moduli of polymeric thin films. *Nat. Mater.* **2004**, *3*, 545–550. [CrossRef] [PubMed]
44. Sidorenko, A.; Krupenkin, T.; Taylor, A.; Fratzl, P.; Aizenberg, J. Reversible switching of hydrogel-actuated nanostructures into complex micropatterns. *Science* **2007**, *315*, 487–490. [CrossRef] [PubMed]
45. Holmes, D.P.; Ursiny, M.; Crosby, A.J. Crumpled surface structures. *Soft Matter* **2008**, *4*, 82–85. [CrossRef] [PubMed]
46. Pandey, A.; Holmes, D.P. Swelling-induced deformations: A materials-defined transition from macroscale to microscale deformations. *Soft Matter* **2013**, *9*, 5524–5528. [CrossRef] [PubMed]
47. Phadnis, A.; Manning, K.C.; Sanders, I.; Burgin, T.P.; Rykaczewski, K. Droplet-train induced spatiotemporal swelling regimes in elastomers. *Soft Matter* **2018**, *14*, 5869–5877. [CrossRef]
48. Stockmayer, W.H. Theory of molecular size distribution and gel formation in branched polymers II. General cross linking. *J. Chem. Phys.* **1944**, *12*, 125–131. [CrossRef]
49. Magalhães, A.S.G.; Almeida Neto, M.P.; Bezerra, M.N.; Ricardo, N.M.; Feitosa, J. Application of FTIR in the determination of acrylate content in poly (sodium acrylate-co-acrylamide) superabsorbent hydrogels. *Química Nova* **2012**, *35*, 1464–1467. [CrossRef]
50. Leitão, R.C.; Moura, C.P.D.; da Silva, L.R.; Ricardo, N.M.; Feitosa, J.; Muniz, E.C.; Fajardo, R.A.; Rodrigues, F.H. Novel superabsorbent hydrogel composite based on poly (acrylamide-co-acrylate)/nontronite: Characterization and swelling performance. *Química Nova* **2015**, *38*, 370–377. [CrossRef]
51. Schott, H. Swelling kinetics of polymers. *J. Macromol. Sci. Part B* **1992**, *31*, 1–9. [CrossRef]
52. Katime, I.; Velada, J.L.; Novoa, R.; Díaz de Apodaca, E.; Puig, J.; Mendizabal, E. Swelling Kinetics of Poly (acrylamide)/Poly (mono-n-alkyl itaconates) Hydrogels. *Polym. Int.* **1996**, *40*, 281–286. [CrossRef]
53. ASTM International. *D638-14 Standard Test Method for Tensile Properties of Plastics*; ASTM International: West Conshohocken, PA, USA, 2014.

Article

UV-VIS Curable PEG Hydrogels for Biomedical Applications with Multifunctionality

Tina Sabel-Grau *, Arina Tyushina, Cigdem Babalik and Marga C. Lensen *

Nanopatterned Biomaterials (Secr. C 1), Department of Chemistry, Technische Universität Berlin, Strasse des 17. Juni 115, 10623 Berlin, Germany; arina1984@web.de (A.T.); cigdem7189@hotmail.de (C.B.)
* Correspondence: Tina@physik.tu-berlin.de (T.S.-G.); Marga@lensenlab.de (M.C.L.)

Abstract: Multifunctional biomedical materials capable of integrating optical functions are highly desirable for many applications, such as advanced intra-ocular lens (IOL) implants. Therefore, poly(ethylene glycol)-diacrylate (PEG-DA) hydrogels are used with different photoinitiators (PI). In addition to standard UV PI Irgacure, Erythrosin B and Eosin Y are used as PI with high sensitivity in the optical range of the spectrum. The minimum PI concentrations for producing new hydrogels with PEG-DA and different PIs were determined. Hydrogel films were obtained, which were applicable for light-based patterning and, hence, the functionalization of surface and volume. Cytotoxicity tests confirm cytocompatibility of hydrogels and compositions. Exploiting the correlation of structure and function allows biomedical materials with multifunctionality.

Keywords: hydrogels; photopolymers; volume holography; photo curing; multifunctional biomedical biomaterials; light-responsive materials

Citation: Sabel-Grau, T.; Tyushina, A.; Babalik, C.; Lensen, M.C. UV-VIS Curable PEG Hydrogels for Biomedical Applications with Multifunctionality. Gels 2022, 8, 164. https://doi.org/10.3390/gels8030164

Academic Editors: Yang Liu and Kiat Hwa Chan

Received: 4 February 2022
Accepted: 3 March 2022
Published: 5 March 2022

Publisher's Note: MDPI stays neutral with regard to jurisdictional claims in published maps and institutional affiliations.

Copyright: © 2022 by the authors. Licensee MDPI, Basel, Switzerland. This article is an open access article distributed under the terms and conditions of the Creative Commons Attribution (CC BY) license (https://creativecommons.org/licenses/by/4.0/).

1. Introduction

In tissue engineering and medical science, hydrogels are particularly suitable as tissue scaffolds due to their tunable properties including water content, swellability, diffusivity and stiffness [1,2]. In fact, many applications are opened up by the explicit control over molecular structure and mechanical properties, such as elasticity, cross-linking degree or surface morphology [3]. Using light to control properties, hydrogels are well-suited scaffolds for light-responsive functionality [4]. Such stimuli-responsive hydrogel materials can change their mechanical properties upon exposure to light. However, relatively little has been studied with respect to the their optical properties and little attention has been paid to their potential photonic functionalities [5]. Control over optical properties and the resulting integration of optical functionality open up new opportunities for multifunctional biomedical materials such as advanced intraocular lens (IOL) implants.

Cataract, the irreversible turbidity of the natural lens of the eye, is one of the most common causes for global blindness and can only be treated by replacing the clouded lens with an artificial IOL implant. Among the state-of-the-art IOLs are modern foldable hydrogel lenses [6]. Persistent problems with IOLs include postoperative calcification [7] and secondary cataract [8]. The processes underlying such postoperative clouding, emerging in vivo from interaction with the biological environment, are still not well understood.

This is where the idea of volume holographic structuring comes into play. Prospective IOLs, based on multifunctional biomedical material with integrated optical functionality, could fulfill their function—i.e., to focus the light onto the retina—with an optically structured volume [9]. As a result, the shape and surface of the IOL remain free and available for other purposes. Thus, subsequent surface modifications remain optional to achieve specific interactions with the biological environment. In order to make this possible, we propose a strategy to combine the optical structuring of the volume and a specific modification of the surface. Therefore, volume holographic structuring can be applied for the integration of three-dimensional optical structures with specific functionality in terms of diffractive properties.

Volume holography is a very interesting field of application for photo-responsive polymers, where diffractive structures are induced by a spatially modulated holographic exposure [10]. Holographic elements such as diffractive structures can accommodate classical optical functions, while at the same time being extremely flat in shape and low in weight. This gives rise to a great potential for replacing classical refractive optical systems or extending them with new functionalities. The prerequisite in each case is the availability of suitable, photo-patternable materials that can exhibit function through structure. An example of such diffractive structures with classical optical function includes holographic lenses.

Poly(ethylene glycol) (PEG)-based hydrogels are generally promising as tissue-engineering scaffolds due to their biocompatibility and intrinsic resistance to protein adsorption and cell adhesion [11]. Furthermore, poly(ethylene glycol)-diacrylate (PEG-DA) hydrogels can be used as model materials for the generation of internal 3D patterns [12]. Acrylate-terminated PEG macromers undergo rapid polymerization in the presence of photoinitiators that generate radicals when exposed to light [13]. This makes PEG-DA hydrogels interesting as scaffolds into which desired bioactivity can be tailored via light-based patterning [12]. In this context, it was shown that hydrogels can also be structured photolithographically using diffusion processes—which are the basis for volume holography.

Additionally, the micropatterning of PEG-based hydrogels with gold nanoparticles allows for the fabrication of functionalized PEG-based hydrogel films [14,15]. The integration—and holographic assembly—of nanoparticles in turn enables the modification of optical properties on the microscale and nanoscale in the form of holographic nanoparticle-polymer composite gratings [16].

In all this, the type of photoinitiator (PI) used is key for the specific photo-response of a certain material. The properties of the PI has strong influence on holographic grating formation in the respective material [17]. It also influences how well certain conditions are met, such as resistance to humidity [18]. Eosin-Y (EY) and Erythrosin B (EB) are amongst the possible PIs applied for holographic grating formation in an AA/PVA photopolymer [19]. EY is used as a PI due to its excellent spectroscopic properties, which makes it suitable for use with light sources in the visible range and safe for living organisms [20]. EB can only be used for free radical polymerization [19].

2. Results and Discussion

2.1. Gel Formation

PEG-DA was mixed with PI (Irgacure, EB and EY, respectively). Films were prepared by photopolymerization with 366 nm for 1 h. In terms of optical transparency, mechanical integrity, flexibility, and stability, the new gels compare well with other gels based on PEG-DA [21–23].

For PEG-DA with Irgacure, EB and EY as PI, a minimum concentration of PI was needed to make gel. With less PI concentration, no hydrogel was formed. The minimum concentration for the different PIs is shown in Table 1. We found a minimum PI concentration for producing the new hydrogels with PEG-DA and different PIs to be 0.025% for Irgacure, 0.1% for EB and 0.5% for EY, respectively.

Table 1. Minimum PI concentration to make gel for the different PIs.

Photoinitiator (PI)	Minimum PI Concentration to Make Gel with PEG-DA
Irgacure 2959	0.025%
EB	0.1%
EY	0.5%

2.2. UV-Vis Spectra

Figure 1 shows the UV-Vis spectra before and after crosslinking for the novel PIs. In general, for all new PEG-DA hydrogels, we find spectra shifted somewhat toward higher wavelengths compared to the pure PIs [24,25], while crosslinking tends to cause a small shift toward lower wavelengths, as already substantiated in the literature [26].

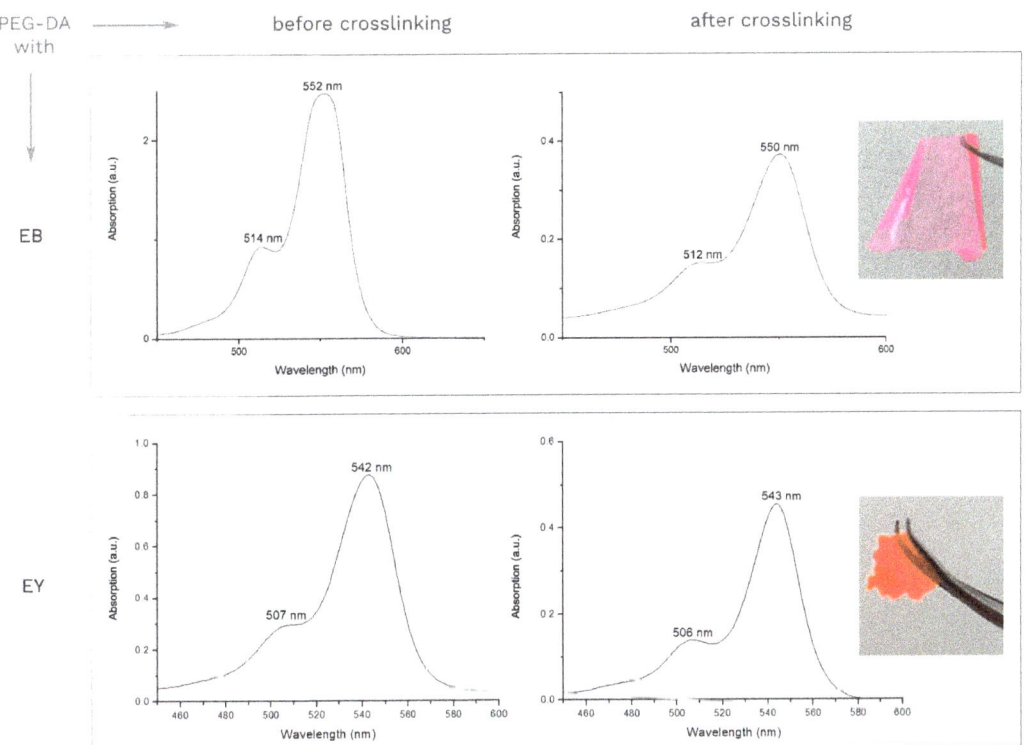

Figure 1. UV-Vis spectra for the novel PIs (EB and EY) with PEG-DA before and after crosslinking, respectively. The hydrogels with new PIs (EB and EY) feature good spectroscopic properties, allowing for use with light sources in the visible range.

While Irgacure, which is an often employed and suitable photoinitiator for biomaterials research, has an absorption maximum in the UV/Vis-spectrum around 300 nm [24], the novel dyes under investigation display a strong absorption of visible light with wavelengths up to 550 nm (see Figure 1).

With EB as PI, the absorption maximum before crosslinking was around 552 nm with a shoulder at 514 nm. After crosslinking, the absorption peak can be observed at 550 nm with a shoulder at 512 nm. The typical color for the hydrogel with EB as PI is pink, as shown in Figure 1.

With EY as PI, the absorption maximum before crosslinking was around 542 nm with a shoulder at 507 nm. After crosslinking, the absorption peak can be observed at 543 nm with a shoulder at 506 nm. The typical color for the hydrogel with EY as PI is orange, as shown in Figure 1.

The major advantage of the new PIs (EB and EY) over the standard Irgacure is their high sensitivity in the optical range of the spectrum, which enables optical patterning—e.g., by volume holography. To counter the disadvantage of strong coloring, the composition could be optimized, e.g., by use of a crosslinker so that the concentrations of EB and EY can be reduced, respectively.

In the next step, the respective optical response must be determined depending on the composition. In some cases (such as with an organic cationic ring-opening polymerization system), competing effects regarding the contribution to the optical grating formation can be observed [27]. It is also known that optical shrinkage can have significant influence on grating formation [28] and that the amount of PEG in a composition affects film shrinkage, as well as its optical properties [29]. Furthermore, we have also observed that photoinitiators may contribute to light-induced modification of optical properties and subsequent pattern formation as well [30].

2.3. Cytotoxicity Tests

A live/dead staining assay has been used to study the cell viability after incubation with PI EB and EY and also EB 0.1% with PEG-DA before and after crosslinking for 24 h. In the live/dead staining assay, dead cells turn up red, while living cells turn up green when observed with a fluorescence microscope. As shown in Figure 2, all cells but one appear to be green, indicating that PIs and hydrogel (PEG-DA with 0.1% PI EB) are cytocompatible and suitable as substrates for studying the behaviour of L929 cells at the biointerface.

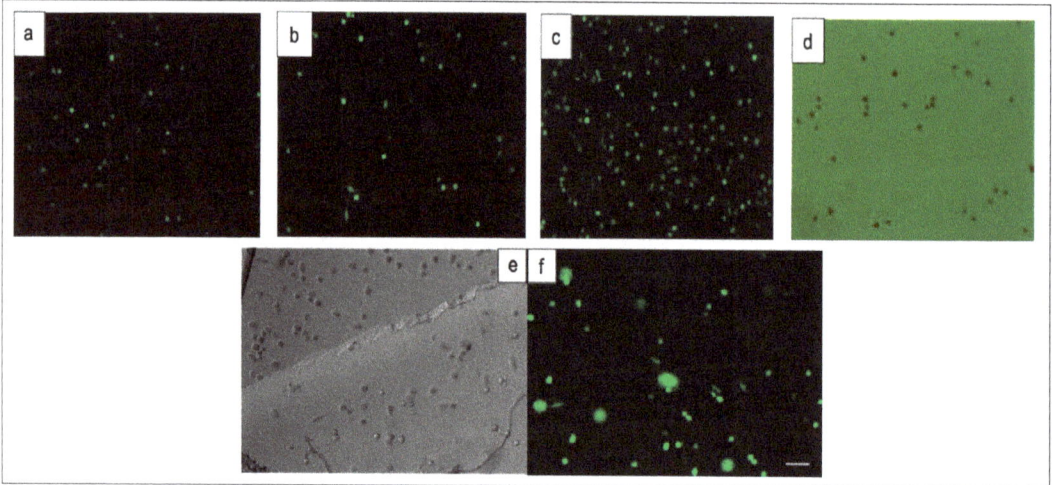

Figure 2. Fluorescent images of cells test by live/dead cell assay with L929 after incubated for 24 h. (**a**) Control cells; (**b**) EB; (**c**) EY; (**d**) EB 0.1% with PEG-DA; (**e**) optical micrograph of cross-linked EB/PEG-DA hydrogel; (**f**) fluorescence image of cross-linked EB/PEG-DA hydrogel. Scale bar depicts 50 µm.

Figure 2 shows fluorescent images of cell tests by live/dead cell assay after being incubated for 24 h with cell line L929. Cell tests shown in Figure 2 confirm cytocompatibility of PEG-DA hydrogels and PIs, hence affirming its aptitude for biomedical applications.

3. Conclusions

In addition to the standard photoinitiator Irgacure, Erythrosin B (EB) and Eosin Y (EY) were used as photoinitiators (PI) in PEG-DA hydrogel. We have determined the minimum PI concentration for producing new hydrogels with PEG-DA for the different PIs respectively. All the PIs are cytocompatible and suitable as substrates for studying the behaviour of L929 cells at the biointerface.

The new PIs (EB and EY) feature good spectroscopic properties, allowing for their application with light sources in the visible range and, thus, for applications in volume holography. Cytotoxicity test with cell line L929 were performed to confirm cytocompatibility of hydrogels and PIs. This opens up many options for PEG-DA hydrogels with PIs as multifunctional biomedical applications.

The strategy to combine optical structuring of the volume and specific modification of the surface is particularly interesting for the design of advanced intraocular lens (IOL) implants: based on a multifunctional biomedical material with integrated optical functionality and operating by the principle 'function by structure', such a new type of IOL is expected to attain enhanced functionality [9]. The optical functionality of an IOL with integrated holographic lens as a diffractive element consists in focusing the light onto the retina. Several holographic elements can be combined in stacks, where the functionality of the individual elements overlap. The selectivity of a stack then results in a superposition of Bragg selectivity of the individual elements [31].

Beyond an enhanced functionality, the transfer of the optical functionality from the surface into the volume of the IOL implant brings further benefits such as the free interface for specific interaction with the biological environment. As the existing problems with conventional IOLs, such as postoperative clouding, emerge in vivo from interactions with the biological environment, they could be better addressed with free-surface IOLs.

The next step towards such a multifunctional optical material is to better understand the processes that underlay optical structuring, such as the interplay of polymerization and diffusion in the case of holographic gratings. Here, the general approach is to understand holographic grating formation as a consequence of photopolymerization and mass transport processes: local polymerization is induced by a light pattern projected into the photosensitive medium. Polymerization proportional to the light intensity results in the induction of a chemical gradient, followed by monomer diffusion and subsequent polymerization. The final grating is formed as a periodic modulation of optical properties, according to the recording light pattern [10].

It now remains to be clarified what role the individual components play in the formation of optical structures in cases of PEG-DA hydrogels with EB and EY. Furthermore, it remains to be examined if other additives—such as crosslinker or dopant, e.g., in the form of azobenzene-functionalized acrylates or gold nanoparticles—have a positive effect on the formation of optical patterns.

4. Materials and Methods
4.1. Preparation of PEG Hydrogels
4.1.1. Chemicals

Poly(ethylene glycol) diacrylate (PEG, Mn 575) and 2-hydroxy-4′-(2-hydroxyethoxy)-2-methylpropiophenone (photoinitiator (PI)—Irgacure 2959), Erythrosin B and Eosin Y were from Sigma-Aldrich Chemie GmbH (Steinheim, Germany). The chemical structures of the hydrogel components are shown in Figure 3.

4.1.2. Synthesis of PEG Based Photopolymer with PIs

PEG-DA was used as precursor. It was mixed with PI (Irgacure, EB and EY, respectively). PI concentration varied between 1% and 6 ppm. The chemical structures of different PIs are shown in Figure 3. For the good mixing of both substances, the mixture was sonicated for around 30 min. At first, the mixture was converted in a cuvette and measured with a UV-Vis spectrometer to obtain spectra before crosslinking. Then, the mixture was dispensed on a glass slide and covered with a thin glass cover slip to achieve a flat and thin hydrogel sample. The glass-sandwich was placed under a UV-light source (366 nm) for 60 min and the glass cover slip was peeled off. A flat, thin standalone hydrogel film was received and also prepared for UV-Vis measurement. Therefore, the hydrogel was placed on a thin glass cover slip and measured with a UV-Vis spectrometer to obtain spectra after crosslinking.

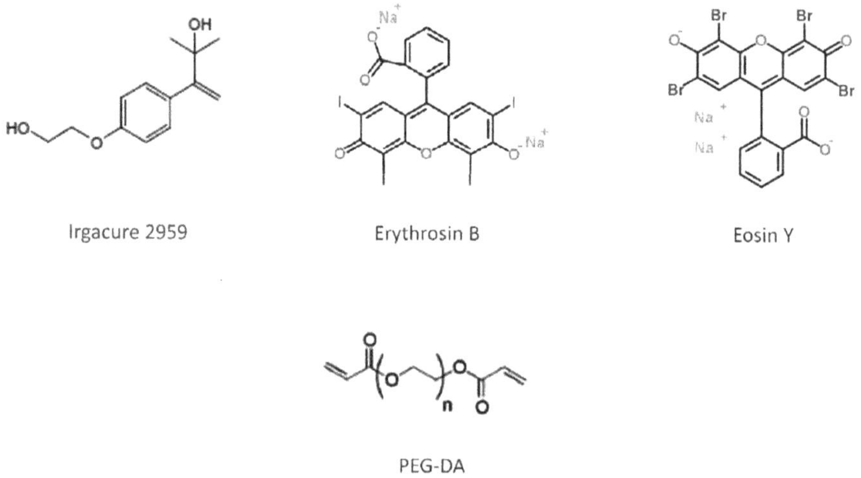

Figure 3. Irgacure 2959, Erythrosin B and Eosin Y are used as photoinitiators (PI); PEG-DA as a precursor for PEG hydrogel.

4.2. Cell Culture

4.2.1. Chemicals

Mouse fibroblast L929 cells were provided by Dr. Lehmann, Fraunhofer Institute for Cell Therapy and Immunology, IZI, Leipzig, Germany. RPMI 1640 medium, Trypsin, Fetal Bovine Serum (FBS) and Penicillin/Streptomycin (PS) were provided by PAA Laboratories GmbH, Austria, and cell culture plates are from SPL Live Sciences Inc., Seoul, Korea. The Incubator CB150 Series was from Binder GmbH, Germany. Phosphate Buffered Saline solution (Dulbecco's PBS) was purchased from Sigma-Aldrich Chemie, GmbH, Germany. The counter chamber was from Marienfeld Superior (Paul Marienfeld GmbH & Co., KG, Lauda-Königshofen, Germany).

4.2.2. Cell Culture Experiments

The mouse fibroblasts L929 cells were cultured in the tissue culture plate in RPMI 1640 medium with the addition of 10% Fetal Bovine Serum (FBS) and 1% Penicillin/Streptomycin (PS) in a cell culture plate in an incubator at controlled temperature (37 °C) and CO_2 atmosphere (5%). The cells were grown in a cell culture plate and cell culture experiments were performed when a confluency of 75% to 95% was reached.

The hydrogel samples were prior washed with water and kept in a PBS solution for around 30 min before cell culture experiments. As soon as a confluency of at least 75% was reached, the cells were washed with PBS, detached by using trypsin and, after the centrifugation process, a new medium was added on the cells and mixed properly. An amount of 10 µL of this cell medium solution was placed on a cell counter chamber in order to count the cell number by using an optical microscope and to achieve a concentration of 40,000 cells/mL. Depending on the counted cell number, the cell solution was mixed with a defined amount of new medium. The samples were placed in a TCPS plate or on the washed and precut hydrogel. The samples were then cultured within these cells for 24 h, at 37 °C in a 5% CO_2 atmosphere in a TCPS.

4.2.3. Live Dead Cytotoxicity Assay

The live dead cytotoxicity assay is a fluorescence-based method for checking the viability of cells. Hereby, the cells are stained with fluoresceindiacetete (FBS) and propidiumiodid (PI) molecules. FBS is dissociated in the cytoplasma of live cells into green fluorescence molecules and, due to the size and charge of the PI molecules, they only can enter into the cell cytoplasm when the cell membranes are damaged and are bound to nucleic acids, which appear then as red fluorescence color. In this manner, the live cells appear as green-stained cells and dead cells appear as red-stained cells in a fluorescent microscope image.

For the live dead assay, a 1:1 v/v solution of PI and FBS in PBS was prepared and added into the cell culture solution in a dark environment. Immediately after the mixtures are prepared, fluorescence images are taken.

4.3. Analytical Instruments

UV-Vis spectra were obtained with Cary 4000 UV-Vis Spectrometer (Agilent Technologies, Santa Clara, CA, USA). The spectral range from 300 to 900 nm was measured at room temperature. Liquid samples (before crosslinking) were converted in a cuvette for measurement. The flat, thin standalone hydrogel films (after crosslinking) were placed on a thin glass cover slip for measurement with a UV-Vis spectrometer.

The results from cell culture experiments were observed via the optical microscope from Carl Zeiss, Germany, and analyses were performed with the AxioVision V4.8.2 software (Carl Zeiss, Oberkochen, Germany).

Author Contributions: Conceptualization, T.S.-G. and M.C.L.; investigation, A.T. and C.B.; writing—original draft preparation, review and editing, T.S.-G.; supervision, M.C.L.; project administration and funding acquisition, T.S.-G. All authors have read and agreed to the published version of the manuscript.

Funding: This research was funded by Deutsche Forschungsgemeinschaft (DFG, German Research Foundation), grant number SA 2990/1-1. The APC was funded by DFG.

Institutional Review Board Statement: Not applicable.

Informed Consent Statement: Not applicable.

Data Availability Statement: Not applicable.

Conflicts of Interest: The authors declare no conflict of interest. The funders had no role in the design of the study; in the collection, analyses or interpretation of data; in the writing of the manuscript; or in the decision to publish the results.

References

1. Luo, Y.; Shoichet, M.S. A photolabile hydrogel for guided three-dimensional cell growth and migration. *Nat. Mater.* **2004**, *3*, 249–253. [CrossRef]
2. Higgins, C.; Aisenbrey, E.; Wahlquist, J.; Heveran, C.; Ferguson, V.; Bryant, S.; Mcleod, R. Enhanced mechanical properties of photo-clickable thiol-ene PEG hydrogels through repeated photopolymerization of in-swollen macromer. *Soft Matter* **2016**, *12*, 9095–9104.
3. Lensen, M.C.; Schulte, V.; Salber, J.; Dietz, M.; Menges, F.; Möller, M. Cellular responses to novel, micropatterned biomaterials. *Pure Appl. Chem.* **2008**, *80*, 2479–2487. [CrossRef]
4. Li, L.; Scheiger, J.M.; Levkin, P.A. Design and Applications of Photoresponsive Hydrogels. *Adv. Mater.* **2019**, *31*, 1807333. [CrossRef] [PubMed]
5. Choi, M.; Choi, J.W.; Kim, S.; Nizamoglu, S.; Hahn, S.K.; Yun, S.H. Light-guiding hydrogels for cell-based sensing and optogenetic synthesis in vivo. *Nat. Photonics* **2013**, *7*, 987–994. [CrossRef] [PubMed]
6. Izak, A.M.; Werner, L.; Pandey, S.K.; Apple, D.J. Calcification of modern foldable hydrogel intraocular lens designs. *Eye* **2003**, *17*, 393–406. [CrossRef]
7. Kanclerz, P.; Yildirim, T.; Khoramnia, R. Microscopic Characteristics of Late Intraocular Lens Opacifications. *Arch. Pathol. Lab. Med.* **2020**, *145*. [CrossRef]
8. Lundgren, B.; Jonsson, E.; Rolfsen, W. Secondary cataract. *Acta Ophthalmol.* **2009**, *70*, 25–28. [CrossRef]
9. Sabel, T.; Lensen, M.C. Function and structure—Combined optical functionality and specific bio- interaction for multifunctional biomedical materials. *J. Med. Mater. Technol.* **2017**, *1*, 10–12.
10. Sabel, T.; Lensen, M.C. Volume Holography: Novel Materials, Methods and Applications. In *Holographic Materials and Optical Systems*; Naydenova, I., Babeva, T., Nazarova, D., Eds.; InTech: Rijeka, Croatia, 2017.
11. Naahidi, S.; Jafari, M.; Logan, M.; Wang, Y.; Yuan, Y.; Bae, H.; Dixon, B.; Chen, P. Biocompatibility of hydrogel-based scaffolds for tissue engineering applications. *Biotechnol. Adv.* **2017**, *35*, 530–544. [CrossRef]
12. Hahn, B.M.S.; Miller, J.S.; West, J.L. Three-Dimensional Biochemical and Biomechanical Patterning of Hydrogels for Guiding Cell Behavior. *Adv. Mater.* **2006**, *18*, 2679–2684. [CrossRef]

13. Nguyen, K.T.; West, J.L. Photopolymerizable hydrogels for tissue engineering applications. *Biomaterials* **2002**, *23*, 4307–4314. [CrossRef]
14. Yesildag, C.; Ouyang, Z.; Zhang, Z.; Lensen, M.C. Micro-Patterning of PEG-Based Hydrogels with Gold Nanoparticles Using a Reactive Micro-Contact-Printing Approach. *Front. Chem.* **2019**, *6*, 1–10. [CrossRef] [PubMed]
15. Yesildag, C.; Tyushina, A.; Lensen, M.C. Nano-Contact Transfer with Gold Nanoparticles on PEG Hydrogels and Using Wrinkled PDMS-Stamps. *Polymers* **2017**, *9*, 199. [CrossRef] [PubMed]
16. Tomita, Y.; Kageyama, A.; Iso, Y.; Umemoto, K.; Kume, A.; Liu, M.; Pruner, C.; Jenke, T.; Roccia, S.; Geltenbort, P.; et al. Fabrication of nanodiamond-dispersed composite holographic gratings and their light and slow-neutron diffraction properties. *Phys. Rev. Appl.* **2020**, *14*, 44056. [CrossRef]
17. Qi, Y.; Li, H.; Fouassier, J.P.; Lalevée, J.; Sheridan, J.T. Comparison of a new photosensitizer with erythrosine B in an AA / PVA-based photopolymer material. *Appl. Opt.* **2014**, *53*, 1052–1062. [CrossRef]
18. Mikulchyk, T.; Martin, S.; Naydenova, I. Investigation of the sensitivity to humidity of an acrylamide-based photopolymer containing N-phenylglycine as a photoinitiator. *Opt. Mater.* **2014**, *37*, 810–815. [CrossRef]
19. Qi, Y.; Gleeson, M.R.; Guo, J.; Gallego, S.; Sheridan, J.T. Quantitative Comparison of Five Different Photosensitizers for Use in a Quantitative Comparison of Five Different Photosensitizers for Use in a Photopolymer. *Phys. Res. Int.* **2012**, *17*, 11.
20. Tomal, W.; Ortyl, J. Water-Soluble Photoinitiators in Biomedical Applications. *Polymers* **2020**, *12*, 1073. [CrossRef]
21. Schulte, V.A.; Diez, M.; Hu, Y.; Möller, M.; Lensen, M.C. Combined Influence of Substrate Stiffness and Surface Topography on the Antiadhesive Properties of Acr-sP(EO-stat-PO) Hydrogels. *Biomacromolecules* **2010**, *11*, 3375–3383. [CrossRef]
22. Lensen, M.C.; Schulte, V.A.; Diez, M. Cell Adhesion and Spreading on an Intrinsically Anti-Adhesive PEG Biomaterial. In *Biomaterials*; Pignatello, R., Ed.; IntechOpen: Rijeka, Croatia, 2011.
23. Kelleher, S.; Jongerius, A.; Löbus, A.; Strehmel, C.; Zhang, Z.; Lensen, M.C. AFM Characterisation of elastically micropatterned surfaces fabricated by Fill-Molding In Capillaries (FIMIC) and investigation of the topographical influence on cell adhesion to the patterns. *Adv. Eng. Mater.* **2012**, *14*, B56–B65. [CrossRef]
24. Kaastrup, K.; Sikes, H.D. Using photo-initiated polymerization reactions to detect molecular recognition. *Chem. Soc. Rev.* **2016**, *45*, 532–545. [CrossRef] [PubMed]
25. Yang, R.; Soper, S.A.; Wang, W. A new UV lithography photoresist based on composite of EPON resins 165 and 154 for fabrication of high-aspect-ratio microstructures. *Sensors Actuators A Phys.* **2007**, *135*, 625–636. [CrossRef]
26. Sung, J.; Lee, D.G.; Lee, S.; Park, J.; Jung, H.W. Crosslinking Dynamics and Gelation Characteristics of Photo- and Thermally Polymerized Poly(Ethylene Glycol) Hydrogels. *Materials* **2020**, *13*, 3277. [CrossRef]
27. Sabel, T.; Zschocher, M. Transition of refractive index contrast in course of grating growth. *Sci. Rep.* **2013**, *3*, 1–7. [CrossRef]
28. Sabel, T. Spatially resolved analysis of Bragg selectivity. *Appl. Sci.* **2015**, *5*, 1064–1075. [CrossRef]
29. Dimitrov, O.; Stambolova, I.; Vassilev, S.; Lazarova, K.; Babeva, T.; Mladenova, R. Surface and Morphological Features of ZrO_2 Sol-Gel Coatings Obtained by Polymer Modified Solution. *Mater. Proc.* **2020**, *2*, 1–8.
30. Sabel, T. Volume hologram formation in SU-8 photoresist. *Polymers* **2017**, *9*, 198. [CrossRef]
31. Akbari, H.; Naydenova, I.; Persechini, L.; Garner, S.M.; Cimo, P.; Martin, S. Diffractive Optical Elements with a Large Angle of Operation Recorded in Acrylamide Based Photopolymer on Flexible Substrates. *Int. J. Polym. Sci.* **2014**, *2014*, 918285. [CrossRef]

Article

The Effect of Crosslinking Degree of Hydrogels on Hydrogel Adhesion

Zhangkang Li [1], Cheng Yu [2], Hitendra Kumar [3], Xiao He [4], Qingye Lu [4], Huiyu Bai [2,*], Keekyoung Kim [1,3,*] and Jinguang Hu [1,4,*]

1. Department of Biomedical Engineering, University of Calgary, 2500 University Dr. NW, Calgary, AB T2N 1N4, Canada
2. Key Laboratory of Synthetic and Biological Colloids, Ministry of Education, School of Chemical and Material Engineering, Jiangnan University, Wuxi 214122, China
3. Department of Mechanical and Manufacturing Engineering, University of Calgary, 2500 University Dr. NW, Calgary, AB T2N 1N4, Canada
4. Department of Chemical and Petroleum Engineering, University of Calgary, 2500 University Dr. NW, Calgary, AB T2N 1N4, Canada
* Correspondence: Authors: hybai@jiangnan.edu.cn (H.B.); keekyoung.kim@ucalgary.ca (K.K.); jinguang.hu@ucalgary.ca (J.H.)

Abstract: The development of adhesive hydrogel materials has brought numerous advances to biomedical engineering. Hydrogel adhesion has drawn much attention in research and applications. In this paper, the study of hydrogel adhesion is no longer limited to the surface of hydrogels. Here, the effect of the internal crosslinking degree of hydrogels prepared by different methods on hydrogel adhesion was explored to find the generality. The results show that with the increase in crosslinking degree, the hydrogel adhesion decreased significantly due to the limitation of segment mobility. Moreover, two simple strategies to improve hydrogel adhesion generated by hydrogen bonding were proposed. One was to keep the functional groups used for hydrogel adhesion and the other was to enhance the flexibility of polymer chains that make up hydrogels. We hope this study can provide another approach for improving the hydrogel adhesion generated by hydrogen bonding.

Keywords: hydrogel; adhesive strength; crosslinking; hydrogen bonding

1. Introduction

In several day-to-day applications, adhering materials to each other requires the use of a glue [1]. In this case, these two materials are called adherends, and the glue is adhesive [2]. Hydrogels are highly porous three-dimensional crosslinked polymer networks consisting of hydrophilic polymers that enable hydrogels to retain large amounts of water [2]. However, the use of a glue is not feasible when an attachment between human tissues and hydrogels is desired. In the case of hydrogel adhesion, hydrogels are used as the adherend and adhesive simultaneously [1]. Therefore, an adhesive hydrogel surface should be designed if the hydrogels are used in biomedicine, such as for wound dressing and wearable devices [3,4]. Hydrogel adhesion is central to many applications in medicine and engineering [5,6]; therefore, an ideal hydrogel with excellent adhesive performance has drawn considerable attention [7]. Hydrogel adhesion usually results from surface energy, intermolecular interactions, and near-surface effects [1,2]. Academic researchers have typically focused on the surface of hydrogels or the interface between hydrogels and substrates to create strong hydrogel adhesion [3,8]. The internal structure of hydrogel can also affect the adhesion of the hydrogel's surface, but a few research studies have explored the relationship between the internal structure of hydrogels and surface adhesion [9,10].

A polymer precursor solution typically consists of different polymer chains. Once these polymer chains connect by crosslinking, the precursor solution will transform into solid hydrogels [11,12]. The crosslinking degree of hydrogels can affect the internal structure of hydrogels, which further results in the change in the mechanical and swelling properties of hydrogels. Usually, a higher crosslinking degree of hydrogels corresponds to a higher compressive and tensile strength and lower equilibrium swelling rate [13,14]. In addition to the mechanical and swelling properties, hydrogel adhesion is one of the key hydrogel properties. It has been widely used in medicine and engineering in recent years [15,16]. Although the hydrogel adhesion inherently relies on engineering the contact surface at soft and hydrated interfaces, we found the adhesion is also affected by the crosslinking degree of hydrogels. Usually, hydrogels can be prepared by different crosslinking methods, mainly including the freezing–thawing cycle, photo-crosslinking and thermal-crosslinking [17,18]. To prove the generality of this finding, polyvinyl alcohol (PVA), polyacrylamide (PAM) and polyvinyl alcohol-bearing styrylpyridinium group (PVA-SbQ) hydrogels, prepared by the freezing–thawing (F-T) cycle, thermal-crosslinking and photo-crosslinking, respectively, were chosen to study the effect of crosslinking on hydrogel adhesion produced by hydrogen bonding in this study. Furthermore, the reason why crosslinking degree could change the hydrogel adhesion is also discussed. We hope this study can provide insights into improving the hydrogel adhesion generated by hydrogen bonding.

2. Results and Discussion

2.1. The Effect of The Crosslinking Degree of PVA Hydrogels on Hydrogel Adhesion, and Mechanical and Swelling Properties

PVA is a conventional raw material for preparing hydrogels because of its satisfactory biocompatibility, biodegradation, and nontoxicity [19,20]. PVA hydrogels can be formed through the well-known F-T method via the physical crosslinking by hydrogen bonds (Figure 1a,b) [21,22]. During the freezing process, water freezes and reduces interaction with hydroxyl groups on PVA chains [23]. As a result, the hydroxyl groups on PVA chains bind to form hydrogen bonding (Figure 1a). Consequently, the PVA solution becomes a hydrogel due to the development of a 3D structure (Figure 1b). With the increased freezing time, there is an increase in the hydrogen bond formation. Hydrogen bonds act as physical crosslinking points in PVA hydrogels. Therefore, the crosslinking degree depends on the freezing time.

There are several methods of testing hydrogel adhesion [1]. Figure 1c shows the modified method for testing the shear adhesive strength of PVA hydrogels used in this study unlike the general method (Figure 1d) [1,24]. The adherends were stuck to the same side of hydrogels to avoid significant errors due to the low adhesive strength of PVA hydrogels. Figure 1e shows that with the increase in the freezing time of PVA hydrogels, the adhesive strength decreased from 1600 Pa to 400 Pa. In addition, the swelling ratio of PVA hydrogels decreased with the increase in crosslinking degree that results from the increased freezing time (Figure 1f). Conversely, the tensile strength, elongation at break and compressive strength of PVA hydrogels improved with longer freezing duration because of the increased crosslinking degree (Figure 1g–i). It is well-known that the crosslinking degree benefits mechanical strength because a higher crosslinking degree results in a denser hydrogel structure [12,25,26]. In addition to mechanical strength, the crosslinking degree affects hydrogel adhesion simultaneously. For PVA hydrogels prepared by the F-T cycle, on the one hand, most of the hydroxyl groups on PVA formed hydrogen bonds used for crosslinking during the F-T process, which resulted in the lack of free hydroxyl groups used for adhesion [7,27]. On the other hand, the increase in the crosslinking degree of the PVA hydrogels limited the accessibility of hydroxyl groups due to the decrease in segment mobility. Consequently, the adhesive property between hydrogels and substrates reduced significantly with increased freezing time. This phenomenon was also applied to composite hydrogels. As shown in Figure 1j, the adhesion of polyvinyl alcohol/cellulose nanocrystal (PVA/CNC) composite hydrogels decreased with the increase in freezing duration.

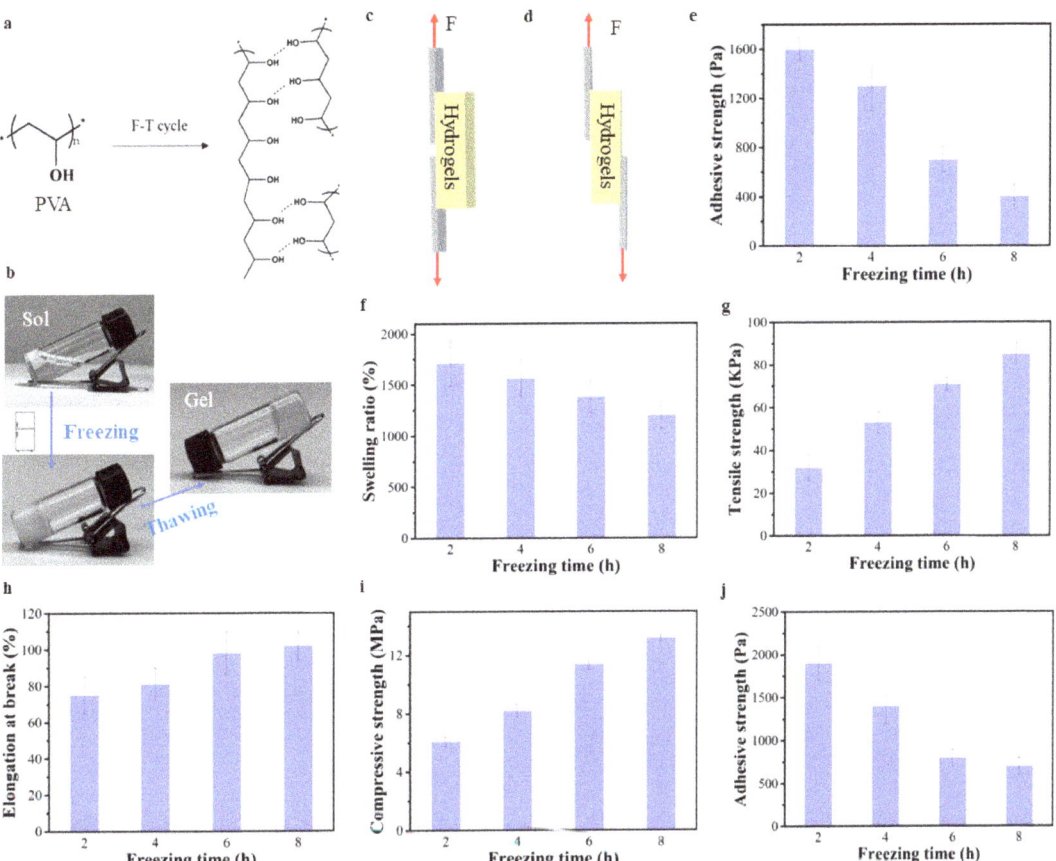

Figure 1. (a) The formation mechanism of PVA hydrogels. (b) The preparation process of PVA hydrogels. (c) The modified method of testing hydrogel adhesion used in this study. (d) The general method of testing hydrogel adhesion. (e) The adhesive strength of PVA hydrogels prepared by different freezing times. (f) The swelling ratio of PVA hydrogels prepared by different freezing times. (g) The tensile strength of PVA hydrogels prepared by different freezing times. (h) The elongation at break of PVA hydrogels prepared by different freezing times. (i) The compressive strength of PVA hydrogels prepared by different freezing times. (j) The adhesive strength of PVA/CNC hydrogels prepared by different freezing times.

2.2. The Effect of The Crosslinking Degree of PAM Hydrogels on Hydrogel Adhesion, and Mechanical and Swelling Properties

Unlike PVA hydrogels prepared by the F-T cycle, PAM hydrogels are usually synthesized by thermal-crosslinking [28,29]. The covalent bonding formed by the polymerization acts as crosslinking points in these types of hydrogels rather than hydrogen bonding (Figure 2a), which results in the sol–gel transition (Figure 2b). The crosslinking degree depends on the crosslinker content and the thermal-crosslinking time. Therefore, PAM hydrogels with different crosslinking degrees were obtained by adjusting crosslinking time and crosslinker content to investigate the relationship between crosslinking degree and hydrogel adhesion. As shown in Figure 2c,d, with the increase in the crosslinking time and crosslinker content, the hydrogel adhesion decreased. This observation implied that a higher crosslinking degree was not helpful for hydrogel adhesion. Unlike the PVA hydrogels, the free functional groups used for adhesion did not decrease in PAM hydrogels [1,2].

The decrease in hydrogel adhesion mainly resulted from the limitation of segment mobility. Although there were a lot of free amino groups on the PAM hydrogels, the amino group could not move to the hydrogel surface to interact with some functional groups on substrates as the crosslinking degree of PAM hydrogels increased. In addition, with the increase in crosslinking time and crosslinker content, the swelling ratio of PAM hydrogels simultaneously decreased (Figure 2e,f). The decreased swelling ratio further corroborated that the crosslinking degree of the hydrogel increased and a denser hydrogel structure was formed [30–32]. With the increase in crosslinking degree caused by increased crosslinking time, the compressive strength of PAM hydrogels gradually increased (Figure 2g). Conversely, while the increase in the crosslinker content improved the crosslinking degree, the compressive strength of PAM hydrogels did not increase significantly (Figure 2h). This is because the higher crosslinker content resulted in the crosslinking of PAM without complete polymerization. Accordingly, the length of PAM polymer chains that make up the hydrogel network were short, which resulted in the slight decrease in compressive strength. Therefore, it is challenging to enhance the mechanical and adhesive properties of hydrogels simultaneously.

2.3. The Effect of The Crosslinking Degree of PVA-SbQ Hydrogels on Hydrogel Adhesion, and Mechanical and Swelling Properties

In addition to the F-T cycle and thermal-crosslinking, photo-crosslinking is another common method for preparing hydrogels. To prove the generality of the effect of crosslinking degree on hydrogel adhesion, the PVA-SbQ hydrogels with different degrees were prepared by adjusting the photo-crosslinking duration. Instead of the conventional F-T cycle or using chemical crosslinkers, PVA-SbQ hydrogels could be formed by a fast and facile photo-crosslinking method via the photodimerization of carbon–carbon double bonds (C=C bonds) on SbQ functional groups (Figure 3a,b) [33,34]. As shown in Figure 3c, the adhesion of a PVA-SbQ hydrogel sharply decreased with the increase in photo-crosslinking time. The increased photo-crosslinking time also reduced the swelling ratio of PVA-SbQ hydrogels, which suggested an increase in the crosslinking degree (Figure 3d) [30–32]. Conversely, the compressive strength and tensile strength steadily increased as the photo-crosslinking duration was increased (Figure 3e,f). As with the PAM hydrogels, the decrease in hydrogel adhesion primarily resulted from the limitation of segment mobility, and the enhanced mechanical properties were caused by the increase in the crosslinking degree. This proved that photo-crosslinked hydrogels also followed the aforementioned relationship between the degree of crosslinking and hydrogel adhesiveness.

2.4. The Strategies for Constructing Strong Hydrogel Adhesion

In the study, three kinds of hydrogels were prepared by the freezing–thawing cycle, thermal-crosslinking and photo-crosslinking, respectively. The hydrogel adhesion was primarily generated by hydrogen bonding because of the amino and hydroxyl groups. In addition to the amino and hydroxyl groups shown in these hydrogels, carboxyl groups can also contribute towards hydrogel adhesion. Therefore, amino, hydroxyl or carboxyl groups are required to construct hydrogel adhesion generated by hydrogen bonding (Figure 4a). However, these functional groups should be free. For instance, the maximum adhesive strength of PVA-SbQ hydrogels is much higher than that of PVA hydrogels (Figure 4b). Although there were many hydroxyl groups in PVA hydrogels, they formed hydrogen bonding used for crosslinking with themselves rather than interacting with the functional groups on the substrate. As a result, these hydroxyl groups are not free and will contribute towards hydrogel adhesion. On the contrary, the fabricated PVA-SbQ hydrogel holds massive free hydroxyl groups as the crosslinking was mainly driven by photodimerization of SbQ groups instead of hydrogen bonds forming between hydroxyl groups. The advantage of hydrogen bond-triggered adhesion is that it is recyclable since the hydrogen bonding formation utilizes dynamic and noncovalent interactions. The cyclic stripping test demonstrated that the adhesive strength of the PVA-SbQ hydrogel did not reduce significantly with the increase in the number of stripping cycles (Figure 4c). In addition to the free functional groups, it was found that the crosslinking degree also affected the hydrogel adhesion. The increase in the crosslinking degree of the hydrogel restricted the mobility of the polymer chains (Figure 4d). Hence, the polymer chains with functional groups were not quickly diffused towards the hydrogel surface to form intimate contact with the substrate. As a result, excessive crosslinking of hydrogels led to a reduced adhesive strength (Figure 4e).

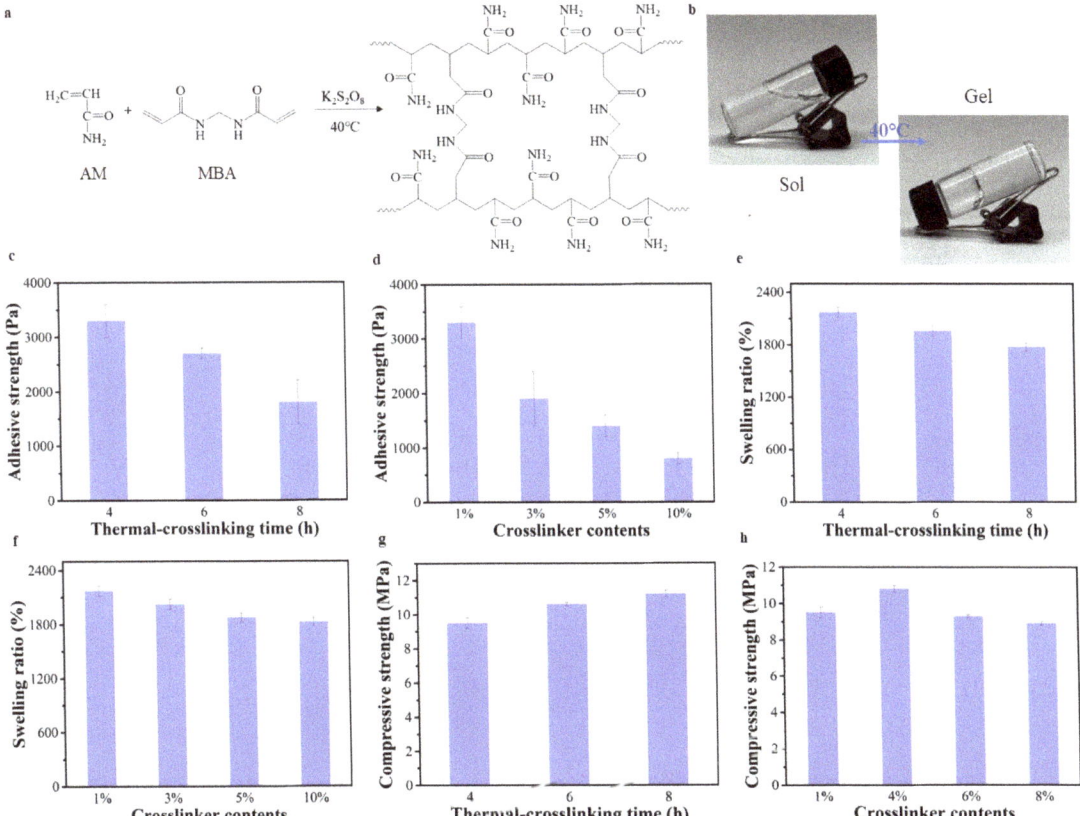

Figure 2. (a) The formation mechanism of PAM hydrogels. (b) The preparation process of PAM hydrogels. (c) The adhesive strength of PAM hydrogels prepared by different thermal-crosslinking times. (d) The adhesive strength of PAM hydrogels prepared with different crosslinker contents. (e) The swelling ratio of PAM hydrogels prepared by different thermal-crosslinking times. (f) The swelling ratio of PAM hydrogels prepared with different crosslinker contents. (g) The compressive strength of PAM hydrogels prepared by different thermal-crosslinking times. (h) The compressive strength of PAM hydrogels prepared with different crosslinker contents.

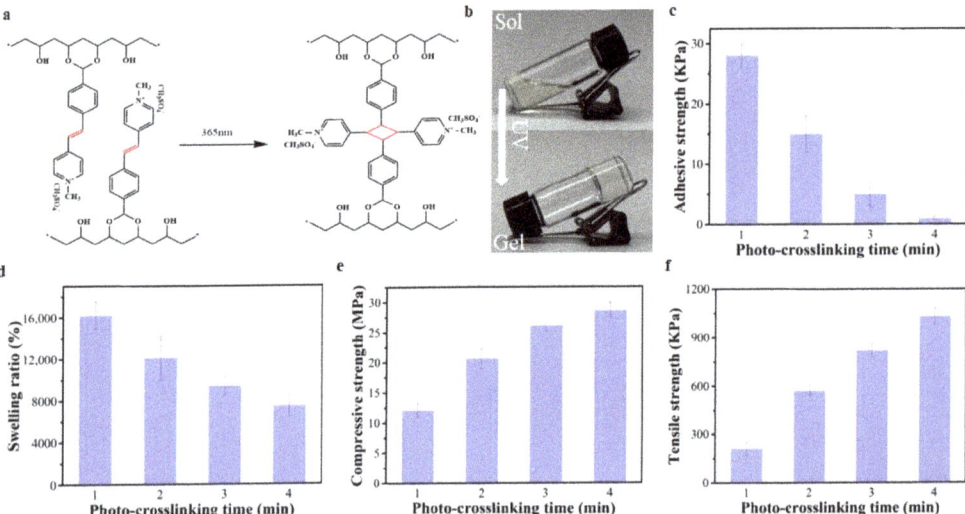

Figure 3. (a) The formation mechanism of PVA-SbQ hydrogels. (b) The preparation process of PVA-SbQ hydrogels. (c) The adhesive strength of PVA-SbQ hydrogels prepared by different photo-crosslinking times. (d) The swelling ratio of PVA-SbQ hydrogels prepared by different photo-crosslinking times. (e) The compressive strength of PVA-SbQ hydrogels prepared by different photo-crosslinking times. (f) The tensile strength of PVA-SbQ hydrogels prepared by different photo-crosslinking times.

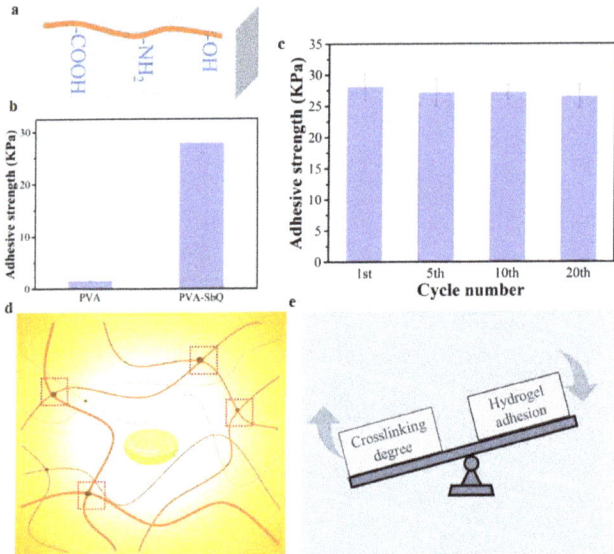

Figure 4. (a) The functional groups required for hydrogel adhesion generated by hydrogen bonds. (b) The maximum adhesive strength of PVA and PVA-SbQ hydrogels. (c) The adhesive strength of PVA-SbQ hydrogels under different stripping cycles. (d) The effect of crosslinking degree of hydrogels on chain movement. (e) The effect of crosslinking degree on hydrogel adhesion.

3. Conclusions

In this study, we investigated PVA, PAM and PVA-SbQ hydrogels fabricated by the freezing–thawing cycle, thermal-crosslinking and photo-crosslinking, respectively. These three types of hydrogels present representative models of hydrogels formed by different crosslinking mechanisms and extend the observations and derived conclusion to a wide class of materials. The hydrogel adhesion capability decreased with the increase in the crosslinking degree of these hydrogels. This is because a higher crosslinking degree would limit segment mobility. As a result, functional groups on polymer chains could not move to the hydrogel surface to interact with the substrate to generate hydrogel adhesion. Therefore, in addition to free functional groups used to interact with substrates, the flexibility of polymer chains that make up hydrogels is vital for hydrogel adhesion. However, the decrease in crosslinking degree in order to increase the availability of free functional groups and higher segment mobility would result in poor mechanical strength of the hydrogels and limit their functionality. Therefore, it is essential as well as challenging to maintain the balance between hydrogel adhesion and mechanical strength. However, a rational design of a hydrogel crosslinked network and free functional groups can allow better control over these hydrogel characteristics.

4. Materials and Methods

4.1. Materials

Polyvinyl alcohol (PVA, Mn ≈ 130,000, 99% hydrolysis), N,N'-methylenebisacrylamide (MBA), potassium persulfate (KPS) and acrylamide (AM) were obtained from Sigma-Aldrich (St. Louis, MO, USA) and used without further purification. Polyvinyl alcohol-bearing styrylpyridinium group (PVA-SbQ) solution was purchased from Shanghai KCKI Printing Technology Co., Ltd., Shanghai, China (polymerization degree was 1700, and the concentration of SbQ was 0.03–0.05 mol kg^{-1}), and cellulose nanocrystal (CNC) aqueous suspension was prepared from cotton linters.

4.2. The Preparation of PVA Hydrogel

First, 8.0 g PVA powder was added to 92 g deionized water to obtain 8 wt% PVA solution. Secondly, the mixtures were put into in a three-mouth flask and then stirred at 95 °C until the PVA powder was completely dissolved. Subsequently, the PVA solution was cooled down to room temperature, until the air bubbles completely disappeared. Lastly, 8 wt% PVA solution was poured into a watch glass and placed in a −20 °C refrigerator for 2, 4, 6 and 8 h respectively to obtain PVA hydrogels with different freezing durations.

4.3. The Preparation of PVA/CNC Hydrogels

First, the CNC aqueous suspension was prepared by acid hydrolysis method. Specific steps are as follows: 3 g cotton fiber was added to 62 mL 65% H_2SO_4 and treated at 50 °C for 45 min under constant stirring. The mixture was centrifuged and dialyzed against deionized water to remove the residual acid. The aqueous suspension was then ultrasonicated to evenly disperse the CNC and break agglomerates. The dispersed CNC was diluted to obtain a homogenous aqueous dispersion.

Next, 8 wt% PVA solution was added to the CNC aqueous suspension at given CNC/PVA mass ratio of 0.01:1. Subsequently, the PVA/CNC solution was stirred at 90 °C for 2 h and then ultrasonicated for additional 20 min. Finally, the PVA/CNC solution was poured into a watch glass and placed in a −20 °C refrigerator for 2, 4, 6 and 8 h, respectively, to obtain PVA/CNC hydrogels with different freezing durations.

4.4. The Preparation of PAM Hydrogels

An 8 wt% AM solution was obtained by dissolving 8 g AM in 92 g deionized water. Subsequently, different weights of MBA powders were added into the 8 wt% AM solution to obtain samples with different compositions, where the mass ratio between MBA and AM were 1:100, 3:100, 5:100 and 10:100, respectively. These mixtures were then stirred continuously for 30 min. After the addition of the MBA and stirring for 30 min, 0.0192 g KPS was added into the solution and stirred for another 10 min to initiate polymerization. Finally, the prepared hydrogel solutions were placed at a temperature of 45 °C for 4 h to obtain PAM hydrogels with different MBA compositions. Among the various hydrogels, the PAM hydrogels with 1% MBA were selected to be placed in 45 °C for 4, 6 and 8 h, respectively, to obtain and study PAM hydrogels with different thermal-crosslinking times.

4.5. The Preparation of PVA-SbQ Hydrogels

PVA-SbQ hydrogels were directly formed from 8 wt% PVA-SbQ solution with different photo-crosslinking times by being exposed to an F300 UV lamp for 1, 2, 3, 4 min, respectively.

4.6. Characterization

Tensile strength of the hydrogels was evaluated by performing tensile tests on an Instron 5967 electronic universal testing machine. The PVA and PVA/SbQ hydrogels were cut into dumbbell-shaped samples with a diameter of 7–9 mm and a thickness of 3–5 mm. Additionally, the compressive tests were carried out on cylindrical hydrogel samples with a diameter of 7–9 mm and a thickness of 3–5 mm by an Instron 5967 electronic universal testing machine at room temperature. The compression rate was fixed at 2 mm/min. Measurements of each sample were repeated at least three times and the acquired data were reported as average value and standard deviation.

To estimate the adhesive strength of hydrogels, the shear adhesive strength tests were performed. The PVA, PVA/CNC, PAM and PVA/SbQ hydrogels were cut into a rectangular spline with a size of 40 mm × 10 mm × 5 mm and then adhered to metal sheets, where the adhesion area on one side was 10 mm × 10 mm and the other side was 10 mm × 20 mm, as shown in Figure 1c. The force response was recorded while keeping one metal sheet fixed and moving another metal sheet at a speed of 20 mm/min using a universal testing machine. In the experiment, 5 samples were tested in each group to obtain the average value and standard deviation. The adhesive strength was calculated by the maximum load divided by the adhesion area.

To test the swelling ratio of hydrogels, the hydrogel samples were immersed in distilled water for 3 days at least at room temperature, until the hydrogel weight did not increase. In the next step, these swollen hydrogels were placed in a 40 °C oven to evaporate all of the water in hydrogels. The weight of swollen hydrogel and dried hydrogels were recorded, respectively. The equilibrium swelling ratio was calculated by the following equation

$$ESR\% = W_s/W_d \times 100 \tag{1}$$

where W_s and W_d are the weight of the swollen hydrogel at room temperature and dried hydrogels, respectively.

Author Contributions: Conceptualization, Z.L. and H.B.; methodology, Z.L. and C.Y.; software, H.K. and X.H.; formal analysis, Z.L.; resources, H.B., Q.L. and K.K.; writing—original draft preparation, Z.L. and J.H.; writing—review and editing, J.H.; supervision, J.H. and K.K. All authors have read and agreed to the published version of the manuscript.

Funding: The authors acknowledge the support from China Scholarship Council scholarship (Z.L.), and the Canada First Research Excellence Fund (CFREF) for its Global Research Initiative in Sustainable Low Carbon Unconventional Resources (J.H.).

Institutional Review Board Statement: Not applicable.

Informed Consent Statement: Not applicable.

Data Availability Statement: Not applicable.

Conflicts of Interest: The authors declare that they have no conflict of interest.

References

1. Yang, J.; Bai, R.; Chen, B.; Suo, Z. Hydrogel adhesion: A supramolecular synergy of chemistry, topology, and mechanics. *Adv. Funct. Mater.* **2020**, *30*, 1901693. [CrossRef]
2. Zhang, W.; Wang, R.; Sun, Z.M.; Zhu, X.; Zhao, Q.; Zhang, T.; Cholewinski, A.; Yang, F.; Zhao, B.; Pinnaratip, R.; et al. Catechol-functionalized hydrogels: Biomimetic design, adhesion mechanism, and biomedical applications. *Chem. Soc. Rev.* **2020**, *49*, 433–464. [CrossRef] [PubMed]
3. Kamoun, E.A.; Kenawy, E.-R.S.; Chen, X. A review on polymeric hydrogel membranes for wound dressing applications: PVA-based hydrogel dressings. *J. Adv. Res.* **2017**, *8*, 217–233. [CrossRef] [PubMed]
4. Liang, Y.; He, J.; Guo, B. Functional hydrogels as wound dressing to enhance wound healing. *ACS Nano* **2021**, *15*, 12687–12722. [CrossRef]
5. Bovone, G.; Dudaryeva, O.Y.; Marco-Dufort, B.; Tibbitt, M.W. Engineering hydrogel adhesion for biomedical applications via chemical design of the junction. *ACS Biomater. Sci. Eng.* **2021**, *7*, 4048–4076. [CrossRef]
6. Caló, E.; Khutoryanskiy, V.V. Biomedical applications of hydrogels: A review of patents and commercial products. *Eur. Polym. J.* **2015**, *65*, 252–267. [CrossRef]
7. Li, Z.; Wang, D.; Bai, H.; Zhang, S.; Ma, P.; Dong, W. Photo-crosslinking strategy constructs adhesive, superabsorbent, and tough PVA-based hydrogel through controlling the balance of cohesion and adhesion. *Macromol. Mater. Eng.* **2020**, *305*, 1900623. [CrossRef]
8. Wang, L.; Zhou, M.; Xu, T.; Zhang, X. Multifunctional hydrogel as wound dressing for intelligent wound monitoring. *Chem. Eng. J.* **2022**, *433*, 134625. [CrossRef]
9. Yang, J.; Steck, J.; Bai, R.; Suo, Z. Topological adhesion II. Stretchable adhesion. *Extreme Mech. Lett.* **2020**, *40*, 100891. [CrossRef]
10. Steck, J.; Kim, J.; Yang, J.; Hassan, S.; Suo, Z. Topological adhesion. I. Rapid and strong topohesives. *Extreme Mech. Lett.* **2020**, *39*, 100803. [CrossRef]
11. Bai, H.; Li, Z.; Zhang, S.; Wang, W.; Dong, W. Interpenetrating polymer networks in polyvinyl alcohol/cellulose nanocrystals hydrogels to develop absorbent materials. *Carbohydr. Polym.* **2018**, *200*, 468–476. [CrossRef] [PubMed]
12. Li, Z.; Bai, H.; Zhang, S.; Wang, W.; Ma, P.; Dong, W. DN strategy constructed photo-crosslinked PVA/CNC/P(NIPPAm-co-AA) hydrogels with temperature-sensitive and pH-sensitive properties. *New J. Chem.* **2018**, *42*, 13453–13460. [CrossRef]
13. Liu, X.; He, X.; Yang, B.; Lai, L.; Chen, N.; Hu, J.; Lu, Q. Dual physically cross-linked hydrogels incorporating hydrophobic interactions with promising repairability and ultrahigh elongation. *Adv. Funct. Mater.* **2021**, *31*, 2008187. [CrossRef]
14. Dodero, A.; Pianella, L.; Vicini, S.; Alloisio, M.; Ottonelli, M.; Castellano, M. Alginate-based hydrogels prepared via ionic gelation: An experimental design approach to predict the crosslinking degree. *Eur. Polym. J.* **2019**, *118*, 586–594. [CrossRef]
15. Liu, X.; Zhang, Q.; Duan, L.; Gao, G. Bioinspired nucleobase-driven nonswellable adhesive and tough gel with excellent underwater adhesion. *ACS Appl. Mater. Interfaces* **2019**, *11*, 6644–6651. [CrossRef] [PubMed]
16. Liu, X.; Zhang, Q.; Gao, Z.; Hou, R.; Gao, G. Bioinspired adhesive hydrogel driven by adenine and thymine. *ACS Appl. Mater. Interfaces* **2017**, *9*, 17645–17652. [CrossRef] [PubMed]
17. Chen, Y.; Li, J.; Lu, J.; Ding, M.; Chen, Y. Synthesis and properties of poly (vinyl alcohol) hydrogels with high strength and toughness. *Polym. Test.* **2022**, *108*, 107516. [CrossRef]
18. Liu, J.; Su, C.; Chen, Y.; Tian, S.; Lu, C.; Huang, W.; Lv, Q. Current understanding of the applications of photocrosslinked hydrogels in biomedical engineering. *Gels* **2022**, *8*, 216. [CrossRef]
19. Jiang, S.; Liu, S.; Feng, W. PVA hydrogel properties for biomedical application. *J. Mech. Behav. Biomed. Mater.* **2011**, *4*, 1228–1233. [CrossRef]
20. Wu, S.; Hua, M.; Alsaid, Y.; Du, Y.; Ma, Y.; Zhao, Y.; Lo, C.; Wang, C.; Wu, D.; Yao, B.; et al. Poly (vinyl alcohol) hydrogels with broad-range tunable mechanical properties via the hofmeister effect. *Adv. Mater.* **2021**, *33*, 2007829. [CrossRef]
21. Hassan, C.M.; Peppas, N.A. Cellular PVA hydrogels produced by freeze/thawing. *J. Appl. Polym. Sci.* **2000**, *76*, 2075–2079. [CrossRef]
22. Zhang, H.; Xia, H.; Zhao, Y. Poly (vinyl alcohol) hydrogel can autonomously self-heal. *ACS Macro Lett.* **2012**, *1*, 1233–1236. [CrossRef] [PubMed]
23. Hassan, C.M.; Peppas, N.A. Structure and morphology of freeze/thawed PVA hydrogels. *Macromolecules* **2000**, *33*, 2472–2479. [CrossRef]
24. Bai, H.; Yu, C.; Zhu, H.; Zhang, S.; Ma, P.; Dong, W. Mussel-inspired cellulose-based adhesive with underwater adhesion ability. *Cellulose* **2021**, *29*, 893–906. [CrossRef]
25. Cui, T.; Sun, Y.; Wu, Y.; Wang, J.; Ding, Y.; Cheng, J.; Guo, M. Mechanical, microstructural, and rheological characterization of gelatin-dialdehyde starch hydrogels constructed by dual dynamic crosslinking. *LWT* **2022**, *161*, 113374. [CrossRef]
26. Lin, X.; Zhao, X.; Xu, C.; Wang, L.; Xia, Y. Progress in the mechanical enhancement of hydrogels: Fabrication strategies and underlying mechanisms. *J. Appl. Polym. Sci.* **2022**, *60*, 2525–2542. [CrossRef]
27. Wang, M.; Bai, J.; Shao, K.; Tang, W.; Zhao, X.; Lin, D.; Huang, S.; Chen, C.; Ding, Z.; Ye, J. Poly (vinyl alcohol) hydrogels: The old and new functional materials. *Int. J. Polym. Sci.* **2021**, *2021*, 2225426. [CrossRef]
28. Sennakesavan, G.; Mostakhdemin, M.; Dkhar, L.K.; Seyfoddin, A.; Fatihhi, S. Acrylic acid/acrylamide based hydrogels and its properties—A review. *Polym. Degrad. Stab.* **2020**, *180*, 109308. [CrossRef]

29. Gombert, Y.; Roncoroni, F.; Sánchez-Ferrer, A.; Spencer, N.D. The hierarchical bulk molecular structure of poly (acrylamide) hydrogels: Beyond the fishing net. *Soft Matter* **2020**, *16*, 9789–9798. [CrossRef]
30. García-García, J.M.; Liras, M.; Garrido, I.Q.; Gallardo, A.; París, R. Swelling control in thermo-responsive hydrogels based on 2-(2-methoxyethoxy) ethyl methacrylate by crosslinking and copolymerization with N-isopropylacrylamide. *Polym. J.* **2011**, *43*, 887–892. [CrossRef]
31. Zhu, Q.; Barney, C.W.; Erk, K.A. Effect of ionic crosslinking on the swelling and mechanical response of model superabsorbent polymer hydrogels for internally cured concrete. *Mater. Struct. Constr.* **2015**, *48*, 2261–2276. [CrossRef]
32. Mahkam, M.; Doostie, L. The relation between swelling properties and cross-linking of hydrogels designed for colon-specific drug delivery. *Drug Deliv. J. Deliv. Target. Ther. Agents* **2005**, *12*, 343–347. [CrossRef] [PubMed]
33. Avramescu, A.; Noguer, T.; Avramescu, M.; Marty, J.-L. Screen-printed biosensors for the control of wine quality based on lactate and acetaldehyde determination. *Anal. Chim. Acta* **2002**, *458*, 203–213. [CrossRef]
34. Soldatkin, A.P.; Montoriol, J.; Sant, W.; Martelet, C.; Jaffrezic-Renault, N. Development of potentiometric creatinine-sensitive biosensor based on ISFET and creatinine deiminase immobilised in PVA/SbQ photopolymeric membrane. *Mater. Sci. Eng. C* **2002**, *21*, 75–79. [CrossRef]

Article

Novel Hydrogel Material with Tailored Internal Architecture Modified by "Bio" Amphiphilic Components—Design and Analysis by a Physico-Chemical Approach

Richard Heger [1,2,*], Martin Kadlec [1,2], Monika Trudicova [1,2], Natalia Zinkovska [1,2], Jan Hajzler [3], Miloslav Pekar [2,*] and Jiri Smilek [2,*]

1. Materials Research Center, Faculty of Chemistry, Brno University of Technology, Purkynova 118, 61200 Brno, Czech Republic; martin.kadlec@vut.cz (M.K.); xctrudicova@fch.vut.cz (M.T.); xczinkovska@fch.vut.cz (N.Z.)
2. Institute of Physical and Applied Chemistry, Faculty of Chemistry, Brno University of Technology, Purkynova 118, 61200 Brno, Czech Republic
3. Institute of Materials Science, Faculty of Chemistry, Brno University of Technology, Purkynova 118, 61200 Brno, Czech Republic; xchajzler@fch.vut.cz
* Correspondence: xchegerr@fch.vut.cz (R.H.); pekar@fch.vut.cz (M.P.); smilek@fch.vut.cz (J.S.)

Abstract: Nowadays, hydrogels are found in many applications ranging from the industrial to the biological (e.g., tissue engineering, drug delivery systems, cosmetics, water treatment, and many more). According to the specific needs of individual applications, it is necessary to be able to modify the properties of hydrogel materials, particularly the transport and mechanical properties related to their structure, which are crucial for the potential use of the hydrogels in modern material engineering. Therefore, the possibility of preparing hydrogel materials with tunable properties is a very real topic and is still being researched. A simple way to modify these properties is to alter the internal structure by adding another component. The addition of natural substances is convenient due to their biocompatibility and the possibility of biodegradation. Therefore, this work focused on hydrogels modified by a substance that is naturally found in the tissues of our body, namely lecithin. Hydrogels were prepared by different types of crosslinking (physical, ionic, and chemical). Their mechanical properties were monitored and these investigations were supplemented by drying and rehydration measurements, and supported by the morphological characterization of xerogels. With the addition of natural lecithin, it is possible to modify crucial properties of hydrogels such as porosity and mechanical properties, which will play a role in the final applications.

Keywords: lecithin; hydrogel; rheology; scanning electron microscopy; drying and swelling; extracellular matrix; mesh size

1. Introduction

Hydrogels are hydrophilic polymers with a three-dimensional network structure that have the ability to absorb a large volume of water due to the presence of hydrophilic moieties, which makes them particularly suitable materials for biomedical applications (e.g., scaffolds) [1]. Selecting the pertinent components for the fabrication of the final hydrogel allows for a functional and applicable material with unique properties (e.g., porosity, biocompatibility, biodegradability) to be obtained. This exact customizable functionality makes these materials appropriate and desirable for a wide range of application areas (tissue engineering, pharmacy, water treatment, material engineering, etc.).

An equally important property of hydrogels is their ability to simulate and mimic biological systems such as the extracellular matrix (ECM), which is, in fact, a structural support network composed of diverse proteins, sugars, and other components. ECM regulates cellular processes including survival, growth, proliferation, migration, and differentiation [2]. Engineering a tailored in vitro environment mimicking the organized structure of ECM is a huge challenge and a desired goal. Since the scaffolds must offer relevant properties sufficient for cellular function, hydrogels have an advantage as potential materials due to their tunable physico-chemical (electrical charge and pore size) and mechanical (stiffness, tensile strength) properties [3]. The majority of hydrogels are also biocompatible, for example, naturally derived polymers such as agarose, alginate, chitosan, collagen, fibrin, gelatin, hyaluronic acid, and dextran as well as biocompatible synthetic gels based on poly(ethylene glycol) (PEG), poly(vinyl alcohol) (PVA), and poly(hydroxyethyl methacrylate) (PHEMA) [4].

Since the 3D network structure of hydrogels is mainly responsible for their mechanical properties and porous microstructure, one of the possibilities of how to modify, upgrade, or tailor properties of hydrogels is to incorporate hydrophobic or micellar domains into the gel structure [5].

Pure hydrophobic association (HA) hydrogels refer to physically crosslinked hydrogels formed by hydrophobic interactions, which account for 5–20% of the total amount of polymer. The bulk of hydrophobic association hydrogels are produced by micellar copolymerization [6]. For instance, Tuncaboylu et al. attempted to improve the low mechanical strength of self-healing hydrogels by creating hybrid hydrogels with strong hydrophobic interactions between hydrophilic polymers mediated by the large hydrophobic moiety of a physical crosslinker (stearyl methacrylate) [7]. The addition of NaCl to the reaction solution during the copolymerization of large hydrophobes (stearyl methacrylate (C18)) with the hydrophilic monomer acrylamide (AAm) in an aqueous solution of sodium dodecyl sulfate (SDS) led to micellar growth and the solubilization of the large hydrophobes within the SDS micelles. Rheological measurements showed that the hydrophobic associations surrounded by surfactant micelles acted as reversible breakable crosslinks responsible for the rapid self-healing of the hydrogels [7].

An alternative approach to enhance the toughness of the hydrogel network is to introduce particles as additional crosslinking points (e.g., latex particles, nanoparticles) [6]. Latex particles (LPs) that are usually prepared via emulsion polymerization ensure effective energy dissipation and provide hydrogels with higher mechanical properties. Gu et al. [8] proposed a method that encompassed the adsorption of the hydrophobic alkyl chains of hydrophobic monomers on the surface of the latex microspheres and their subsequent stabilization in the presence of surfactants, thus forming hydrophobic association centers as the first physical crosslinking points. Moreover, anionic sulfate radicals (originating from the dissociation of the persulfate) were attracted toward the cationic chains of latex microspheres (obtained via surfactant-free emulsion copolymerization of styrene with a vinylidene comonomer bearing a cationic side group) and formed secondary physical crosslinking centers. The incorporation of cationic latex microspheres led to an improvement in the tensile and compression strength of the modified hydrogel compared with pure hydrophobic association hydrogel.

Since inorganic nanoparticles have a high specific surface area, their incorporation into the hydrogel network could also improve its mechanical behavior relating to surface structure and charging [6]. At the same time, the introduction of calcium carbonate nanoparticles [9], hydroxyapatite [10], kaolin [11], and laponite particles [12] could also induce hydrogel adhesion.

On the other hand, the embodiment of polymeric nanoparticles provides the ability to encapsulate both hydrophobic and hydrophilic substances [6]. In addition, Arno et al. investigated how particle morphology (e.g., particle shape, size, and surface) affected the adhesion and mechanical properties of the resultant calcium-alginate hydrogels [13]. The authors demonstrated that 2D platelets substantially improved both the adhesion between hydrogel surfaces and the material's mechanical strength when blended into the polymeric network compared to their 0D spherical or 1D cylindrical counterparts.

The properties of hydrogels, as mentioned previously, can be adapted not only through the appropriate choice of materials and crosslinking techniques, but also by modifying the internal structure of the gel by using a structure modifier such as lecithin during the preparation process. It should be remembered that lecithin is a typical amphiphilic phospholipid mixture primarily containing distearoylphosphatidylcholine, which possesses good biocompatibility and capability to enhance the bioavailability of co-administered drugs [14]. Lecithin in water systems can self-assemble into array of liquid-crystalline structures depending on the amount of water and temperature. The most likely structures formed under normal working laboratory conditions are lamellar liquid-crystalline structures [15]. Moreover, varying the ratio of lecithin in the multi-component hydrogel system may further improve the applicability and functionality of designed gels. The transport and mechanical properties of materials are given by their internal structure and can be greatly affected by its rearrangement.

Among the different types of lecithin-based systems, the most common platforms in this area are liposomes and microemulsions [16]. Liposomes are an example of soft phospholipid nanoparticles with typical diameters of around 100 nm [17]. Due to their closed vesicular structure, hydrophilic active compounds could be embedded into their internal water compartments, while hydrophobic compounds could be loaded into the bilayer of the liposome. In most cases, lecithin-based liposomal hydrogels are used as carriers; nevertheless, such systems still have certain disadvantages such as a slow and uncontrolled process of drug release [18]. In contrast, lecithin microemulsion-based gels or organogels have some advantages over liposomal hydrogels such as an easier preparation procedure, an absence of organic solvents, and higher storage stability due to the thermodynamic stability of microemulsions [19]. The matrix of lecithin microemulsion-based gels is composed of lecithin, which acts as a surfactant as well as a gelling agent in the presence of a nonpolar organic solvent (external phase) or a polar agent, which is usually water.

Substantial research is focused on modifying the internal structures of hydrogels, however, to the best of our knowledge, there has previously been no systematic study investigating the preparation and targeted modification of the internal structures of biocompatible hydrogels that focused on the use of natural amphiphilic substances and their crucial (e.g., mechanical) application properties.

Thus, this work focuses on the effect of the structure modifier lecithin (as stated before, the lecithin is able to self-organize into liquid-crystalline structures) and its concentration on the resultant mechanical properties of differently crosslinked hydrogels. The results of this work could provide a deeper understanding of the interactions between lecithin and the hydrogel network, and, alternatively, between lecithin and model drugs. Lecithin aggregates in hydrogels can also be viewed as a model of phospholipid structures (like cell membranes) occurring in real tissues, and thus as a model of their potential impact on the rheological or transport properties of the extracellular matrix.

2. Results and Discussion

On the basis of the prior experience of our team and in an attempt to investigate the effect of different crosslinking strategies on the final properties of hydrogels, the following materials were selected: agarose as a physically crosslinked hydrogel, alginate crosslinked by polyvalent ions as an ionically crosslinked hydrogel, and PVA-chitosan as a chemically crosslinked hydrogel.

As stated in Section 4, for each type of crosslinking, four different samples were investigated. Three samples with lecithin additions at different concentrations (0.5, 1, and 2 wt.%) were labeled according to their lecithin concentration (i.e., "0.5", "1" and "2"). The fourth sample was a reference sample without lecithin, simply marked as "R". The lecithin concentrations were selected on the basis of preliminary experiments focused mainly on estimating the maximum amount of lecithin that could be incorporated into the hydrogel matrix.

2.1. Physical Crosslinking

Agarose was a representative of the physically crosslinked hydrogel matrix, whose properties were affected by lecithin content. Hydrogel samples after preparation as well as samples after the drying and rehydration procedure were studied (schematic figure of the preparation procedure can be seen in the Supplementary Materials Figure S1).

2.1.1. Rheology

Amplitude sweep results for physically crosslinked hydrogels obtained under an applied oscillatory strain of 1 Hz suggest that differences in lecithin concentration have, from a viscoelastic property point of view, a minimal influence on the hydrogel structure after preparation, especially with respect to the width of the linear viscoelastic region (as can be seen in Figure 1a). The storage as well as the loss modulus gradually increased with increasing lecithin concentration, which might be due to the overall higher dry content of the hydrogels. The effect of lecithin concentration on the viscoelastic properties of agarose hydrogels was also minimal in the linear viscoelastic region (LVR), which is the range of the values of storage modulus where the hydrogel is able to resist the applied oscillatory strain and can thus indicate the strength of non-covalent hydrogel nodes. Probably, the strength of the physically crosslinked hydrogel is provided mainly by non-covalent weak interactions (H-bonding) between the chains of agarose. Lecithin only had a small effect on the viscoelastic properties of 1 wt.% aq. agarose. The obtained values marking the end of the LVR were very similar for all samples physically crosslinked (Table 1). The values reported in the tables were either obtained by rheology software (TRIOS TA Instruments) analyses (cross-over point, average moduli values in LVR) or calculated. The end of LVR was obtained by comparing the average value of storage modulus in LVR with each point, where the deviation greater than 5% marked the end of the LVR. The mesh size calculations are described in Section 4.2. The cross-over point ($G' = G''$), the point at which the hydrogel was irreversibly damaged, was very similar for all samples.

Table 1. Values for physically crosslinked agarose hydrogels after preparation obtained from strain and frequency sweep tests before drying.

Lecithin Concentration (wt.%)	Cross-Over Point		Average Moduli Values in LVR		End of LVR	Mesh Size
	G' (Pa)	Strain (%)	G' (Pa)	G'' (Pa)	Strain (%)	Mesh (nm)
0 (R)	157.5 ± 4.1	425.8 ± 2.2	3299 ± 277	366 ± 28	2.5 ± 1.0	13.3 ± 0.1
0.5	207.9 ± 2.1	414.1 ± 4.5	4576 ± 12	551 ± 15	1.8 ± 0.0	13.4 ± 0.4
1	194.9 ± 10.7	433.2 ± 10.2	4002 ± 81	461 ± 4	1.8 ± 0.0	12.7 ± 0.1
2	224.5 ± 0.0	468.0 ± 2.9	4880 ± 27	529 ± 8	1.8 ± 0.0	12.9 ± 0.3

The same amplitude sweep tests were performed on samples dried to the xerogel form and again rehydrated. The amount of absorbed water had a significant effect on these samples. As can be seen from Figure 1b as well as from the dry matter content experiments (Section 2.1.2), the samples with the highest lecithin content were able to reabsorb the largest amount of water (twice as much water as the sample without lecithin). This was also reflected in the amplitude sweep results because the moduli values for these hydrogels decreased proportionately. The reference sample had the highest moduli values, whereas the lowest values were observed for the samples with the greatest lecithin concentrations. The moduli values were somewhat larger than those for the samples studied after preparation (Table 2), mainly due to the elevated values of the swelling degrees of the systems after drying and rehydration in comparison with those of the just prepared hydrogels. Lecithin, therefore, favored water absorption. For the physically crosslinked hydrogels, even the cross-over point was affected, and samples with higher lecithin concentrations shifted the cross-over point to higher strain values. This could be the effect of the attractive interactions between lecithin and the polysaccharide chains, leading to the reinforcement of the hydrogels obtained after their drying and rehydration. In the initially prepared hydrogels, lecithin was dispersed to a greater extent in a liquid medium without this (strong) effect. This could be explained by the H-bonding between polysaccharide chains and lecithin, which are more significant for the rehydrated hydrogels because of the absence of water (in xerogel), which could not interfere. The same could be observed for the cross-over point, which again gradually increased with lecithin concentration.

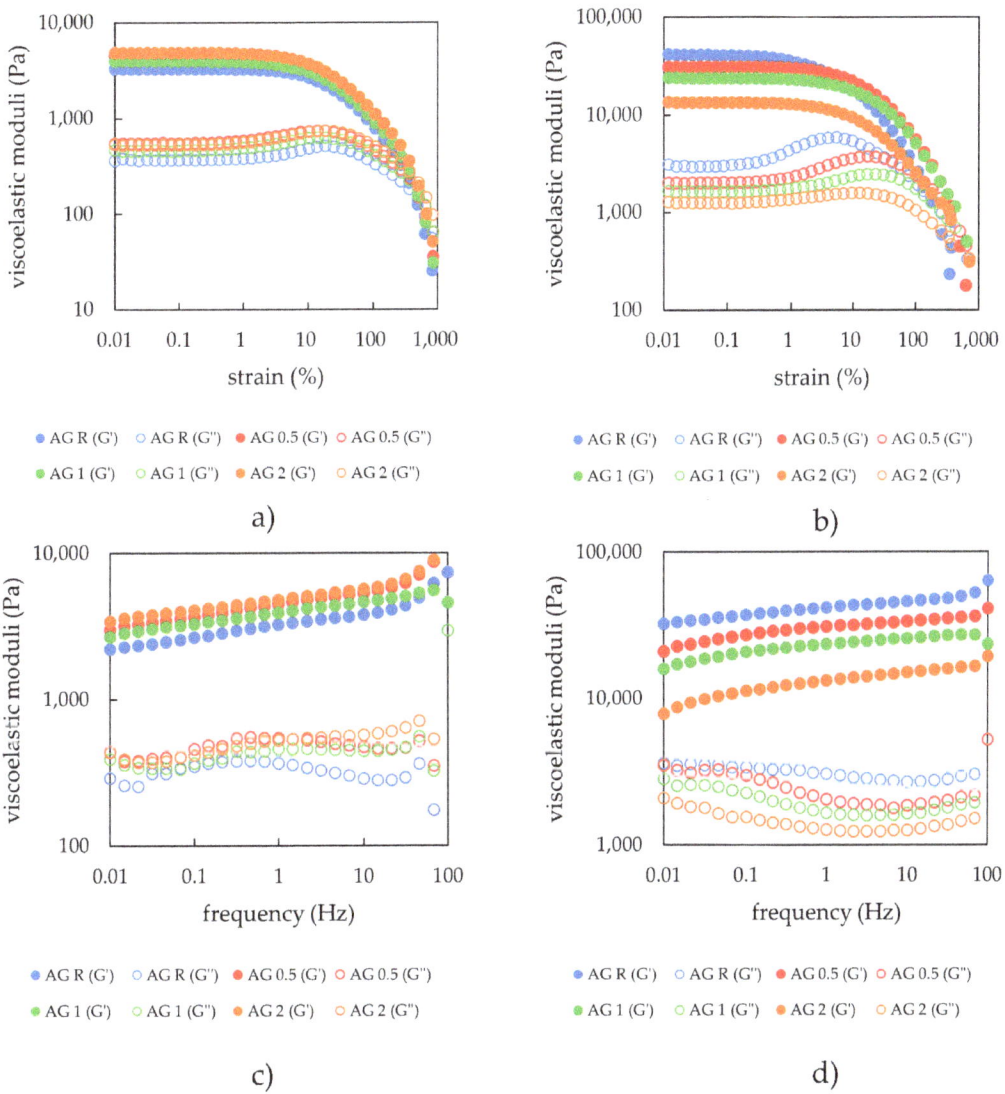

Figure 1. (**a**) Strain sweep of agarose hydrogels with different lecithin concentrations (0, 0.5, 1, and 2 wt.%) after preparation; (**b**) strain sweep of agarose hydrogels with different lecithin concentrations (0, 0.5, 1, and 2 wt.%) after drying and rehydration of the xerogels; (**c**) frequency sweep of agarose hydrogels with different lecithin concentrations (0, 0.5, 1, and 2 wt.%) after preparation; (**d**) frequency sweep of agarose hydrogels with different lecithin concentrations (0, 0.5, 1, and 2 wt.%) after drying and rehydration of the xerogels.

Table 2. Values obtained from strain and frequency sweep tests for physically crosslinked agarose hydrogels after drying and rehydration.

Lecithin Concentration (wt.%)	Cross-Over Point		Average Moduli Values in LVR		End of LVR	Mesh Size
	G' (Pa)	Strain (%)	G' (Pa)	G'' (Pa)	Strain (%)	Mesh (nm)
0 (R)	1814 ± 340.6	250.4 ± 131.7	41,386 ± 10,517	2977 ± 707	0.3 ± 0.1	7.6 ± 1.1
0.5	1005.7 ± 142.9	718.5 ± 129.3	31,216 ± 980	2010 ± 4	1.2 ± 0.5	7.6 ± 0.2
1	542.7 ± 0.0	1148.7 ± 0.0	23,829 ± 3	1642 ± 118	1.4 ± 0.3	8.2 ± 0.1
2	350.8 ± 35.5	1257.6 ± 12.2	13,506 ± 1217	1256 ± 122	0.9 ± 0.0	9.0 ± 0.3

Frequency sweep test results are presented in Figure 1 and show that the shape of the rheograms for all hydrogel samples was very similar. The storage modulus was dominant, which means that the samples act as a fully crosslinked gel material with a fully crosslinked internal structure. The trend of the moduli values was the same as that observed for the amplitude sweep tests and therefore indicates that the lecithin addition increased the values of the storage moduli as well as of the loss moduli, which was well correlated with the higher dry matter content, as previously stated. With increasing oscillation frequency values, the moduli values increased, which means that the hydrogel samples were not completely relaxed, and the degree of relaxation was influenced by the type of crosslinking. Practically, the average relaxation time of the hydrogel network exceeds the period associated with the progressively increasing frequency of the applied oscillatory deformations. The values of the mesh size of the internal structure of the hydrogels calculated from the frequency sweep tests using Equations (3) and (4) are recorded in Tables 1 and 2. The results for the freshly prepared hydrogels showed the same trend as other rheological data (i.e., that the mesh size does not differ substantially between the concentrations), whereas for the rehydrated xerogels, a slight increase could be observed at higher lecithin concentrations, which can be explained by lecithin fitting itself into the pores and thus increasing its size. This could be explained by lecithin forming lamellar liquid-crystalline structures in absorbed water along with the already mentioned H-bonding between the polysaccharide chains and lecithin. When comparing the absolute values of mesh sizes for freshly prepared and rehydrated hydrogels, we can see that the pores decreased in size after rehydration.

Based on the results of the strain and frequency sweeps performed onto the freshly prepared agarose hydrogels, it can be seen that lecithin, as an amphiphilic natural component, does not lead to a substantially modified viscoelastic behavior of these physically crosslinked hydrogels in the range of lecithin concentrations used (see Figure 1a,c). Agarose, which forms a thermoreversible physical hydrogel in an aqueous medium in the form of a natural linear polysaccharide, was not expected to interact significantly with amphiphilic lecithin. Thus, it was not expected that agarose could significantly interact with amphiphilic lecithin. Lecithin thus serves only as a filler, and does not interfere significantly with the internal structure of the hydrogel. Therefore, lecithin plays an important role in the rehydration of dried samples. Thus, an increasingly higher content of lecithin in the structure of such type of hydrogels causes the viscoelastic moduli storage and loss moduli to gradually decrease. Practically, the presence of lecithin affects the ability of agarose xerogels (hydrogel after drying) to reabsorb water (i.e., to swell) (see Figures 1b,d and 2). The final viscoelastic properties of hydrogels are definitely affected by the amount of dispersion medium (water) after the swelling of xerogels. If the addition of lecithin, as the modifier of the internal architecture of hydrogels, is able to change the swelling properties, it will also definitely change the viscoelastic properties due to the different amount of water. From the applicative point of view, this finding is absolutely essential, given that by choosing a suitable concentration of additive (lecithin), we were able to prepare hydrogels with the required properties (especially viscoelastic) tailored to a specific purpose.

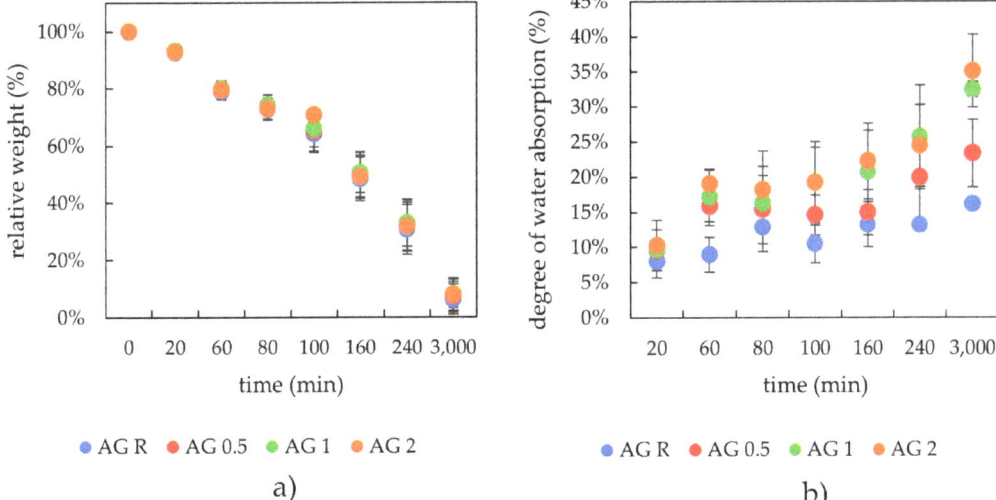

Figure 2. Drying (**a**) and rehydration (**b**) of the physically crosslinked agarose hydrogels with different contents of lecithin.

2.1.2. Drying and Rehydration Measurements

The amounts of water and dry matter associated with the studied gels are two of the most important parameters for hydrogel characterization and future applicability. Dry matter affects the behavior of the final material. The same is true for the water inside the hydrogel, which significantly affects, for example, the transport properties. As stated in Section 1, these parameters predetermine the applicative nature of the final system.

The results of the drying kinetics of physically crosslinked hydrogels can be seen in Figure 2a. At the start of these experiments, all weights of the hydrogels (2 ± 0.2 g) and xerogels were comparably the same. It can be seen that the lecithin addition had no influence on the drying kinetics. The most likely explanation is that water retained by lecithin is not bound as tightly as water hydrating agarose. Conversely, during the swelling process, hydrogel with lecithin easily draws water (more easily than the agarose hydrogel solely) and this resulted in the lecithin-agarose samples showing a higher swelling ability with corresponding lower moduli (Figure 1). The swelling experiments demonstrated the influence of lecithin on the swelling capacity. Therefore, the lecithin structures insert themselves into the hydrogel pores and support the water intake. The kinetics of the swelling process was very similar for all samples, with a peculiarity noted at the onset of the experiment, where the samples richer in lecithin (1 and 2 wt.%) revealed a greater rate of water absorption. Additionally, the same systems (agarose with 1 and 2 wt.% of lecithin) were able to absorb the largest amount of water.

2.1.3. Morphological Characterization of Xerogels

Morphological characterization was performed on dried samples; therefore, the results may not correspond to the results obtained from methods where hydrogels are studied in native form (specifically, rheology). From the results obtained by scanning electron microscopy (SEM), the effect of lecithin addition could be observed in sectional view. The surfaces of these xerogels were smooth and with no visible pores on the micrometer scale. In sectional view, the lecithin-free xerogel exhibited a layered structure of polymer fibers with no visible interferences (see Figure 3). The same layered morphology was also observed for xerogels of agarose with different contents of lecithin even though there were regions of fusion of adjacent layers. Overall, the general morphology, practically devoid of pores as revealed by SEM, is most likely due to a compact structure resulting via the air drying procedure applied to hydrogels to finally obtain xerogels.

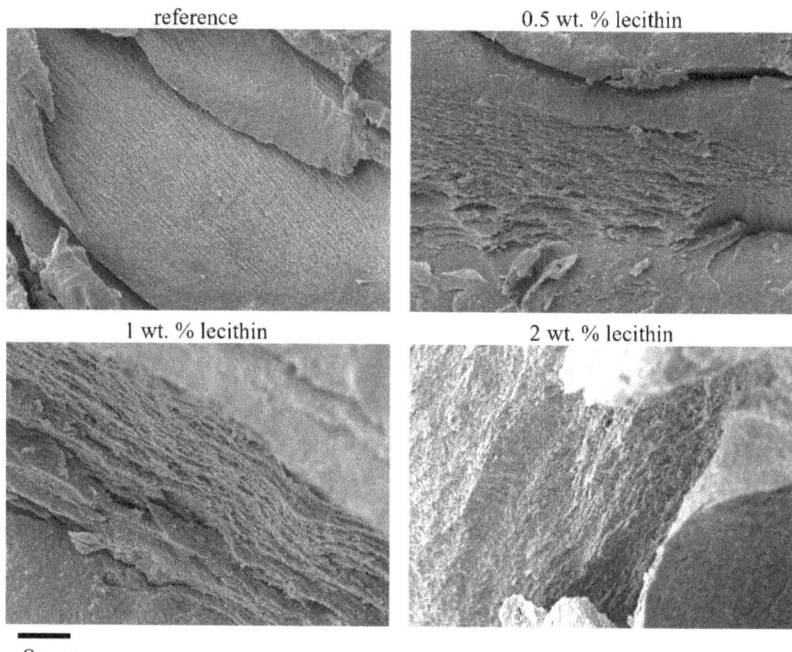

Figure 3. Physically crosslinked agarose xerogels with different lecithin contents observed in sectional view by SEM. Magnification 5000×.

For these xerogels, gas sorption measurements were also performed (Table 3). The low values of the specific surface suggest a lack of the pore structure of xerogels, with a slight dependence on the compactness of layered morphology of these systems in dry state. Even if the results of gas sorption are in line with those of SEM investigation, the gas sorption method is not quite a suitable technique for determining the structure of these xerogels.

Table 3. Specific surface area for physically crosslinked agarose xerogels with the addition of lecithin determined by gas sorption.

Concentration of Lecithin (wt.%)	Specific Surface Area (m^2/g)
0 (R)	3.4
0.5	1.0
1	1.9
2	2.1

2.2. Ionic Crosslinking

Sodium alginate crosslinked by the calcium chloride in the two to one weight ratio was a representative of the ionically crosslinked hydrogel matrix, where the negatively charged poly(guluronic) acid units of alginate (-COO$^-$) interact with the polyvalent ions (Ca^{2+}) to form a bond (schematic figure of the preparation procedure can be seen in Supplementary Materials, Figure S2). The final properties were also affected by lecithin addition. Hydrogel samples, both after preparation and dried and rehydrated, were studied by rheology, drying, and rehydration as well as morphological characterization.

2.2.1. Rheology

Ionically crosslinked hydrogels also underwent amplitude sweep tests. What is immediately observable is the decreasing trend of moduli for the freshly prepared samples as lecithin content increase (see Figure 4a). One of the reasons for this is the water intake during gelling, which increases for samples with ascending lecithin concentration (Section 2.2.2), the amphiphilic component playing a major role in the preparation of ionically crosslinked hydrogels. Larger lecithin addition also modified some characteristics of the hydrogels (see Table 4). The average moduli values in LVR steadily decreased after lecithin addition, thus making the gel softer. The most likely explanation is that after the crosslinking of alginate by calcium ions, free calcium chloride is still present in the system and is able to interact with the added lecithin micelles due to its dissociated form. Higher lecithin content causes a competitive interaction and as a result, lecithin displaces the calcium ions in the crosslinked alginate. Further lecithin could interact with the alginate via quaternary ammonia or with the calcium ions via negatively charged phosphate residues. For the moduli decrease, we could suggest that newly formed nodes are weaker and, in a lesser amount compared with the original alginate gel. Such competitive interactions were observable even during sample preparation, where the precipitate was visible on the surface of the solution. They were also confirmed by viscosity measurements, where the solution of calcium chloride and lecithin had higher viscosity values than expected, based on the viscosity of lecithin in water and of calcium chloride in water (figure is available in Supplementary Materials Figure S3). Other rheological data were very similar for the samples and, as stated earlier, the biggest differences were in the moduli values, thus in the hydrogel strength.

Table 4. Values for ionically crosslinked alginate hydrogels after preparation obtained from strain and frequency sweep tests before drying.

Lecithin Concentration (wt.%)	Cross-Over Point		Average Moduli Values in LVR		End of LVR	Mesh Size
	G' (Pa)	Strain (%)	G' (Pa)	G'' (Pa)	Strain (%)	Mesh (nm)
0 (R)	150.4 ± 9.1	260.4 ± 18.6	1667 ± 192	165 ± 23	1.7 ± 0.0	10.9 ± 0.4
0.5	158.6 ± 12.5	275.3 ± 24.4	2138 ± 480	245 ± 66	1.3 ± 0.0	11.0 ± 0.7
1	110.8 ± 1.2	260.8 ± 5.4	1052 ± 1	104 ± 0	1.6 ± 0.3	13.8 ± 1.9
2	65.3 ± 17.9	278.2 ± 9.0	468 ± 15	41 ± 0	2.1 ± 0.4	17.3 ± 1.5

The rehydrated samples followed a similar trend with respect to the moduli values, where these values decreased with increasing lecithin concentration. Average moduli values in LVR reported in the table below (Table 5) were higher than those presented in Table 4 because the rehydrated samples were not able to reabsorb the same amount of water as the freshly prepared hydrogels. Such behavior could be due to a compact arrangement favored by non-covalent interactions (mainly ionic interactions induced by Ca^{2+} ions onto both alginate and lecithin components) during the drying process.

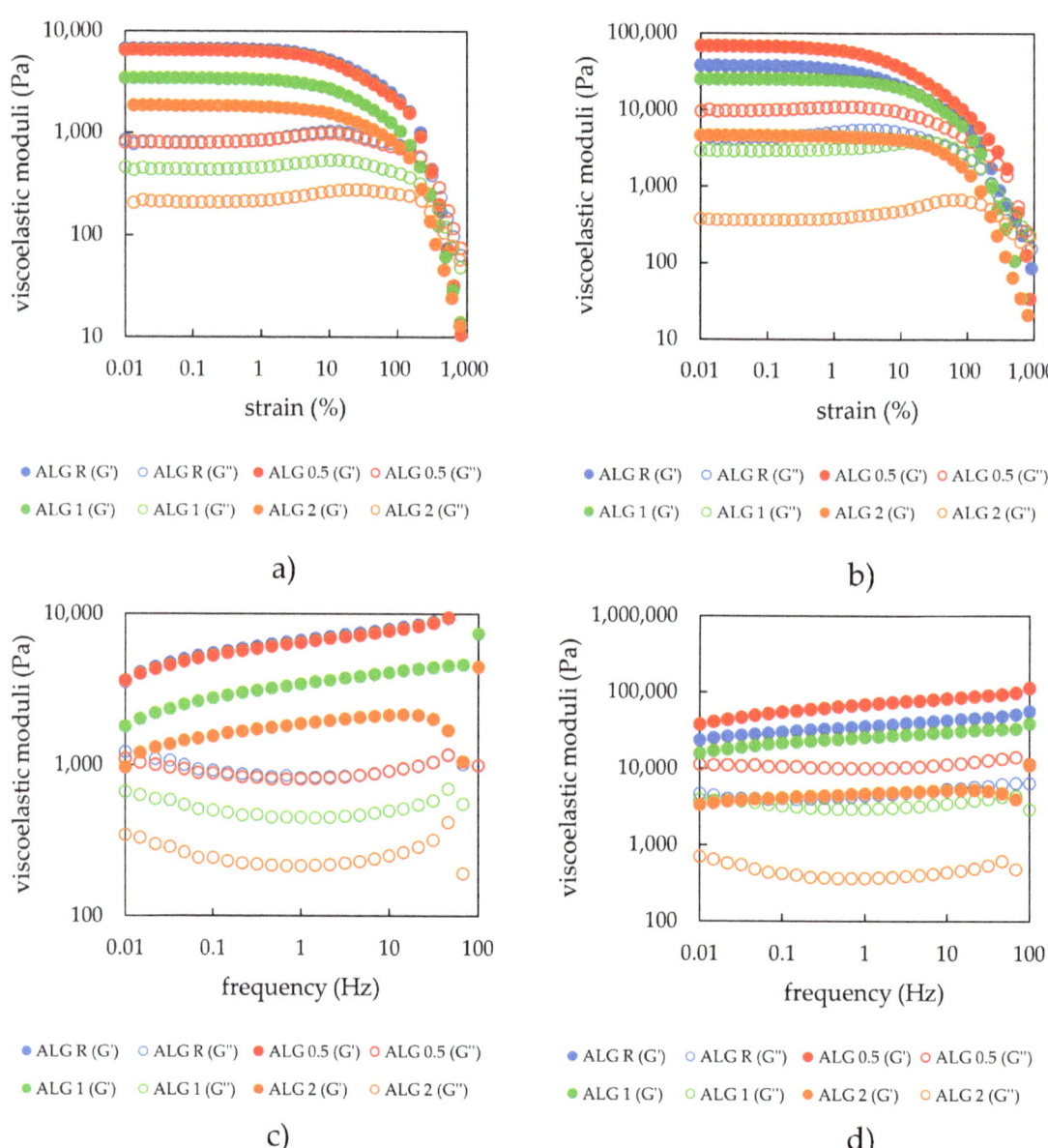

Figure 4. (**a**) Strain sweep of alginate hydrogels with the addition of different lecithin concentrations (0, 0.5, 1, and 2 wt.%) after preparation; (**b**) strain sweep of alginate hydrogels with different lecithin concentrations (0, 0.5, 1, and 2 wt.%) after drying and rehydration of the xerogels (frequency applied 1 Hz); (**c**) frequency sweep of alginate hydrogels with different lecithin concentrations (0, 0.5, 1, and 2 wt.%) after preparation; (**d**) frequency sweep of alginate hydrogels with different lecithin concentrations (0, 0.5, 1, and 2 wt.%) after drying and rehydration of the xerogels.

Table 5. Values for ionically crosslinked alginate hydrogels after drying and rehydration obtained from the strain and frequency sweep tests.

Lecithin Concentration (wt.%)	Cross-Over Point		Average Moduli Values in LVR		End of LVR	Mesh Size
	G' (Pa)	Strain (%)	G' (Pa)	G'' (Pa)	Strain (%)	Mesh (nm)
0 (R)	479.2 ± 129.7	210.8 ± 119.7	26,342 ± 13,355	3191 ± 1346	1.6 ± 0.4	4.6 ± 1.4
0.5	894.9 ± 612.4	522.6 ± 51.8	68,513 ± 17,434	9861 ± 1533	0.6 ± 0.6	12.3 ± 2.1
1	1179.5 ± 106.7	209.1 ± 37.3	25,386 ± 741	2912 ± 45	1.2 ± 0.2	8.3 ± 1.8
2	553.5 ± 24.3	189.5 ± 17.4	4599 ± 500	1842 ± 1447	2.4 ± 0.0	7.6 ± 0.5

The rheograms obtained during the frequency sweep tests (expressed as viscoelastic moduli on applied frequency) (Figure 4c) obeyed the same order as those that resulted from the amplitude sweep tests (storage and loss moduli as a function of oscillatory applied strain of 1 Hz) (Figure 4a) for all the studied alginate and alginate-lecithin hydrogels. The calculated mesh size from the rheological (frequency sweep) measurement for freshly prepared ionically crosslinked alginate hydrogels indicated the effect of lecithin on the structural properties of these hydrogels. The higher addition of lecithin causes a higher mesh size (more than 50% if the hydrogels without/with 2 wt.% of lecithin is compared). The effect of lecithin concentration was also not observed for dried and rehydrated hydrogels. Although ionically crosslinked hydrogels have the ability to reabsorb the dispersion medium and again create a network internal structure by water intake, the internal structure of these hydrogels is probably damaged by the air-drying process. Moreover, swelled hydrogels differ in mesh size values in comparison with freshly prepared (e.g., hydrogels with 2 wt.% of lecithin had a mesh size of 17.3 nm while the mesh size of the hydrogels with the same concentration of lecithin after swelling was 7.6 nm). Therefore, the effect of lecithin on the mesh size of hydrogels repeatedly prepared by drying and swelling in water medium was negligible.

2.2.2. Drying and Rehydration Measurements

The drying curves for the alginate-lecithin systems were very similar almost irrespective of the lecithin content, in contrast to the drying dependence obtained for the freshly prepared hydrogels of alginate solely (Figure 5a). The different kinetics regarding the rate of water loss during the drying step could be due to the way lecithin fills the hydrogel pores and holds water within, and also due to the favorable electrostatic Ca^{2+}-lecithin interactions, which influence the hydrogel structure and thus enable it to better hold water. As for the swelling after drying, it can be observed that the samples with higher lecithin concentrations were able to absorb water more rapidly and to a higher capacity, which is again due to the modified hydrogel network due to the presence of lecithin.

2.2.3. Morphological Characterization of Xerogels

SEM images taken for xerogels prepared by ionic crosslinking show the effect of lecithin on the surface morphology of the samples (see Figure 6). Surface morphology of lecithin-free samples and of those with 0.5 wt.% lecithin exhibited a roughness due to the many micrometer-sized crystals of $CaCl_2$ resulted after air-drying. Instead, the surface of xerogels with 1 and 2 wt.% lecithin is practically devoid of crystalline aggregates, with some degree of roughness, which led to a more compact structure of these mixed systems in their dry state. The morphological characteristics microscopically revealed are in accordance with the decreasing tendency of the specific surface values (from gas sorption measurements, Table 6) as the lecithin content rose. On the other hand, the lack of $CaCl_2$ crystalline aggregates for the systems with a higher lecithin content (1 and 2 wt.%) could be related to Ca^{2+} consumption in favorable electrostatic interactions with lecithin anions, which means that the crystalline structures observed in the case of alginate xerogels without lecithin and for those with 0.5 wt.% lecithin could be due to the excess of $CaCl_2$ contained in these explored samples.

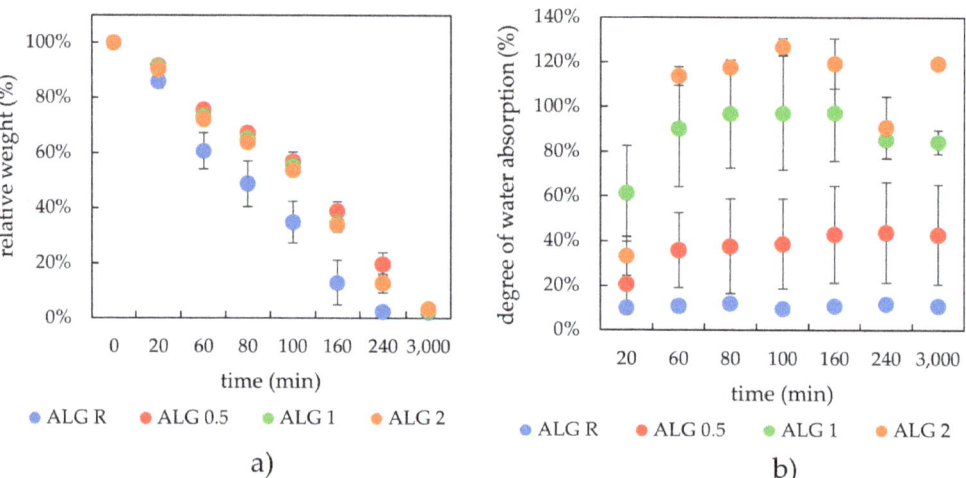

Figure 5. Drying (**a**) and rehydration (**b**) of ionically crosslinked alginate hydrogels with different lecithin content.

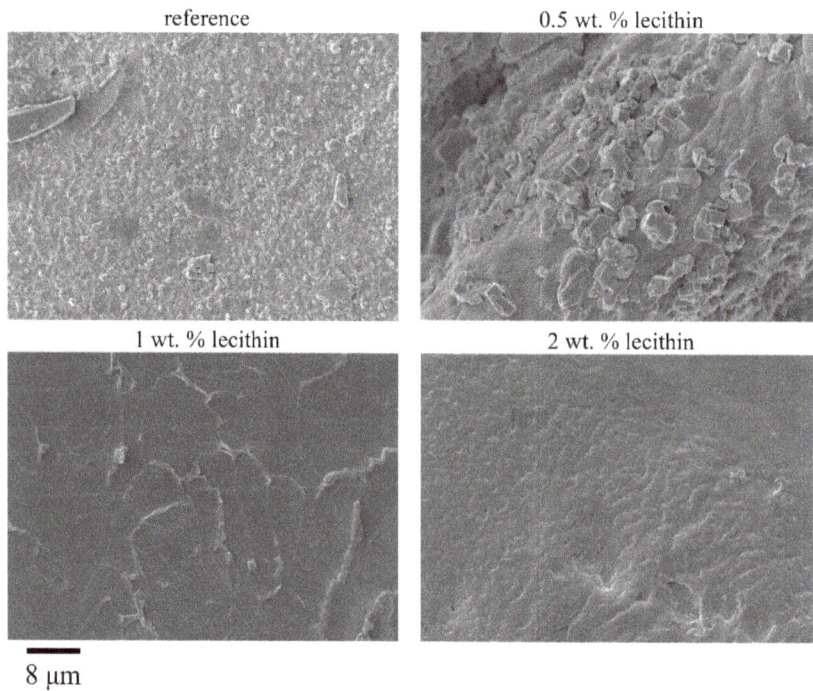

Figure 6. Surface morphologies of ionically crosslinked alginate xerogels with the addition of lecithin revealed by SEM. Magnification 5000×.

Table 6. Specific surface area for ionically crosslinked alginate xerogels with the addition of lecithin determined by gas sorption.

Concentration of Lecithin (wt.%)	Specific Surface Area (m^2/g)
0 (R)	9.1
0.5	6.3
1	5.9
2	4.7

2.3. Chemical Crosslinking

Poly(vinyl alcohol) and chitosan crosslinked by the epichlorohydrin was a representative of the chemically crosslinked hydrogel matrix. Epichlorohydrin reacts with either the hydroxyl group of PVA or amino group of chitosan to form a highly reactive intermediate. This intermediate product reacts with another hydroxyl (PVA) or amino group (chitosan) to form the crosslinked structure. Study of these hydrogels, both in their freshly prepared state, after air-drying at 40 °C and their subsequent rehydration and as xerogels, showed some physico-mechanical properties altered by the lecithin content (schematic figure of the preparation procedure can be seen in the Supplementary Materials Figure S4).

2.3.1. Rheology

For chemically crosslinked hydrogels, the amplitude sweep results showed that the addition of lecithin modified the rheological properties of hydrogels (see Figure 7a). However, the highest lecithin concentration did not lead to further changes in the mechanical properties. The same can be said after comparing the data points (see Table 7). At the same time, a higher content of lecithin decreased the values marking the end of the LVR as well as the strength of the hydrogels and the cross-over point values. The results are acceptable after taking into account the preparation and final state of the hydrogel. An important step of the preparation procedure is drying of the liquid mixture, which leads to crosslinking of the nodes and its subsequent rehydration. If lecithin is present, the rehydration is improved.

Table 7. Values for chemically crosslinked PVA-chitosan hydrogels obtained from strain and frequency sweep tests before drying.

	Cross-Over Point		Average Moduli Values in LVR		End of LVR	Mesh Size
Lecithin Concentration (wt.%)	G' (Pa)	Strain (%)	G' (Pa)	G'' (Pa)	Strain (%)	Mesh (nm)
0 (R)	1665.3 ± 43.2	53.8 ± 8.2	8629 ± 304	398 ± 4	1.6 ± 0.3	13.6 ± 0.7
0.5	1005.5 ± 32.4	49.4 ± 18.4	6644 ± 1503	307 ± 44	1.2 ± 0.9	13.8 ± 0.6
1	666.6 ± 5.4	40.2 ± 3.2	4545 ± 129	377 ± 68	0.6 ± 0.1	12.7 ± 0.1
2	631.6 ± 24.7	39.1 ± 4.7	4398 ± 195	421 ± 5	0.7 ± 0.1	12.9 ± 0.1

The same experiments were performed for hydrogel samples dried and rehydrated. The dried and rehydrated hydrogels with lecithin assembled into the pores ended up with modified properties (see Figure 7b), specifically, an increase in moduli values and a decrease in the values marking the cross-over point, in contrast to the reference sample. As can be seen in Figure 7b and Table 8, the presence of lecithin makes the hydrogels obtained after the drying–rehydration step much more deformation resistant, characterized by much higher values of strain at the cross-over point. At the same time, for these mixed rehydrated hydrogels, lecithin, irrespective of its content, exerted a larger influence in the enhancement of the hydrogels' strength (average moduli values in LVR) when compared to the rehydrated systems physically and ionically crosslinked.

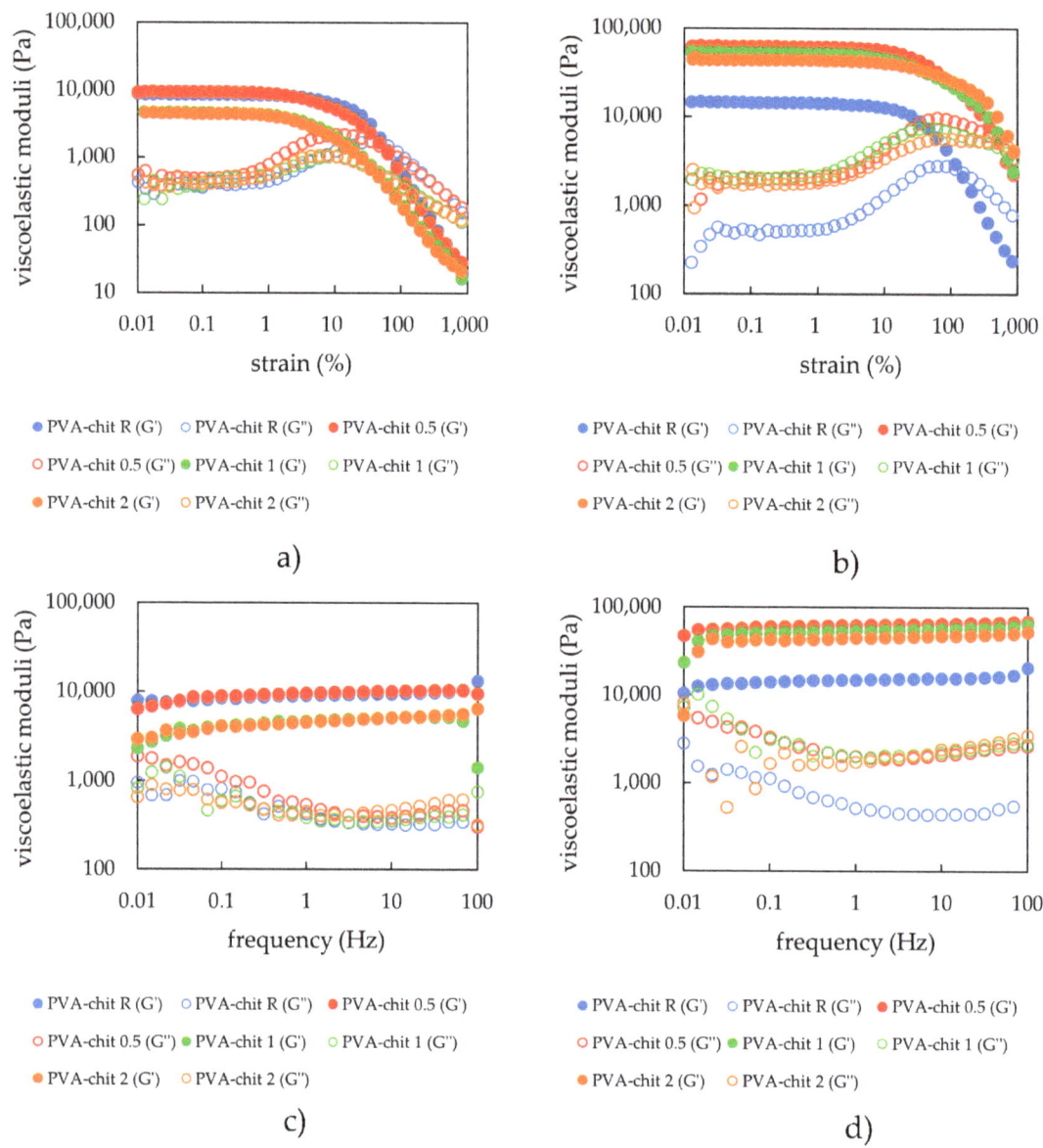

Figure 7. (**a**) Strain sweep of PVA-chitosan hydrogels with the different lecithin concentrations (0, 0.5, 1, and 2 wt.%) after preparation; (**b**) strain sweep of PVA-chitosan hydrogels with different lecithin concentrations (0, 0.5, 1, and 2 wt.%) after drying and rehydration of the xerogels (frequency applied–1 Hz); (**c**) frequency sweep of PVA-chitosan hydrogels with different lecithin concentrations (0, 0.5, 1, and 2 wt.%) after preparation; (**d**) frequency sweep of PVA-chitosan hydrogels with different lecithin concentrations (0, 0.5, 1, and 2 wt.%) after drying and rehydration of the xerogels.

Table 8. Values for chemically crosslinked PVA-chitosan hydrogels after drying and rehydration obtained from strain and frequency sweep tests.

Lecithin Concentration (wt.%)	Cross-Over Point		Average Moduli Values in LVR		End of LVR	Mesh Size
	G' (Pa)	Strain (%)	G' (Pa)	G'' (Pa)	Strain (%)	Mesh (nm)
0 (R)	2470.0 ± 494.7	138.5 ± 13.5	14,514 ± 1413	532 ± 33	3.2 ± 0.0	11.6 ± 0.3
0.5	7122.4 ± 633.3	379.1 ± 233.0	62,099 ± 6505	1928 ± 65	5.0 ± 1.0	7.1 ± 0.1
1	4964.6 ± 275.8	502.2 ± 277.5	52,833 ± 10,153	2089 ± 246	3.0 ± 1.7	6.2 ± 1.3
2	4074.2 ± 182.3	900.1 ± 97.5	43,685 ± 3177	1761 ± 211	5.9 ± 2.3	8.1 ± 0.0

The frequency and amplitude sweep results indicated the same tendency discussed above (see comparatively Figure 7). Thus, a critical lecithin concentration is necessary to modify the properties of this type of chemically crosslinked hydrogels (according to the results lying between 0.5 and 1 wt.%); also, there is a maximum concentration above which further modifications do not occur (differences between 1 and 2 wt.% are negligible). The significant difference in the chemically crosslinked hydrogels (comparing to the physically and ionically crosslinked) is the relaxation phenomenon characterized by much longer relaxation times in contrast to covalently crosslinked systems. Covalently crosslinked hydrogels exhibit almost constant values of storage moduli over the whole range of the applied frequencies. The same trend was also observed for the dried and rehydrated samples. Again, for all samples, the storage modulus prevailed in comparison to the loss modulus. The mesh sizes of these samples (Tables 7 and 8) were not affected by the content of lecithin, a result that can be explained by the character of covalent crosslinking, which is stronger than physical and ionic crosslinking. On the other hand, the same trend of decreasing mesh sizes after rehydration could be observed.

2.3.2. Drying and Rehydration Measurements

As can be seen from Figure 8, the drying and swelling kinetics were not significantly altered by the addition of lecithin. Only a marginal influence was observed for samples with the highest lecithin concentrations, which were able to absorb the most water. This generally smaller influence of lecithin can be explained by the structure of chemically crosslinked hydrogels, which are characterized by a high enough crosslinking density and, consequently, by a smaller pore size morphology. The structure is more organized due to the stronger covalent bonds. The water absorption for this kind of hydrogel possessing stronger covalent cross linkages was very fast and occurred almost immediately during the first minutes of the swelling experiments.

2.3.3. Morphological Characterization of Xerogels

Results on the structural characterization of chemically crosslinked xerogels were similar to those for physically crosslinked hydrogels. The surface morphology of these xerogels looked smooth with no visible pores. In sectional view, SEM images revealed clear layered structures, with an interlayer roughness increasing with lecithin content (Figure 9), which in turn led to a gradual ascension of the value of specific surface (Table 9). Despite this fact, an apparently less corrugated surface observed for lecithin-free hydrogels had a higher specific surface area (Table 9), which might be explained by a greater compactness associated with the layered structure of the mixed xerogels.

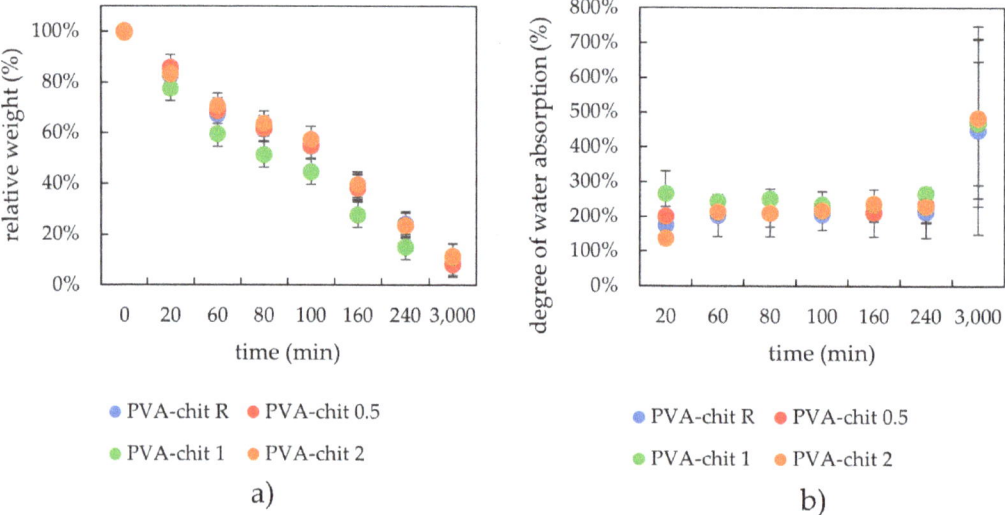

Figure 8. Drying (**a**) and rehydration (**b**) of chemically crosslinked PVA-chitosan hydrogels with different lecithin content.

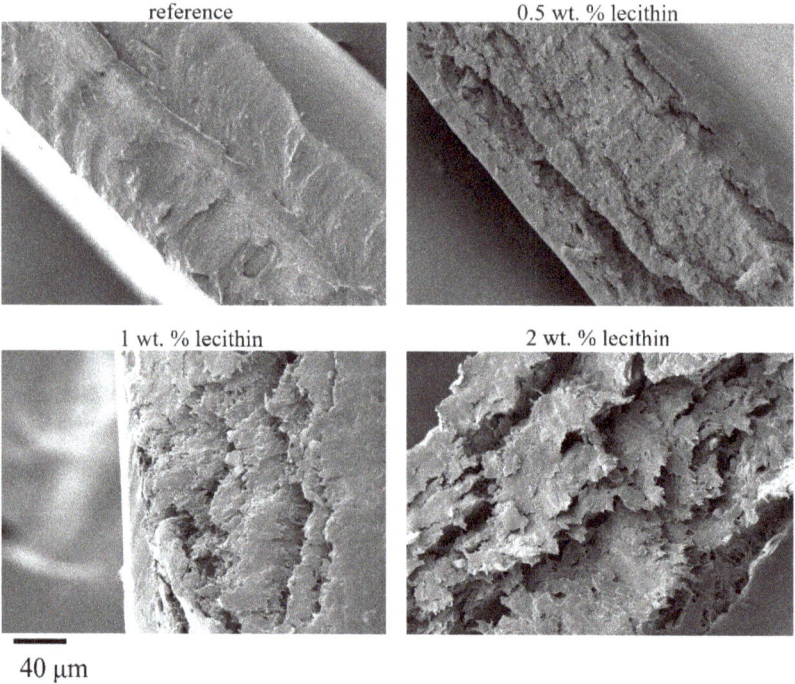

Figure 9. SEM images of chemically crosslinked PVA-chitosan xerogels with the addition of lecithin taken from the sectional view. Magnification 1000×.

Table 9. Specific surface area for chemically crosslinked PVA-chitosan xerogels with different lecithin content determined by gas sorption.

Concentration of Lecithin (wt.%)	Specific Surface Area (m^2/g)
0 (R)	2.9
0.5	0.8
1	1.2
2	1.6

3. Conclusions

This work studied the influence of lecithin (L-α-phosphatidylcholine) on three differently crosslinked hydrogels (physically crosslinked agarose, alginate ionically crosslinked by calcium ions, and a mixture of PVA and chitosan chemically crosslinked by epichlorohydrin). The bulk of this work was to study differences between the gels investigated immediately after preparation and the corresponding rehydrated xerogels (prepared by swelling). By choosing the lecithin content, we were able to modify some of the mechanical properties of the hydrogels with a modified internal structure, especially in the case of the rehydrated ones. In this regard, the addition of lecithin had the strongest influence in enhancing the strength of chemically crosslinked PVA-chitosan gels, which is partially consistent with the mesh size and by the amount of water absorbed into their structure after being previous air-dried. Apart from the rheological data and those obtained from the kinetics of water loss during hydrogel dehydration, these conclusions were supported by the scanning electron microscopy and gas sorption experiments performed on the xerogels. For this type of material, even though gas sorption appears to be inappropriate, however, it serves to confirm the non-porous structure of the xerogels.

In this work, we determined that the addition of phospholipid lecithin into the hydrogel matrix can alter their mechanical properties, which might be highly beneficial knowledge for the use of such hydrogels in particular applications. However, the transport properties also need to be investigated. Therefore, further transport experiments are required, which are absolutely crucial for a better understanding of such hydrogel materials and how they can be used in final applications.

4. Materials and Methods

Hydrogels with distinct gelation mechanisms (physical, ionic, chemical crosslinking) [20] were studied. As an example of a physically crosslinked matrix, the linear thermoreversible polysaccharide agarose (Agarose E, Condalab, Madrid, Spain) at 1 wt.%, was used [21]. As an example of an ionically crosslinked matrix, sodium alginate (Sigma-Aldrich, Prague, Czech Republic) at 2 wt.% crosslinked by calcium chloride (Lach-Ner, Neratovice, Czech Republic) at a two to one weight ratio was chosen [22]. For chemically crosslinked hydrogels, poly(vinyl alcohol) (Sigma-Aldrich, Prague, Czech Republic) mixed with chitosan (low molecular weight, Sigma-Aldrich, Prague, Czech Republic) and crosslinked by epichlorohydrin (Sigma-Aldrich, Prague, Czech Republic) was employed [23]. L-α-Phosphatidylcholine (lecithin) was incorporated into all hydrogel samples before gelation at three different weight percentage concentrations (Sigma-Aldrich, Czech Republic, Prague).

The materials and their concentrations and ratios were selected on the basis of data previously reported [20–24] and can be seen in the table below (Table 10).

Table 10. Concentrations of each individual component in the final hydrogel form (agarose, sodium alginate, calcium chloride, PVA, chitosan, and lecithin).

Physically Crosslinked Hydrogels		
Sample	Agarose (wt.%)	Lecithin (wt.%)
AG R	1	0
AG 0.5	1	0.5
AG 1	1	1
AG 2	1	2
Ionically Crosslinked Hydrogels		
Sample	Sodium Alginate (wt.%)	Calcium Chloride (mol·dm^3)	Lecithin (wt.%)
ALG R	2	0.1	0
ALG 0.5	2	0.1	0.5
ALG 1	2	0.1	1
ALG 2	2	0.1	2
Chemically Crosslinked Hydrogels		
Sample	PVA (wt.%)	Chitosan (wt.%)	Lecithin (wt.%)
PVA R	7.8	2.5	0
PVA 0.5	7.8	2.5	0.5
PVA 1	7.8	2.5	1
PVA 2	7.8	2.5	2

4.1. Water Loss during Drying and Rehydration Measurements

The ability to hold, release, and absorb water was tested by different approaches. Water loss was monitored by means of simple drying tests. All samples were dried either in the laboratory dryer at 40 °C and regularly weighed, or in a semi-automatic moisture analyzer (IR-35, Denver Instrument, Denver, CO, USA), where the weight was recorded automatically. The relative weight of the hydrogel (x) during drying was calculated using the following formula:

$$x = \frac{m_t}{m_0} \cdot 100 \qquad (1)$$

where m_t is the weight of the gel at time t, and m_0 is the weight of the hydrogel in the swollen state.

Often very small weight losses of water from the hydrogel samples made using drying scales more difficult. For this reason, drying kinetics were mostly studied using the combination of laboratory driers and analytical scales, upon which samples were weighed every twenty minutes. After the samples were dried to the xerogel form, they were inserted into a water bath, where they were kept until they reached their maximum water absorption capacity. The degree of water absorption (m_a) was calculated by:

$$m_a = \frac{m_t}{m_x} \cdot 100 \qquad (2)$$

where m_t is the weight of the hydrogel at time t, and m_x is the weight of the xerogel. The hydrogel samples were regularly weighed on analytical scales to study their swelling kinetics.

4.2. Rheology

Hydrogels are semi-solid materials that exhibit distinctive mechanical characteristics lying between those of solids and liquids. Therefore, rheology is indeed an appropriate technique for studying their behavior [25–29]. The mechanical properties of the prepared hydrogels were determined by rheological characterization using a rotational rheometer (Discovery HR-2, TA Instruments) employing cross-hatched 20 mm plate–plate geometry to avoid potential sensor wall-slippage during measurement. The complex rheological procedure consisted of strain sweep and frequency sweep tests. The strain sweep test is a useful tool for obtaining information about samples if fluid-like or gel-like behavior under different values of applied strain prevails. In addition, it is possible to determine the region where the deformation is non-destructive (the linear viscoelastic region-LVR) as well as the behavior of the sample when the LVR strain limit is exceeded. The other mentioned test, the frequency sweep test, serves the purpose of describing hydrogel behavior in the non-deformation range (LVR) and provides information about different crosslinking sites (if applicable) in the internal structure of the hydrogel. Both tests were carried out on freshly prepared samples and rehydrated ones. The rehydrated samples were first dried to constant mass in the laboratory dryer for two days at a constant temperature of 40 °C and further rehydrated for three days in distilled water. Freshly prepared agarose and alginate samples were measured within a gap of 1000 μm. The gap for rehydrated samples varied according to the thickness of the gel, which depended on its swelling capacity, 500 μm for agarose gels and 1000 μm for alginate gels. PVA-chitosan hydrogels (both fresh and rehydrated) were measured within a gap of 200 μm due to the limited thickness of the prepared hydrogel foils. Prior to each applied test, samples were allowed to temper and rest for 180 s after loading into the measuring gap.

To obtain a suitable value of constant amplitude strain for the linear viscoelastic region (LVR), which was an essential parameter for ongoing frequency sweep tests, strain sweep tests were conducted first within the amplitude strain range of 0.01–1000% under a constant frequency of oscillation of 1 Hz in at least two repetitions, using a freshly loaded sample for each test. From these measurements, a strain of 0.1% was chosen as a suitable value of deformation for ongoing frequency tests, because this strain value lays within the LVR for all fresh and rehydrated samples. The range of oscillating frequencies for the frequency sweep tests was set to 0.01–100 Hz. Like the former strain sweep tests, the frequency sweep tests were also conducted in at least two repetitions. A summary of settings for both rheology tests is presented in Table 11.

Table 11. Summary of settings for rheology measurements (conditioning step, amplitude sweep, and frequency sweep).

Conditioning Step	
Temperature	25 °C
Time	180 s
Amplitude Sweep		**Frequency Sweep**	
temperature	25 °C	temperature	25 °C
strain	0.01–1000%	strain	0.1%
points per decade	8	points per decade	6
frequency	1 Hz	frequency	0.01–100 Hz

Routine techniques that are usable for the characterization of the internal structures of many materials (e.g., scanning electron microscopy) have some limitations in the study of hydrogels. One of the most limiting factors is that the structures of hydrogels are mostly studied in a dried state. The internal structures of a hydrogel in the presence of water and in the absence of water must certainly differ. Moreover, the preparation of the hydrogel in its dried state is also critical because the dispersion medium (water) must be removed (mostly by evaporation or by sublimation if lyophilization is used). Unfortunately, both of these processes (evaporation as well as sublimation) have a significant impact on the final xerogel morphology. Simply, the fragile internal structure of the hydrogel may be critically damaged by the removal of the dispersion medium. Thus, such a resulting structure (specifically, the porous structure) revealed by scanning electron microscopy often has low informative value with respect to the internal structure of the hydrogel in its swollen state. Therefore, an alternative way to determine the pore size (and then obtain information about the internal structure of the hydrogel) must be found. An interesting solution to this problem is offered by the rheological characterization of the hydrogel, which involves the calculation of the mesh size.

Mesh size, as one of the most critical parameters in hydrogel characterization, was calculated by means of relaxation spectra (relaxation moduli G and relaxation time λ) from the frequency sweep oscillation measurements in accordance with the Maxwell model [30]. The frequency sweep (viscoelastic moduli as a function of oscillation frequency) was interpolated by continuous relaxation spectra in TRIOS software (TA Instruments, New Castle, DE, USA).

Typical relaxation spectra can be found in the Supplementary Materials (Figure S5). On the basis of previous rheological investigation [25], it was concluded that the optimal number of Maxwell elements was 4, in order to fit the frequency sweep measurements of the hydrogels. Four relaxation moduli were obtained from continuous relaxation spectra analyses. The sum of relaxation moduli was calculated in order to determine the crosslinking density [31] (see Equation (3), where ρ_x represents the crosslinking density (mol·m^{-3})) and provides information on the density of the junction in the swollen hydrogel form. G (Pa) is the sum of 4 relaxation moduli, R (J·mol^{-1}·K^{-1}) represents the universal gas constant, and T is the thermodynamic temperature in Kelvins.

$$\rho_x = \frac{G}{RT} \quad (3)$$

If all criteria are met (in particular, frequency sweep measurements are realized in the linear viscoelastic region and the mechanical properties of hydrogels with different crosslinking are consistent with rubber elasticity theory [32]), finally the mesh size can be calculated using Equation (4), where ξ is the mesh size (unit: m) and N_A represents Avogadro's number.

$$\xi = \sqrt[3]{\frac{6}{\pi \rho_x N_A}} \quad (4)$$

4.3. Morphological Characterization of Xerogels

Since the structure affects properties that are crucial for hydrogel applications, determining the hydrogel morphology is one of the most important characterizations. There are many direct (microscopy) and indirect (scattering-based) methods to characterize hydrogel morphology [33]. Several direct visualization techniques (light microscopy, laser scanning confocal microscopy, and micro-computed tomography) that can handle swollen hydrogels have considerable disadvantages (e.g., limited resolution) [34]. On the other hand, commonly used scanning electron microscopy includes a critical step (i.e., the inevitable solidification of the sample using drying or freezing, during which the collapse of the structure or the creation of artifacts can occur) [35,36]. Kaberova et al. [37] tested the usability of scanning electron microscopy and concluded that the results from this method should always be confirmed by microscopy techniques applicable for gels in their swollen state.

For the characterization of dry samples, the specific surface area (the Brunauer–Emmett–Teller (BET) approach) is typically determined. The specific surface area is not suitable for characterizing hydrogels because of the already mentioned artifacts that appear during the preparation of dried samples. However, it can be used, for example, for the characterization of materials used in a dried state and that can form hydrogels (adsorbent) [38], or for the confirmation of reversible porosity [39].

The structure of the xerogels was studied in this work. Specifically, scanning electron microscopy and gas sorption were chosen as suitable techniques for determining the internal architecture of xerogels. Since the mechanical properties were studied for hydrogels right after preparation and also for swollen hydrogels after dehydration, it seemed convenient to investigate the structural properties of the hydrogels in these forms. Since this form is a dry form, it was possible to avoid deformation of the structure caused by the preparation of hydrogels for scanning electron microscopy.

4.3.1. Scanning Electron Microscopy

To determine changes in hydrogel structure, xerogels of all prepared samples were subjected to direct visualization using scanning electron microscopy. The samples were dried in a laboratory dryer at 40 °C. A few small specimens were taken from each studied sample to maintain objective observation. These specimens were subsequently gold-coated in a sputtering device (POLARON) and investigated using a ZEISS EVO LS 10 scanning electron microscope.

Both the surface morphologies and sectional images of samples were recorded. Observations were realized in secondary electron (SE) mode and the accelerating voltage was set to 5 kV to avoid charging of the samples.

4.3.2. Gas Sorption

A NOVA 2200e high-speed gas sorption analyzer (Quantachrome Instruments) was used to determine the specific surface area. The samples were weighed into a measuring cell (0.05–0.1 g). The measuring cell was placed in a degassing station, where the degassing process was carried out at 75 °C for 20 h. After cooling, the degassed sample was weighed to four decimal places. The samples were placed in a measuring station. The adsorption and desorption isotherms were measured under liquid nitrogen (77 K) from 0.05–0.95 of the relative pressure P/P_0. The obtained data were processed by NovaWin software and specific surface area was calculated by the multi-point BET method.

Supplementary Materials: The following supporting information can be downloaded at: https://www.mdpi.com/article/10.3390/gels8020115/s1, Figure S1: Preparation procedure of physically crosslinked agarose hydrogels; Figure S2: Preparation procedure of ionically crosslinked alginate hydrogels; Figure S3: Dynamic viscosity measurements for combinations of solutions of lecithin, $CaCl_2$ and alginate; Figure S4: Preparation procedure of chemically crosslinked PVA-chitosan hydrogels; Figure S5: Typical relaxation spectra for mesh size calculations, TRIOS software (TA Instruments).

Author Contributions: Conceptualization, R.H. and J.S.; Methodology, R.H., M.K., M.T. and J.H.; Validation, R.H., M.K. and M.T.; Formal analysis, R.H. and J.S.; Investigation, R.H., M.K., M.T., J.H. and N.Z.; Data curation, R.H., M.T. and M.K.; Writing—original draft preparation, R.H., M.K., M.T. and N.Z.; Writing—review and editing, J.S. and M.P.; Visualization, R.H. and N.Z.; Project administration, R.H.; Funding acquisition, R.H.; Supervision, J.S. All authors have read and agreed to the published version of the manuscript.

Funding: This research project was supported by the project Quality Internal Grants of BUT (KInG BUT), Reg. No. CZ.02.2.69/0.0/0.0/19_073/0016948, which is financed from the Operational Program: Research, Development, and Education.

Institutional Review Board Statement: Not applicable.

Informed Consent Statement: Not applicable.

Data Availability Statement: The data used in this study are available on request from the corresponding author.

Conflicts of Interest: The authors declare no conflict of interest.

References

1. Aswathy, S.; Narendrakumar, U.; Manjubala, I. Commercial hydrogels for biomedical applications. *Heliyon* **2020**, *6*, e03719. [CrossRef] [PubMed]
2. Kular, J.K.; Basu, S.; Sharma, R.I. The extracellular matrix: Structure, composition, age-related differences, tools for analysis and applications for tissue engineering. *J. Tissue Eng.* **2014**, *5*, 2041731414557112. [CrossRef] [PubMed]
3. Geckil, H.; Xu, F.; Zhang, X.; Moon, S.; Demirci, U. Engineering hydrogels as extracellular matrix mimics. *Nanomedicine* **2010**, *5*, 469–484. [CrossRef]
4. Gadjanski, I. Recent advances on gradient hydrogels in biomimetic cartilage tissue engineering. *F1000Research* **2017**, *6*, 2158. [CrossRef]
5. Pekař, M. Hydrogels with Micellar Hydrophobic (Nano)Domains. *Front. Mater.* **2015**, *1*, 35. [CrossRef]

6. Zhang, Y.; Chen, Q.; Dai, Z.; Dai, Y.; Xia, F.; Zhang, X. Nanocomposite adhesive hydrogels: From design to application. *J. Mater. Chem. B* **2021**, *9*, 585–593. [CrossRef]
7. Tuncaboylu, D.C.; Argun, A.; Algi, M.P.; Okay, O. Autonomic self-healing in covalently crosslinked hydrogels containing hydrophobic domains. *Polymer* **2013**, *54*, 6381–6388. [CrossRef]
8. Gu, S.; Duan, L.; Ren, X.; Gao, G.H. Robust, tough and anti-fatigue cationic latex composite hydrogels based on dual physically cross-linked networks. *J. Colloid Interface Sci.* **2017**, *492*, 119–126. [CrossRef]
9. Li, A.; Jia, Y.; Sun, S.; Xu, Y.; Minsky, B.B.; Stuart, M.A.C.; Cölfen, H.; von Klitzing, R.; Guo, X. Mineral-Enhanced Polyacrylic Acid Hydrogel as an Oyster-Inspired Organic–Inorganic Hybrid Adhesive. *ACS Appl. Mater. Interfaces* **2018**, *10*, 10471–10479. [CrossRef]
10. Cui, C.; Wu, T.; Gao, F.; Fan, C.; Xu, Z.; Wang, H.; Liu, B.; Liu, W. An Autolytic High Strength Instant Adhesive Hydrogel for Emergency Self-Rescue. *Adv. Funct. Mater.* **2018**, *28*, 1804925. [CrossRef]
11. Fan, X.; Wang, S.; Fang, Y.; Li, P.; Zhou, W.; Wang, Z.; Chen, M.; Liu, H. Tough polyacrylamide-tannic acid-kaolin adhesive hydrogels for quick hemostatic application. *Mater. Sci. Eng. C* **2020**, *109*, 110649. [CrossRef] [PubMed]
12. Rajabi, N.; Kharaziha, M.; Emadi, R.; Zarrabi, A.; Mokhtari, H.; Salehi, S. An adhesive and injectable nanocomposite hydrogel of thiolated gelatin/gelatin methacrylate/Laponite® as a potential surgical sealant. *J. Colloid Interface Sci.* **2020**, *564*, 155–169. [CrossRef] [PubMed]
13. Arno, M.C.; Inam, M.; Weems, A.C.; Li, Z.; Binch, A.L.A.; Platt, C.I.; Richardson, S.M.; Hoyland, J.A.; Dove, A.P.; O'Reilly, R.K. Exploiting the role of nanoparticle shape in enhancing hydrogel adhesive and mechanical properties. *Nat. Commun.* **2020**, *11*, 1420. [CrossRef] [PubMed]
14. Zhang, H.; Hao, R.; Ren, X.; Yu, L.; Yang, H.; Yu, H. PEG/lecithin–liquid-crystalline composite hydrogels for quasi-zero-order combined release of hydrophilic and lipophilic drugs. *RSC Adv.* **2013**, *3*, 22927–22930. [CrossRef]
15. Shchipunov, Y.A. Lecithin. In *Encyclopedia of Surface and Colloid Science*, 3rd ed.; Somasundaran, P., Ed.; CRC Press: Boca Raton, FL, USA, 2015; Volume 3, pp. 3674–3693.
16. Elnaggar, Y.S.; El-Refaie, W.M.; El-Massik, M.A.; Abdallah, O.Y. Lecithin-based nanostructured gels for skin delivery: An update on state of art and recent applications. *J. Control. Release* **2014**, *180*, 10–24. [CrossRef] [PubMed]
17. Thompson, B.R.; Zarket, B.C.; Lauten, E.H.; Amin, S.; Muthukrishnan, S.; Raghavan, S.R. Liposomes Entrapped in Biopolymer Hydrogels Can Spontaneously Release into the External Solution. *Langmuir* **2020**, *36*, 7268–7276. [CrossRef]
18. Li, D.; An, X.; Mu, Y. A liposomal hydrogel with enzyme triggered release for infected wound. *Chem. Phys. Lipids* **2019**, *223*, 104783. [CrossRef]
19. Talaat, S.M.; Elnaggar, Y.S.R.; Abdalla, O.Y. Lecithin Microemulsion Lipogels Versus Conventional Gels for Skin Targeting of Terconazole: In Vitro, Ex Vivo, and In Vivo Investigation. *AAPS PharmSciTech* **2019**, *20*, 161. [CrossRef]
20. Maitra, J.; Shukla, V.K. Cross-linking in Hydrogels—A Review. *Am. J. Polym. Sci.* **2014**, *4*, 25–31. [CrossRef]
21. Trudicova, M.; Smilek, J.; Kalina, M.; Smilkova, M.; Adamkova, K.; Hrubanova, K.; Krzyzanek, V.; Sedlacek, P. Multiscale Experimental Evaluation of Agarose-Based Semi-Interpenetrating Polymer Network Hydrogels as Materials with Tunable Rheological and Transport Performance. *Polymers* **2020**, *12*, 2561. [CrossRef]
22. Kuo, C.K.; Ma, P.X. Ionically crosslinked alginate hydrogels as scaffolds for tissue engineering: Part 1. Structure, gelation rate and mechanical properties. *Biomaterials* **2001**, *22*, 511–521. [CrossRef]
23. Garnica-Palafox, I.M.; Sánchez-Arévalo, F.M.; Velasquillo, C.; García-Carvajal, Z.; Garcia-Lopez, J.; Ortega-Sánchez, C.; Ibarra, C.; Luna-Barcenas, G.; Solís-Arrieta, L. Mechanical and structural response of a hybrid hydrogel based on chitosan and poly(vinyl alcohol) cross-linked with epichlorohydrin for potential use in tissue engineering. *J. Biomater. Sci. Polym. Ed.* **2014**, *25*, 32–50. [CrossRef] [PubMed]
24. Mendes, A.C.L.; Shekarforoush, E.; Engwer, C.; Beeren, S.; Gorzelanny, C.; Goycoolea, F.M.; Chronakis, I.S. Co-assembly of chitosan and phospholipids into hybrid hydrogels. *Pure Appl. Chem.* **2016**, *88*, 905–916. [CrossRef]
25. Smilek, J.; Jarábková, S.; Velcer, T.; Pekař, M. Compositional and Temperature Effects on the Rheological Properties of Polyelectrolyte–Surfactant Hydrogels. *Polymers* **2019**, *11*, 927. [CrossRef]
26. Mourycová, J.; Datta, K.K.R.; Procházková, A.; Ploténá, M.; Enev, V.; Smilek, J.; Másilko, J.; Pekař, M. Facile synthesis and rheological characterization of nanocomposite hyaluronan-organoclay hydrogels. *Int. J. Biol. Macromol.* **2018**, *111*, 680–684. [CrossRef]
27. Derkach, S.R.; Ilyin, S.O.; Maklakova, A.A.; Kulichikhin, V.G.; Malkin, A.Y. The rheology of gelatin hydrogels modified by κ-carrageenan. *LWT-Food Sci. Technol.* **2015**, *63*, 612–619. [CrossRef]
28. López-Marcial, G.R.; Zeng, A.Y.; Osuna, C.; Dennis, J.; García, J.M.; O'Connell, G.D. Agarose-Based Hydrogels as Suitable Bioprinting Materials for Tissue Engineering. *ACS Biomater. Sci. Eng.* **2018**, *4*, 3610–3616. [CrossRef]
29. Gila-Vilchez, C.; Bonhome-Espinosa, A.B.; Kuzhir, P.; Zubarev, A.; Duran, J.D.G.; Lopez-Lopez, M.T. Rheology of magnetic alginate hydrogels. *J. Rheol.* **2018**, *62*, 1083–1096. [CrossRef]
30. Gradzielski, M.; Hoffmann, I. Polyelectrolyte-surfactant complexes (PESCs) composed of oppositely charged components. *Curr. Opin. Colloid Interface Sci.* **2018**, *35*, 124–141. [CrossRef]
31. Pescosolido, L.; Feruglio, L.; Farra, R.; Fiorentino, S.; Colombo, I.; Coviello, T.; Matricardi, P.; Hennink, W.E.; Vermonden, T.; Grassi, M. Mesh size distribution determination of interpenetrating polymer network hydrogels. *Soft Matter* **2012**, *8*, 7708–7715. [CrossRef]

32. Flory, P.J. *Principles of Polymer Chemistry*; Cornell University Press: New York, NY, USA, 1953.
33. Raghuwanshi, V.S.; Garnier, G. Characterisation of hydrogels: Linking the nano to the microscale. *Adv. Colloid Interface Sci.* **2019**, *274*, 102044. [CrossRef] [PubMed]
34. Suchý, T.; Šupová, M.; Bartoš, M.; Sedláček, R.; Piola, M.; Soncini, M.; Fiore, G.B.; Sauerová, P.; Kalbáčová, M.H. Dry versus hydrated collagen scaffolds: Are dry states representative of hydrated states? *J. Mater. Sci. Mater. Med.* **2018**, *29*, 20. [CrossRef] [PubMed]
35. Muthulakshmi, L.; Pavithra, U.; Sivaranjani, V.; Balasubramanian, N.; Sakthivel, K.M.; Pruncu, C.I. A novel Ag/carrageenan–gelatin hybrid hydrogel nanocomposite and its biological applications: Preparation and characterization. *J. Mech. Behav. Biomed. Mater.* **2021**, *115*, 104257. [CrossRef]
36. Marmorat, C.; Arinstein, A.; Koifman, N.; Talmon, Y.; Zussman, E.; Rafailovich, M. Cryo-Imaging of Hydrogels Supermolecular Structure. *Sci. Rep.* **2016**, *6*, 25495. [CrossRef] [PubMed]
37. Kaberova, Z.; Karpushkin, E.; Nevoralová, M.; Vetrík, M.; Šlouf, M.; Dušková-Smrčková, M. Microscopic Structure of Swollen Hydrogels by Scanning Electron and Light Microscopies: Artifacts and Reality. *Polymers* **2020**, *12*, 578. [CrossRef]
38. Bhagat, S.D.; Kim, Y.-H.; Yi, G.; Ahn, Y.-S.; Yeo, J.-G.; Choi, Y.-T. Mesoporous SiO_2 powders with high specific surface area by microwave drying of hydrogels: A facile synthesis. *Microporous Mesoporous Mater.* **2008**, *108*, 333–339. [CrossRef]
39. Weber, J.; Bergström, L. Mesoporous Hydrogels: Revealing Reversible Porosity by Cryoporometry, X-ray Scattering, and Gas Adsorption. *Langmuir* **2010**, *26*, 10158–10164. [CrossRef]

Article

Factors Affecting the Time and Process of CMC Drying Using Refractance Window or Conductive Hydro-Drying

Rubén D. Múnera-Tangarife [1,2], Efraín Solarte-Rodríguez [3], Carlos Vélez-Pasos [2] and Claudia I. Ochoa-Martínez [2,*]

[1] Grupo GIEPRONAL, School of Basic Sciences, Technology and Engineering, Universidad Nacional Abierta y a Distancia, Palmira 763531, Colombia; ruben.munera@unad.edu.co
[2] Grupo GIPAB, School of Food Engineering, Universidad del Valle, Santiago de Cali 760001, Colombia; carlos.velez@correounivalle.edu.co
[3] Quantum Optics Research Group, Department of Physics, Universidad del Valle, Santiago de Cali 760001, Colombia; efrain.solarte@correounivalle.edu.co
* Correspondence: claudia.ochoa@correounivalle.edu.co

Citation: Múnera-Tangarife, R.D.; Solarte-Rodríguez, E.; Vélez-Pasos, C.; Ochoa-Martínez, C.I. Factors Affecting the Time and Process of CMC Drying Using Refractance Window or Conductive Hydro-Drying. Gels 2021, 7, 257. https://doi.org/10.3390/gels7040257

Academic Editors: Yang Liu and Kiat Hwa Chan

Received: 15 November 2021
Accepted: 9 December 2021
Published: 11 December 2021

Publisher's Note: MDPI stays neutral with regard to jurisdictional claims in published maps and institutional affiliations.

Copyright: © 2021 by the authors. Licensee MDPI, Basel, Switzerland. This article is an open access article distributed under the terms and conditions of the Creative Commons Attribution (CC BY) license (https://creativecommons.org/licenses/by/4.0/).

Abstract: Intensive research on biodegradable films based on natural raw materials such as carboxymethyl cellulose (CMC) has been performed because it enables the production of transparent films with suitable barrier properties against oxygen and fats. Considering the importance of the production of this type of film at the industrial level, a scalable and continuous drying method is required. Refractance window-conductive hydro drying (RW-CHD) is a sustainable and energy-efficient method with high potential in drying this kind of compound. The objective of this study was to evaluate the factors (CMC thickness, heating water temperature, and film type) and radiation penetration depth that affect drying time and energy consumption. It was found that drying time decreased with increasing temperature and decreasing thickness. Similarly, energy consumption decreased with decreasing temperature and thickness. However, the drying time and energy consumed per unit weight of product obtained were equivalent when drying at any of the thicknesses evaluated. Film type had little effect on time and energy consumption compared to the effects of temperature and CMC thickness. The radiation penetration depth into the CMC was determined to be 1.20 ± 0.19 mm. When the thickness was close to this value, the radiation energy was better utilized, which was reflected in a higher heating rate at the beginning of drying.

Keywords: carboxymethyl cellulose; drying; RW; CHD; radiation penetration depth

1. Introduction

The limited access to non-renewable resources for packaging materials has turned the focus to biopolymers. In the last decade, intensive research on biodegradable films based on natural raw materials such as carboxymethyl cellulose (CMC) has been published, as it has non-toxic and non-allergic effects and produces a transparent film with suitable barrier properties against oxygen and fats [1,2]. These films are used for food packaging, adhesives, biocomposites, and hydrogel films.

Antosik et al. [1] prepared CMC-based films as carriers for a pressure-sensitive adhesive. The polysaccharide solution was poured into a polystyrene mold and dried for 24 h at 60 °C. Hasheminya et al. [2] prepared films with Kefiran, essential oil, and CMC; the final step of the process was to pour the solution onto glass plates and dry at 25 °C for 72 h. In addition, CMC was used for its binding capacity in tea waste bioplastics; the drying process took place in an oven for 3 h at 70 °C [3]. Cheng et al. [4] prepared a composite film using CMC, konjac glucomannan, and palm olein to improve the moisture barrier properties. The solutions were dried at room temperature for approximately 20 h. Li et al. [5] produced a starch-CMC-based film; samples were dried at room temperature for 12 h. Another example of based CMC films is hydrogels. They can offer new opportunities to design efficient packaging materials when prepared with CMC and polyvinylpyrrolidone [6].

These works used the casting method to dry the films at laboratory scale. Considering the importance of the production of this type of film at the industrial level, a drying method that is scalable, continuous, and energy efficient is required.

Refractance Window® (RW) drying is a technique developed by MCD Technologies, Inc. (Tacoma, WA, USA) [7,8]. This technique is called conductive hydro-drying (CHD) for high thicknesses [9]. The material to be dried is placed on a film that is in contact with water (95–97 °C) at atmospheric pressure. The water is recirculated below its boiling point to provide the thermal energy necessary for drying. According to some authors, this film is transparent to infrared radiation [7,8,10]. The thermal energy of the water is transmitted to the food through the plastic film by conduction and radiation, and moisture is carried by air flowing over the food layer [9]. This technology has had substantial growth in the last five years (75% of the publications on the subject have been in this period) due to its advantages in quality and cost.

One of the advantages of RW-CHD drying is the drying time, which is relatively short when compared to other drying techniques, such as solar drying, tray drying, or freeze drying. In tray drying, the products need to be dried from 3 to 5 h at high temperature whereas the processing time in freeze drying varies from 18 to 24 h [11]. Ochoa-Martínez et al. [12] found that, for the RW technique, the moisture content decreases rapidly to a value below 5% wb in about 30 min for 1-mm samples and 60 min for 2-mm samples. In contrast, it took about 240 min to obtain similar results with the tray drying technique. Baeghbali et al. [13] obtained a drying time of 20–24 h for freeze drying whereas the observed time was 5–7 min for RW for different products. Nindo et al. [14] reported residence times (h) of drying for asparagus puree for freeze drying (18–24), tray drying (2.5–5.5), as well as spouted bed (1.2–2.3), microwave (0.5–1.6), and RW (0.074) technologies. Longer drying times adversely affect product quality. The faster drying rate in RW-CHD has been attributed to the radiative heat transfer that occurs between the heating water and the food [14,15]. In CHD, there is, additionally, a significant effect of conduction heat transfer [16].

On the other hand, compared with conventional dryers, RW dryers have high thermal efficiency (i.e., two-fold higher than freeze dryers) [8]. Baeghbali et al. [13] compared the overall energy efficiency from different drying techniques for pomegranate juice drying. The highest energy efficiency (31.56%) was shown by the RW dryer, followed by a spray dryer (12.92%) and a freeze dryer (1.12%). Additionally, Baeghbali et al. [17] reported that the overall energy used for carrot puree drying was 0.375–0.525 kW for RW dryers, which was lower than the 70–84 kW used in the case of freeze drying. Similar results were reported by Bernaert et al. [18], where the energy efficiency of RW was three-fold and 40-fold higher than spray drying and freeze drying, respectively.

In addition to the above advantages, the drying equipment, energy consumption, and operation costs are lower than those of tray, freeze, drum, and spray drying methods. Additionally, it has negligible CO_2 emissions [19]. For example, the cost of the WR drier can be between one-third and one-half the cost of freeze drying [8,15,20]. Compared to other drying systems such as drum drying or convection drying, production costs can be up to 70% lower [21,22].

The effect of process variables (sample thickness and heating water temperature) on physicochemical properties has been studied for numerous foods, especially fruits and vegetables and, to a lesser extent, meats, dairy products, and others. Mahanti et al. [23] conducted an extensive review on this subject. Moderate food temperatures and short drying times reduce the deterioration of food characteristics. Other important process variables are the type and thickness of the film used, which affect the percentage of radiation transmission. These variables have not been studied. Most of the relevant works have used polyethylene terephthalate (PET) films. Other authors [9,16,24] performed calculations to determine the relative contribution of heat transfer mechanisms in RW-CHD drying. In no case was the penetration of infrared radiation observed, which is a critical aspect affecting heat transfer in this technology. There are no available data in the literature about this variable for infrared wavelengths, except that of Ginzburg [25] for various food materials and Almeida et al. [26] for potatoes in near-infrared drying. In addition, heat fluxes emitted by water, water–film, and water–film–CMC systems have not been measured to characterize RW-CHD drying.

In RW-CHD, comprehension of the factors affecting drying time and energy consumption, such as the thickness of the sample, the temperature of the heating water, and the type and thickness of the film, is important. Only the first two have been studied for numerous products, especially fruits and vegetables and, to a lesser extent, meats, dairy products, and others. This technology has not been applied in drying gels such as CMC. Because of its high moisture content and long processing time, it is essential to use highly efficient drying methods for the drying of CMC films that can be used as biodegradable packaging materials.

The main objective of the work was to evaluate the factors affecting the drying time and energy consumption in the RW-CHD drying of CMC. The radiation penetration depth into CMC and the effect of operating conditions (CMC thickness, heating water temperature, and film type and thickness) on drying time and energy consumption were determined.

In this work, it was found that the drying time decreased with increasing temperature and decreasing thickness. Similarly, energy consumption decreased with decreasing temperature and thickness. However, the drying time and energy input per unit weight of product obtained were equivalent when drying at 1.5 or 3.5 mm. That is important for deciding the amount of sample to be dried (product thickness) as a function of operating cost and final product quality when exposed to high temperatures for a longer time. Compared to temperature and CMC thickness, film type had a minor effect on time and energy consumption. This was due to the films' high radiation transparency (low resistance to heat flow), regardless of type and thickness. On the other hand, the radiation penetration depth in CMC was determined to be 1.20 ± 0.19 mm. This was independent of the type and thickness of the film and the temperature of the heating water since it is a property of the material; no effect of temperature was observed because the wavelength range at the temperatures studied is very small. For thicknesses close to this value, more use is made of the radiation energy, which was reflected in a higher heating rate at the beginning of drying (3 min).

2. Results and Discussion

2.1. Drying Kinetics

Figure 1 shows the drying kinetics of CMC with thicknesses of 1.5, 2.5, and 3.5 mm for the three films (PP, LDPE, and PET) at 90 and 70 °C. As expected, as the sample thickness decreased and the drying temperature increased, the water loss was higher for every film.

Considering that the air conditions in contact with the sample during drying were $60.4 \pm 6.5\%$ relative humidity and 26.9 ± 1.7 °C, the equilibrium moisture content for CMC is 0.17 kg water/kg dry solid (15% wb) according to the sorption isotherm reported by Torres et al. [27]. In the kinetics shown, values close to this equilibrium value can be observed.

An initial part with a constant slope was observed in these curves, corresponding to the constant drying rate period, up to an approximate moisture value of 5 kg water/kg ss (83% wb). The high drying rate in this period was notorious, unlike tray drying, which is very short with a low drying rate [12]. During this period, there was water saturation on the surface and easy replacement of evaporated water due to the low resistance to mass transfer in the interior. In RW-CHD, the sample exhibited rapid heating (in the first seconds of drying), maintaining the continuous transport of water inside the sample to the surface, in contrast to the tray drying, where the sample heating was slow, and heat transfer was the limiting mechanism. Moisture removal from the surface to the air occurred by convective diffusion, the driving force of which is the water concentration difference. Parrouffe et al. [28] found that infrared radiation had no significant influence on reducing drying time beyond the constant rate period due to the low moisture content at the surface, which generally has a low penetration depth or a low absorption. The drying rate in the constant rate period (kg water/kg ss·min) corresponding to the slope of the curves in the linear region is presented in Figure 2.

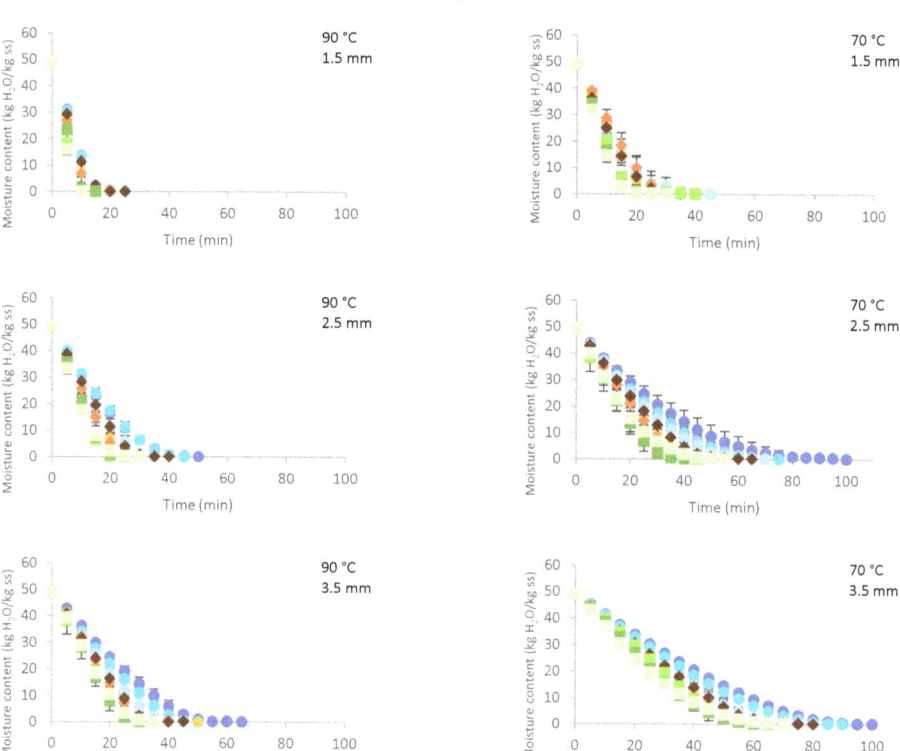

Figure 1. Effect of drying time, water temperature, and CMC thickness on moisture content (●PP 0.83, ●PP 0.55, PP 0.38, ◆LDPE 0.15, ◆LDPE 0.10, ◆LDPE 0.03, ■PET 0.25, ■PET 0.18, PET 0.08).

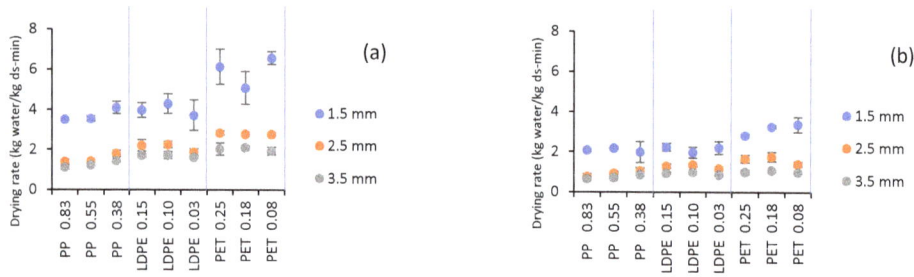

Figure 2. Effect of type and thickness of plastic films and CMC thickness on the drying rate: (**a**) 90 °C and (**b**) 70 °C.

In the analysis of variance, it was observed that the main factors (type-film thickness, CMC thickness, and heating water temperature) and their interactions had a significant effect on the drying rate ($p < 0.05$). According to the F-Ratio value, the relative importance of CMC thickness (1548.63) related to temperature (429.83) and film type-thickness (73.59) was highlighted. Tukey tests for temperature, CMC thickness, and film thickness type are presented in Tables 1–3.

Table 1. Effect of water temperature on CMC drying kinetics.

Temperature (°C)	Drying Rate (kg Water/kg ds·min)	Drying Flux (kg Water/h·m^2)	Time (min)
70	1.53 [a]	3.38 [a]	52.37 [a]
80	2.14 [b]	4.73 [b]	37.52 [b]
90	2.78 [c]	6.05 [c]	29.33 [c]

Values followed by the same letter are not significantly different ($p < 0.05$, Tukey's test).

Table 2. Effect of CMC thickness on drying kinetics.

CMC Thickness (mm)	Drying Rate (kg Water/kg ds·min)	Drying Flux (kg Water/h·m^2)	Time (min)
1.5	3.49 [a]	4.96 [a]	22.67 [a]
2.5	1.70 [b]	4.66 [b]	42.58 [b]
3.5	1.26 [c]	4.55 [b]	53.97 [c]

Values followed by the same letter are not significantly different ($p < 0.05$, Tukey's test).

Table 3. Effect of type and thickness of plastic film on CMC drying kinetics.

Film Type-Thickness (mm)	Drying Rate (kg Water/kg ds·min)	Drying Flux (kg Water/h·m^2)	Time (min)
PP 0.83	1.55 [d]	3.46 [e]	55.86 [a]
PP 0.55	1.70 [cd]	3.75 [de]	48.85 [b]
PP 0.38	1.90 [bcd]	4.13 [cd]	42.80 [c]
LDPE 0.15	2.06 [bc]	4.68 [b]	38.31 [cd]
LDPE 0.10	2.10 [b]	4.85 [b]	37.66 [cd]
LDPE 0.03	1.93 [bcd]	4.59 [bc]	39.42 [cd]
PET 0.25	2.68 [a]	5.77 [a]	30.36 [e]
PET 0.18	2.66 [a]	5.62 [a]	30.56 [e]
PET 0.08	2.76 [a]	5.64 [a]	33.86 [de]

Values followed by the same letter are not significantly different ($p < 0.05$, Tukey's test).

According to the Tukey analysis presented in Table 1, the mean value of the drying rate was significantly different for the three temperatures and the three CMC thicknesses studied. The rate increased with increasing heating water temperature and decreasing CMC thickness (higher values were obtained for a CMC thickness of 1.5 mm) and higher differences for PET films. The mean value of the drying rate increased by 81.7% when increasing the temperature from 70 to 90 °C and was 177.0% when decreasing the CMC thickness from 3.5 to 1.5 mm. Similar behavior was observed by Zotarelli et al. [24] in mango drying, with an increase of 40% when decreasing the thickness from 2 to 3 mm.

Table 2 shows that the three thicknesses of each film evaluated had a similar mean drying rate value. The highest rate was obtained for the PET films, while the other two materials presented lower values that were statistically equal.

Although the tests with PET for a sample thickness of 1.5 mm showed a higher drying rate in the constant period (moisture content dropped from 98% to 80% wb), statistical analysis of the results showed that the film type had little effect on time compared to the temperature effect and CMC thickness. According to Tsilingiris [29], a slight change in chemical composition, thickness, and measurement processes should produce substantial changes in the optical properties of the film.

Figure 3 shows the drying flux, which corresponds to the water evaporation capacity of the equipment per unit of drying area (kg water/h·m^2). These values were between 2 and 9 kg/h·m^2, similar to those obtained by Zotarelli et al. [24] (between 2.67 and 10.75 kg/h·m^2) when drying mango pulp (thickness of 2, 3, and 5 mm) and heating water of 75, 85 and 95 °C. In addition, Nindo et al. [14] obtained a value of 10 kg/h·m^2 when drying pumpkin puree, with thicknesses between 0.4 and 0.6 mm in the pilot plant and of 3.1, 3.9, and 4.6 kg/h·m^2 at the industrial level.

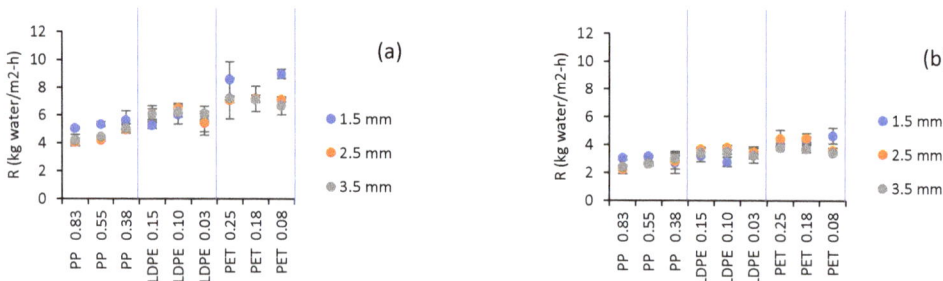

Figure 3. Effect of type and thickness of plastic films and CMC thickness on the drying flux (R): (**a**) 90 °C and (**b**) 70 °C.

All factors and their interactions affected the drying flux ($p < 0.5$). According to the F-ratio, the water temperature (700.62) had the most significant effect. Film thickness (92.80) and CMC thickness (17.84) had far less significant effects. According to Tukey's test (Table 1), the three temperatures had effects that differed in terms of significance. The mean value of the drying flux increased with increasing temperature. A similar effect was observed in the drying of mango pulp [24]. By increasing the temperature from 70 to 90 °C, the drying flux increased by 79%.

On the other hand, Tukey's test for CMC thickness (Table 2) showed that the mean value of the drying flux was higher and statistically different for CMC thickness of 1.5 mm. The mean drying flux obtained with 2.5 and 3.5 mm CMC thicknesses was not significantly different. The mean drying flux value increased by 9.0% when reducing the thickness from 3.5 to 1.5 mm. Zotarelli et al. [24] observed similar behavior when decreasing the thickness of mango pulp.

The Tukey test (Table 3) for film type and thickness showed that the mean value of the drying flux was statistically similar for the three thicknesses of each of the films and different for each type of film. The highest drying flux was obtained with PET, followed by LDPE and PP.

Figure 4 shows the drying time required to reach a final moisture content of 0.312 kg water/kg ss (this corresponds to the highest final moisture content of all treatments). Statistically, it was observed that all factors had a significant effect ($p < 0.05$) on drying time (min). According to the F-ratio, the most significant effect was CMC thickness (993.90), followed by temperature (540.26). Shende and Datta [30] showed the same behavior for mango drying. Film type-thickness (93.34) showed a minimal effect. Azizi et al. [31] also found that increasing the thickness of PET film from 0.1 to 0.3 mm does not have any significant effect on the drying time of kiwifruit irrespective of water temperature and sample thickness.

In studies of fruit drying using PET, drying times that were comparable to those presented in Figure 4 were obtained. For cornelian cherry pulp with a thickness of 1 mm at 90, 95, and 98 °C, the drying time was between 15 and 20 min [32]; a time of 20 min was obtained for mango pulp with a thickness of 2 mm at 95 °C [30]; and a time of 40 min was recorded for tomato pulp with a thickness between 1 and 1.5 mm at 90 °C and 60 min at 75 °C [22]. The increase in drying time with increasing food thickness was also observed in the drying of papaya puree [33] and mango slices [12].

According to Tukey's test, significantly different means were obtained for the temperatures and CMC thicknesses evaluated (Tables 1 and 2). As expected, the mean value of drying time decreased with increasing temperature and decreasing thickness. A decrease from 90 to 70 °C increased the drying time by 78.6%; an increase in thickness from 1.5 to 3.5 mm increased the drying time by 138.1%. However, there was a direct relationship between time and the amount of final product obtained, which in turn was related to CMC thickness ($R^2 = 0.9992$) (Table 4), meaning that it was equivalent to drying with 1.5 or 3.5 mm in terms of the final amount obtained. This is important for deciding on the amount of sample to dry in terms of final product quality and operating costs. According to Table 3, for PP, the mean value of drying time was significantly different for each of its thicknesses and higher than for the other films evaluated (the evaluated thicknesses of PP were very high compared to those of LDPE and PET). The lowest average value of the drying time was obtained with the PET films (regardless of their thickness), which agrees with the higher drying rate obtained.

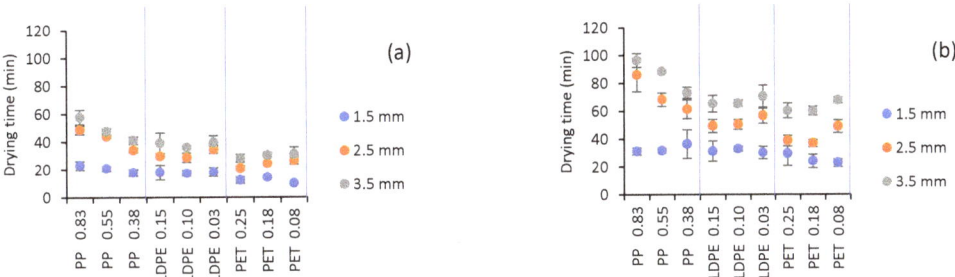

Figure 4. Effect of type and thickness of plastic films and CMC thickness on the drying time: (**a**) 90 °C and (**b**) 70 °C.

Table 4. Relation between the dry product and the sample thickness.

Thickness (mm)	Drying Time (min)	Wet Sample Weight (g)	Dry Sample Weight (g)	Drying Index (g Dry Sample/h)
1.5	22.7	33.3 ± 1.2	0.9 ± 0.0	2.6 ± 0.9
2.5	42.6	64.8 ± 3.8	1.7 ± 0.1	2.6 ± 0.8
3.5	54.0	84.9 ± 3.0	2.2 ± 0.1	2.7 ± 0.8

When the yield analysis (g/h-m^2) was conducted for the conditions studied (Table 5), it was observed that the highest values were obtained for 3.5 mm food thickness and 90 °C. The highest yield was obtained for PET 0.25. The best performances for LDPE and PP were observed with LDPE 0.1 and PP 0.38, respectively. In addition to the yield, it was necessary to consider the temperatures to which the product was subjected during drying, which affected the quality of the final product.

Table 5. Effects of process factors on the drying yields (g/h-m^2).

Temperature (°C)	70			80			90		
Thickness (mm)	1.5	2.5	3.5	1.5	2.5	3.5	1.5	2.5	3.5
PP 0.83	61.5	45.2	51.0	62.2	66.3	70.4	81.8	78.3	86.5
PP 0.55	60.4	54.7	55.9	89.3	72.4	73.1	95.0	89.1	101.5
PP 0.38	49.1	58.0	64.1	83.3	80.2	88.1	101.2	107.0	113.1
LDPE 0.15	60.7	76.0	74.1	72.1	96.4	99.8	97.8	123.0	120.8
LDPE 0.10	56.5	73.6	71.5	80.1	97.2	106.0	106.4	132.9	132.6
LDPE 0.03	64.8	68.5	70.3	94.8	105.4	106.2	108.8	110.8	126.6
PET 0.25	62.1	90.8	83.7	122.0	120.6	120.2	150.7	157.2	165.7
PET 0.18	73.9	91.0	75.8	118.2	123.8	105.2	129.5	139.8	147.2
PET 0.08	79.7	69.3	67.9	125.3	93.1	95.3	180.5	128.6	148.6

2.2. Temperature Profiles

Figure 5 shows the temperature profiles of the main components of the system (water, water–film interface, film–CMC interface, and CMC–air interface) for some of the conditions studied.

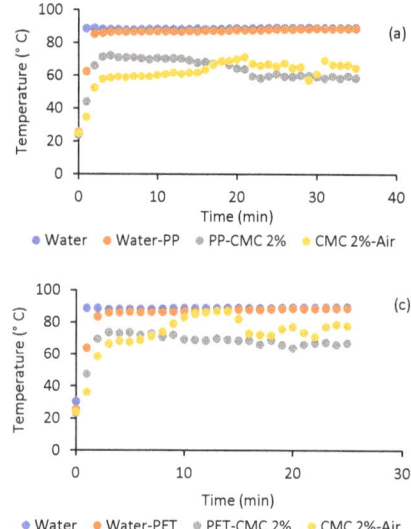
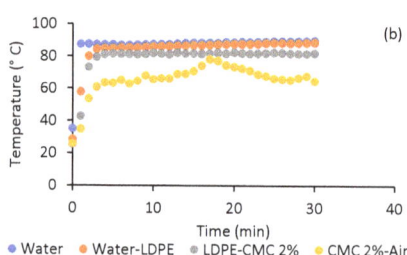

Figure 5. Temperature profiles of CMC-2.5 mm drying at water temperature of 90 °C: (**a**) PP 0.38 mm, (**b**) LDPE 0.15 mm, and (**c**) PET 0.18 mm.

The interfaces of PP-CMC and PET-CMC films were rapidly heated by radiation and conduction (within the first three minutes of drying) to approximately 10–20 °C below the source temperature. During drying, their temperature decreased by 5–10 °C as the CMC absorbed energy from the film to equilibrate. In general, they may not heat up to the source temperature due to their high transmissivity (70–90 %) and low conductivity (PP: 0.23–0.26 W/mK; PET: 0.16–0.20 W/mK) [34]. On the other hand, when using LDPE, the interface temperature with CMC increased rapidly. Due to its high conductivity (0.25–0.33 W/mK) [34], it reached a value close to the heating water (about 5 °C below the water temperature) and remained constant during drying. The film interface temperature that did not depend on the food was always constant, while that which was in contact with the food was a function of the food temperature.

The energy absorbed by the sample was used to heat (sensible heat) and to evaporate the water (latent heat). In general, the CMC temperature remained below the water temperature due to evaporative cooling and the effect exerted by the air temperature (26.9 ± 1.7 °C). On the other hand, the CMC initially absorbed energy from the film and warmed up to equilibrium with the temperature of the film–CMC interface (the film lost energy and the food gained energy); subsequently, heating of the sample was observed, and finally, they equilibrated at the same temperature. An increase in the sample temperature up to water temperature was only observed in the drying with LDPE-0.03 due to the low film thickness and high conductivity. The sample temperature rose to that of the water in tests carried out at 70 °C with 1.5 mm thickness, regardless of the film used.

2.3. Heat Flux Emitted by Water, Water–Film, and Water–Film–CMC Systems

The radiative heat flux emitted by the water at 70, 80, and 90 °C and the radiative heat flux transmitted through the film (water–film system) determined during the first 50 s were measured (Figure 6). It was observed that the heat flux was significantly higher for times longer than 50 s (these data are not presented). The heating of the film after this time (Figure 5) increased its emissivity, causing an increase in heat flux. Statistically, a high-temperature effect was observed.

According to Figure 6, all films showed high transmittance to radiation, with the heat flux being similar to that of the water. These results agree with those presented by

Tsilingiris [29] for PP and LDPE films. That confirms that the films were transparent to radiation to a high degree, which allowed the heating of the sample, as discussed in the previous section.

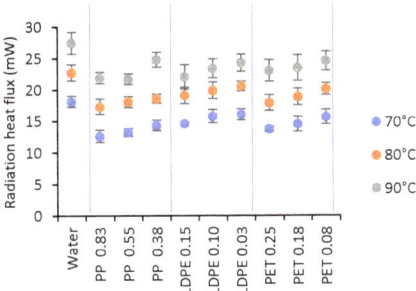

Figure 6. Radiation heat transfer for water and water-drying film systems.

On the other hand, the radiation heat flux leaving the CMC during the first 70 s of drying is presented in Figure 7 for some of the conditions studied. It was observed that the radiative heat flux was higher for heating water at 90 °C when compared to 80 or 70 °C. It was also observed that it increased with time; initially, this heat flux was low due to the high radiation absorption of the food; subsequently, the flux increased due to the combined effect of the radiation emission of the food upon heating and the lower absorption.

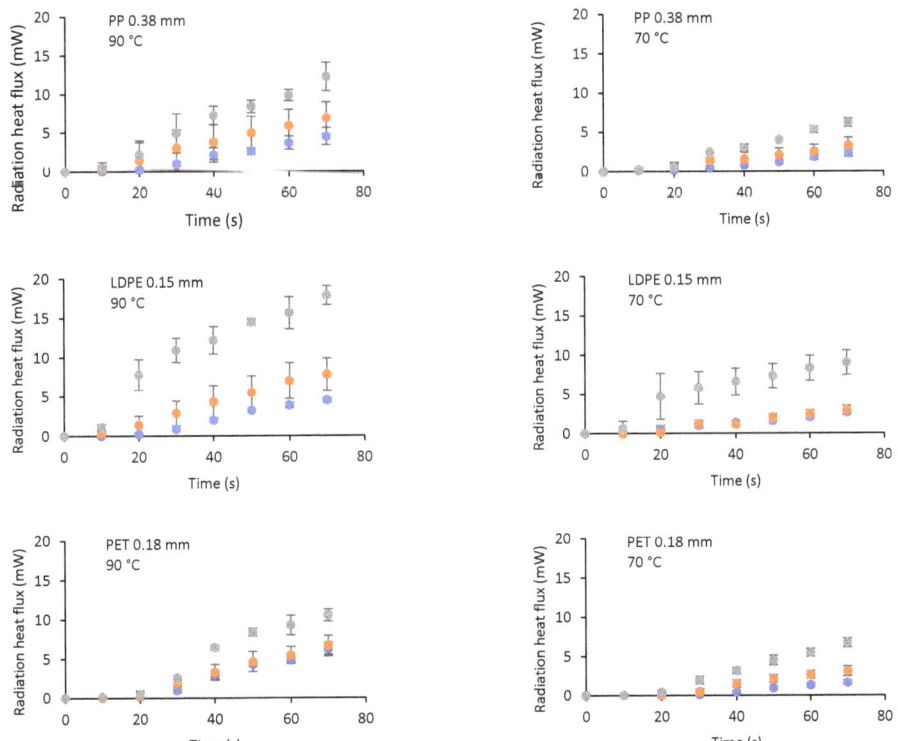

Figure 7. Effect of the type of plastic film, water temperature, and CMC thickness (●1.5 mm, ●2.5 mm y ●3.5 mm) on the radiation heat flux leaving the CMC.

2.4. Penetration Depth of the Infrared Radiation in CMC

The penetration depth is defined as the depth at which the radiation intensity decays by 37% (1/e) of its initial value [35,36]. In other words, 63% of the incident radiation is used [11,30]. Depending on the material, electromagnetic waves can travel inside the food or be absorbed on its surface [37]. Using (Equation (1)), the parameters I_0 and μ^{-1} were obtained (Table 6).

$$I = I_0 e^{-\mu x} \quad (1)$$

where I is the radiation passing through the solid (water–film–food system) of thickness x, I_0 is the initial radiation reaching the solid, and μ is the attenuation coefficient.

Table 6. Beer–Lambert Law parameters for radiant heat flow.

Water Temperature (°C)	70			80			90		
Film Type-Thickness (mm)	μ^{-1}	Ln (Io)	R^2	μ^{-1}	Ln (Io)	R^2	μ^{-1}	Ln (Io)	R^2
PP 0.83	1.33	2.35	0.94	1.35	2.69	0.96	0.73	1.37	1.00
PP 0.55	1.10	2.54	1.00	1.34	2.78	0.96	0.83	1.20	0.99
PP 0.38	1.05	2.57	0.98	0.73	3.04	0.99	0.89	1.13	0.98
LDPE 0.15	1.21	2.71	0.93	1.31	2.91	1.00	0.93	1.08	0.95
LDPE 0.10	1.22	2.79	0.99	1.23	2.97	1.00	0.74	1.36	0.98
LDPE 0.03	1.26	2.69	0.96	1.27	2.96	0.99	0.66	1.52	0.94
PET 0.25	1.33	2.33	0.91	1.41	2.73	0.96	0.88	1.13	0.97
PET 0.18	0.70	2.73	1.00	0.95	2.71	0.97	0.84	1.19	0.86
PET 0.08	1.16	2.67	0.99	1.33	2.94	0.99	0.80	1.25	0.98

The ANOVA and Tukey's test showed no significant effect of temperature or film type-thickness, so the values obtained for all penetration depths were averaged (Table 6). The average value of the radiation penetration depth in 2% CMC was 1.20 ± 0.19 mm. Although the penetration depth is known to depend on the nature of the material and temperature of the emitting source (wavelength) [37], the value obtained depended only on the material (CMC 2%) in the range of temperatures evaluated (70, 80, and 90 °C). It was, as expected, independent of the type and thickness of the films evaluated.

The total emissive power includes the energy of all wavelengths in the radiation spectrum [37]. The wavelength at which the maximum emissive power of radiation occurs, λ_{max}, depends on the emitter temperature (T, K) and is given by Wien's displacement law, presented in Equation (2) [38].

$$\lambda_{max} = \frac{2898}{T} \quad (2)$$

According to Equation (2), for heating water temperatures between 70 and 90 °C, the wavelengths at which the maximum emissive power is obtained are 8.4 μm and 8.0 μm, respectively. This slight difference in wavelengths for the studied temperature range explains the lack of an effect of this factor on the penetration depth.

There is little information available in the literature on the penetration depth of infrared radiation [26,37]. For foods with high humidity (>80), penetration depth values between 1 and 7 mm were found, depending on the spectral peaks of the radiation source [25]. For potatoes with humidity between 67 and 82%, radiation penetration depth values between 0.45 and 2.85 mm were obtained [26], and for carrots, 1.5 mm was reported [25].

The amount of infrared radiation that a vegetable or animal sample can absorb depends on the wavelength of the emitter and the sample composition [37]. Foods are complex mixtures of biochemical molecules, biological polymers, inorganic salts, and water. Water shows high absorption at wavelengths of 3.0, 4.7, 6.0, and between 12 and 15.3 μm [39]. These values do not coincide in the wavelength range where the highest emissive power of water is obtained (8.4 μm and 8.0 μm at the temperatures used). On the other hand, carbohydrates (i.e., CMC) show two strong absorption bands at 3 μm and between 7 and 10 µm [39], coinciding with the emission bands of water. Accordingly, resonant absorption occurs in carbohydrates, which heat the water in the CMC. Subsequently, a thermal heating effect of the water occurs due to the dipole orientation in the electromagnetic field of radiation.

2.5. Energy Consumption

Figure 8 shows the results of energy consumption. According to the analysis of variance, film thickness, CMC thickness, and the heating water temperature had a significant effect on energy consumption ($p < 0.05$). According to the F-ratio, the most significant factor was CMC thickness (907.56), while the heating water temperature (101.8) and film thickness (72.56) had a much smaller effect.

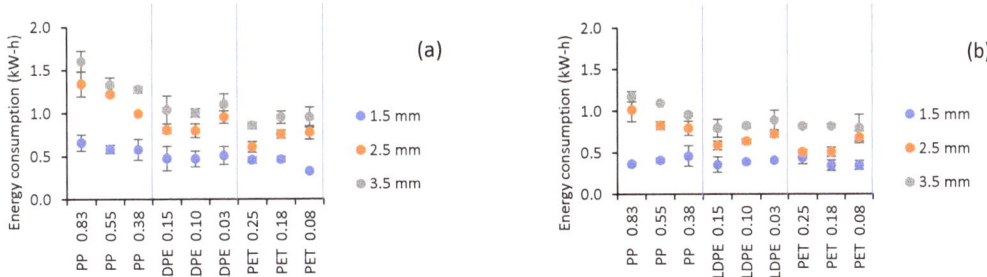

Figure 8. Effects of type and thickness of plastic films, CMC thickness, and water temperature ((**a**) 90 °C and (**b**) 70 °C) on energy consumption.

Tukey's test showed significant differences for both thickness and temperature. Energy consumption increased by 125.0%, thereby increasing the CMC thickness ($p < 0.05$) from 1.5 to 3.5 and 26.9% with the increase in temperature from 70 to 90 °C. This behavior was expected considering the greater amount of water to evaporate at the higher thickness and the more significant amount of energy required with increasing temperature.

The Tukey test for film thickness and type showed that the mean energy consumption value was higher for PP 0.83 mm, followed by PP 0.55 and 0.38 mm. There was no significant difference in the mean value of energy consumption ($p < 0.05$) among the other films, probably due to a combined effect of transmissivity and conductivity.

In the present work, the energy consumption varied between 0.44 and 0.99 kWh depending on the thickness of the sample. On the other hand, Baeghbali et al. [13] reported a consumption of 4.31 ± 0.82 kWh (using a pilot-scale continuous dryer in the drying of pomegranate juice). When analyzing the specific energy consumption ratio (SER) [19], which in turn was related to the CMC thickness (Table 7), a direct relationship was found ($R^2 = 0.9989$). This means that the energy consumption, in terms of the final amount of product obtained for the three thicknesses studied, was similar. As mentioned in the time calculation, this information is essential for deciding on the amount of sample to dry depending on the final product quality and operating costs. Menon et al. [19], in their review of energy efficiency in drying technologies, reported SER values between 0.13 and 1.9 kWh/kg for super-heated steam drying, between 0.7 and 37.1 kWh/kg for microwave drying, and between 1.41 and 3.11 kWh/kg for impinging heat drying.

Table 7. Specific energy consumption (SER).

CMC Thickness (mm)	Energy Consumption (kWh)	SER (kWh/kg Water)
1.5	0.44 ± 0.09	13.6 ± 2.8
2.5	0.79 ± 0.21	12.5 ± 2.8
3.5	0.99 ± 0.20	12.0 ± 2.3

3. Materials and Methods

3.1. Carboxymethylcellulose (CMC)

CMC powder (Gelycel F1 3500, specification 10031) with 4.14% humidity (Agenquímicos, Santiago de Cali, Colombia) was used. It was diluted in distilled water to obtain a 2% solution (98% moisture) using a blender (Samurai, Innova, Colombia) until a uniform mixture was obtained. The solution was left at rest for 24 h to release all air bubbles. Metal frames of 9.5 cm × 29.5 cm were used to obtain CMC sheets of the desired thickness (1.5, 2.5, and 3.5 mm).

3.2. Equipment

A stationary RW-CHD dryer (HS-50, Ceirobots, Santiago de Cali, Colombia) (Figure 9) was used. Two drying trays (10.3 cm × 30.3 cm), the bottoms of which were composed of plastic film, were in contact with the hot water. CMC was placed on the plastic films. The water was heated by two resistors (14.7 A each) in a 20 L tank, and the temperature was maintained through a recirculation system. A fan was used to remove the water vapor produced during drying. Three films of different thicknesses were evaluated, each according to commercial availability: polypropylene (PP) of 0.83, 0.55, and 0.38 mm; low density polyethylene (LDPE) of 0.15, 0.10, and 0.03 mm; and polyethylene terephthalate (PET) of 0.25, 0.18, and 0.08 mm.

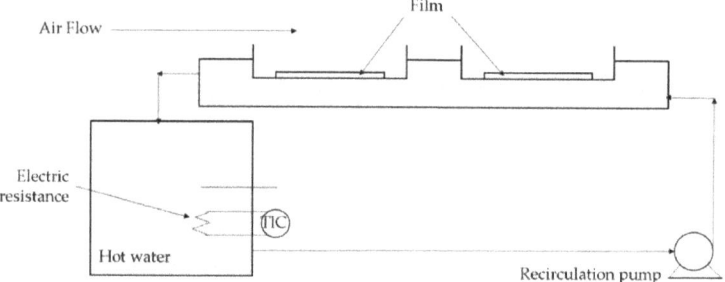

Figure 9. Schematic diagram of an RW-CHD dryer.

3.3. Experimental Design

The factors evaluated were the type of film with different thicknesses (PP-0.83, PP-0.55, PP-0.38, LDPE-0.15, LDPE-0.10, LDPE-0.03, PET-0.25, PET-0.18, PET-0.08), the thickness of the CMC (1.5, 2.5, and 3.5 mm), and the temperature of the heating water (70, 80, and 90 °C). A 9 × 3 × 3 complete randomized factorial design was used. All tests were performed in triplicate for a total of 243 treatments. The response variables evaluated were moisture content and drying time, temperature profiles, radiation heat flux, radiation penetration depth, and energy consumption. In addition, the radiant heat flux for the water and water–film system was determined at 70, 80, and 90 °C.

3.4. Response Variables

3.4.1. Moisture Content and Drying Time

Moisture content (db) was determined every 5 min. The drying tray was re-removed from the dryer, and the bottom dried and weighed. A balance (Ohaus®, Ad-venturerTM, Shanghai, China, accurate to 0.01 g) was used. The initial and final moisture content of the sample was determined by oven drying (Thermo Scientific, Heratherm OGS60, Langenselbold Germany) at 70 °C for 24 h [32]. The drying time required to reach a moisture content of 0.312 kg water/kg ss of CMC was determined to compare treatments (this value was selected considering that it was the minimum value obtained in some treatments).

3.4.2. Temperature Profiles

The temperatures of the hot water, the water–film interface, and the film-water-food interface were measured every 5 s using J-type thermocouples (Omega, 5TC-TT-J-30-72, diameter 0.25 mm, Norwalk, CT, USA) and a data acquisition system (Comark Instruments, model Diligence Evg N3014, Norwich, UK), with an accuracy of 0.2 °C. A thermographic camera (Flir, Model E4, Wilsonville, OR, USA) was used to measure the temperature of the food–air interface.

3.4.3. Radiative Heat Flow

The radiative heat flux emitted by the water, the water–film system, and the water–film–food system was measured for all experimental design conditions. A potentiometer (Molectron Detector Inc., PowerMax 5200, Portland, OR, USA) with a radiation measurement scale between 0 and 10 W (12 scale ranges; the scale from 0 to 100 mW was used) and an accuracy of 0.1 mW was used. This equipment used a pyroelectric sensor to measure the radiation heat flux in a spectral range of 0.25 to 11 µm (from ultraviolet to infrared). The diameter of the measuring sensor was 19 mm. The sensor was placed 4 cm from the radiant-heat-flux-emitting surface. The measurements were performed before the heating of the film or CMC, and the radiative heat flux was calculated by dividing the heat flux by the sensor area (0.000284 m^2). Measurements were taken between 30 and 70 s. Before 30 s, the system had not stabilized, and after 80 s, conduction heat transfer was likely to occur due to heating of the film. All these measurements were taken on the drying film without food.

3.4.4. Radiation Penetration Depth

The radiation penetration depth (μ^{-1} or $x_{0.37}$) is defined as the depth at which the radiation intensity decays by 37% (1/e) of its initial value [35,36]. Using the radiation attenuation law, also known as the Beer–Lambert Law [29,40] (Equation (1)), the values of the intercept ($ln(I_0)$) and the slope ($-\mu$) were determined. For this, the experimental values of I vs. x (1.5-, 2.5-, and 3.5-mm thickness CMC, for heating water at 70, 80, and 90 °C) were used.

The radiation penetration depth was calculated for a radiation heat flux at 30 s, thus considering only the radiation emitted by the hot water passing through the film and the food. Data corresponding to 10 and 20 s were not considered due to the system's instability at the beginning of the measurement.

3.4.5. Energy Consumption

Energy consumption was determined using a Peacefair meter, model PZEM-022 (China), for alternating current in the voltage range of 80 to 260 V, 50/60 Hz, up to 100 A. The sensing element of this equipment was placed at the power input of the drying equipment.

3.5. Statistical Analysis

Statgraphics Centurion 19 software (Statgraphics Technologies, Inc., version 19.2.01, 2021, The Plains, VA, USA) was used to perform analysis of variance and determine the dependence between process variables and response variables, with a 95% confidence level ($p < 0.05$). It was also used to determine the independence between means using Tukey's test.

Author Contributions: Conceptualization, R.D.M.-T. E.S.-R., C.V.-P. and C.I.O.-M.; methodology, R.D.M.-T.; validation, R.D.M.-T.; formal analysis, R.D.M.-T., E.S.-R., C.V.-P. and C.I.O.-M.; investigation, R.D.M.-T., E.S.-R., C.V.-P. and C.I.O.-M.; resources, R.D.M.-T., E.S.-R., C.V.-P. and C.I.O.-M.; writing—original draft preparation, R.D.M.-T., E.S.-R., C.V.-P. and C.I.O.-M.; writing—review and editing, R.D.M.-T., E.S.-R., C.V.-P. and C.I.O.-M.; funding acquisition, R.D.M.-T., C.V.-P. and C.I.O.-M. All authors have read and agreed to the published version of the manuscript.

Funding: This research was funded by Universidad del Valle Convocatoria interna 2018 and by Universidad Nacional Abierta y a Distancia—UNAD, Proyecto PIE_G_34_18ECBTI.

Institutional Review Board Statement: Not applicable.

Informed Consent Statement: Not applicable.

Data Availability Statement: The data supporting reported results can be found in Múnera-Tangarife, R.D. (2021) "Evaluación de los factores que afectan la transferencia de calor en el secado por ventana de refractancia-hidrosecado" (Ph.D. thesis, Universidad del Valle, Colombia), or they are available upon request from the corresponding author.

Conflicts of Interest: The authors declare no conflict of interest. The funders had no role in the design of the study; in the collection, analyses, or interpretation of data; in the writing of the manuscript, or in the decision to publish the results.

References

1. Antosik, A.K.; Wilpiszewska, K.; Czech, Z. Carboxymethylated Polysaccharide-Based Films as Carriers for Acrylic Pressure-Sensitive Adhesives. *Int. J. Adhes. Adhes.* **2017**, *73*, 75–79. [CrossRef]
2. Hasheminya, S.-M.; Mokarram, R.R.; Ghanbarzadeh, B.; Hamishekar, H.; Kafil, H.S.; Dehghannya, J. Influence of Simultaneous Application of Copper Oxide Nanoparticles and Satureja Khuzestanica Essential Oil on Properties of Kefiran–Carboxymethyl Cellulose Films. *Polym. Test.* **2019**, *73*, 377–388. [CrossRef]
3. Liu, M.; Arshadi, M.; Javi, F.; Lawrence, P.; Davachi, S.M.; Abbaspourrad, A. Green and Facile Preparation of Hydrophobic Bioplastics from Tea Waste. *J. Clean. Prod.* **2020**, *276*, 123353. [CrossRef]
4. Cheng, L.H.; Abd Karim, A.; Seow, C.C. Characterisation of Composite Films Made of Konjac Glucomannan (KGM), Carboxymethyl Cellulose (CMC) and Lipid. *Food Chem.* **2008**, *107*, 411–418. [CrossRef]
5. Li, Y.; Shoemaker, C.F.; Ma, J.; Shen, X.; Zhong, F. Paste Viscosity of Rice Starches of Different Amylose Content and Carboxymethylcellulose Formed by Dry Heating and the Physical Properties of Their Films. *Food Chem.* **2008**, *109*, 616–623. [CrossRef]
6. Gregorova, A.; Saha, N.; Kitano, T.; Saha, P. Hydrothermal Effect and Mechanical Stress Properties of Carboxymethylcellulose Based Hydrogel Food Packaging. *Carbohydr. Polym.* **2015**, *117*, 559–568. [CrossRef]
7. Magoon, R.E. Method and Apparatus for Drying Fruit Pulp and the Like. U.S. Patent 4,631,837A, 30 December 1986.
8. Nindo, C.I.; Tang, J. Refractance Window Dehydration Technology: A Novel Contact Drying Method. *Dry. Technol.* **2007**, *25*, 37–48. [CrossRef]
9. Ortiz-Jerez, M.J.; Ochoa-Martínez, C.I. Heat Transfer Mechanisms in Conductive Hydro-Drying of Pumpkin (*Cucurbita maxima*) Pieces. *Dry. Technol.* **2015**, *33*, 965–972. [CrossRef]
10. Kudra, T.; Mujumdar, A.S. *Advanced Drying Technologies*, 2nd ed.; Taylor & Francis: Boca Raton, FL, USA, 2009; ISBN 978-1-4200-7387-4.
11. Abonyi, B.I.; Tang, J.; Edwards, C.G. *Evaluation of Energy Efficiency and Quality Retention for the Refractance WindowTM Drying System*; Washington State University: Tacoma, WA, USA, 1999; p. 38.
12. Ochoa-Martínez, C.I.; Quintero, P.T.; Ayala, A.A.; Ortiz, M.J. Drying Characteristics of Mango Slices Using the Refractance WindowTM Technique. *J. Food Eng.* **2012**, *109*, 69–75. [CrossRef]
13. Baeghbali, V.; Niakousari, M.; Farahnaky, A. Refractance Window Drying of Pomegranate Juice: Quality Retention and Energy Efficiency. *LWT-Food Sci. Technol.* **2016**, *66*, 34–40. [CrossRef]
14. Nindo, C.I.; Feng, H.; Shen, G.Q.; Tang, J.; Kang, D.H. Energy Utilization and Microbial Reduction in a New Film Drying System. *J. Food Process. Preserv.* **2003**, *27*, 117–136. [CrossRef]
15. Clarke., P.T. Refractance WindowTM—"Down Under". In Proceedings of the Proceedings on the 14th International Drying Symposium, São Paulo, Brasil, 22–25 August 2004; pp. 813–820.
16. Ortiz-Jerez, M.J.; Gulati, T.; Datta, A.K.; Ochoa-Martínez, C.I. Quantitative Understanding of Refractance WindowTM Drying. *Food Bioprod. Process.* **2015**, *95*, 237–253. [CrossRef]
17. Baeghbali, V.; Niakosari, M.; Kiani, M. Design, Manufacture and Investigating Functionality of a New Batch Refractance Window System. In Proceedings of the 5th International Conference on Innovations in Food and Bioprocess Technology, Pathumthani, Thailand, 7–9 December 2010. [CrossRef]
18. Bernaert, N.; Van Droogenbroeck, B.; Van Pamel, E.; De Ruyck, H. Innovative Refractance Window Drying Technology to Keep Nutrient Value during Processing. *Trends Food Sci. Technol.* **2019**, *84*, 22–24. [CrossRef]
19. Menon, A.; Stojceska, V.; Tassou, S.A. A Systematic Review on the Recent Advances of the Energy Efficiency Improvements in Non-Conventional Food Drying Technologies. *Trends Food Sci. Technol.* **2020**, *100*, 67–76. [CrossRef]
20. Nindo, C.I.; Tang, J.; Cakir, E.; Powers, J.R. *Potential of Refractance Window Technology for Value Added Processing of Fruits and Vegetables in Developing Countries*; American Society of Agricultural and Biological Engineers: Portland, OR, USA, 2006.
21. Kaspar, K.L.; Park, J.S.; Mathison, B.D.; Brown, C.R.; Massimino, S.; Chew, B.P. Processing of Pigmented-Flesh Potatoes (*Solanum tuberosum* L.) on the Retention of Bioactive Compounds: Potato Processing on Nutrient Retention. *Int. J. Food Sci. Technol.* **2012**, *47*, 376–382. [CrossRef]

22. Abul-Fadl, M.M.; Ghanem, T.H. Effect of Refractance-Window (RW) Drying Method on Quality Criteria of Produced Tomato Powder as Compared to the Convection Drying Method. *World Appl. Sci. J.* **2011**, *15*, 953–965.
23. Mahanti, N.K.; Chakraborty, S.K.; Sudhakar, A.; Verma, D.K.; Shankar, S.; Thakur, M.; Singh, S.; Tripathy, S.; Gupta, A.K.; Srivastav, P.P. Refractance WindowTM-Drying vs.Other Drying Methods and Effect of Different Process Parameters on Quality of Foods: A Comprehensive Review of Trends and Technological Developments. *Future Foods* **2021**, *3*, 100024. [CrossRef]
24. Zotarelli, M.F.; Carciofi, B.A.M.; Laurindo, J.B. Effect of Process Variables on the Drying Rate of Mango Pulp by Refractance Window. *Food Res. Int.* **2015**, *69*, 410–417. [CrossRef]
25. Ginzburg, A.S. *Application of Infrared Radiation in Food Processing. Chemical and Process Engineering Series*; Leonard Hill: London, UK, 1969.
26. Almeida, M.; Torrance, K.E.; Datta, A.K. Measurement of Optical Properties of Foods in Near- and Mid-Infrared Radiation. *Int. J. Food Prop.* **2006**, *9*, 651–664. [CrossRef]
27. Torres, M.D.; Moreira, R.; Chenlo, F.; Vázquez, M.J. Water Adsorption Isotherms of Carboxymethyl Cellulose, Guar, Locust Bean, Tragacanth and Xanthan Gums. *Carbohydr. Polym.* **2012**, *89*, 592–598. [CrossRef] [PubMed]
28. Parrouffe, J.M.; Dostie, M.; Navarri, P.; Andrieu, J.; Mujumdar, A.S. Heat and Mass Transfer Relationship in Combined Infrared and Convective Drying. *Dry. Technol.* **1997**, *15*, 399–425. [CrossRef]
29. Tsilingiris, P.T. Comparative Evaluation of the Infrared Transmission of Polymer Films. *Energy Convers. Manag.* **2003**, *44*, 2839–2856. [CrossRef]
30. Shende, D.; Datta, A.K. Optimization Study for Refractance Window Drying Process of Langra Variety Mango. *J. Food Sci. Technol.* **2020**, *57*, 683–692. [CrossRef] [PubMed]
31. Azizi, D.; Jafari, S.M.; Mirzaei, H.; Dehnad, D. The Influence of Refractance Window Drying on Qualitative Properties of Kiwifruit Slices. *Int. J. Food Eng.* **2017**, *13*, 20160201; [CrossRef]
32. Tontul, I.; Eroğlu, E.; Topuz, A. Convective and Refractance Window Drying of Cornelian Cherry Pulp: Effect on Physicochemical Properties. *J. Food Process. Eng.* **2018**, *41*, e12917. [CrossRef]
33. Ocoró-Zamora, M.; Ayala-Aponte, A. Influence of Thickness on the Drying of Papaya Puree (*Carica papaya* L.) through Refractance WindowTM Technology. *DYNA* **2013**, *80*, 147–154.
34. Múnera-Tangarife, R.D. *Evaluación de Los Factores Que Afectan La Transferencia de Calor En El Secado Por Ventana de Refractancia-Hidrosecado*; Universidad del Valle: Santiago de Cali, Colombia, 2021.
35. Richardson, P. (Ed.) *Thermal Technologies in Food Processing*; Woodhead Publishing in Food Science and Technology; CRC Press: Boca Raton, FL, USA; Woodhead: Cambridge, UK, 2001; ISBN 978-1-85573-558-3.
36. Sun, D.-W. *Thermal Food Processing: New Technologies and Quality Issues*; CRC Press: Boca Raton, FL, USA, 2012; ISBN 978-1-4398-7679-4.
37. Pawar, S.B.; Pratape, V.M. Fundamentals of Infrared Heating and Its Application in Drying of Food Materials: A Review: Mapping of Infrared Drying of Foods. *J. Food Process. Eng.* **2017**, *40*, e12308. [CrossRef]
38. Krishnamurthy, K.; Khurana, H.K.; Soojin, J.; Irudayaraj, J.; Demirci, A. Infrared Heating in Food Processing: An Overview. *Comp. Rev. Food Sci Food Saf.* **2008**, *7*, 2–13. [CrossRef]
39. Sandu, C. Infrared Radiative Drying in Food Engineering: A Process Analysis. *Biotechnol. Prog.* **1986**, *2*, 109–119. [CrossRef] [PubMed]
40. Siegel, R.; Howell, J.R.; Mengüc, M.P. *Thermal Radiation Heat Transfer*; CRC Press: Boca Raton, FL, USA, 2016; ISBN 978-1-4987-5774-4.

Article

Development of Novel Superabsorbent Hybrid Hydrogels by E-Beam Crosslinking

Ion Călina [1], Maria Demeter [1,*], Anca Scărișoreanu [1,*] and Marin Micutz [2]

[1] National Institute for Lasers Plasma and Radiation Physics, 409 Atomiștilor, 077125 Măgurele, Romania; calina.cosmin@inflpr.ro

[2] Department of Physical Chemistry, University of Bucharest, 4-12 Regina Elisabeta Blvd., 030018 Bucharest, Romania; micutz@gw-chimie.math.unibuc.ro

* Correspondence: maria.dumitrascu@inflpr.ro (M.D.); anca.scarisoreanu@inflpr.ro (A.S.)

Abstract: In this study, several superabsorbent hybrid hydrogel compositions prepared from xanthan gum (XG)/sodium carboxymethylcellulose (CMC)/graphene oxide (GO) were synthesized by e-beam radiation crosslinking. We studied and evaluated the effects of GO content from the chemical structure of the hydrogels according to: sol-gel analysis, swelling degree, diffusion of water, ATR-FTIR spectroscopy, network structure, and dynamic mechanical analysis. The gel fraction and swelling properties of the prepared hydrogels depended on the polymer compositions and the absorbed dose. The hybrid XGCMCGO hydrogels showed superabsorbent capacity and reached equilibrium in less than 6 h. In particular, the XGCMCGO (70:30) hydrogel reached the highest swelling degree of about 6000%, at an irradiation dose of 15 kGy. The magnitude of the elastic (G′) and viscous (G″) moduli were strongly dependent on the absorbed dose. When the degree of crosslinking was higher, the G′ parameter was found to exceed 1000 Pa. In the case of the XGCMCGO (80:20) hydrogel compositions, the Mc and ξ parameters decreased with the absorbed dose, while crosslinking density increased, which demonstrated that we obtained a superabsorbent hydrogel with a permanent structure.

Keywords: hydrogel; superabsorbent; e-beam; swelling; crosslinking

1. Introduction

In the specialized scientific literature, there are reports on a series of composite hydrogels that have various characteristic properties and can have many applications: delivering or controlled releasing of drugs [1–3], adsorbents for water purification [4–6], adsorbents of dyestuffs in various fields [7], and energy-storage devices [8,9]. Most hydrogels containing graphene oxide (GO) in their composition were obtained using chemical crosslinking agents.

Huang et al. developed some superabsorbent hydrogels based on different GO concentrations (GO/poly (acrylic acid-co-acrylamide)) by in situ radical solution polymerization, and obtained hydrogels with a swelling capacity of up to 1100% [10,11].

Hydrogel crosslinking using ionizing radiation is a well-established method that is clean and rapid, and that ensures the obtaining of sterile products with a permanent and homogeneous network structure.

By varying the absorbed dose, the degree of crosslinking, on which the swelling degree strongly depends, can be controlled. Since it has many hydrophilic groups on its surface, GO is an excellent material that can be used in mixtures with several polymers in order to obtain new hydrogels with superabsorbent properties [12–14]. Natural polymers such as CMC and XG are more susceptible to degradation when they are irradiated in aqueous solutions due to the indirect effect of water. The effect of e-beam irradiation on XG, which is used as an ingredient in the food industry to determine irradiation stability, has been investigated by Li et al. [15]. The study showed that the viscosity of irradiated aqueous

XG-based solutions was reduced as the irradiation dose increased (5–50 kGy), suggesting degradation or depolymerizations of polysaccharide macromolecules.

Comprehensives studies on the γ-radiation of XG in solid and aqueous states were performed by Şen and Hayrabolulu et al. They showed that the effect of γ-radiation in air for XG is chain scission, which is effective when the dose rate is decreased due to the enhanced oxidative degradations during irradiation [16,17]. The same research group developed superabsorbent hydrogels based on XG using γ-radiation crosslinking synthesis at very low absorbed dose, such as 1–2 kGy [18].

Another study investigated how the irradiation dose (0–30 kGy) influenced the viscosity of CMC solutions, and developed methods to control the degradation [19].

Most often, the crosslinking by irradiation of natural polymers is accompanied by the degradation of the polymer chain. Despite this major disadvantage, natural polymers are preferred over the petroleum-derived synthetic polymers for ecological reasons, but also due to the fact that natural polymers have inherent properties of biocompatibility and biodegradability. Thus, materials produced for biomedical purposes, in addition to being easier to process, are much more easily accepted by cells and tissues.

One of the reasons for using graphene oxide (GO) to create hybrid hydrogels with natural polymers such as xanthan gum (XG) and caboxymethylcellulose (CMC) in their composition is primarily related to obtaining a hydrogel with improved mechanical properties compared to the initial polymers. It is well known that natural polymers have low mechanical properties, although they have excellent absorption properties due to their highly hydrophilic nature.

GO induces elasticity of polymer chains, thermal and chemical stability. It is a material suitable for increasing the mechanical strength of natural polymers and due to the ease with which GO can be dispersed in a polymer mixture increasing the miscibility of these compounds [20,21].

A particularly important role that GO can play in the composition of some hydrogels is related to the lyophilization process when the hydrogels lose their water content. GO hydrogels obviously maintain their chemical composition and still have high porosity, a large specific surface area, and a high affinity for various molecules. Recent studies have shown an increased drug-loading capacity and prolonged release when using GO-based hydrogel systems, making them suitable for advanced drug-delivery applications [22]. In our previous study, superabsorbent hydrogels incorporated with GO were obtained by e-beam irradiation in the absence of air, at low radiation doses (0.5–3 kGy), from a high concentration of polymers in aqueous solution (paste-like condition) [23]. The compositions of the hydrogels above were characterized by a lower swelling capacity and a higher crosslinking density.

The present study sought to develop and characterize a new hybrid hydrogel with superabsorbent properties composed exclusively of biodegradable polymers (XG and CMC), incorporated with GO, to obtain a hydrogel with the best structural, rheological, and swelling properties, and that could be used in the field of biomedical engineering [24] and in hygienic products.

2. Results and Discussion

2.1. Sol-Gel Analysis

Figure 1 shows the variation of the gel fraction (GF) depending on the absorbed dose of the XGCMCGO hydrogels with compositions of 50:50, 80:20, and 70:30. For these hydrogel compositions, it was observed that the highest percentage of the GF was obtained at the lowest absorbed dose: 78% for XGCMCGO (50:50) and 73% for XGCMCGO (80:20); while for XGCMCGO (70:30), at the lowest dose, it obtained the lowest percentage of the GF, namely 60%. The XGCMCGO (70:30) hydrogel showed an almost constant value of GF over the entire range of applied radiation doses.

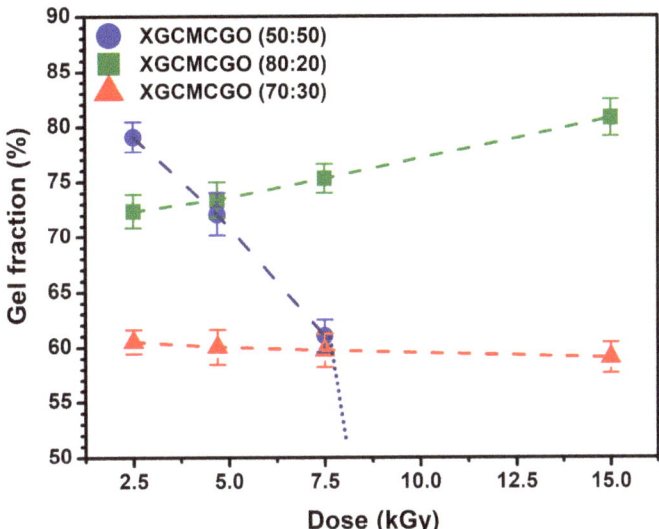

Figure 1. Gel fraction as a function of the absorbed dose during e-beam irradiation for the XGCMCGO hydrogels.

For the hydrogel samples having the compositions of 50:50 and 70:30, the percentage of GF decreased with an increasing absorbed dose. The exception to this rule was the hydrogel with equal concentrations of polymers (XGCMCGO (50:50)), for which the determination of the GF was no longer possible at doses above 7.5 kGy, due to the advanced degradation of the samples after irradiation.

For the XGCMCGO (80:20) hydrogel, the percentage of the GF increased with an increasing absorbed dose; the highest percentage of the GF was 78% at an irradiation dose of 15 kGy.

The obtained values were comparable to those of other studies regarding the e-beam crosslinking of similar hydrogel compositions. For example, Sung et al. prepared CMC/GO hydrogels by e-beam irradiation at 30 kGy and obtained a gel fraction of 67.3% [20].

Another study, performed by Said et al., which investigated the formation of CMC/AA hydrogels in aqueous solutions under the effect of e-beam irradiation, showed that the gel fraction increased quickly with a dose of 50 kGy, reaching a value of 70%, then in the dose range of 50–100 kGy, showed a slight increase [25].

The rate of hydrogel formation depends on the p_0/q_0 ratio of the polymer. If the crosslinking process predominates over the degradation process, an insoluble gel is formed as a detriment to the degradation process [26]. When polymers are subjected to ionizing radiation, crosslinking and main chain scission occur simultaneously. The quantitative estimation of the extent of crosslinking and degradation can be made using the Charlesby–Rosiak equation. The p_0/q_0 ratios calculated using the Charlesby–Rosiak equation for the XGCMCGO hydrogels are presented in Table 1.

Table 1. Sol-gel parameters calculated according the Charlesby–Rosiak Equation (3).

Hydrogel	p_0/q_0	Dg (kGy)
XGCMCGO (50:50)	1.04	0.23
XGCMCGO (80:20)	0.64	0.24
XGCMCGO (70:30)	0.66	0.79

The lowest value of the p_0/q_0 of 0.64 was obtained for the hydrogel composition of 80:20, which contained the highest XG concentration. The XGCMCGO (50:50) and

XGCMCGO (70:30) hydrogels showed p_0/q_0 ratios of 1.04 and 0.66, respectively. It is well known in the radiation-chemistry field that when various polymeric composition subjected to a treatment with ionizing radiation show a p_0/q_0 less than 2, the crosslinking processes predominated [27]. In the present study, these values were slightly higher compared to those obtained for hydrogels prepared with high concentrations of acrylic acid (70% AA) in data presented in our previous study, where the p_0/q_0 ratio was equal to 0.28 [23].

The gelation dose (D_g) is another parameter that shows the minimum dose required to obtain an insoluble gel during irradiation. D_g values decreased as the concentration of polymers and crosslinking agent ($N'N$—methylenebis(acrylamide), NMBA) increased. We observed that the gelation occurred at a very low dose (0.23 kGy) for the XGCMCGO (50:50) hydrogel.

The radiation-chemical yield of crosslinking (G(X)) decreased with an increase in the irradiation dose for all the XGCMCGO hydrogel compositions. The maximum value of this parameter was 132.2 µmol/J for the XGCMCGO (70:30) hydrogel at a dose of 2.5 kGy. For the radiation-chemical yield of scission (G(S)), the maximum value was 252.5 µmol/J for the XGCMCGO (50:50) hydrogel at a dose of 2.5 kGy (Table 2).

Table 2. Radiation-chemical yield of crosslinking G(X) and chain scission G(S) calculated for the XGCMCGO hydrogels.

Dose (kGy)	Radiation-Chemical Yields (µmol/J)	XGCMCGO (50:50)	XGCMCGO (80:20)	XGCMCGO (70:30)
2.5	G(X)	121.4	113.7	132.2
	G(S)	252.5	100.1	121.6
4.7	G(X)	54.4	65.8	73.5
	G(S)	113.2	57.9	67.6
7.5	G(X)	33.6	46.8	53.4
	G(S)	69.8	41.2	49.1
15	G(X)	-	27.6	27.9
	G(S)	-	24.3	25.6

As shown in the results presented in Table 2, G(S) > G(X) for the XGCMCGO (50:50) hydrogel, with approximately double the value. Therefore, it was obvious that in this case, the degradation processes predominated.

On the other hand, we observed that G(X) > G(S) for the XGCMCGO (80:20) and XGCMCGO (70:30) hydrogels. In these cases, the crosslinking predominated.

2.2. Swelling Degree

To characterize the network structure and determine the effective crosslinking density of the XGCMCGO hydrogels, the swelling properties in deionized water (DI) were first investigated.

Figure 2 shows the variation of the swelling degree (SD) at equilibrium, as a function of time and the absorbed dose, for the XGCMCGO hydrogels. All hydrogels showed good swelling ability. The maximum SD obtained at a dose of 15 kGy was ~6000%, which classified these XGCMCGO (70:30) hydrogels as "superabsorbent hydrogels". In the case of hydrogels with low concentrations of CMC, XG, AA, and NMBA, the SD had a value of 2250% for the XGCMCGO (50:50) hydrogel obtained at 4.7 kGy. At irradiation doses greater than 5 kGy, this hydrogel was degraded in deionized water (DI) in less than 2 h. For the XGCMCGO (80:20) hydrogel, the SD showed a maximum value of 2300% at a dose of 7.5 kGy.

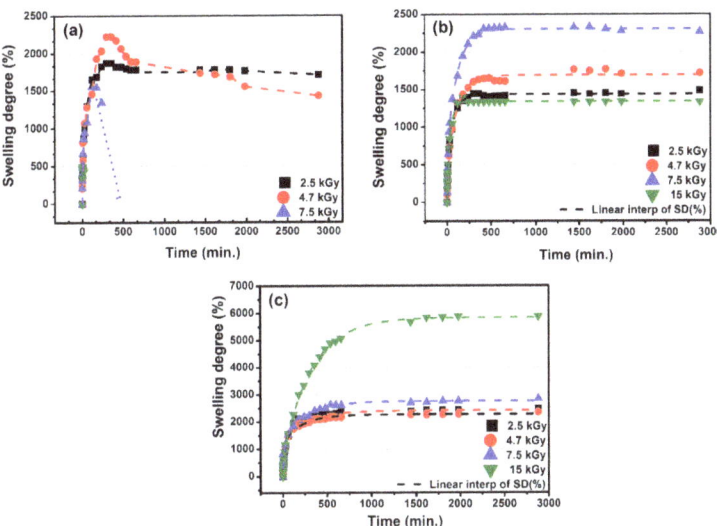

Figure 2. The swelling degree of the hydrogels in DI water: (**a**) XGCMCGO (50:50); (**b**) XGCMCGO (80:20); (**c**) XGCMCGO (70:30).

Moreover, we can specify that these hydrogels had the ability to reach equilibrium in less than 6 h and were stable in deionized water; of these, the most suitable proved to be the XGCMCGO (70:30) hydrogel. Previous studies of superabsorbent hydrogels based on CMC/AA/montmorillonite obtained by γ-irradiation showed a swelling degree of 16,000% [28]. Compared to our study, the above study used a high concentration of polymers; i.e., 10% CMC and 30% AA. Sultana et al. synthesized copolymer hydrogels from acrylamide/CMC by γ-irradiation, and they obtained an SD value in DI water of 795% [29]. We have shown that by using low concentrations of polymers and GO and appropriate mass ratios between them, with a moderate radiation dose rate, stable hydrogels with different degrees of swelling were obtained.

Superabsorbent hydrogels based on xanthan gum/polyacrylic acid/graphene oxide were prepared as absorbers for removing methylene blue dye from the water. For these hydrogels, the maximum swelling value in the neutral medium was found to be 2100% for a GO concentration of 1% [30]. Superabsorbent resin based on acrylic acid/CMC/GO showed the highest swelling capacity with 0.6% of GO in hydrogel composition, about 750 g g^{-1} in distilled water. The above study demonstrated that the incorporation of a moderate amount of GO in a CMC-based hydrogel can improve the water capacity of the material [31]. The CMC/GO composite superabsorbents prepared by e-beam irradiation at 10 kGy showed a swelling capacity of 140 g g^{-1} in distilled water [32]. After analyzing the data, we concluded that the XGCMCGO hydrogels presented in this study, in addition to having the lowest concentration of 0.1% GO, had the best swelling properties in a neutral environment.

2.3. Diffusion of Water

The analysis of water diffusion mechanisms in swollen polymeric systems has received considerable attention in recent years due to the important applications of swollen hydrogels in biomedical and pharmaceutical engineering. When a hydrogel is in contact with water, water enters the hydrogel by diffusion and the hydrogel network expands, resulting in the swelling of the hydrogel. Diffusion involves the migration of water into pre-existing or dynamically formed spaces between the macromolecular chains of the hydrogel [33]. To follow this process, Equation (7) was applied to the initial stage of swelling process, up to 60% of maximum swelling (equilibrium state) [34]. Table 3 shows the values of water diffusion mechanisms characteristic to the XGCMCGO hydrogels.

Table 3. Parameters k, n, and D for the XGCMCGO hydrogels.

XGCMCGO (50:50)						
Dose (kGy)	k	n	R^2	D (cm^2/s)	R^2	Mechanism
2.5	−3.42	0.29	0.91	0.02	0.98	Fickian
4.7	−3.41	0.42	0.96	0.03	0.98	Fickian
7.5	−3.47	0.47	0.99	0.03	0.98	Fickian
15	−3.08	0.28	0.94	0.03	0.98	Fickian
XGCMCGO (80:20)						
Dose (kGy)	k	n	R^2	D (cm^2/s)	R^2	Mechanism
2.5	−3.59	0.52	0.96	0.02	0.92	Non—Fickian
4.7	−3.44	0.45	0.97	0.03	0.95	Fickian
7.5	−3.53	0.59	0.99	0.03	0.93	Non—Fickian
15	−2.98	0.38	0.97	0.04	0.99	Fickian
XGCMCGO (70:30)						
Dose (kGy)	k	n	R^2	D (cm^2/s)	R^2	Mechanism
2.5	−3.15	0.44	0.98	0.03	0.98	Fickian
4.7	−2.94	0.52	0.99	0.04	0.97	Non—Fickian
7.5	−2.93	0.45	0.98	0.04	0.99	Fickian
15	−2.94	0.46	0.99	0.03	0.99	Fickian

A value of $n \leq 0.5$ indicated a Fickian diffusion mechanism. In this case, the rate of solvent diffusion was much lower compared to the macromolecular chain relaxation of the polymer. A value of n in the range $0.5 < n < 1$ indicated a non-Fickian transport, in which the diffusion of the solvent into the hydrogel structure was rapid compared to the relaxation rate of the macromolecular chain of the polymer [35].

Figure 3a–c show graphically, in logarithmic scale, the values of the parameter F = f(t). Figure 3d–f show graphically F = f (t 0.5). The values of the coefficients n, k, and D were calculated from the slope and intersection of the lines resulting from the swelling kinetics. The calculated values for the diffusion exponents n were less than 0.5 for the XGCMCGO (50:50) hydrogel at all irradiation doses, thus indicating that the solvent transport mechanism was of a Fickian type.

The XGCMCGO (80:20) hydrogel also presented a Fickian diffusion mechanism, but only in the cases of irradiation doses of 4.7 and 15 kGy. For the other irradiation doses (2.5 kGy and 7.5 kGy) the diffusion mechanism was non-Fickian. When the diffusion was non-Fickian, the relaxation time of the macromolecular chain and the diffusion had the same order of magnitude. As the solvent diffused into the hydrogel, the rearrangement of the macromolecular chains did not occur immediately.

The mechanism of water diffusion of the XGCMCGO (70:30) hydrogel was of a Fickian type; the exception was the hydrogel crosslinked with 4.7 kGy, in which case a value of n greater than 0.5 was obtained, thus exhibiting a non-Fickian diffusion behavior. In another similar study of a hybrid hydrogel obtained from acrylamide and XG, a value of n equal to 0.48 was reported [36]. Another study showed the obtaining of a GO-based hydrogel in which the inorganic component was incorporated in different concentrations. For this, values of n were obtained in the range of 0.5–1, which indicated a non-Fickian diffusion, which is specific to crosslinked hydrogels [37].

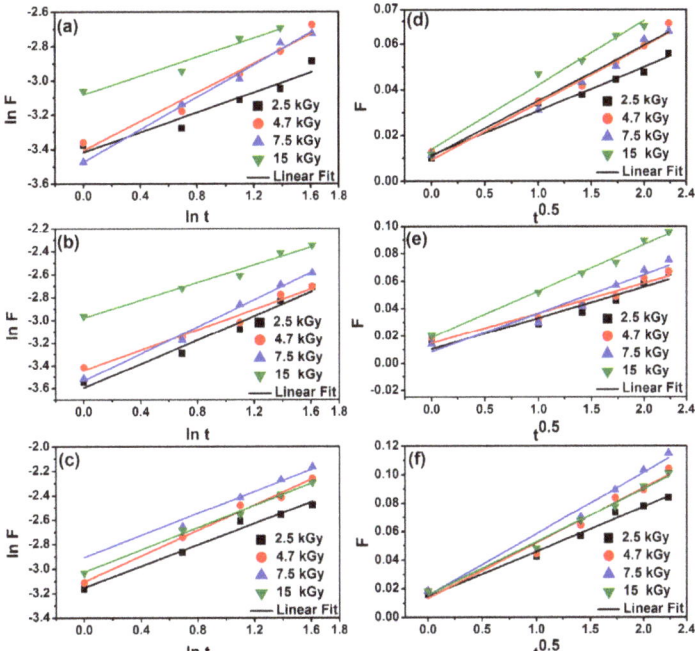

Figure 3. Swelling kinetics curves for the hydrogels: (**a**) XGCMCGO (50:50); (**b**) XGCMCGO (80:20); (**c**) XGCMCGO (70:30). Plots of F versus t0.5 for the hydrogels: (**d**) XGCMCGO (50:50); (**e**) XGCMCGO (80:20); (**f**) XGCMCGO (70:30).

The values of the diffusion coefficient (D) were found in the range of 0.02–0.04 cm^2/s, depending on the irradiation dose and the composition of the hydrogel. Compared to other studies, these values were much lower, suggesting the formation of high-molecular-weight hydrogels [38], as can be seen in the rheological measurements. For example, following a study on hydrogels obtained from CMC/acrylamide/GO by radical polymerization, the values obtained for the diffusion coefficient were in the range of 0.66–1.26 cm^2/s [39].

2.4. ATR-FTIR Spectroscopy

The characteristic FTIR spectra for the natural polymers (XG and CMC) and GO are shown in Figure 4. The FTIR spectrum characteristic of XG showed: 3308 cm^{-1} (O–H groups); 2917 cm^{-1} (CH$_3$ and CH$_2$ functional groups); 1700–1730 cm^{-1} (C = O); 1607 cm^{-1} (C = O); 1405 cm^{-1} (C–O group); and 1025 cm^{-1} (C–O group). The FTIR spectra of CMC showed: 3303 cm^{-1} (O–H group); 2923 cm^{-1} (C–H stretching vibrations); 1586 cm^{-1} (COO– group); 1420 cm^{-1} (CH$_2$); 1320 cm^{-1} (O-H groups); and 1030 cm^{-1} (C–O group). Regarding the FTIR spectrum of GO, it showed: 3231 cm^{-1} (O–H); 1732 cm^{-1} (C = O and C–H groups); 1620 cm^{-1} (C = C); 1275 cm^{-1} (C–OH); and 1053 cm^{-1} (C–O).

Figure 5a,b shows the characteristic FTIR spectra for nonirradiated polymeric blends with various ratios (XG:CMC) without GO (left side) and containing GO (right side). According to the acquired FTIR spectra, the displacement of the characteristic bands located in the range of 3600–600 cm^{-1} toward smaller wavenumbers corresponded to the formation of hydrogen bonds due to the miscibility of the polymers. In addition, increases in the band intensities in this area were proportional to the concentrations of natural polymers (XG and CMC) included in the polymeric blend.

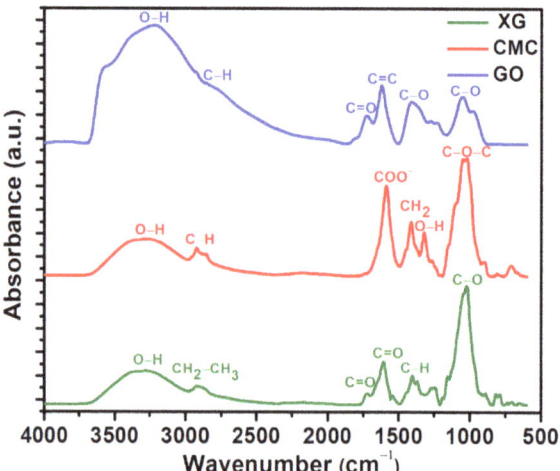

Figure 4. The FTIR spectra of pure XG, CMC, and pure GO.

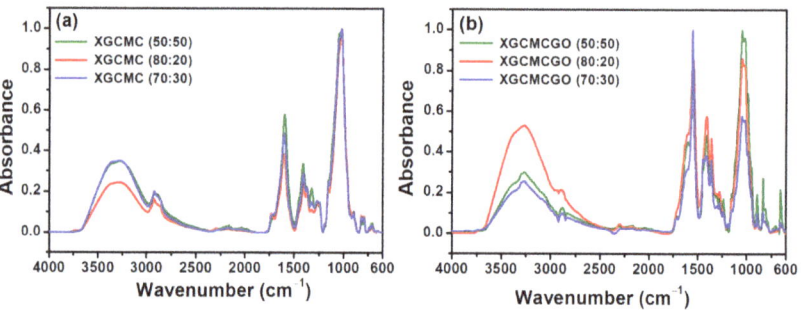

Figure 5. (**a**) The FTIR spectra of XGCMC without GO dried hydrogel; (**b**) the FTIR spectra of XGCMCGO dried hydrogel.

For the XGCMCGO unirradiated polymeric blend, the hydrophilic functional groups (–COOH, –C = O, –OH, and –C–O–C) of GO played a key role in improving the compatibility between the polymer matrix and GO.

The FTIR-ATR spectra, corresponding to XGCMCGO hydrogels irradiated at doses in the range of 2.5–15 kGy, are shown in Figure 6. It can be seen that the intensity of the absorption bands varied with the irradiation dose.

The intensity of specific absorption bands in the range of 3300–3000) cm^{-1} for XGCM-CGO (50:50) increased with the absorbed dose. The position of the bands assigned to the groups: C–H varied from 2929 cm^{-1} to 2940 cm^{-1}; C = O varied from 1614 cm^{-1} to 1637 cm^{-1}; and COO– varied from 1538 cm^{-1} to 1542 cm^{-1}, after e-beam irradiation at a dose of 15 kGy.

In the case of the XGCMCGO (80:20) hydrogel, the value of the band assigned to the O–H group decreased with the irradiation dose from 3268 cm^{-1} to 3253 cm^{-1}. The intensity of the bands assigned to CH groups increased significantly for the irradiated polymeric blends. The decrease of the wavenumber from 1558/1412/1193/1054 cm^{-1}, due to the application of a dose of 15 kGy, to 1557/1403/1157/1053 cm^{-1}, respectively, may have been caused by the different degrees of crosslinking of the hydrogel. As the absorbed dose increased, the crosslinking density also increased (see Table 4).

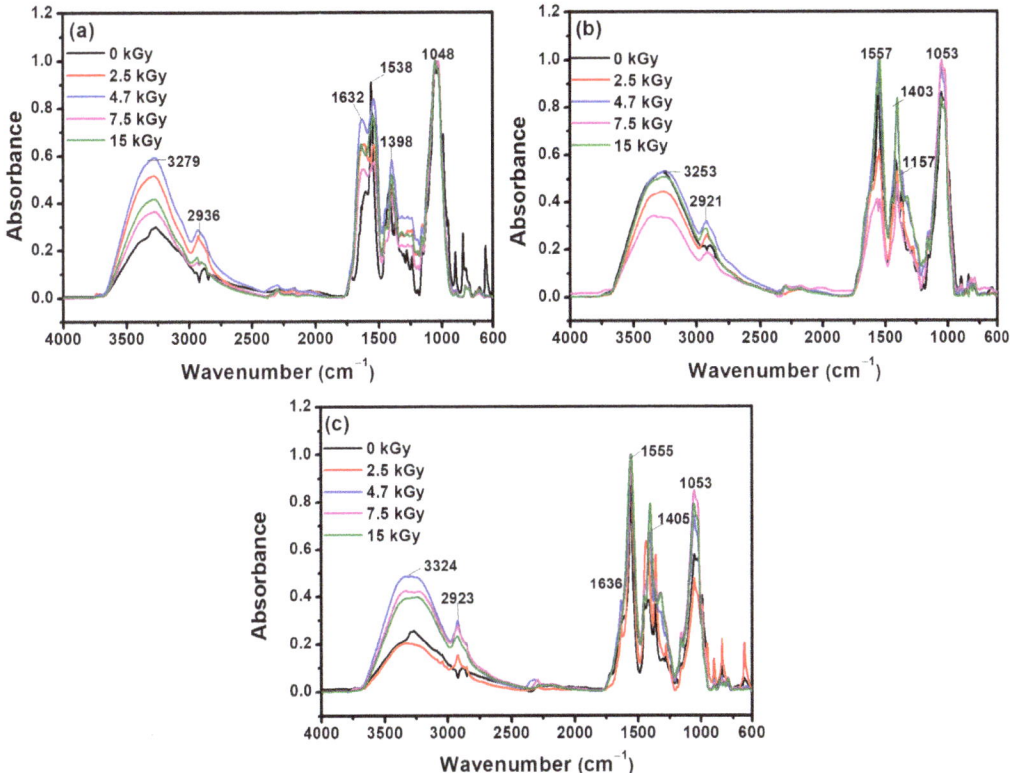

Figure 6. The FTIR spectra of unirradiated and irradiated dried hydrogels: (**a**) XGCMCGO (50:50); (**b**) XGCMCGO (80:20); (**c**) XGCMCGO (70:30).

Table 4. The elastic modulus (G′), the molecular weight between two crosslinks of the hydrogel chains (M_C), the crosslink density (ν_e), and network mesh size (ξ) of the XGCMCGO hydrogels.

Dose (kGy)	G′ (Pa)	M_C (kg/mol)	ν_e (mol/m^3)	ξ (nm)
XGCMCGO (50:50)				
2.5	869	132.7	0.76	140.1
4.7	712	157.3	0.64	158.4
7.5	646	160.3	0.63	170.6
15	200	-	-	-
XGCMCGO (80:20)				
2.5	912	210.7	0.49	138.5
4.7	933	195.9	0.52	132.8
7.5	1020	172.1	0.59	128.8
15	1052	146.5	0.70	123.6
XGCMCGO (70:30)				
2.5	1016	250.4	0.41	145.9
4.7	840	241.3	0.43	171.8
7.5	795	207.8	0.49	191.7
15	766	200.1	0.52	210.6

For the XGCMCGO (70:30) hydrogels, the intensity of the absorption bands in the range of 3400–2800 cm^{-1} increased with a dose of 7.5 kGy. At this dose, the position corresponding to the OH group shifted to smaller wavenumbers, from 3333 to 3324 cm^{-1}. The band assigned to the COO–group was shifted from 1556 cm^{-1} to 1550 cm^{-1} when the hydrogel was irradiated with a dose of 15 kGy. A similar trend was observed for the other bands in the range of 1500–800 cm^{-1}. The intensity of the characteristic bands for the XGCMCGO (70:30) hydrogel increased when it was irradiated at doses up to 7 kGy due to the fact that the crosslinking density decreased, a direct consequence of increased swelling. These results were supported by the data obtained from swelling experiments (~6000% at a dose of 15 kGy).

2.5. Characterization of the Network Structure

The characterization of the network structure of a hydrogel is a complex procedure due to the many types of networks encountered in the practice; namely: regular, irregular, and weakly or strongly crosslinked networks.

Basic structural parameters for the XGCMCGO hydrogels such as the elastic modulus (G'), average molecular weight between two successive crosslinks (M_C), crosslink density (ν_e), and mesh size (ξ) were calculated according to the equations presented in Section 4.7. C_n was taken as a weighted average of the characteristic ratios of the polymers: XG = 271 [40], CMC = 10 [41] and AA = 6.7 [42]. M_r is the monomeric unit of XG, CMC, and AA, taken as an average (XG = 933 g/mol, CMC = 234 g/mol, and AA = 72 g/mol) [43].

Using the values of the elastic modulus (G') determined by rheological analysis for each hydrogel composition and based on the theory of rubber elasticity, we determined M_C. The network parameters calculated for the XGCMCGO hydrogels are presented in Table 4.

The values of the M_C parameter for all hydrogel compositions were in the range of 132.7–250.4 kg/mol. For the XGCMCGO (80:20) and (70:30) hydrogel compositions, the corresponding Mc values decreased with an increase in the absorbed dose; while for the XGCMCGO (50:50) hydrogel compositions, the corresponding M_C values increased. This increase demonstrated once again the degradation of this polymeric mixture under the action of ionizing radiation.

For the XGCMCGO (50:50) and (70:30) hydrogel compositions, the corresponding ξ values increased with an increase in the irradiation dose; while for the XGCMCGO (80:20) hydrogel compositions, the corresponding ξ values decreased with an increase in the irradiation dose. The decrease of the parameter ξ may have been due to the increase in the crosslinking density. Therefore, a large number of active or free radical centers were formed on the polymer chains, and a large number of small chains were formed, which led to crosslinking and the formation of a more compact network structure with a smaller mesh size. Figure 7 shows the appearance of the polymer mixture before and after irradiation with e-beams.

For this reason, ν_e is one of the most important structural parameters for characterizing a class of hydrogels that have superabsorbent properties. According to the values obtained, it was shown that this parameter increased as the absorbed dose increased in the cases of the XGCMCGO (80:20) and (70:30) hydrogels. The behavior was completely different in the case of XGCMCGO (50:50), where the values for ν_e decreased with an increase in the absorbed dose. All the values obtained for ν_e were in the range of 0.41–0.76 mol/m^3, and showed the obtaining of hydrogels with a moderately crosslinked network structure that allowed the absorption of an increased amount of fluids.

Knowing the hydrogel pore size is important to understanding the mechanism for transport of bioactive substances in the macromolecular network of such materials. For example, controlling the rate of a drug diffusion is essential, because it reflects the space available for a drug molecule to diffuse inside or outside the swollen hydrogel network [44].

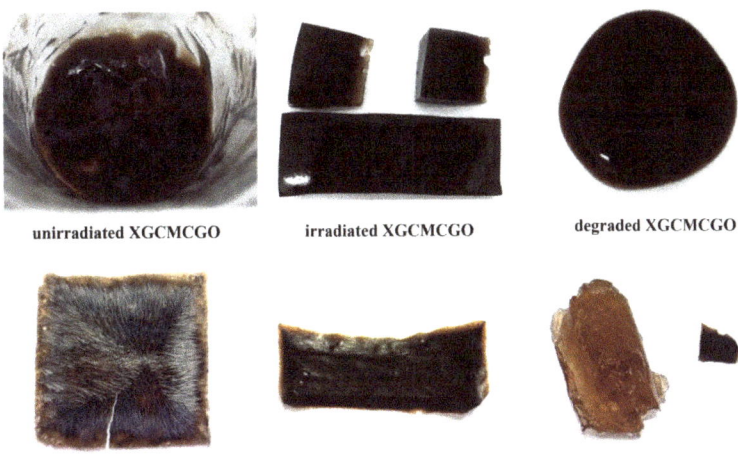

Figure 7. XGCMCGO hydrogels in different hypostases.

The values of ξ obtained for the XGCMCGO hydrogels were in the range of 123.6–210.6 nm. These values were similar to those obtained by other authors who developed hydrogels with applicability in tissue engineering, with ξ values previous reported in the range of 139–258 nm [45].

2.6. Rheological Analysis

The hydrogels were evaluated to determine the viscoelastic parameters G′ and G″. The elastic modulus (G′) shows the elasticity of the crosslinked bonds and the total elasticity of the material, while the loss or viscous modulus (G″) provides a perspective on the viscosity of the material. The ratio between G′ and G″ indicates the total resistance to deformation of the material [46].

The mechanical properties of hydrogels are important when selecting material for biomedical and other applications. They depend on the composition of the hydrogel and its water content. Hydrogels with a higher water content generally have a better permeability and biocompatibility [47]. However, there are some disadvantages, as a high degree of swelling is accompanied by a decrease in mechanical strength. For many applications, the combination of a high swelling degree and good mechanical properties is very important. Many approaches have been used to improve the mechanical properties of hydrogels, including the copolymerization of hydrophilic and hydrophobic monomers, increasing the crosslinking density, and varying the polymerization conditions [48].

Figure 8 shows the variation of the elastic and viscous (G′ and G″) moduli as a function of angular frequency (ω) and the absorbed dose for the XGCMCGO hydrogels. We observed that G′ was greater than G″ for all the hydrogel compositions. This is a critical requirement for hydrogels, as a G′ greater than the G″ suggests that the hydrogel has a higher elastic behavior [49].

Rheological analysis confirms the crosslinking or degradation processes that occur depending on the composition of the polymer mixture. Moreover, the effect of irradiation at various irradiation doses is highlighted, and the G′ and G″ is dependent on the composition of the hydrogel and the absorbed dose. A rheological analysis was performed on swollen hydrogel samples. The determined values of G′ and G″ are shown in Table 4.

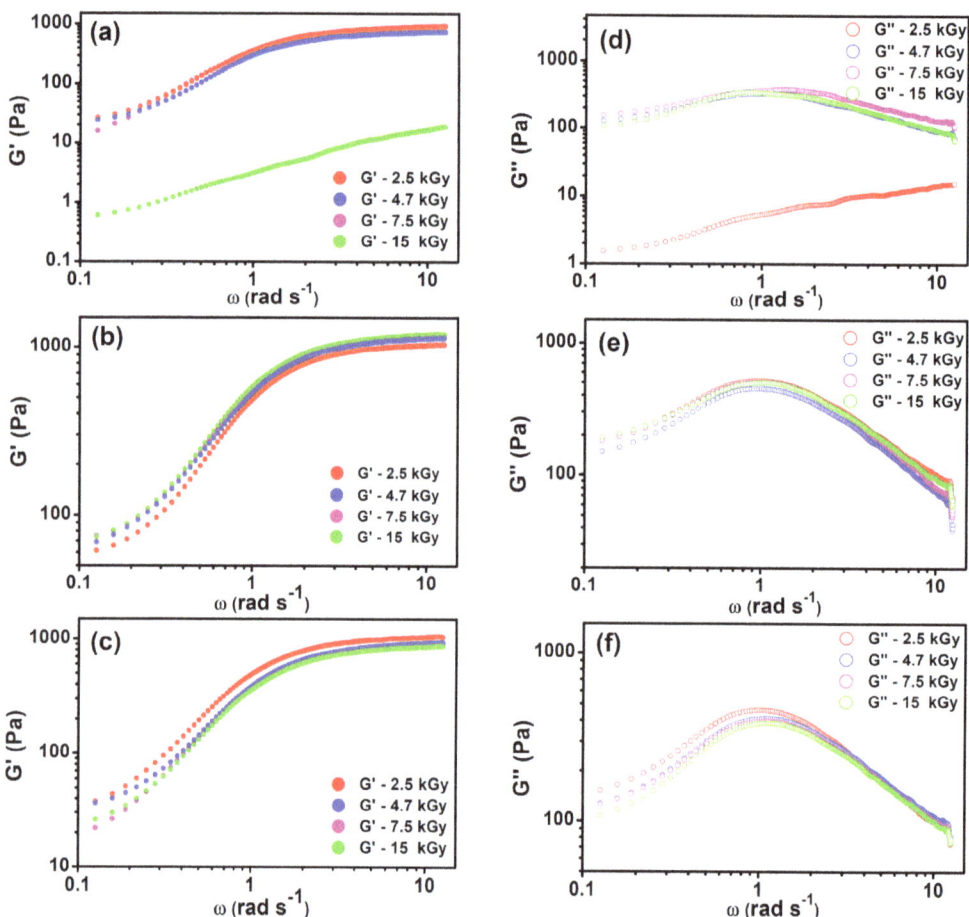

Figure 8. Elastic modulus (G′), as a function of the angular frequency (ω), of the (**a**) XGCMCGO (50:50), (**b**) XGCMCGO (80:20), and (**c**) XGCMCGO (70:30) hydrogels and their evolutions as a function of the absorbed dose. Viscous modulus (G″), as a function of the angular frequency (ω), of the (**d**) XGCMCGO (50:50), (**e**) XGCMCGO (80:20), and (**f**) XGCMCGO (70:30) hydrogels, and their evolutions as a function of the absorbed dose.

For the XGCMCGO (80:20) hydrogel, the G′ increased with an increase in the absorbed dose, with a maximum value for G′ = 1052 Pa at 15 kGy. The XGCMCGO (50:50) hydrogel had a G′ = 869 Pa at 2.5 kGy. When this polymeric blend was irradiated with 15 kGy, the G′ decreased very drastically down to 12 Pa. In the case of the XGCMCGO (70:30) hydrogel, G′ = 913 Pa was obtained at 2.5 kGy, and at 15 kGy, G′ = 766 Pa.

The G′ decreased with the absorbed dose in the cases of the XGCMCGO (50:50) and (70:30) hydrogels. We concluded that the decrease in G′ reflected the reduction of the crosslinking density (ν_e), while the decrease in G″ was due to the inhibition of the viscous behavior, showing a complete degradation of the polymer blend.

The XGCMCGO (80:20) hydrogel had a very well defined elastic behavior depending on the absorbed dose. In a recent study, a superabsorbent hydrogel based on CMC/starch/GO presented a value of G′ = 8050 Pa [50]. In another study, hydrogels obtained from CMC by redox polymerization for soft-tissue healing applications had a value of G′ = 814 Pa [45]. As shown by the rheological analysis and the sol-gel analysis, the XGCMCGO (80:20) hydrogel had the best properties.

3. Conclusions

We prepared novel and complex XGCMCGO hydrogels by e-beam crosslinking to be used as a potential substrate in biomedical engineering. Irradiation of XGCMC polymeric blends with e-beams proved to be a suitable technique, especially when GO was incorporated into their composition. Irradiation of polymeric mixtures with a wide range of doses allowed us to obtain hydrogels with different properties.

XGCMCGO hybrid hydrogels had better mechanical strength compared to the pure CMC hydrogel, and a better flexibility and improved swelling behavior than the simple XG hydrogel.

The gel fraction and the swelling properties of the prepared hydrogels depended on the composition of the polymers and the absorbed dose.

The hydrogels showed superabsorbent capacity and reached equilibrium in less than 6 h, especially the XGCMCGO (70:30) hydrogel, which reached the highest swelling degree of about 6000%, at 15 kGy. The crosslinking process predominated compared to the degradation process.

By characterizing the network structure, we observed that the hybrid hydrogels with high XG concentration showed the best structural properties. The crosslinking density increased with the absorbed dose.

The mesh size obtained from the experimental data of rheological analysis was in the range of 123–210 nm; these values were comparable to those of several hydrogels.

The interaction between the hydrogel components was highlighted by FT-IR analysis, and the increase in absorption band intensities for the characteristic functional group, as well as their shifting toward lower wavenumbers, were correlated with the crosslinking degree, which decreased at a dose of 15 kGy.

For all hydrogel compositions, $G' > G''$, thus suggesting that the hydrogels had a higher elastic behavior. The G' had values in a wide range of 12–1052 Pa, with the maximum value obtained at a dose of 15 kGy for the XGCMCGO (80:20) hydrogel composition.

Therefore, the prepared XGCMCGO superabsorbent hydrogels belong to a class of environmentally friendly materials, and might have potential practical applications in many areas, such as in biomedical engineering and hygienic products.

4. Materials and Methods

4.1. Materials

Sodium carboxymethylcellulose (CMC, Mw = 2.5×10^5 g/mol), $N'N$-methylene-bis-acrylamide (NMBA 99%, Mw = 154.17 g/mol), acrylic acid anhydrous 99% containing MEHQ as inhibitor (AA, Mw = 72.06 g/mol), and NaOH were purchased from Merck KGaA, Darmstadt, Germany. A commercial xanthan gum (XG—food grade, produced by Jungbunzlauer, Wien, Austria) in powder form with a molecular weight of Mw = 1.6×10^6 g/mol was used. Ultra-highly concentrated single-layer graphene oxide (6.2 g/L) was purchased from Graphene Laboratory Inc., New York, NY, USA.

4.2. Synthesis of Hydrogels and E-Beam Irradiation

In this experiment, three different hydrogels based on XGCMCGO with different content of XG and CMC in the presence of AA, NaOH, and NMBA were prepared. XG (5 wt %) and CMC (2 wt %) were dissolved in DI water at room temperature. After complete solubilization of the XG and CMC, they were mixed with each other in different compositions. For each XG:CMC ratio, 75 mL of the mixture was prepared. The ratios of XG:CMC in the mixtures were 50:50; 80:20, and 70:30. For a better understanding of the experiments, sample compositions are presented in Table 5.

Table 5. Sample composition details.

Hydrogel	Chemical Composition (%, w/v)					
	XG	CMC	GO	AA	NaOH	NMBA
XGCMCGO (50:50)	2.1	0.8	0.1	0.8	0.4	2
XGCMCGO (80:20)	3.3	0.3	0.1	2.8	1.4	1.5
XGCMCGO (70:30)	2.9	0.5	0.1	5.9	2.9	0.5

Then, 15 mL of each homogeneous solution of XGCMCGO was packed in a hermetically sealed polyethylene zip bag and subjected to e-beam irradiation at predetermined doses (2.5–15 kGy). The e-beam sample irradiation was performed in air at room temperature (25 °C) using a linear electron accelerator (National Institute for Laser, Plasma and Radiation Physics, Măgurele, Romania) at a fixed beam energy of 6 MeV (average beam current of 10 µA, pulse length of 3.75 µs, pulse repetition rate of 50 Hz, and average dose rate of 1 kGy/min) [51]. The dosimetry was performed using graphite calorimeters.

4.3. Sol-Gel Analysis

The hydrogel samples were dried in a vacuum oven to a constant weight and then immersed in DI water for 48 h at room temperature (25 °C). After 48 h, the swollen hydrogels were removed from the water and dried at 30 °C to a constant weight. The gel fraction (GF) and soluble fraction (s) were calculated as follows:

$$G(\%) = \left(\frac{W_d}{W_i}\right) \quad (1)$$

$$s = 1 - G \quad (2)$$

where W_i is the initial weight of dried sample after irradiation, W_d is the weight of the dried insoluble part of sample after immersion for 48 h, and s is soluble fraction of the polymer. All measurements were carried out in triplicate for each sample, and all values were expressed as mean value and standard deviation of three independent samples.

The gelation doses (the doses necessary to produce the first insoluble gel fraction) and the degradation vs. crosslinking ratios for the XGCMCGO hydrogels were calculated using a customized computer program for sol-gel analysis (Gelsol95), which was based on the Charlesby–Rosiak formula [26]:

$$s + \sqrt{s} = \frac{p_0}{q_0} + \left(2 - \frac{p_0}{q_0}\right)\left(\frac{D_v + D_g}{D_v + D}\right) \quad (3)$$

where p_0 is the degradation density (i.e., average number of main chain scissions per monomer unit and per unit dose), q_0 is the crosslinking density (i.e., fraction of monomer units crosslinked per unit dose), D is the absorbed dose (kGy), D_g is the gelation dose (kGy), and D_v is the virtual dose (kGy) (the dose necessary to transform the real sample into a sample with the molecular weight distribution of $M_w/M_n = 2$).

The radiation yields of crosslinking and degradation (scission) were calculated using the following equations:

$$G(X) = \frac{4.9 \cdot 10^2 \cdot c}{M_C \cdot D \cdot \rho} \quad (4)$$

$$G(S) = G(X) \cdot 2\frac{p_0}{q_0} \quad (5)$$

where G(X) is the radiation yield of crosslinking (expressed as number of moles of crosslinking bonds per Joule), G(S) is the radiation yield of chain scission (mol/J), M_C (kg/mol) is the average molecular weight between two successive crosslinks, c (g/L) is the polymer concentration in irradiated solution, D is absorbed dose (J/kg), and ρ (kg/m^3) is the polymer density [52].

4.4. Swelling Degree

The swelling properties of hydrogels were explored by placing the dried hydrogels in DI water at room temperature for 48 h to reach swelling equilibrium. At specified times, the swelled hydrogels were taken out of the distilled water, blotted with paper, weighed, and immersed again.

The swelling degree (SD(%)) was calculated as a function of the dry (W_d) and swollen (W_s) hydrogel weights using Equation (6) [53]:

$$SD(\%) = \frac{(W_s - W_d)}{W_d} \cdot 100 \tag{6}$$

4.5. Diffussion of Water

The most basic law of Fick's was used for the explanation of swelling kinetics and diffusion of the polymeric structures. The following equation was used to determine the nature of diffusion of water into hydrogels [54]:

$$F = \frac{M_t}{M_\infty} = kt^n \tag{7}$$

where F is the fraction of swelling due to the water uptake, M_t is the adsorbed water at time t, M_∞ is the adsorbed water at equilibrium, k is a proportionality constant, and n is the diffusional exponent. The first 60% of the water uptake data were fitted to Equation (7), and the corresponding values of k and n were obtained.

4.6. ATR-FTIR Spectroscopy

The changes in chemical structure of the crosslinked hydrogels were investigated. ATR-FTIR spectra of unirradiated and irradiated samples were taken with a PerkinElmer Spectrum 100 FTIR Spectrometer. The samples for FTIR analysis were first dried in a vacuum oven at 30 °C for 72 h. The samples were subjected to wavenumbers ranging from 4000 to 600 cm^{-1} at ambient temperature and a resolution of 4 cm^{-1}, averaged from 50 scans/sample.

4.7. Characterization of Network Structure

Network parameters of the XGCMCGO hydrogels, such as the average molecular weight between two crosslinks (Mc), crosslinking density (ve), and mesh size (ξ), were determined by using the swelling and rheological measurements. Using elastic modulus (G′) values determined from rheological measurements and based on the rubber elasticity theory, Mc could be determined using the following equation [55]:

$$M_C = \frac{A\rho RT(v_{2r})^{2/3}(v_{2s})^{1/3}}{G'} \tag{8}$$

where R is the universal gas constant (8.314 m^3 Pa/molK), T is the absolute experimental temperature (298.15 °K), v_{2r} is the polymer volume fraction after e-beam crosslinking, v_{2s} is the polymer volume fraction of the crosslinked hydrogel in swollen state, ρ (kg/m^3) is the polymer density, and the factor A equals 1 for an affine network and 1−2/ϕ for a phantom network.

The effective crosslink density (v_e) of the hydrogels was calculated using Equation (9):

$$v_e = \frac{\rho}{M_C} \tag{9}$$

The polymer volume fractions (v_{2r} and v_{2s}) were determined using Equation (10):

$$v_{2r(s)} = \frac{\left[1 + \left(w_{2r(s)} - 1\right) \cdot \rho_{hydrogel}\right]^{-1}}{\rho_{solvent}} \tag{10}$$

where $\rho_{hydrogel}$ and $\rho_{solvent}$ are the densities of the hydrogel and solvent (kg/m^3), respectively; and $w_{2r(s)}$ is the weight of the hydrogel after e-beam crosslinking after swelling (g). The weight swelling ratio of hydrogels after crosslinking (w_{2r}) was calculated as: w_{2r} = hydrogel mass after irradiation/hydrogel dry mass. The weight swelling ratio of hydrogels after swelling (w_{2s}) was calculated as: w_{2s} = hydrogel mass after swelling/hydrogel dry mass.

The mesh size of the polymer network (ξ) was determined using Equation (11) [56]:

$$\xi = v_{2s}^{-1/3} \cdot \left[C_n \left(\frac{2M_C}{M_r} \right) \right]^{-1/2} \cdot l \tag{11}$$

where C_n is the Flory characteristic ratio, M_r is the average molecular weight of the repeating unit, and l is the carbon–carbon bond length (0.154 nm).

4.8. Dynamic Rheological Measurements

Dynamic rheological measurements of the hydrogels were performed by employing an MFR 2100 Micro Fourier Rheometer (GBC, Australia) equipped with a home-made temperature control jacket connected to a Lauda E100 circulating water bath.

The operating parameters of the instrument during rheological investigation were as follows: gap between plates—400 µm; displacement amplitude—0.03 µm (to fall into the linear viscoelasticity domain); frequency domain—0.005–2.000 Hz (with a step of 0.005 Hz, which led to angular frequencies, in rad/s, of 2π times higher than the corresponding frequencies taken in Hz); equilibration time for each of the isothermal measurements—20 min; and 30 scans per rheogram.

The dynamic rheological parameters of storage modulus (G′) and loss modulus (G″) were determined to evaluate the stability of the hydrogel network. All rheological measurements were performed in triplicate at the same constant temperature of 23 °C, and all values were expressed as mean value and standard deviation.

Author Contributions: Conceptualization, I.C., M.D. and A.S.; methodology, I.C. and M.D.; investigation, I.C., M.D., A.S. and M.M.; data curation, I.C., M.D. and M.M.; writing—original draft preparation, I.C.; writing—review and editing, M.D. and A.S.; supervision, M.D. and A.S.; visualization, M.M. All authors have read and agreed to the published version of the manuscript.

Funding: This research was funded by the Romanian Ministry of National Education and Research, by Installations and Special Objectives of National Interest, and by the Nucleu LAPLAS VI Program (Contract No. 16N/08.02.2019). The APC was funded by Installations and Special Objectives of National Interest.

Conflicts of Interest: The authors declare no conflict of interest.

References

1. Qiu, Y.; Park, K. Environment-sensitive hydrogels for drug delivery. *Adv. Drug Deliv. Rev.* **2001**, *53*, 321–339. [CrossRef]
2. Miyata, T.; Uragami, T.; Nakamae, K. Biomolecule-sensitive hydrogels. *Adv. Drug Deliv. Rev.* **2002**, *54*, 79–98. [CrossRef]
3. Gupta, P.; Vermani, K.; Garg, S. Hydrogels: From controlled release to pH-responsive drug delivery. *Drug Discov. Today* **2002**, *7*, 569–579. [CrossRef]
4. Guo, S.; Dong, S. Graphene nanosheet: Synthesis, molecular engineering, thin film, hybrids, and energy and analytical applications. *Chem. Soc. Rev.* **2011**, *40*, 2644–2672. [CrossRef]
5. Chen, Y.; Chen, L.; Bai, H.; Li, L. Graphene oxide–chitosan composite hydrogels as broad-spectrum adsorbents for water purification. *J. Mater. Chem. A* **2013**, *1*, 1992–2001. [CrossRef]
6. Gao, H.; Sun, Y.; Zhou, J.; Xu, R.; Duan, H. Mussel-Inspired Synthesis of Polydopamine-Functionalized Graphene Hydrogel as Reusable Adsorbents for Water Purification. *ACS Appl. Mater. Interfaces* **2013**, *5*, 425–432. [CrossRef]
7. Wang, J.; Su, S.; Qiu, J. Biocompatible swelling graphene oxide reinforced double network hydrogels with high toughness and stiffness. *New J. Chem.* **2017**, *41*, 3781–3789. [CrossRef]
8. Peng, S.; Han, X.; Li, L.; Zhu, Z.; Cheng, F.; Srinivansan, M.; Adams, S.; Ramakrishna, S. Unique Cobalt Sulfide/Reduced Graphene Oxide Composite as an Anode for Sodium-Ion Batteries with Superior Rate Capability and Long Cycling Stability. *Small* **2016**, *12*, 1359–1368. [CrossRef]

9. Xu, H.; Adolfsson, K.; Xie, L.; Hassanzadeh, S.; Pettersson, T.; Hakkarainen, M. Zero-Dimensional and Highly Oxygenated Graphene Oxide for Multifunctional Poly(lactic acid) Bionanocomposites. *ACS Sustain. Chem. Eng.* **2016**, *4*. [CrossRef]
10. Huang, Y.; Zeng, M.; Ren, J.; Wang, J.; Fan, L.; Xu, Q. Preparation and swelling properties of graphene oxide/poly(acrylic acid-co-acrylamide) super-absorbent hydrogel nanocomposites. *Colloids Surf. A Physicochem. Eng. Asp.* **2012**, *401*, 97–106. [CrossRef]
11. Lei, H.; Xie, M.; Zhao, Y.; Zhang, F.; Xu, Y.; Xie, J. Chitosan/sodium alginate modificated graphene oxide-based nanocomposite as a carrier for drug delivery. *Ceram. Int.* **2016**, *42*. [CrossRef]
12. Geim, A.K.; Novoselov, K.S. The rise of graphene. *Nat. Mater.* **2007**, *6*, 183–191. [CrossRef]
13. Balandin, A.A.; Ghosh, S.; Bao, W.; Calizo, I.; Teweldebrhan, D.; Miao, F.; Lau, C.N. Superior Thermal Conductivity of Single-Layer Graphene. *Nano Lett.* **2008**, *8*, 902–907. [CrossRef]
14. Lee, C.; Wei, X.; Kysar, J.; Hone, J. Measurement of the Elastic Properties and Intrinsic Strength of Monolayer Graphene. *Science* **2008**, *321*, 385–388. [CrossRef]
15. Li, Y.; Ha, Y.; Wang, F.; Li, Y. Effect of Irradiation on the Molecular Weight, Structure and Apparent Viscosity of Xanthan Gum in Aqueous Solution. *Adv. Mater. Res.* **2011**, *239–242*, 2632–2637. [CrossRef]
16. Sen, M.; Hayrabolulu, H.; Taskin, P.; Torun, M.; Cutrubinis, M.; Güven, O. Radiation induced degradation of xanthan gum in the solid state. *Radiat. Phys. Chem.* **2015**, *124*. [CrossRef]
17. Hayrabolulu, H.; Cutrubinis, M.; Güven, O.; Sen, M. Radiation induced degradation of xanthan gum in aqueous solution. *Radiat. Phys. Chem.* **2017**, *144*. [CrossRef]
18. Hayrabolulu, H.; Demeter, M.; Cutrubinis, M.; Şen, M. Radiation synthesis and characterization of xanthan gum hydrogels. *Radiat. Phys. Chem.* **2021**, *188*, 109613. [CrossRef]
19. Choi, J.; Lee, H.-S.; Kim, J.-H.; Lee, K.W.; Lee, J.; Seo, S.J.; Kang, K.W.; Byun, M.W. Controlling the radiation degradation of carboxymethylcellulose solution. *Polym. Degrad. Stab.* **2008**, *93*, 310–315. [CrossRef]
20. Sung, Y.; Kim, T.-H.; Lee, B. Syntheses of carboxymethylcellulose/graphene nanocomposite superabsorbent hydrogels with improved gel properties using electron beam radiation. *Macromol. Res.* **2016**, *24*, 143–151. [CrossRef]
21. Lee, S.; Lee, H.; Sim, J.H.; Sohn, D. Graphene oxide/poly(acrylic acid) hydrogel by γ-ray pre-irradiation on graphene oxide surface. *Macromol. Res.* **2014**, *22*, 165–172. [CrossRef]
22. Pinelli, F.; Nespoli, T.; Rossi, F. Graphene Oxide-Chitosan Aerogels: Synthesis, Characterization, and Use as Adsorbent Material for Water Contaminants. *Gels* **2021**, *7*, 149. [CrossRef]
23. Calina, I.; Demeter, M.; Vancea, C.; Scarisoreanu, A.; Meltzer, V. E-beam radiation synthesis of xanthan-gum/carboxymethylcellulose superabsorbent hydrogels with incorporated graphene oxide. *J. Macromol. Sci. Part A* **2018**, *55*, 260–268. [CrossRef]
24. Das, B.; Prasad, K.E.; Ramamurty, U.; Rao, C. Nano-indentation studies on polymer matrix composites reinforced by few-layer graphene. *Nanotechnology* **2009**, *20*, 125705. [CrossRef]
25. Said, H.M.; Alla, S.G.A.; El-Naggar, A.W.M. Synthesis and characterization of novel gels based on carboxymethyl cellulose/acrylic acid prepared by electron beam irradiation. *React. Funct. Polym.* **2004**, *61*, 397–404. [CrossRef]
26. Olejniczak, J.; Rosiak, J.; Charlesby, A. Gel/dose curves for polymers undergoing simultaneous crosslinking and scission. *Int. J. Radiat. Appl. Instrum. Part C Radiat. Phys. Chem.* **1991**, *37*, 499–504. [CrossRef]
27. Mozalewska, W.; Czechowska-Biskup, R.; Olejnik, A.K.; Wach, R.A.; Ulański, P.; Rosiak, J.M. Chitosan-containing hydrogel wound dressings prepared by radiation technique. *Radiat. Phys. Chem.* **2017**. [CrossRef]
28. Salmawi, K.; El-Naggar, A.; Ibrahim, S. Gamma Irradiation Synthesis of Carboxymethyl Cellulose/Acrylic Acid/Clay Superabsorbent Hydrogel. *Adv. Polym. Technol.* **2016**, *37*, 515–521. [CrossRef]
29. Sultana, S.; Islam, M.R.; Dafader, N.C.; Haque, M.E. Preparation of carboxymethyl cellulose/acrylamide copoly-mer hydrogel using gamma radiation and investigation of its swelling behavior. *J. Bangladesh Chem. Soc.* **2012**, *25*, 132–138. [CrossRef]
30. Hosseini, S.M.; Shahrousvand, M.; Shojaei, S.; Khonakdar, H.A.; Asefnejad, A.; Goodarzi, V. Preparation of superabsorbent eco-friendly semi-interpenetrating network based on cross-linked poly acrylic acid/xanthan gum/graphene oxide (PAA/XG/GO): Characterization and dye removal ability. *Int. J. Biol. Macromol.* **2020**, *152*, 884–893. [CrossRef]
31. Wang, Z.; Ning, A.; Xie, P.; Gao, G.; Xie, L.; Li, X.; Song, A. Synthesis and swelling behaviors of carboxymethyl cellulose-based superabsorbent resin hybridized with graphene oxide. *Carbohydr. Polym.* **2017**, *157*, 48–56. [CrossRef] [PubMed]
32. Kim, B.; Kim, T.-H.; Lee, B. Optimal synthesis of carboxymethylcellulose-based composite superabsorbents. *Korean J. Chem. Eng.* **2021**, *38*, 215–225. [CrossRef]
33. Saraydın, D.; Isıkver, Y.; Sahiner, N.; Güven, O. The Influence of Preparation Methods on the Swelling and Network Properties of Acrylamide Hydrogels with Crosslinkers. *J. Macromol. Sci. Part A* **2004**, *41*, 419–431. [CrossRef]
34. Karadağ, E.; Saraydın, D.; Güven, O. Influence of some crosslinkers on the swelling of acrylamide-crotonic acid hydrogels. *Turkish J. Chem.* **1997**, *21*, 151–161.
35. Colombo, P.; Bettini, R.; Santi, P.; Peppas, N.A. Swellable matrices for controlled drug delivery: Gel-layer behaviour, mechanisms and optimal performance. *Pharm. Sci. Technolo. Today* **2000**, *3*, 198–204. [CrossRef]
36. Karadağ, E.; Ödemiş, H.; Kundakçi, S.; Üzüm, Ö.B. Swelling characterization of acrylamide/zinc acrylate/xanthan gum/sepiolite hybrid hydrogels and Its application in sorption of janus green B from aqueous solutions. *Adv. Polym. Technol.* **2016**, *35*, 248–259. [CrossRef]

37. Dai, H.; Zhang, Y.; Ma, L.; Zhang, H.; Huang, H. Synthesis and response of pineapple peel carboxymethyl cellulose-g-poly (acrylic acid-co-acrylamide)/graphene oxide hydrogels. *Carbohydr. Polym.* **2019**, *215*, 366–376. [CrossRef]
38. Siouffi, A.; Guiochon, G. Chromatography: Thin-Layer (PLANAR) | Theory of Thin-Layer (Planar) Chromatography. *Encycl. Sep. Sci.* **2000**, 915–930. [CrossRef]
39. Varaprasad, K.; Jayaramudu, T.; Sadiku, E.R. Removal of dye by carboxymethyl cellulose, acrylamide and graphene oxide via a free radical polymerization process. *Carbohydr. Polym.* **2017**, *164*, 186–194. [CrossRef]
40. Borsali, R.; Pecora, R. *Soft-Matter Characterization*; Springer Science & Business Media: Berlin/Heidelberg, Germany, 2008; ISBN 140204464X.
41. Robinson, G.; Ross-Murphy, S.B.; Morris, E.R. Viscosity-molecular weight relationships, intrinsic chain flexibility, and dynamic solution properties of guar galactomannan. *Carbohydr. Res.* **1982**, *107*, 17–32. [CrossRef]
42. Gudeman, L.F.; Peppas, N.A. pH-sensitive membranes from poly (vinyl alcohol)/poly (acrylic acid) interpenetrating networks. *J. Membr. Sci.* **1995**, *107*, 239–248. [CrossRef]
43. Fekete, T.; Borsa, J.; Takács, E.; Wojnarovits, L. Synthesis of carboxymethylcellulose/acrylic acid hydrogels with superabsorbent properties by radiation-initiated crosslinking. *Radiat. Phys. Chem.* **2015**, *124*. [CrossRef]
44. Wong, R.; Ashton, M.; Dodou, K. Effect of Crosslinking Agent Concentration on the Properties of Unmedicated Hydrogels. *Pharmaceutics* **2015**, *7*, 305–319. [CrossRef]
45. Varma, D.M.; Gold, G.T.; Taub, P.J.; Nicoll, S.B. Injectable carboxymethylcellulose hydrogels for soft tissue filler applications. *Acta Biomater.* **2014**, *10*, 4996–5004. [CrossRef]
46. Kalia, S.; Choudhury, A.R. Synthesis and rheological studies of a novel composite hydrogel of xanthan, gellan and pullulan. *Int. J. Biol. Macromol.* **2019**, *137*, 475–482. [CrossRef]
47. Micic, M.; Suljovrujic, E. Network parameters and biocompatibility of p(2-hydroxyethyl methacrylate/itaconic acid/oligo(ethylene glycol) acrylate) dual-responsive hydrogels. *Eur. Polym. J.* **2013**, *49*, 3223–3233. [CrossRef]
48. Lorenzo, E.; Katime, I. Some Mechanical Properties of Poly[(acrylic acid)-co-(itaconic acid)] Hydrogels. *Macromol. Mater. Eng.* **2003**, *288*, 607–612. [CrossRef]
49. Lejardi, A.; Hernandez, R.; Gonzalez, M.C.; Santos, J.; Etxeberria, A.; Sarasua, J.-R.; Mijangos, C. Novel hydrogels of chitosan and poly(vinyl alcohol)-g-glycolic acid copolymer with enhanced rheological properties. *Carbohydr. Polym.* **2014**, *103*, 267–273. [CrossRef]
50. Braihi, A. Viscoelastic and Rheological Properties of Carboxymethyl Cellulose/Starch/Graphite Oxide as Superabsorbent Hydrogel Nano Composites (SHNCs). *Int. J. Mater. Sci. Appl.* **2015**, *4*, 30–36. [CrossRef]
51. Manaila, E.; Craciun, G.; Ighigeanu, D.; Lungu, B.; Dumitru, M.; Stelescu, M. Electron Beam Irradiation: A Method for Degradation of Composites Based on Natural Rubber and Plasticized Starch. *Polymers* **2021**, *13*, 1950. [CrossRef]
52. Suhartini, M.; Yoshii, F.; Nagasawa, N.; Mitomo, H. Radiation Yield and Radicals Produced in Irradiated Poly (Butylene Succinate). *Atom Indon.* **2011**, *30*. [CrossRef]
53. Nagasawa, N.; Yagi, T.; Kume, T.; Yoshii, F. Radiation crosslinking of carboxymethyl starch. *Carbohydr. Polym.* **2004**, *58*, 109–113. [CrossRef]
54. Ritger, P.L.; Peppas, N.A. A simple equation for description of solute release I. Fickian and non-fickian release from non-swellable devices in the form of slabs, spheres, cylinders or discs. *J. Control. Release* **1987**, *5*, 23–36. [CrossRef]
55. Mahmudi, N.; Sen, M.; Rendevski, S.; Güven, O. Radiation synthesis of low swelling acrylamide based hydrogels and determination of average molecular weight between cross-links. *Nucl. Instrum. Methods Phys. Res. Sect. B Beam Interact. Mater. Atoms* **2007**, *265*, 375–378. [CrossRef]
56. Canal, T.; Peppas, N.A. Correlation between mesh size and equilibrium degree of swelling of polymeric networks. *J. Biomed. Mater. Res.* **1989**, *23*, 1183–1193. [CrossRef]

Article
Instantaneous Degelling Thermoresponsive Hydrogel

Noam Y. Steinman and Abraham J. Domb *

The Alex Grass Center for Drug Design and Synthesis and Center for Cannabis Research, Faculty of Medicine, Institute of Drug Research, School of Pharmacy, The Hebrew University of Jerusalem, Jerusalem 91120, Israel; noam.steinman@mail.huji.ac.il
* Correspondence: avid@ekmd.huji.ac.il

Abstract: Responsive polymeric hydrogels have found wide application in the clinic as injectable, biocompatible, and biodegradable materials capable of controlled release of therapeutics. In this article, we introduce a thermoresponsive polymer hydrogel bearing covalent disulfide bonds. The cold aqueous polymer solution forms a hydrogel upon heating to physiological temperatures and undergoes slow degradation by hydrolytic cleavage of ester bonds. The disulfide functionality allows for immediate reductive cleavage of the redox-sensitive bond embedded within the polymer structure, affording the option of instantaneous hydrogel collapse. Poly(ethylene glycol)-b-poly(lactic acid)-S-S-poly(lactic acid)-b-poly(ethylene glycol) (PEG-PLA-SS-PLA-PEG) copolymer was synthesized by grafting PEG to PLA-SS-PLA via urethane linkages. The aqueous solution of the resultant copolymer was a free-flowing solution at ambient temperatures and formed a hydrogel above 32 °C. The immediate collapsibility of the hydrogel was displayed via reaction with NaBH$_4$ as a relatively strong reducing agent, yet stability was displayed even in glutathione solution, in which the polymer degraded slowly by hydrolytic degradation. The polymeric hydrogel is capable of either long-term or immediate degradation and thus represents an attractive candidate as a biocompatible material for the controlled release of drugs.

Keywords: PEG-PLA; thermoresponsive hydrogel; redox-sensitive

Citation: Steinman, N.Y.; Domb, A.J. Instantaneous Degelling Thermoresponsive Hydrogel. *Gels* **2021**, *7*, 169. https://doi.org/10.3390/gels7040169

Academic Editor: Yang Liu

Received: 4 October 2021
Accepted: 12 October 2021
Published: 14 October 2021

Publisher's Note: MDPI stays neutral with regard to jurisdictional claims in published maps and institutional affiliations.

Copyright: © 2021 by the authors. Licensee MDPI, Basel, Switzerland. This article is an open access article distributed under the terms and conditions of the Creative Commons Attribution (CC BY) license (https://creativecommons.org/licenses/by/4.0/).

1. Introduction

Biodegradable polymers are ubiquitous across the pharmaceutical industry. These materials often constitute the primary platform for the delivery of therapeutic agents, as their slow degradation in vivo allows for sustained drug release over an extended period of time without the need for subsequent removal of the delivery vehicle. 'Smart' polymers capable of responding to external stimuli were developed to afford targeted drug release based on the presence of the relevant stimulus. These materials bear functional groups capable of a quick response to small changes in temperature, pH, or light, which lead to a physical change in the polymer that triggers drug release [1].

Polymeric hydrogels describe polymers capable of absorbing large amounts of water to form a gel due to crosslinking of the polymer chains. Smart thermoresponsive hydrogels were developed to undergo a sol–gel transition in response to temperature variation [2]. This feature enables a liquid solution to be injected into a physiological environment and form a gel in situ at the point of injection due to the temperature change. The molecular structures of such polymers determine their gelling temperatures as well as important features such as biocompatibility [3,4].

Copolymers of poly (lactic-co-glycolic acid) (PLGA) and poly (ethylene glycol) (PEG) were described at length as biocompatible and biodegradable polymers capable of forming thermoresponsive hydrogels in water [5]. Triblock copolymers, either with the morphology PLA-PEG-PLGA or PEG-PLGA-PEG, are soluble in water and reversibly form hydrogels upon heating [6–8]. The gelling behavior of the hydrogels formed from these copolymers is affected by overall polymer molecular weight, the ratio between polymer blocks, and the concentration of the polymer in solution [3,9–11]. Hence, the rational design of each element of the polymer structure is critical to achieving desired gel properties.

Inspired by the crucial role of cysteine–cysteine bonds and cleavage thereof in biological processes, the disulfide (S–S) bonds were exploited in a variety of functional materials due to their redox responsiveness. Materials containing disulfide bonds may undergo specific cleavage under reductive conditions, particularly upon exposure to the reductive intracellular space, rendering redox-responsive functionality to polymers. Disulfide bonds were incorporated in polyurethanes to afford self-healing properties [12], in electrochemical polymers and devices [13,14], and in drug carriers to render reduction-specific crosslinking or drug binding [15–17] for the controlled release of antitumor drugs due to high concentrations of reducing agents in the tumor microenvironment [18], to reduce toxicity in gene delivery platforms [19–22], and in supramolecular polymer applications [23]. The opportunity to incorporate disulfide bonds in biocompatible thermoreversible hydrogels may render these materials dual-responsiveness to both temperature and reduction.

In this work, we describe a triblock PEG-PLA-PEG copolymer bearing one disulfide bond per molecule. The thermoresponsiveness of the polymer solution afforded a fully water-soluble material at cool temperatures, which formed a gel between 32 and 40 °C. The polymer possesses a cleavable disulfide bond capable of cleavage upon exposure to strong reducing agents, thereby rendering immediate hydrogel collapse. This proof of concept was displayed by the cleavage of the disulfide bond in the presence of hydride. A synthetic analog without a disulfide bond was stable as a hydrogel even under harsh reductive conditions. The work here represents, to the best of our knowledge, the first report of a thermoresponsive hydrogel capable of controlled instantaneous collapse, in this example upon exposure to reducing conditions (Figure 1).

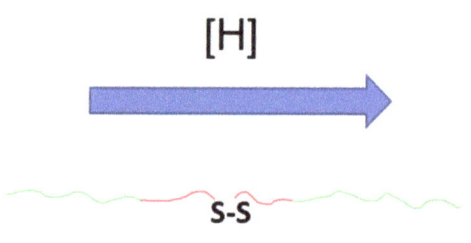

Figure 1. PEG-PLA-PEG triblock copolymers with embedded disulfide bonds immediately collapse upon addition of reducing agent NaBH$_4$. Disulfide-bearing polymer hydrogels afford controlled hydrogel collapsibility due to S–S bond cleavage.

2. Results

2.1. Rational Design

Jeong et al. first reported the synthesis of triblock copolymers containing a central poly (lactic acid) (PLA) block bearing two sidechains of poly (ethylene glycol) (PEG) [24]. The synthetic method reported there comprised of two steps. First, poly(ethylene glycol) methyl ether Mn = 550 (mPEG 550) was used as a macroinitiator of lactide polymerization (via ring-opening polymerization in the presence of stannous octoate catalyst) to form a diblock copolymer. Two hydroxyl termini of two distinct PLA blocks were subsequently linked by reaction with diisocyanate, forming urethane linkages (Figure 2A). Whereas the original research focused on polymers with sol–gel properties, which required cooling to move from sol form to gel morphology, subsequent reports focused on the development of sol–gel transitions, which occurred upon heating of the polymer solution. The enhancement of sol-to-gel properties from relying on cooling (gel at low temperatures and sol at high temperatures) to being responsive to heating (sol at room temperature and gel at elevated temperature) was crucial, as the practical application of such gels relies on the incorporation of active agents in a room-temperature solution, which upon injection and exposure to physiological temperatures forms a semi-solid gel capable of extended release of the active agent. These sol–gel thermosensitive hydrogels were based on an amorphous PLA polymer block (as opposed to crystalline poly (L-lactic acid) polymer blocks employed in the original work). Gel properties, biodegradability, and drug release were all subsequently optimized, leading to over two decades of research and application of injectable sol–gel biodegradable polymer solutions [6,11,25].

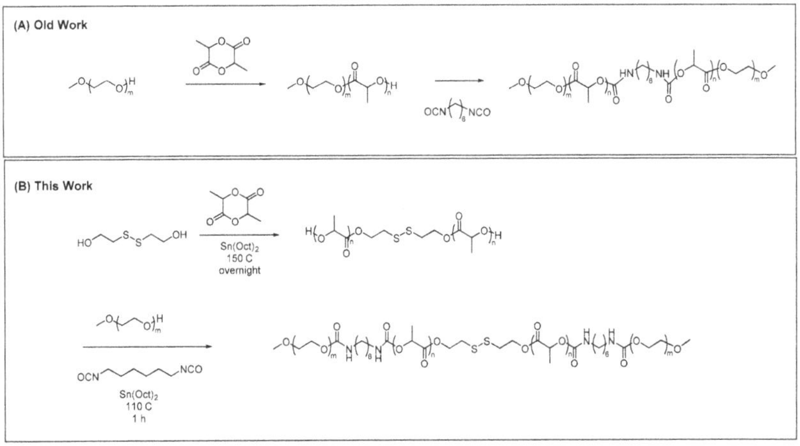

Figure 2. (**A**) Synthesis by Jeong et al. of a PEG-PLA-PEG triblock copolymer with urethane linkages [24]. (**B**) Synthesis of PEG-PLA-PEG with urethane linkages between polymer blocks and one embedded disulfide bond.

The PEG-PLA-PEG triblock copolymer described here is an analog of the original work performed by Jeong and co-workers. Its chemical structure bears the modular hydrophobic and hydrophilic units at the appropriate ratios for effecting sol–gel transitions near physiological temperatures, with the added benefit of one embedded disulfide bond. In doing so, the 'smart' hydrogel was imbued with secondary responsiveness to reduction agents in addition to its thermal responsivity. The hydrogel resultant thereof was therefore capable of injection as a liquid at ambient temperatures, gelling upon moderate heating, and subsequent collapse of the 3D gel structure upon exposure to strong reducing agents [26]. The following is a description of the synthesis and characterization of the novel triblock copolymer with a comprehensive study of the gel properties of aqueous solutions prepared thereof.

2.2. Synthesis

In a similar fashion as the original work on PEG-PLA-PEG triblock copolymers, several synthetic steps were employed in order to incorporate a disulfide bond into the triblock copolymer structure. In the first synthetic step, 2-hydroxyethyl disulfide was used as the initiator of DL-lactide polymerization. In doing so, an amorphous DL-poly(lactic acid) (DL-PLA) was obtained bearing one embedded disulfide bond (PLA(SS)). In order to complete the triblock copolymer synthesis, mPEG with hydroxyl termini were linked to either terminus of the amorphous PLA diol by a 2:2:1 molar reaction between mPEG diisocyanate and PLA(SS), affording urethane bridges between PEG and PLA polymer blocks with an overall PEG-PLA-SS-PLA-PEG copolymer structure (Figure 2B).

2.3. Characterization

The formation of PLA(SS) and PEG-PLA-SS-PLA-PEG polymers was confirmed spectroscopically by 1H NMR. Polymer molecular weights (MW) were estimated by size-exclusion chromatography (SEC). The presence of a disulfide bond was confirmed by elemental analysis (CHNS).

2.3.1. ^{1}H NMR

The structures of both PLA(SS) and PEG-PLA-SS-PLA-PEG were confirmed by 1H NMR (Figure 3) by observing all available proton peaks. For PLA(SS), in addition to characteristic multiplets for lactic acid at 5.2 and 1.5 ppm [3], a downfield shift of 2-hydroxyethyl disulfide CH_2 protons was observed upon polymerization in the PLA(SS) spectrum, indicating the formation of the ester bond at the termini of the disulfide reagent. No other peaks were observed in the spectrum, indicating a pure substance.

Upon linkage to PEG, no chemical shifts were observed for peaks from PLA(SS), confirming the stability of the PLA block of the copolymer to the reaction conditions. Urethane linkages were confirmed by a peak at δ 3.05 with 2x integration relative to the CH_2 peak at the disulfide, as expected. A characteristic PEG singlet was observed at δ 3.55, and all integrations were calculated relative to CH_3 of mPEG. No other peaks were observed in the spectrum, indicating a pure substance (Figure 3).

2.3.2. Molecular Weight Determination

Weight average (Mw) and number average (Mn) molecular weights of PLA(SS) and PEG-PLA-SS-PLA-PEG were estimated by SEC by comparing to polystyrene standards. Polydispersity (PDI) values were calculated from these measured values. Mn was further calculated by ^{1}H NMR peak integrations using known integration values (CH_2 of 2-hydroxyethyl disulfide for PLA(SS), CH_3 of mPEG for triblock PEG-PLA-PEG copolymer). Mn values calculated by NMR closely reflected the expected chain lengths based on feed ratios of starting materials, indicating full conversion from PLA(SS) to triblock copolymer with PEG. Estimated values provided by SEC indicated increased polymer MW upon copolymerization with PEG by urethane bridge, and only a slight increase in PDI values indicated polymers with high purity (Table 1).

Table 1. Molecular weight determination of disulfide PLA and disulfide-containing PEG-PLA-PEG copolymer.

Polymer	PLA(SS)	PEG-PLA-SS-PLA-PEG
Expected MW [1,2]	1.53	1.96
Mn (^{1}H NMR) [1,3]	1.46	3.11
Mn (SEC) [1,4]	2.65	4.24
Mw (SEC) [1,4]	2.85	5.97
PDI [5]	1.07	1.41

[1] Values given in kDa. [2] Based on feed ratios of starting materials. [3] Calculated based on peak integrations of known value. [4] Calculated from size-exclusion chromatography (SEC). [5] Calculated from SEC as Mw/Mn.

2.4. Rheometry Studies

The PEG-PLA-SS-PLA-PEG triblock copolymer formed a free-flowing solution in double-distilled water at room temperature (25% w/w). Upon incubation at physiological temperature, the solution formed a hydrogel which did not flow upon inversion of the test tube [3,4]. In order to evaluate the flow behavior of the copolymer solution and determine the temperature at which the sol–gel transition occurred, the viscosity of copolymer solutions was measured as a function of temperature (Figure 4). Upon heating, an increase in viscosity was observed beginning at ~32 °C with a maximum 70-fold increase in viscosity at 36 °C. The high viscosity values were stable through physiologically relevant temperatures. Upon cooling of the hydrogel, the opposite trend was observed, indicating reversibility of the thermoresponsive system. This flow behavior indicates a polymeric solution capable of injectability at ambient temperatures (<30 °C) with temperature-induced gelling occurring upon exposure to physiological temperatures.

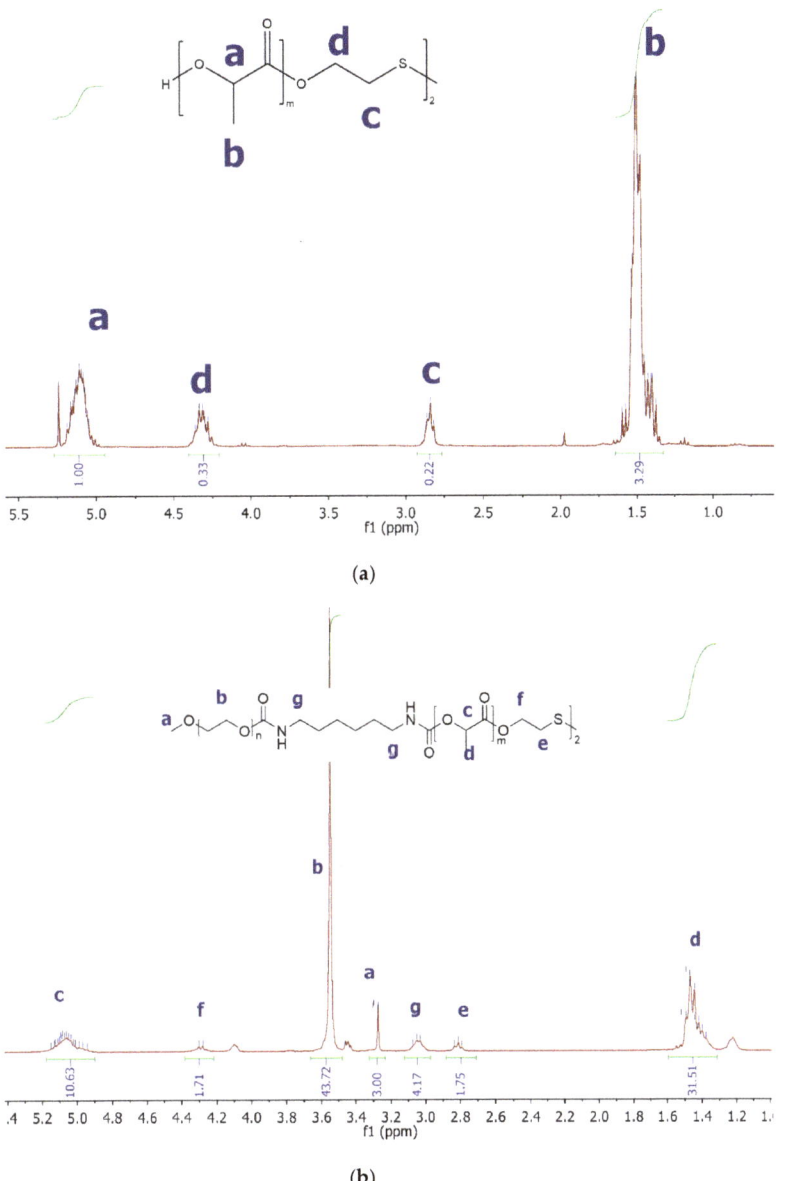

Figure 3. (a) 1H NMR spectra of PLA(SS) with peak assignments.; (b) 1H NMR spectra of PEG-PLA-SS-PLA-PEG with peak assignments.

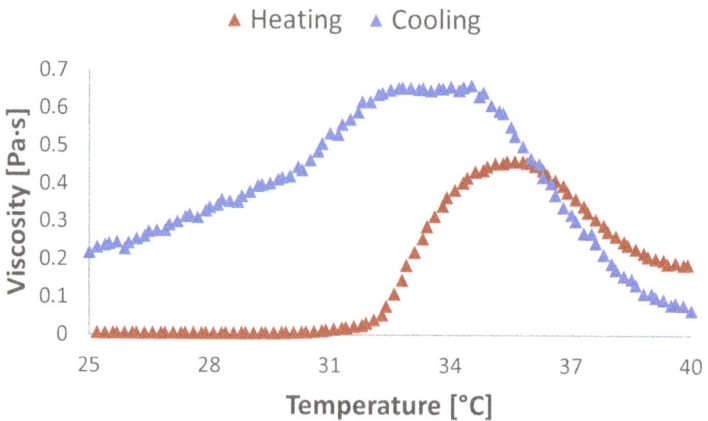

Figure 4. Heating (red) and cooling (blue) viscosity curves of 25% w/w PEG-PLA-SS-PLA-PEG aqueous solutions.

2.5. Degradation Studies

Degradation of injectable implants such as triblock copolymer hydrogels is an important parameter for the determination of the hydrogel's potential applicability to the clinic [27]. The stability of the hydrogel under physiological conditions affects the therapeutic window and controlled release profile of incorporated drugs, and biodegradation of the hydrogel matrix excludes the necessity for post-injection removal of the material, allowing it to slowly degrade and be eliminated by the body.

In the case of the PEG-PLA-PEG copolymer with an embedded disulfide bond, two modes of degradation are possible. One degradable bond is the repeating ester bond throughout the PLA polymer block, known to be susceptible to hydrolytic cleavage [28]. Secondly, the disulfide bond may degrade upon exposure to reducing agents. Hydrolytic degradation of the ester bonds was expected to occur slowly over time, allowing for slow degradation of the hydrogel matrix. The disulfide bond, however, was expected to remain intact under in vitro physiological conditions, degrading only upon exposure to reducing agent. Due to its position in the center of the polymer chain, cleavage of the disulfide bond should lead to an immediate release of water retained in the hydrogel matrix due to rupture of the triblock copolymer structure. In order to investigate these effects, degradation of PEG-PLA-SS-PLA-PEG triblock copolymer was studied in water or in the presence of reducing agents such as sodium borohydride ($NaBH_4$) or L-glutathione.

In order to confirm the redox reactivity of the disulfide bond embedded in the PEG-PLA-PEG triblock copolymer structure, $NaBH_4$, a strong hydride-based reducing agent, was added to the hydrogel matrix. The hydride was expected to reduce the disulfide bond, affording diblock PEG-PLA copolymers with thiol end groups (Figure 5). Indeed, a reduction in polymer molecular weight to half was observed upon reaction with hydride, indicating the cleavage of the reduction-sensitive disulfide bond in the triblock copolymer structure (Figure 5). A chemical analog of the triblock copolymer was prepared using 1,6-hexanediol as the initiator of PLA synthesis, resulting in PEG-PLA-PEG triblock copolymer without a disulfide bond (experimental details and spectral data available as Supplementary Materials, Figures S1 and S2). The molecular weight of this polymer was only slightly reduced upon exposure to the same hydride reagent, thereby confirming the role of the disulfide bond in a molecular weight reduction in the original polymer (Figure 5).

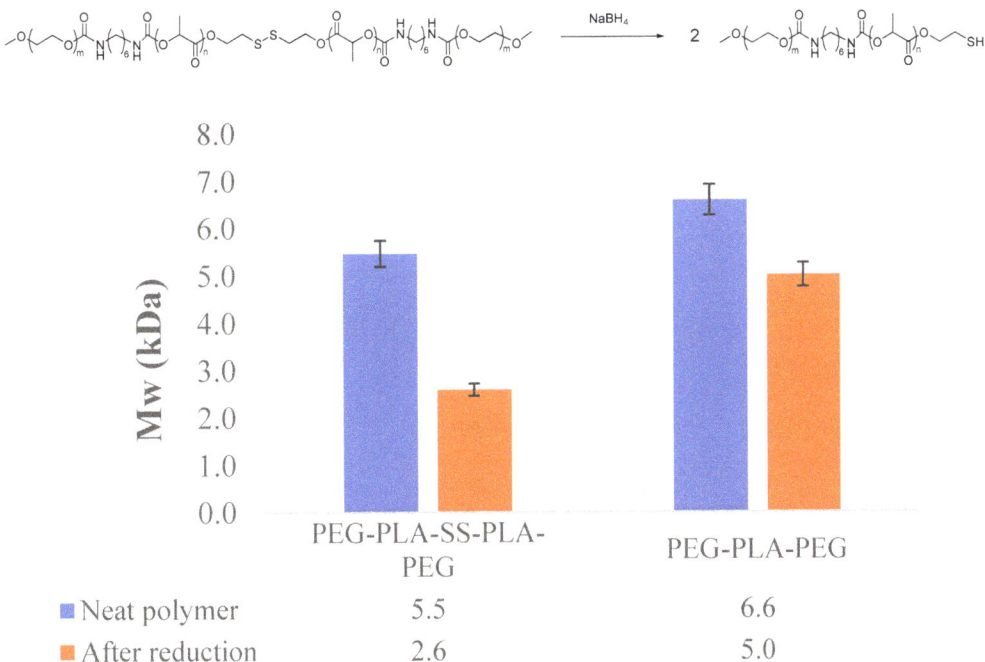

Figure 5. Sensitivity of the embedded disulfide bond in PEG-PLA-PEG triblock copolymer was confirmed by reaction with NaBH$_4$ and subsequent reduction in molecular weight by half. A chemical analog lacking the disulfide bond retained most of its molecular weight under the same conditions, indicating the role of disulfide bond cleavage in molecular weight reduction. Error bars represent the standard deviation of experiments performed in triplicate.

In order to evaluate the in vitro degradation of PEG-PLA-SS-PLA-PEG hydrogels, polymer molecular weight was monitored for two weeks at physiological temperature (Figure 6). The hydrogel remained stable for this period, after which the molecular weight was sufficiently reduced to inhibit the water retention capabilities of the hydrogel, resulting in the release of water from the 3D hydrogel structure. A hydrogel containing 1% L-glutathione displayed no effect on the degradation rate of the polymer. We hypothesize that the rigidity of the hydrogel prevented the thiol reducing agent from reducing the disulfide bond embedded within the triblock copolymer structure. The near-constant PDI and slow Mw reduction in the polymer under either condition indicates a hydrolytically degradable polymeric hydrogel with stability in the presence of thiol reducing agents (Figure 6).

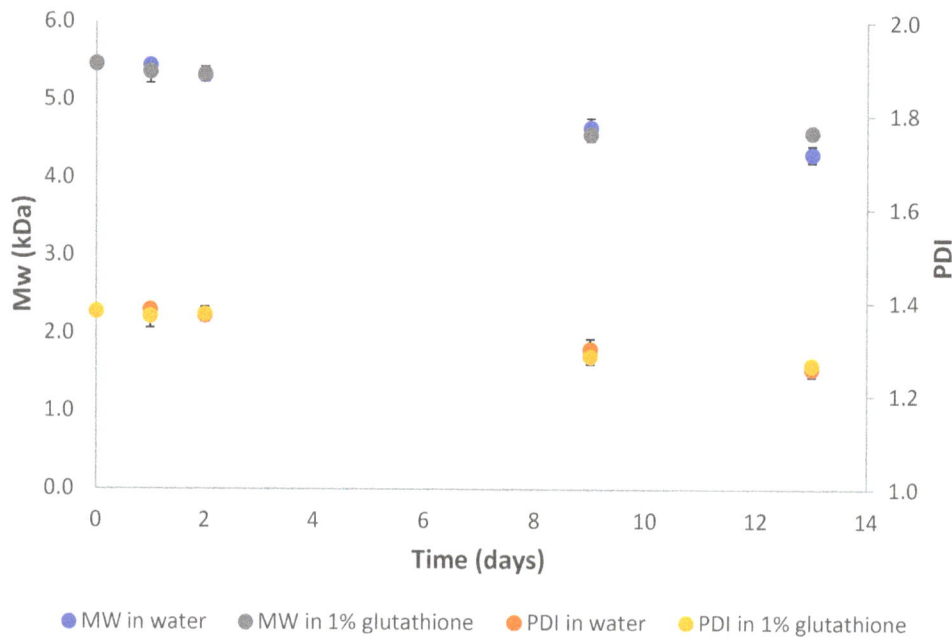

Figure 6. Degradation of PEG-PLA-SS-PLA-PEG hydrogel in distilled water or 1% glutathione solution displayed a slow hydrolytic degradation for two weeks with near-constant PDI values.

3. Discussion

Polymer systems that undergo phase transitions in response to environmental stimuli such as temperature and pH have been widely investigated for drug delivery and tissue engineering applications [29]. The aqueous copolymer solution presented in this work represents a programmable biodegradable responsive polymer gel in which its gelling properties can be additionally controlled by a cleavable S–S bond triggered to cleave and change the gel properties. The polymer is a non-crosslinked triblock copolymer of polyethylene glycol (PEG) blocks and a hydroxy acid aliphatic polyester block that possesses thermoresponsive properties where the polymer is fully water soluble at temperatures between 0 °C and about 30 °C and gels at body temperature (37 °C). In addition to hydrolytically degradable ester bonds, the polymer possesses, as mentioned above, a cleavable bond that breaks upon application of an external reductive trigger that is faster than the hydrolysis of the ester bonds of the polyester block. The cleavage of this bond results in the collapse of the gel to dissolution in water with no remaining gel properties (Figure 7). This class of polymer may have medical use as a device for drug carrying or cell support and delivery. For drug or cell delivery, the rate of de-gelation can be programmed so that after its formation at body temperature, the gel may gradually or instantly lose/loosen its gel properties to allow faster release from the gel of the entrapped cells or drugs. The change in gel properties, after initial gelation at the site of action, can be tailored according to the need for a specific application, either immediately or gradually, based on the presence of reducing agents. Of specific interest is the delivery of therapeutic stem cells carrying anticancer cargo at cancerous tissue. The cells may be delivered to the site of treatment as a solution that instantly gels on the target tissue due to temperature change and starts to lose its gel properties to allow the cells to move towards cancerous tissue due to the increased neoplastic concentration of reducing agents [18].

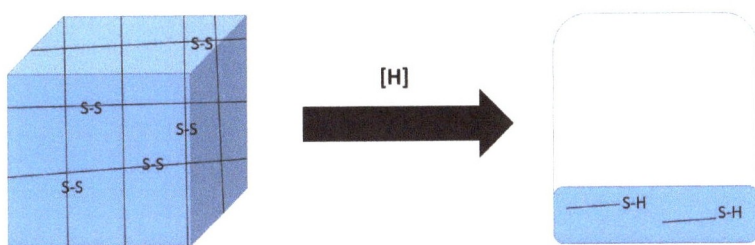

Figure 7. Three-dimensional structure of a physically crosslinked hydrogel may be instantaneously collapsed by exposure to reducing conditions.

Several important parameters are involved in the development of these systems, including robust synthetic methods, defined stimuli-responsive gelation, and degradation for application in vivo. The physical crosslinking mechanism of PEG-PLA-PEG copolymers hydrogel formation permits the additional advantage of reversibility, as no chemical bonds must be cleaved in order to re-obtain the polymer in solution [30]. However, to date, there have been no reports of non-crosslinked thermoresponsive polymer gels that possess all of the following qualities: (1) are water-soluble below physiological temperatures, (2) form a semi-solid continuous gel at physiological temperatures, and (3) possess breakable bonds that permit an immediate programmable change in gel properties following exposure to an external trigger. The PEG-PLA-PEG triblock copolymers with embedded disulfide bond described here represent such material, with gelling properties suitable for physiological application and susceptibility to water leakage upon exposure to reducing agents. The proof of concept afforded by cleavage with a hydride reagent represents an interesting option for the development of reduction-sensitive hydrogels capable of instantaneous collapse upon exposure to reducing agents.

4. Conclusions

Thermoresponsive hydrogels capable of immediate collapse upon exposure to strong reducing agents represent an interesting injectable platform for the controlled release of drugs. At ambient temperatures, the polymer dissolved in water (25%) exists as a free-flowing solution that forms a hydrogel upon heating. The disulfide bond embedded in the polymer structure induces a controlled hydrogel collapse by the addition of $NaBH_4$. The hydrogel is otherwise stable for a period of over two weeks. Hydrolytic degradation of ester bonds results over time in hydrogel collapse into safe by-products. The hydrogels described here have the potential to be applied as delivery vehicles for the slow release of therapeutics with immediate release induced upon contact with reducing agents.

5. Materials and Methods

5.1. Materials

Stannous octoate, hexamethylene diisocyanate, and poly(ethylene glycol) monomethyl ether were purchased from Sigma Aldrich (Rehovot, Israel). Lactide was purchased from Purac Biochem BV (Gorinchem, The Netherlands). Solvents were purchased from Bio-Lab Ltd. (Jerusalem, Israel).

5.2. General Methods

Chemical reactions were performed in dry glassware under N_2 gas. ^{1}H NMR spectra were obtained on a Varian 300 MHz spectrometer with $CDCl_3$ as the solvent and tetramethylsilane as the shift reference. The molecular weights of the polymers were estimated using a gel permeation chromatography (GPC) system consisting of a Waters 1515 Isocratic HPLC pump with a Waters 2410 Refractive Index Detector and a Rheodyne (Cotati, CA, USA) injection valve with a 20 mL loop (Waters, Milford, MA, USA). Samples were eluted with $CHCl_3$ through a linear Styragel HR4E column (7.8 × 300 mm i.d.; Waters) at a

5.3. Synthesis

PEG-PLA-PEG triblock copolymer with disulfide bond was prepared in two steps. First, 50 µL of a 10% solution of stannous octoate in dichloromethane (DCM) was added to a melt of 2-hydroxyethyl disulfide (0.13 g, 0.82 mmol) and D,L-lactide (2.4 g, 16 mmol). The solvent was allowed to evaporate, and the vial was purged with N_2. The mixture was stirred at 120 °C for 2 h, followed by overnight stirring at 150 °C. The crude polymer was taken up in DCM and precipitated into ether to afford PLA disulfide PLA(SS) as a pure substance. ^{1}H NMR (300 MHz, CDCl$_3$, δ): 5.24–5.05 (m, LA), 4.36–4.28 (m, CH$_2$ α-ester), 2.87–2.82 (m, CH$_2$ α-disulfide), 1.59–1.37 (m, LA). Anal. Calcd: C, 46.74; H, 5.91; S, 3.80. Found: C, 48.00; H, 5.70; S, 4.42.

In the next step, hexamethylene diisocyanate (0.11 mL, 0.68 mmol) was added to a melt of PLA(SS) (1.0 g, 0.70 mol) and poly(ethylene glycol) methyl ether Mn = 550 (0.34 mL, 0.68 mmol). One drop of stannous octoate was added, and the mixture was stirred at 110 °C for one hour. The crude product was dissolved in cold water, filtered, and lyophilized to afford PEG-PLA(SS)-PEG as a pure substance. Proton nuclear magnetic resonance (^{1}H NMR) (300 MHz, CDCl$_3$, δ): 5.15–4.94 (m, LA), 4.30–4.28 (m, CH$_2$ α-ester), 4.11–4.08 (m, PEG α-urethane), 3.55 (s, PEG), 3.28 (s, CH$_3$), 3.07–3.03 (m, CH$_2$ α-urethane), 2.83–2.79 (t, J = 6 Hz, CH$_2$ α-disulfide), 1.50–1.38 (m, LA). Anal. Calcd: C, 50.53; H, 7.12; N, 2.05; S, 1.88. Found: C, 51.09; H, 7.22; N, 1.99; S, 2.27.

5.4. Rheological Studies

Sol–gel transitions were measured on 25% w/w polymer solutions in double-distilled water (DDW) by rotational tests on a Physica MCR 101 rheometer (Anton Paar, Austria) as a function of temperature from 25 to 45 °C.

5.5. Degradation Studies

25% w/w polymer solutions were prepared in DDW or in 1% solutions of either reduced L-glutathione or NaBH$_4$. Solutions were incubated at 37 °C, and molecular weight was monitored by GPC for two weeks.

Supplementary Materials: The following are available online at https://www.mdpi.com/article/10.3390/gels7040169/s1, Experimental details: Figure S1: ^{1}H NMR spectrum of PLA analog; Figure S2: ^{1}H NMR spectrum of PEG-PLA-PEG analog.

Author Contributions: Conceptualization, N.Y.S. and A.J.D.; methodology, N.Y.S.; investigation, N.Y.S.; resources, A.J.D.; data curation, N.Y.S.; writing—original draft preparation, N.Y.S.; writing—review and editing, A.J.D.; supervision, A.J.D.; funding acquisition, A.J.D. All authors have read and agreed to the published version of the manuscript.

Funding: This research did not receive any specific grant from funding agencies in the public, commercial, or not-for-profit sectors.

Institutional Review Board Statement: Not applicable.

Informed Consent Statement: Not applicable.

Data Availability Statement: Not applicable.

Conflicts of Interest: The authors declare no conflict of interest.

References

1. Priya James, H.; John, R.; Alex, A.; Anoop, K.R. Smart polymers for the controlled delivery of drugs—A concise overview. *Acta Pharm. Sin. B* **2014**, *4*, 120–127. [CrossRef]
2. Jeong, B.; Kim, S.W.; Bae, Y.H. Thermosensitive sol–gel reversible hydrogels. *Adv. Drug Deliv. Rev.* **2012**, *64*, 154–162. [CrossRef]

3. Steinman, N.Y.; Haim-Zada, M.; Goldstein, I.A.; Goldberg, A.H.; Haber, T.; Berlin, J.M.; Domb, A.J. Effect of PLGA block molecular weight on gelling temperature of PLGA-PEG-PLGA thermoresponsive copolymers. *J. Polym. Sci. Part A Polym. Chem.* **2019**, *57*, 35–39. [CrossRef]
4. Steinman, N.Y.; Bentolila, N.Y.; Domb, A.J. Effect of Molecular Weight on Gelling and Viscoelastic Properties of Poly (caprolactone)–b-Poly (ethylene glycol)–b-Poly (caprolactone)(PCL–PEG–PCL) Hydrogels. *Polymers* **2020**, *12*, 2372. [CrossRef]
5. Maeda, T. Structures and Applications of Thermoresponsive Hydrogels and Nanocomposite-Hydrogels Based on Copolymers with Poly (Ethylene Glycol) and Poly (Lactide-Co-Glycolide) Blocks. *Bioengineering* **2019**, *6*, 107. [CrossRef]
6. Jeong, B.; Bae, Y.H.; Kim, S.W. In situ gelation of PEG-PLGA-PEG triblock copolymer aqueous solutions and degradation thereof. *J. Biomed. Mater. Res.* **2000**, *50*, 171–177. [CrossRef]
7. Shim, M.S.; Lee, H.T.; Shim, W.S.; Park, I.; Lee, H.; Chang, T.; Kim, S.W.; Lee, D.S. Poly(D,L-lactic acid-co-glycolic acid)-b-poly(ethylene glycol)-b-poly (D,L-lactic acid-co-glycolic acid) triblock copolymer and thermoreversible phase transition in water. *J. Biomed. Mater. Res.* **2002**, *61*, 188–196. [CrossRef]
8. Lee, D.S.; Shim, M.S.; Kim, S.W.; Lee, H.; Park, I.; Chang, T. Novel thermoreversible gelation of biodegradable PLGA-block-PEO-block-PLGA triblock copolymers in aqueous solution. *Macromol. Rapid Commun.* **2001**, *22*, 587–592. [CrossRef]
9. Qiao, M.; Chen, D.; Ma, X.; Liu, Y. Injectable biodegradable temperature-responsive PLGA–PEG–PLGA copolymers: Synthesis and effect of copolymer composition on the drug release from the copolymer-based hydrogels. *Int. J. Pharm.* **2005**, *294*, 103–112. [CrossRef]
10. Jeong, B.; Bae, Y.H.; Kim, S.W. Biodegradable thermosensitive micelles of PEG-PLGA-PEG triblock copolymers. *Colloids Surf. B Biointerfaces* **1999**, *16*, 185–193. [CrossRef]
11. Jeong, B.; Bae, Y.H.; Kim, S.W. Thermoreversible gelation of PEG– PLGA– PEG triblock copolymer aqueous solutions. *Macromolecules* **1999**, *32*, 7064–7069. [CrossRef]
12. Jian, X.; Hu, Y.; Zhou, W.; Xiao, L. Self-healing polyurethane based on disulfide bond and hydrogen bond. *Polym. Adv. Technol.* **2018**, *29*, 463–469. [CrossRef]
13. Naskar, K.; Dey, A.; Maity, S.; Ray, P.P.; Ghosh, P.; Sinha, C. Biporous Cd (II) Coordination Polymer via in Situ Disulfide Bond Formation: Self-Healing and Application to Photosensitive Optoelectronic Device. *Inorg. Chem.* **2020**, *59*, 5518–5528. [CrossRef] [PubMed]
14. Naoi, K.; Kawase, K.i.; Mori, M.; Komiyama, M. Electrochemistry of Poly (2, 2′-dithiodianiline): A New Class of High Energy Conducting Polymer Interconnected with S-S Bonds. *J. Electrochem. Soc.* **1997**, *144*, L173. [CrossRef]
15. Liu, L.; Liu, P. Synthesis strategies for disulfide bond-containing polymer-based drug delivery system for reduction-responsive controlled release. *Front. Mater. Sci.* **2015**, *9*, 211–226. [CrossRef]
16. Zhang, P.; Wu, J.; Xiao, F.; Zhao, D.; Luan, Y. Disulfide bond based polymeric drug carriers for cancer chemotherapy and relevant redox environments in mammals. *Med. Res. Rev.* **2018**, *38*, 1485–1510. [CrossRef]
17. Wang, Y.-C.; Wang, F.; Sun, T.-M.; Wang, J. Redox-responsive nanoparticles from the single disulfide bond-bridged block copolymer as drug carriers for overcoming multidrug resistance in cancer cells. *Bioconj. Chem.* **2011**, *22*, 1939–1945. [CrossRef]
18. Wong, D.Y.-K.; Hsiao, Y.-L.; Poon, C.-K.; Kwan, P.-C.; Chao, S.-Y.; Chou, S.-T.; Yang, C.-S. Glutathione concentration in oral cancer tissues. *Cancer Lett.* **1994**, *81*, 111–116. [CrossRef]
19. Hoon Jeong, J.; Christensen, L.V.; Yockman, J.W.; Zhong, Z.; Engbersen, J.F.J.; Jong Kim, W.; Feijen, J.; Wan Kim, S. Reducible poly(amido ethylenimine) directed to enhance RNA interference. *Biomaterials* **2007**, *28*, 1912–1917. [CrossRef]
20. Taranejoo, S.; Chandrasekaran, R.; Cheng, W.; Hourigan, K. Bioreducible PEI-functionalized glycol chitosan: A novel gene vector with reduced cytotoxicity and improved transfection efficiency. *Carbohydr. Polym.* **2016**, *153*, 160–168. [CrossRef]
21. Davoodi, P.; Srinivasan, M.P.; Wang, C.-H. Synthesis of intracellular reduction-sensitive amphiphilic polyethyleneimine and poly(ε-caprolactone) graft copolymer for on-demand release of doxorubicin and p53 plasmid DNA. *Acta Biomater.* **2016**, *39*, 79–93. [CrossRef]
22. Wang, L.-H.; Wu, D.-C.; Xu, H.-X.; You, Y.-Z. High DNA-Binding Affinity and Gene-Transfection Efficacy of Bioreducible Cationic Nanomicelles with a Fluorinated Core. *Angew. Chem. Int. Ed.* **2016**, *55*, 755–759. [CrossRef]
23. Yang, H.; Bai, Y.; Yu, B.; Wang, Z.; Zhang, X. Supramolecular polymers bearing disulfide bonds. *Polym. Chem.* **2014**, *5*, 6439–6443. [CrossRef]
24. Jeong, B.; Bae, Y.H.; Lee, D.S.; Kim, S.W. Biodegradable block copolymers as injectable drug-delivery systems. *Nature* **1997**, *388*, 860–862. [CrossRef]
25. Jeong, B.; Bae, Y.H.; Kim, S.W. Drug release from biodegradable injectable thermosensitive hydrogel of PEG–PLGA–PEG triblock copolymers. *J. Control Release* **2000**, *63*, 155–163. [CrossRef]
26. Li, X.; Wang, Y.; Chen, J.; Wang, Y.; Ma, J.; Wu, G. Controlled release of protein from biodegradable multi-sensitive injectable poly (ether-urethane) hydrogel. *ACS Appl. Mater. Interfaces* **2014**, *6*, 3640–3647. [CrossRef] [PubMed]
27. Doppalapudi, S.; Jain, A.; Khan, W.; Domb, A.J. Biodegradable polymers—An overview. *Polym. Adv. Technol.* **2014**, *25*, 427–435. [CrossRef]
28. Ramot, Y.; Haim-Zada, M.; Domb, A.J.; Nyska, A. Biocompatibility and safety of PLA and its copolymers. *Adv. Drug Deliv. Rev.* **2016**, *107*, 153–162. [CrossRef] [PubMed]

29. Hogan, K.J.; Mikos, A.G. Biodegradable thermoresponsive polymers: Applications in drug delivery and tissue engineering. *Polymer* **2020**, *211*, 123063. [CrossRef]
30. Ghahremankhani, A.A.; Dorkoosh, F.; Dinarvand, R. PLGA-PEG-PLGA Tri-Block Copolymers as In Situ Gel-Forming Peptide Delivery System: Effect of Formulation Properties on Peptide Release. *Pharm. Dev. Technol.* **2008**, *13*, 49–55. [CrossRef]

Article

Transforming Commercial Copper Sulfide into Injectable Hydrogels for Local Photothermal Therapy

Xiaoran Wang [1,†], Zizhen Yang [1,†], Zhaowei Meng [1,*] and Shao-Kai Sun [2,*]

1. Department of Nuclear Medicine, Tianjin Medical University General Hospital, Tianjin 300052, China; mwxiaoran@126.com (X.W.); yang_zizhen@163.com (Z.Y.)
2. School of Medical Imaging, Tianjin Medical University, Tianjin 300203, China
* Correspondence: zmeng@tmu.edu.cn (Z.M.); shaokaisun@tmu.edu.cn (S.-K.S.); Tel.: +86-186-2203-5159 (Z.M.); +86-22-8333-6093 (S.-K.S.)
† These authors contribute equally to the work.

Abstract: Photothermal therapy (PTT) is a promising local therapy playing an increasingly important role in tumor treatment. To maximize PTT efficacy, various near-infrared photoabsorbers have been developed. Among them, metal sulfides have attracted considerable interest due to the advantages of good stability and high photothermal conversion efficiency. However, the existing synthesis methods of metal-sulfide-based photoabsorbers suffer from the drawbacks of complicated procedures, low raw material utilization, and poor universality. Herein, we proposed a flexible, adjustable strategy capable of transforming commercial metal sulfides into injectable hydrogels for local PTT. We took copper sulfide (CuS) as a typical example, which has intense second-window near-infrared absorption (1064 nm), to systematically investigate its in vitro and in vivo characteristics. CuS hydrogel with good syringeability was synthesized by simply dispersing commercial CuS powders as photoabsorbers in alginate-Ca^{2+} hydrogel. This synthesis strategy exhibits the unique merits of an ultra-simple synthesizing process, 100% loading efficiency, good biocompatibility, low cost, outstanding photothermal capacity, and good universality. The in vitro experiments indicated that the hydrogel exhibits favorable photothermal heating ability, and it obviously destroyed tumor cells under 1064 nm laser irradiation. After intratumoral administration in vivo, large-sized CuS particles in the hydrogel highly efficiently accumulated in tumor tissues, and robust local PTT was realized under mild laser irradiation (0.3 W/cm^2). The developed strategy for the synthesis of CuS hydrogel provides a novel way to utilize commercial metal sulfides for diverse biological applications.

Keywords: commercial copper sulfide (CuS); alginate; hydrogel; photothermal therapy (PTT); near-infrared II windows

1. Introduction

Local therapy has recently attracted an increasing amount of attention in tumor treatment due to the advantages of solid selectivity, controllability, and minor systemic side effects [1,2]. Local therapy mainly includes local ablation (radiofrequency ablation [3–5], irreversible electroporation [6–8], high-intensity focused ultrasound [9], local laser ablation [10], cryoablation [11–13], and chemical ablation [14,15]), local photothermpy (PTT [16,17] and photodynamic therapy [16,18,19]), local radioisotope therapy [20,21], local radiotherapy [22–24], and local chemotherapy [25]. An increasing number of studies have shown that local therapy can be applied to many clinical situations, such as in the preservation of tissue and function [26], situations without medical indication for surgery [27], the local treatment of metastatic tumors [28,29], and the salvage treatment of recurrent tumors [30]. Thus, local therapy is sometimes essential to improve the quality of life of patients with tumors, as well as survival time. At the same time, some studies have reported the success of the combination of local therapy and other treatment methods, such as immunotherapy [31,32], which further illustrates the broad prospect of local therapy.

As a kind of burgeoning local therapy, PTT uses photoabsorbers to transform near-infrared (NIR) light energy to heat energy, which induces tumor cell necrosis [16,33]. NIR-based PTT is mainly conducted in two biological windows: the first NIR window (NIR-I) and the second NIR window (NIR-II). The wavelength range of NIR-I is from 750 nm to 1000 nm, and the range of NIR-II is from 1000 nm to 1350 nm [34]. Compared with the widely studied NIR-I light, NIR-II light has a stronger penetrating ability and a higher thermal safe power because of its low absorption and scattering in tissue, which has a substantial superiority in curing deeper tumor tissues [34–37]. In addition to the flourishing development of PTT in fundamental studies [38,39], it is very encouraging that PTT based on local administration has also successfully entered a clinical trial for the treatment of prostate cancer [1], which demonstrates the great potential of PTT-based local therapy in clinical transformation.

To date, plenty of biomaterials have been developed as photoabsorbers, such as organic dyes [40–42], organic nanoparticles [43–45], noble metal materials [46–48], carbon materials [49,50], black phosphorus [51,52], and metal oxide and sulfides [53,54]. Among them, metal sulfides, such as CuS, Bi_2S_3, WS_2, CoS, NiS, and FeS, have high photothermal conversion efficiency because of the surface plasmon resonance effect [55,56], and they have been widely used in PTT in recent years [57–62]. In particular, CuS, which possesses strong absorption in the NIR-II bio-window, has been extensively used in NIR-II PTT [63,64]. The current photoabsorbers are mainly obtained using either bottom-up methods or top-down methods, such as coupling thermal oxidation etching and liquid exfoliation to form a solvent-dispersible system [65,66]. However, these methods face several common problems, such as complex steps, high time and energy costs, low raw material utilization, and lacking universal strategy [67].

To avoid the use of metal sulfides using complex synthesis methods, commercial metal sulfides are an excellent choice. The advantage of commercial metal sulfides is that they are mature industrial products with reasonable quality control and low cost, but their disadvantages lie in the raw materials having large particles, being insoluble in water, and not being able to be used for biological applications. Recently, our group proposed a smart "turning solid into gel" strategy [68] by dispersing solid materials in alginate–Ca^{2+} hydrogel (ACH), which can transform solid materials into an injectable hydrogel, making the solid materials bioavailable. Therefore, it is fascinating to develop versatile commercial metal-sulfide-based hydrogels as novel photoabsorbers without complex synthesis.

Herein, we introduced a simple and powerful ACH platform to load commercial CuS as a representative sample for local tumor NIR-II PTT (Figure 1). The ultra-simple synthesis, 100% loading efficiency, good biocompatibility, low cost, outstanding photothermal capacity, and extreme flexibility allow this platform to provide more options for highly efficient PTT. The CuS hydrogel (CSH) can be simply obtained through mixing and stirring steps. In vitro experiments indicated that CSH exhibits good syringeability and intense NIR-II absorption (1064 nm). Then, CSH was employed for in vivo PTT studies. The results confirm that this hydrogel not only performs well in killing tumor cells under mild laser irradiation but that it also shows low toxicity in vitro and in vivo. To the best of our knowledge, this is the first time that commercial CuS was elegantly employed for highly efficient PTT in vivo.

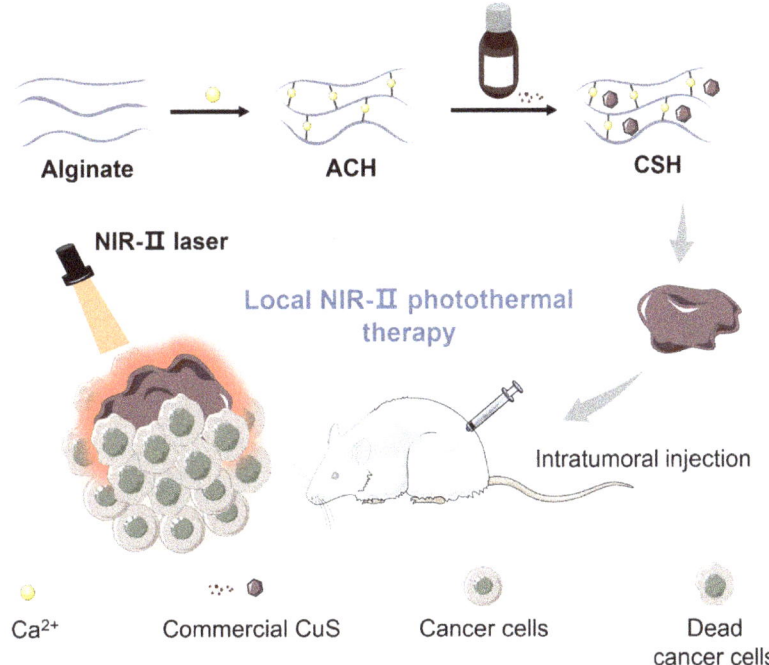

Figure 1. Schematic representation of the synthesis of CSH as a PTT agent for local NIR-II PTT in vivo.

2. Results and Discussion

2.1. Synthesis and Characterization of CSH

Firstly, alginate solution and Ca^{2+} were mixed to produce ACH within 1 min based on their strong coordination interaction (Figure S1). Then, CSH was obtained by dispersing commercial CuS powder into ACH. To investigate the loading capacity of ACH, increasing concentrations of CSH were employed. The maximum loading capacity was 480 mg CuS/mL (Figure 2A). The long-term stability of CSH was also monitored (Table S1 and Figure S2). All concentrations of CSH were stable for more than 4 days, and concentrations of 90 mg CuS/mL and below were still stable after 14 days. Considering that excellent syringeability is essential to potential biological applications, we investigated the maximum loading capacity of CSH capable of fluently being injected with different diameters of syringe needles. The results showed that the maximum injectable concentrations for 0.45, 0.5, 0.6, and 1.2 mm syringe needles were 120, 240, 480, and 480 mg CuS/mL, respectively, and a "TMU" pattern could be written by a 0.45 mm syringe with 20 mg CuS/mL CSH (Figure 2B), which proved its excellent syringeability due to the shear-dependent and reversible gel–sol transition (Figure S3) [69].

Rheological experiments showed that the storage moduli (G') of ACH and CSH were higher than their loss moduli (G"), demonstrating that ACH and CSH were in a gel state with a relatively weak mechanical strength and flexible shape, which made them easily injectable (Figure S4). As the essential components of CSH, the swelling ratio and degradation behavior of ACH were further investigated. The swelling test showed that ACH could reach swelling equilibrium in 10 min in PBS (pH = 7.4). The swelling ratio of ACH was as high as 13,342.6% (Figure S5), which suggested that the internal cross-linking points of ACH were relatively few, the cross-linking density was low, and the water absorption capacity was strong. According to the ACH degradation curve (Figure S6), the degradation rate of ACH in PBS (pH = 7.4) was 51% after 7 days, which showed its excellent degradability.

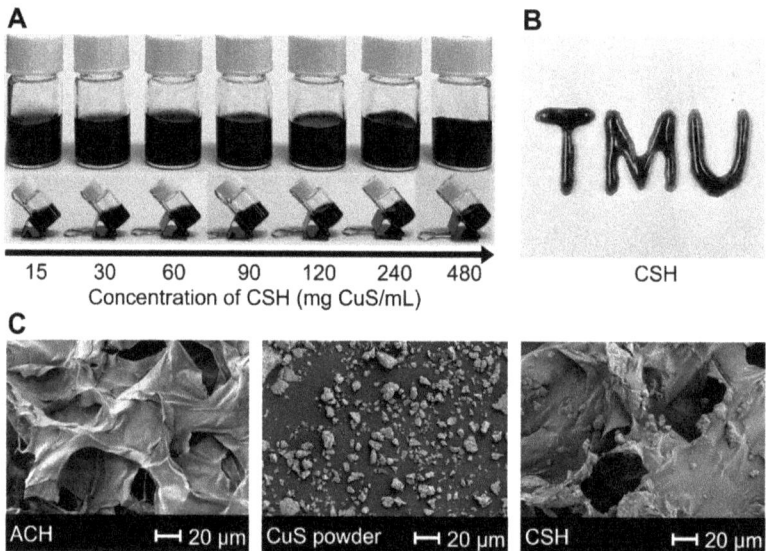

Figure 2. (**A**) Standing and oblique photos of different concentrations of CSH taken immediately after the preparation. (**B**) "TMU" formed by CSH (20 mg CuS/mL) through a 0.45 mm syringe needle. (**C**) SEM images of ACH, CuS particles, and CSH (20 mg CuS/mL).

The scanning electron microscope (SEM) images of ACH, CSH, and commercial CuS powder were characterized, and they indicated that the CuS particles were dispersed in the ACH with a porous structure (Figure 2C).

2.2. Photothermal Performance of CSH In Vitro

To evaluate the photothermal efficiency of CSH in vitro, different concentrations of CSH were treated with NIR-II laser irradiation (1064 nm, 1 W/cm^2) for 5 min, and an infrared thermal camera was used to record the temperature elevations. Under NIR-II laser irradiation, CSH showed good photothermal capacity (Figure 3A). The temperature enhancement of CSH with different concentrations increased from 17.3 °C to 38.1 °C, while the temperature increase of ACH and PBS was just 6.4 °C. The thermal images also demonstrate the outstanding photothermal ability of CSH (Figure 3B). After undergoing the heating–cooling process three times, the heating capacity of CSH did not significantly change (Figure 3C), which indicates that CSH has good photothermal stability under NIR-II laser irradiation. Therefore, not only can the prepared CSH efficiently transform NIR laser energy to heat energy, but it can also remain stable after repeated laser illumination.

2.3. Cytotoxicity and Cellular Uptake of CSH

CSH, which had a great photothermal efficacy under 1064 nm irradiation, super-large loading capacity, and excellent stability, was capable of being used for further studies. To evaluate its cytotoxicity, different concentrations of CSH were added to 4T1 cells in 96-well plates, and the cells were continued to be cultured for 24 h. Then, the cell viabilities were calculated through a standard MTT assay [3-(4,5-Dimethylthiazol-2-yl)-2,5-diphenyltetrazolium bromide, MTT]. The cell viability was as high as 88.8% after incubation with a high concentration of CSH (1 mg CuS/mL), which indicated the low cytotoxicity of CSH (Figure 4A). The cellular uptake experiment proved that CuS particles in CSH could not be uptaken by cells due to their big size (Figure S7), which illustrates that the mechanism of PTT based on CHS is deduced heat conduction instead of the direct interaction of cell and CSH.

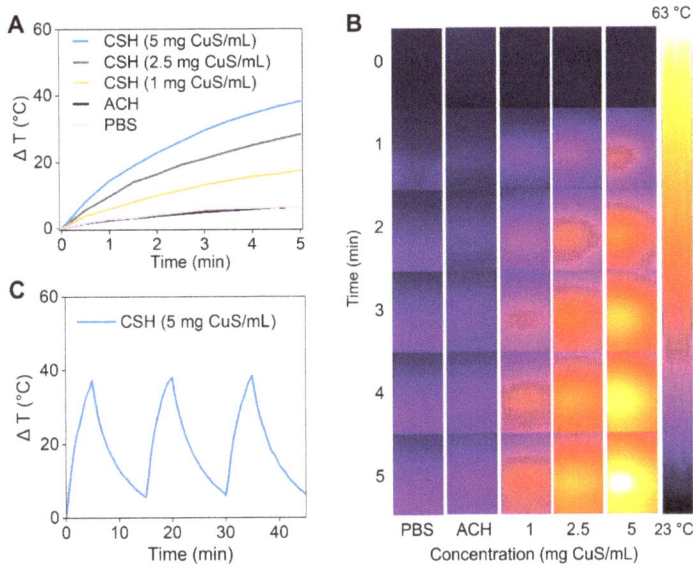

Figure 3. (A) Photothermal heating curves of PBS, ACH, and CSH irradiated by NIR-II laser (1064 nm, 1 W/cm^2). (B) The thermal images of PBS, ACH, and CSH under NIR-II laser irradiation taken by an infrared thermal camera. (C) The photothermal stability of CSH.

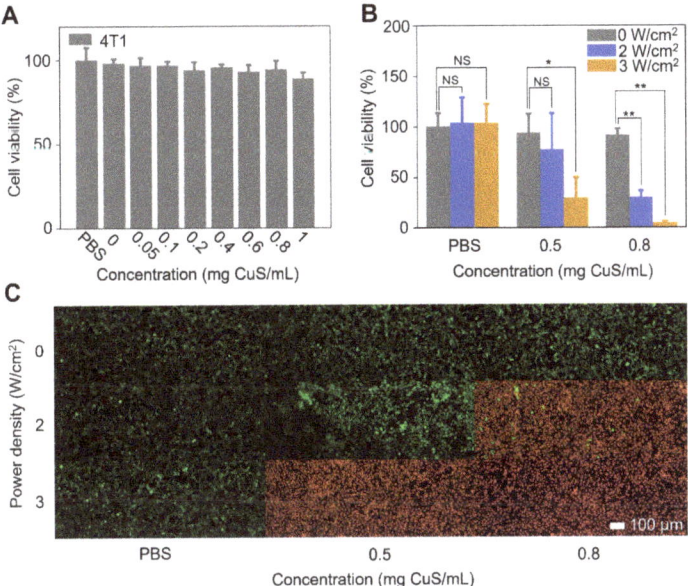

Figure 4. (A) Cytotoxicity of 4T1 cells incubated with different concentrations of CSH or PBS. (B) Viability of 4T1 cells incubated with PBS or CSH (0.5, 0.8 mg CuS/mL) and irradiated by NIR-II laser (1064 nm: 0, 2, or 3 W/cm^2) (* $p < 0.05$, ** $p < 0.01$). (C) Dead/live cell staining test of 4T1 cells treated with PBS or CSH (0.5, 0.8 mg CuS/mL) and irradiated by NIR-II laser (1064 nm: 0, 2, or 3 W/cm^2).

2.4. In Vitro PTT of CSH

Due to the low cytotoxicity of CSH, PTT of 4T1 cells using the hydrogel was investigated with an MTT assay and live and dead cell staining. As the standard procedure, 4T1 cells were cultured in a 96-well plate at 37 °C for 24 h, and different concentrations of CSH (0.5 and 0.8 mg CuS/mL) or PBS were added and incubated with the cells at 37 °C. After 1 h, the cells were irradiated by a 1064 nm laser (0, 2, or 3 W/cm^2) for 5 min. The 4T1 cell viabilities showed CSH-concentration- and laser-power density-dependent deceases. After being treated with both 0.8 mg CuS/mL of CSH and 1064 nm laser irradiation (3 W/cm^2), 4T1 cell viability dropped to less than 4%. However, the viability of 4T1 cells treated with only CSH or laser irradiation remained approximately 100% (Figure 4B). The fluorescent images of live and dead cells, which were stained by calcein acetoxymethyl ester (calcein AM) and propidium iodide (PI), respectively, also showed that the 4T1 cells were significantly destructed after the combined treatments (Figure 4C). These results prove that CSH has an excellent photothermal effect on tumor cells with 1064 nm laser irradiation.

2.5. Intratumoral Retention Test of CSH

In order to assess the intratumoral retention ability of CSH, computer tomography (CT) scans were carried out in vitro and in vivo under a voltage of 120 kV (clinical use). Although CSH was considered to have a weak CT value attenuation (Figure 5A,B), after intratumoral injection, it could still be found at the tumor site with CT scans. During the 2 days of CT monitoring that followed, no significant change was found in the CT value or in the morphology of CSH (Figure 5C,D), proving its good retention ability at tumor sites.

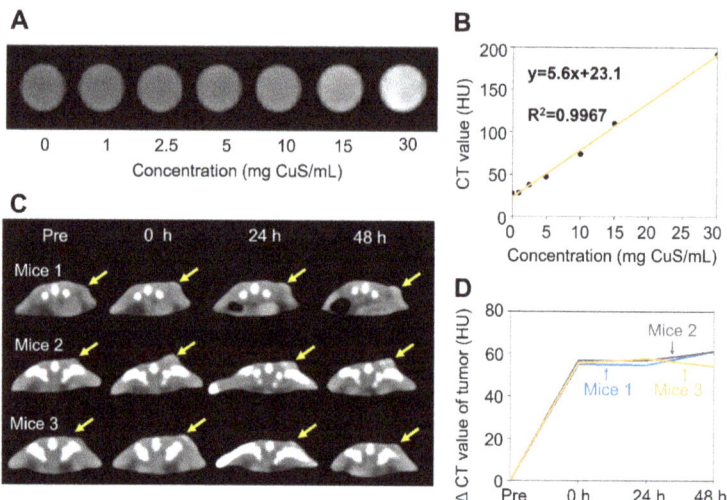

Figure 5. (**A**) CT images of CSH at different concentrations. (**B**) CT value curve of CSH. (**C**) CT scan of BALB/c mice before and 0, 24, and 48 h after being intratumorally injected with 20 mg CuS/mL of CSH (*n* = 3). (**D**) CT value (Hounsfield, HU) changing curves on the tumor site of BALB/c mice.

2.6. In Vivo PTT of CSH

To minimize the damage to surrounding tissues, a mild laser power (0.3 W/cm^2) was employed for in vivo PTT. CSH with a concentration of 20 mg CuS/mL, which can cause a significant temperature rise in vitro, was used to guarantee effective tumor ablation (Figure S8). To evaluate the in vivo photothermal tumor therapy efficacy of CSH in the NIR-II bio-window (1064 nm), BALB/c mice were grouped according to different treatments (*n* = 5) as follows: (1) only PBS; (2) only CSH; (3) PBS + laser; and (4) CSH + laser. In comparison with the control, there was a noticeable temperature increase at the tumor site after being injected with CSH and irradiated with a 1064 nm laser (Figure 6A,B).

Tumor sizes were measured every 2 days to evaluate the anti-tumor capacity. The results showed that the tumor growth of mice treated with both CSH and laser irradiation was effectively inhibited, and the tumors were eliminated after PTT (Figure 6C). There was tumor recurrence in only one mouse in the combined treatment group, and the recurred tumor was significantly smaller than that in the other groups. In contrast, the tumors in the other groups multiplied, and the final tumor volumes after 15 days of growth were about 12.9, 14.1, and 12.5 times larger than the initial tumor volume in groups 1, 2, and 3, respectively (Figure 6D). The tumors were dissected and photographed on the 15th day (Figure 6E). The dissected tumors were weighed, and the ratio of tumor weight to mouse body weight in each group was calculated (Figure 6F), which further revealed that the tumors were obviously suppressed by CSH-based PTT. These results illustrate that CSH can wreck tumors entirely due to its high thermal efficiency in vivo.

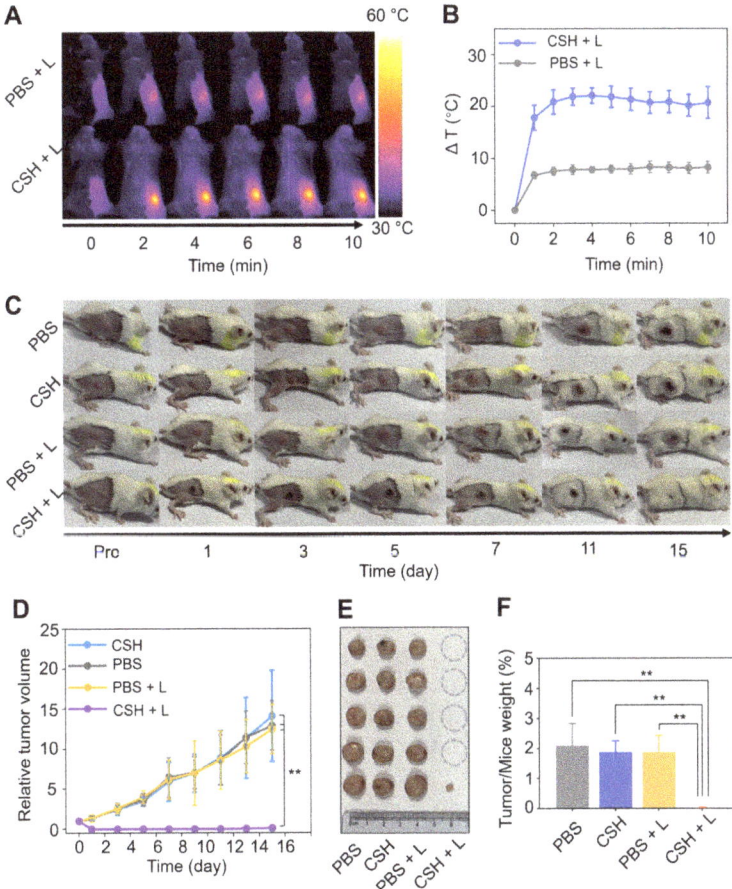

Figure 6. (**A**) Thermal images of tumor-bearing mice treated with only PBS or both CSH and NIR-II laser irradiation. (**B**) Photothermal heating curves of tumor sites taken with an infrared thermal camera. (**C**) Tumor-monitoring photography of mice in various groups. (**D**) Relative tumor volume curves of mice in different groups (n = 5 in each group, ** $p < 0.01$). (**E**) Excised tumors from the mice on the 15th day of the observation period. (**F**) Ratio of final tumor weight to final body weight of mice in different groups (** $p < 0.01$).

2.7. In Vivo Toxicity of CSH

To assess the systemic toxicity of CSH in vivo, the weight monitoring, blood biochemistry analysis, and H&E staining of major organs of the mice were accomplished. The weight monitoring results displayed no noticeable difference in body weight among the mice with various treatments (Figure 7A). The blood biochemistry analysis indicated that the liver and kidney function indexes of the mice were entirely within the normal range (Figure 7B), and no evident inflammatory lesion or organ damage was found in all major organs of the mice (Figure 7C). All of the above results confirm that CSH has good biocompatibility, which makes it a promising PTT agent with good biosafety and photothermal efficacy in vivo.

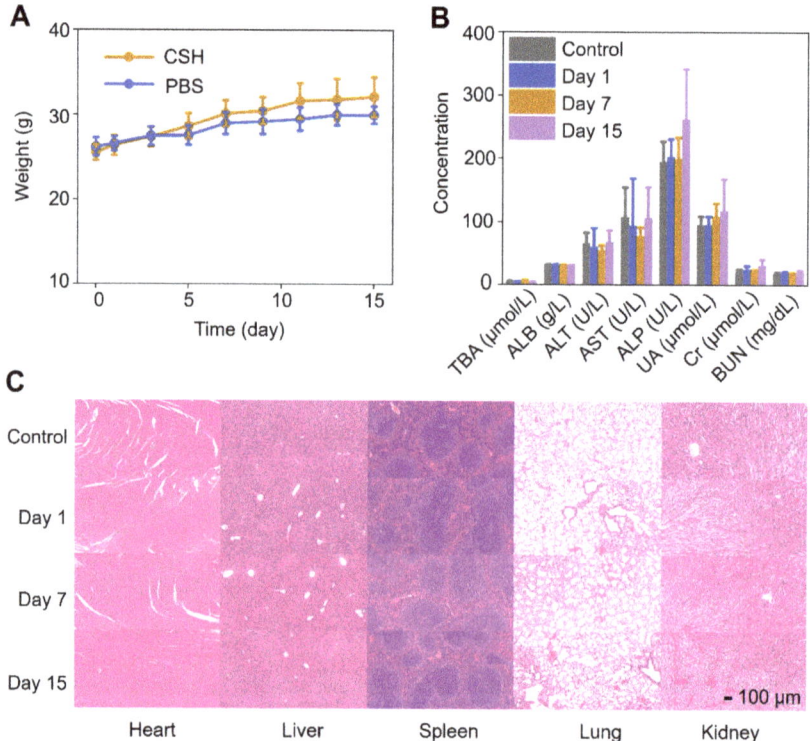

Figure 7. (**A**) Weight-changing curves of mice after being subcutaneously injected with PBS and 20 mg CuS/mL of CSH (n = 5 in each group). (**B**) Blood biochemical indexes of mice measured after being treated with PBS for 15 days and CSH (20 mg CuS/mL) for 1, 7, and 15 days (n = 5 in each group). (**C**) H&E-stained images of major organs of Kunming mice acquired after being treated with PBS for 15 days and CSH (20 mg CuS/mL) for 1, 7, and 15 days.

3. Conclusions

In conclusion, according to the "turning solid into gel" strategy, a robust metal sulfide hydrogel system was established to load commercial metal sulfide powders for high-efficiency tumor PTT. As a representative metal sulfide, commercial CuS powder was studied in depth. The obtained CSH was verified to have good stability, favorable syringeability, potent photothermal efficacy, and excellent retention capability at the injection site. Due to the deeper tissue penetration of NIR-II light, further studies were investigated using 1064 nm laser irradiation. The follow-up experimentations in vitro and in vivo showed the CSH to have negligible toxicity and a high photothermal killing effect on tumor cells under the irradiation of the 1064 nm laser. Therefore, as a new method of photothermal agent preparation, transforming commercial sulfides into injectable hydrogels can help to save costs, improve accuracy, and raise efficiency without worrying about toxicity, all of which give it great hope for clinical transformation.

4. Materials and Methods

4.1. Materials

$CaCl_2$ and sodium alginate (200 ± 20 mPa s) were obtained from Aladdin Biochemical Technology Co., Ltd. (Shanghai, China). CuS was purchased from Sigma-Aldrich trade Co., Ltd. (Shanghai, China). Fetal Bovine Serum (FBS) was provided by Lanzhou Minhai Bio-Engineering Co., Ltd. Dulbecco's Modified Eagle Medium (DMEM) was obtained from ThermoFisher Instrument Co., Ltd. (Suzhou, China). MTT was bought from Aladdin Biochemical Technology Co., Ltd. (Shanghai, China). Calcein AM and PI were provided by Dojindo Chemical L.L.C. (Shanghai, China). DMSO was purchased from Concord Technology Co., Ltd. (Tianjin, China). Ultrapure water was bought from Wahaha Group Co., Ltd. (Hangzhou, China).

4.2. Synthesis of ACH and CSH

Typically, 0.5 mL of sodium alginate (10 mg/mL) and 0.05 mL of $CaCl_2$ (10 mg/mL) were mixed with 0.45 mL H_2O to prepare ACH. Then, commercial CuS powder was added to ACH, and the system was stirred for 15 min to obtain CSH.

4.3. Stability Assessment of CSH

The homogeneous stability of CSH was evaluated for 2 weeks. Briefly, different concentrations of CSH (15, 30, 60, 90, 120, 240, 480 mg CuS/mL) were placed in vials, respectively. If the stability time was less than 10 min or the sulfide could not be dispersed in ACH, the mixture was regarded as overloaded. The stable concentrations of CSH were monitored for 14 days and photographed at different time points (10 min, 20 min, 30 min, 1 h, 2 h, 3 h, 4 h, 6 h, 8 h, 10 h, 12 h, 14 h, 16 h, 18 h, 20 h, 24 h, and then every day). During the monitoring period, if the hydrogel was found to be layered, it was regarded as precipitation.

4.4. Syringeability of CSH

The syringeability of CSH was evaluated using four sizes of syringe needles (26 G, 0.45 mm; 25 G, 0.5 mm; 23 G, 0.6 mm; 18 G, 1.2 mm). Different concentrations of CSH were extruded through the various sizes of syringe needles. For every size of syringe, the maximum injectable concentration was recorded, and a "TMU" (an abbreviation of "Tianjin Medical University") was written by its maximum injectable concentration. Then, a "TMU" was formed by CSH (20 mg CuS/mL) through a 0.45 mm syringe needle, which was used for in vivo PTT.

4.5. Characterization

Rheology experiments of ACH and CSH (20 mg CuS/mL) were conducted on a DHR-2 rheometer (TA Instruments), with a strain amplitude of 1% and an angular frequency of 10 rad/s for dynamic oscillatory time sweep measurements.

The swelling ratio and degradation behavior of ACH were investigated according to a previous study [70]. To calculate the swelling ratio, lyophilized ACH was weighed (recorded as M_0), immersed in PBS (pH = 7.4), and incubated in an incubator shaker at a shaking speed of 100 rpm at 37 °C. The swelled ACH was removed, and the surface water was wiped out. Then, the collected ACH was weighed at a specific time interval (recorded as Mt). The swelling ratio (%, w/w) was calculated using the equation $(M_t - M_0)/M_0 \times 100$. The experiments were carried out in triplicate to obtain an average value. The degradation behavior of ACH was assessed in PBS (pH = 7.4). The lyophilized ACH was weighed (recorded as Mi) and completely immersed in PBS, and then it was degraded in an incubator shaker at 37 °C and 100 rpm. After different time intervals (1, 3, or 7 days), ACH was washed with ultrapure water to remove PBS, freeze-dried, and weighed (recorded as M_f). The degradation rate (%, w/w) was calculated using the equation $(M_i - M_f)/M_i \times 100$.

Field-emission scanning electron microscopy (FE-SEM) images of ACH, CuS powder, and CSH (20 mg CuS/mL) were acquired under a 2 kV accelerating voltage on a Gemini SEM 300 (ZEISS, Germany) microscope.

4.6. Photothermal Performance In Vitro

In order to evaluate the photothermal efficacy of CSH in vitro, PBS or different concentrations of CSH (0, 1, 2.5, and 5 mg CuS/mL) with a volume of 1 cm^3 were placed in cuvettes with a base area of 1 cm^2. Then, cuvettes were irradiated with a 1064 nm laser (1 W/cm^2) for 5 min, and temperature elevations were recorded using an infrared thermal camera. In order to test its photothermal stability, CSH (5 mg CuS/mL, 1 mL) was put into a cuvette and irradiated using NIR-II (1064 nm) laser with a power density of 1 W/cm^2 for 5 min, and then the system was cooled for 10 min to bring the temperature close to room temperature; the process was repeated three times.

4.7. Cell Culture and Animals

The growth and metastasis of 4T1 cells in BALB/c mice are similar to those of human breast cancer, making the cells a relatively classical and widely used cell line to test the therapeutic effects on tumors [71]. Therefore, the 4T1 cell line was used to study CSH in vitro and in vivo. 4T1 cells were cultured in a culture medium with 90% DMEM and 10% FBS. Cells were cultured in a humidified incubator (5% CO_2 and 37 °C), and the culture medium was refreshed at 1–2 day intervals. Kunming mice and BALB/c mice were purchased from Beijing HFK Bioscience Co., Ltd. (Beijing, China). All animal experiments were performed according to the protocols established by the Animal Care and Use Committee of Tianjin Medical University, and all experimental operations were approved by the Animal Care and Use Committee.

4.8. Cytotoxicity and Cellular Uptake Assay

To determine the potential cytotoxic effects of CSH, 4T1 cells (1×10^4 per well) were cultured in 96-well plates with 200 µL of cell culture medium per well for 24 h. Then, after the exchange of the cell medium, PBS or different concentrations of CSH (0, 0.05, 0.1, 0.2, 0.4, 0.6, 0.8, 1 mg CuS/mL) were added to the wells. After 24 h incubation, cell viabilities were evaluated via a standard MTT test. After the cells were washed with PBS, new cell medium and MTT (10 µL, 5 mg/mL) were added and incubated with the cells for 4 h; then, the supernatant was discarded, and 120 µL of DMSO per well was added. Finally, the wells' absorptions at 490 nm were measured using a microplate reader.

The cellular uptake mechanism of CSH was also investigated. In brief, 4T1 cells (1×10^4 per well) were cultured in 96-well plates. After 24 h, PBS or different concentrations of CSH (0, 0.05, 0.1, 0.2, 0.4, 0.6, 0.8, 1 mg CuS/mL) were added and co-incubated with the cells for another 24 h. Then, the cells were washed with PBS, and 120 µL of PBS per well was added. Finally, the cells were observed under a microscope.

4.9. In Vitro Photothermal Cytotoxicity Study

The photothermal cell killing ability of CSH under 1064 nm laser irradiation was evaluated using the MTT assay. 4T1 cells (1.4×10^4 per well) were incubated in a 96-well plate for 24 h. After being washed with PBS, the 4T1 cells were treated with PBS or different concentrations of CSH (0.5 or 0.8 mg CuS/mL) for 1 h, and they were irradiated with varying densities of power of 1064 nm laser (0, 2, or 3 W/cm^2) for 5 min. Then, cell viabilities were measured using the MTT assay, and the absorption of each well at 490 nm was recorded using a microplate reader.

4.10. Live/Dead Cells Staining Test

To further investigate the PTT efficacy in vitro, 4T1 cells (1.4×10^4 per well) were incubated in 96-well plates for 24 h, and CSH (0.5, 0.8 mg CuS/mL) or PBS was added and co-incubated with 4T1 cells for 1 h. Then, 4T1 cells were exposed to NIR-II laser (1064 nm, 0, 2, or 3 W/cm^2) for 5 min, thoroughly washed with PBS twice, and stained with calcein AM and PI. Fluorescent images were recorded with an inverted luminescence microscope.

4.11. Intratumoral Retention Test of CSH

In vitro and in vivo CT scans were carried out via a clinical X-ray CT (SOMATOM Force, Siemens healthineers, Erlangen, Germany) under a clinical voltage (120 kV) [72]. CSHs with different concentrations (0, 1, 2.5, 5, 10, 15, 30 mg CuS/mL) were prepared, and then CT images of CSH were collected. For in vivo CT imaging, 50 µL of CSH (20 mg CuS/mL) was intratumorally injected into BALB/c mice ($n = 3$). Then, the mice were scanned pre-injection and after injection at different time points (0 h, 24 h, and 48 h). CT values were measured using Radiant DICOM Viewer software.

4.12. Anti-Tumor Assessment In Vivo

To ensure biosafety, a mild laser power (0.3 W/cm^2) and CSH with a concentration of 20 mg CuS/mL were used for in vivo PTT. To verify the in vitro heating effect, CSH (20 mg CuS/mL) or PBS was made into 50 µL droplets, and they were irradiated using a 1064 nm laser (0.3 W/cm^2) for 5 min. Thermal images of them were taken, and photothermal heating curves were obtained. Then, to explore the anti-tumor ability of CSH with 1064 nm laser irradiation, tumor-bearing BALB/c mice were divided into 4 groups ($n = 5$) as follows: (1) only PBS, (2) only CSH, (3) PBS + laser, and (4) CSH + laser. Mice in Group 1 were intratumorally injected with PBS (50 µL). Mice in Group 2 were intratumorally injected with CSH (20 mg CuS/mL, 50 µL). Mice in Group 3 were intratumorally injected with PBS and exposed to 1064 nm laser irradiation (0.3 W/cm^2) for 10 min. Mice in Group 4 were intratumorally injected with CSH (20 mg CuS/mL, 50 µL) and exposed to 1064 nm laser irradiation (0.3 W/cm^2) for 10 min. The hyperthermia effect on tumor site was carefully recorded using an infrared thermal camera. Then, tumor sizes were measured and recorded every 2 days. Tumor volume was calculated using the following formula: $V = a \times b^2/2$, where a and b mean the longest and shortest diameters, respectively. The relative volume of the tumors was the ratio of the day's volume to the initial volume. Photos of tumors in all groups were taken every 2 days, and the tumors were removed and weighed on day 15 after the treatment.

4.13. Statistics

The differences between groups were studied using one-way ANOVA, and "p" value < 0.05 was considered as statistically significant. All analyses were conducted using GraphPad Prism 8.0.2 software.

4.14. In Vivo Biosafety Analysis

To evaluate the biosafety of CSH in vivo, weight monitoring, blood biochemistry analysis, and H&E staining were conducted on Kunming mice. To monitor the body weight change, CSH (50 µL, 20 mg CuS/mL) or PBS was subcutaneously injected into Kunming mice ($n = 5$, respectively), and their body weights were recorded every two days until the 15th day. For blood biochemistry analysis and H&E staining, Kunming mice were subcutaneously injected with PBS ($n = 5$) or CSH (50 µL, 20 mg CuS/mL) ($n = 15$). Mice in the hydrogel-injected group were dissected on the 1st, 7th, and 15th days ($n = 5$ every time), and mice in the PBS group were dissected on the 15th day. After the mice were dissected, their major organs (i.e., heart, lung, spleen, liver, and kidney) were removed and stained with hematoxylin and eosin, and blood samples were collected. The blood samples were centrifuged at 3000 rpm for 10 min to separate and collect the supernatant serum. Then, the blood biochemistry biomarkers were analyzed, which included albumin (ALB), total bile acid (TBA), aspartate aminotransferase (AST), alanine aminotransferase (ALT), and alkaline phosphatase (ALP) for liver function assessment, and uric acid (UA), urea nitrogen (BUN), and serum creatinine (Cr) for kidney function evaluation.

Supplementary Materials: The following supporting information can be downloaded at: https://www.mdpi.com/article/10.3390/gels8050319/s1, Figure S1: Photos of ACH stirred for different times after mixing alginate solution and Ca^{2+} solution; Figure S2: Standing and oblique photos of different concentrations of CSH taken on day 7 and day 14; Figure S3: The maximum injectable concentration of CSH through various size of syringes and "TMU" written by them; Figure S4: Rheological properties of ACH and CSH (20 mg CuS/mL); Figure S5: Curve of swelling ratio of ACH in PBS (pH = 7.4); Figure S6: Degradation curve of ACH in PBS (pH = 7.4); Figure S7: Photos of 4T1 cells incubated with different concentrations of CSH or PBS for 24 h; Figure S8: Thermal images and photothermal heating curves of CSH (20 mg CuS/mL) and PBS under 1064 nm laser irradiation in vitro; Table S1: The stability of different concentrations of CSH.

Author Contributions: Conceptualization, S.-K.S. and Z.M.; methodology and validation, X.W. and Z.Y.; writing—original draft preparation, X.W. and Z.Y.; writing—review and editing, S.-K.S. and Z.M. All authors have read and agreed to the published version of the manuscript.

Funding: This work was supported by the following foundations: the National Natural Science Foundation of China (21874101, 21934002, and 82071982 to S.-K.S.; 81571709 and 81971650 to Z.M.); the Natural Science Foundation of Tianjin City (19JCJQJC63700 to S.-K.S.); Young Elite Scientists Sponsorship Program by Tianjin (TJSQNTJ-2018-08 to S.-K.S.); Key Project of Tianjin Science and Technology Committee Foundation grant (16JCZDJC34300 to Z.M.), Tianjin Medical University General Hospital New Century Excellent Talent Program (to Z.M.); Young and Middle-aged Innovative Talent Training Program from the Tianjin Education Committee (to Z.M.); Talent Fostering Program (the 131 Project) from the Tianjin Education Committee; and the Tianjin Human Resources and Social Security Bureau (to Z.M.).

Institutional Review Board Statement: The animal study protocol was approved by the Ethics Committee of Tianjin Medical University (SYXK-2019-0004).

Informed Consent Statement: Not applicable.

Data Availability Statement: Not applicable.

Conflicts of Interest: The authors declare no conflict of interest.

References

1. Rastinehad, A.R.; Anastos, H.; Wajswol, E.; Winoker, J.S.; Sfakianos, J.P.; Doppalapudi, S.K.; Carrick, M.R.; Knauer, C.J.; Taouli, B.; Lewis, S.C.; et al. Gold nanoshell-localized photothermal ablation of prostate tumors in a clinical pilot device study. *Proc. Natl. Acad. Sci. USA* **2019**, *116*, 18590–18596. [CrossRef]
2. Cai, X.; Liu, B. Aggregation-Induced Emission: Recent Advances in Materials and Biomedical Applications. *Angew. Chem. Int. Ed.* **2020**, *59*, 9868–9886. [CrossRef]
3. Kim, T.H.; Koh, Y.H.; Kim, B.H.; Kim, M.J.; Lee, J.H.; Park, B.; Park, J.-W. Proton beam radiotherapy vs. radiofrequency ablation for recurrent hepatocellular carcinoma: A randomized phase III trial. *J. Hepatol.* **2021**, *74*, 603–612. [CrossRef]
4. Lee, S.; Kang, T.W.; Song, K.D.; Lee, M.W.; Rhim, H.; Lim, H.K.; Kim, S.Y.; Sinn, D.H.; Kim, J.M.; Kim, K.; et al. Effect of Microvascular Invasion Risk on Early Recurrence of Hepatocellular Carcinoma After Surgery and Radiofrequency Ablation. *Ann. Surg.* **2021**, *273*, 564–571. [CrossRef]
5. Zhou, W.; Gao, Y.; Tong, Y.; Wu, Q.; Zhou, Y.; Li, Y. Anlotinib enhances the antitumor activity of radiofrequency ablation on lung squamous cell carcinoma. *Pharmacol. Res.* **2021**, *164*, 105392. [CrossRef]
6. Ting, F.; Tran, M.; Böhm, M.; Siriwardana, A.; Van Leeuwen, P.J.; Haynes, A.M.; Delprado, W.; Shnier, R.; Stricker, P.D. Focal irreversible electroporation for prostate cancer: Functional outcomes and short-term oncological control. *Prostate Cancer Prostatic Dis.* **2016**, *19*, 46–52. [CrossRef]
7. Martin, R.C.G., II; Kwon, D.; Chalikonda, S.; Sellers, M.; Kotz, E.; Scoggins, C.; McMasters, K.M.; Watkins, K. Treatment of 200 Locally Advanced (Stage III) Pancreatic Adenocarcinoma Patients With Irreversible Electroporation Safety and Efficacy. *Ann. Surg.* **2015**, *262*, 486–494. [CrossRef]
8. Zhao, J.; Wen, X.; Tian, L.; Li, T.; Xu, C.; Wen, X.; Melancon, M.P.; Gupta, S.; Shen, B.; Peng, W.; et al. Irreversible electroporation reverses resistance to immune checkpoint blockade in pancreatic cancer. *Nat. Commun.* **2019**, *10*, 899. [CrossRef]

9. Feijoo, E.R.; Sivaraman, A.; Barret, E.; Sanchez-Salas, R.; Galiano, M.; Rozet, F.; Prapotnich, D.; Cathala, N.; Mombet, A.; Cathelineau, X. Focal High-intensity Focused Ultrasound Targeted Hemiablation for Unilateral Prostate Cancer: A Prospective Evaluation of Oncologic and Functional Outcomes. *Eur. Urol.* **2016**, *69*, 214–220. [CrossRef]
10. Natarajan, S.; Raman, S.; Priester, A.M.; Garritano, J.; Margolis, D.J.; Lieu, P.; Macairan, M.L.; Huang, J.; Grundfest, W.; Marks, L.S. Focal Laser Ablation of Prostate Cancer: Phase I Clinical Trial. *J. Urol.* **2016**, *196*, 68–75. [CrossRef]
11. Llovet, J.M.; De Baere, T.; Kulik, L.; Haber, P.K.; Greten, T.F.; Meyer, T.; Lencioni, R. Locoregional therapies in the era of molecular and immune treatments for hepatocellular carcinoma. *Nat. Rev. Gastroenterol. Hepatol.* **2021**, *18*, 293–313. [CrossRef]
12. Nault, J.-C.; Sutter, O.; Nahon, P.; Ganne-Carrie, N.; Seror, O. Percutaneous treatment of hepatocellular carcinoma: State of the art and innovations. *J. Hepatol.* **2018**, *68*, 783–797. [CrossRef]
13. Thompson, R.H.; Atwell, T.; Schmit, G.; Lohse, C.M.; Kurup, A.N.; Weisbrod, A.; Psutka, S.P.; Stewart, S.B.; Callstrom, M.R.; Cheville, J.C.; et al. Comparison of Partial Nephrectomy and Percutaneous Ablation for cT1 Renal Masses. *Eur. Urol.* **2015**, *67*, 252–259. [CrossRef]
14. Ahmed, M.; Brace, C.L.; Lee, F.T., Jr.; Goldberg, S.N. Principles of and Advances in Percutaneous Ablation. *Radiology* **2011**, *258*, 351–369. [CrossRef] [PubMed]
15. Ahmed, M.; Solbiati, L.; Brace, C.L.; Breen, D.J.; Callstrom, M.R.; Charboneau, J.W.; Chen, M.-H.; Choi, B.I.; de Baere, T.; Dodd, G.D., III; et al. Image-guided Tumor Ablation: Standardization of Terminology and Reporting Criteria-A 10-Year Update. *Radiology* **2014**, *273*, 241–260. [CrossRef]
16. Li, X.; Lovell, J.F.; Yoon, J.; Chen, X. Clinical development and potential of photothermal and photodynamic therapies for cancer. *Nat. Rev. Clin. Oncol.* **2020**, *17*, 657–674. [CrossRef]
17. Liu, Y.; Bhattarai, P.; Dai, Z.; Chen, X. Photothermal therapy and photoacoustic imaging via nanotheranostics in fighting cancer. *Chem. Soc. Rev.* **2019**, *48*, 2053–2108. [CrossRef]
18. Liu, C.; Cao, Y.; Cheng, Y.; Wang, D.; Xu, T.; Su, L.; Zhang, X.; Dong, H. An open source and reduce expenditure ROS generation strategy for chemodynamic/photodynamic synergistic therapy. *Nat. Commun.* **2020**, *11*, 1735. [CrossRef]
19. Xu, C.; Pu, K. Second near-infrared photothermal materials for combinational nanotheranostics. *Chem. Soc. Rev.* **2021**, *50*, 1111–1137. [CrossRef]
20. Aaronson, D.S.; Yamasaki, I.; Gottschalk, A.; Speight, J.; Hsu, I.C.; Pickett, B.; Roach, M.; Shinohara, K. Salvage permanent perineal radioactive-seed implantation for treating recurrence of localized prostate adenocarcinoma after external beam radiotherapy. *BJU Int.* **2009**, *104*, 600–604. [CrossRef]
21. Chao, Y.; Xu, L.; Liang, C.; Feng, L.; Xu, J.; Dong, Z.; Tian, L.; Yi, X.; Yang, K.; Liu, Z. Combined local immunostimulatory radioisotope therapy and systemic immune checkpoint blockade imparts potent antitumour responses. *Nat. Biomed. Eng.* **2018**, *2*, 611–621. [CrossRef] [PubMed]
22. Formenti, S.C.; Rudqvist, N.-P.; Golden, E.; Cooper, B.; Wennerberg, E.; Lhuillier, C.; Vanpouille-Box, C.; Friedman, K.; de Andrade, L.F.; Wucherpfennig, K.W.; et al. Radiotherapy induces responses of lung cancer to CTLA-4 blockade. *Nat. Med.* **2018**, *24*, 1845–1851. [CrossRef]
23. Kerkmeijer, L.G.W.; Groen, V.H.; Pos, F.J.; Haustermans, K.; Monninkhof, E.M.; Smeenk, R.J.; Kunze-Busch, M.; de Boer, J.C.J.; van Zijp, J.V.D.V.; van Vulpen, M.; et al. Focal Boost to the Intraprostatic Tumor in External Beam Radiotherapy for Patients With Localized Prostate Cancer: Results From the FLAME Randomized Phase III Trial. *J. Clin. Oncol.* **2021**, *39*, 787–796. [CrossRef]
24. Rodriguez-Ruiz, M.E.; Vanpouille-Box, C.; Melero, I.; Formenti, S.C.; Demaria, S. Immunological Mechanisms Responsible for Radiation-Induced Abscopal Effect. *Trends Immunol.* **2018**, *39*, 644–655. [CrossRef]
25. Norouzi, M.; Nazari, B.; Miller, D.W. Injectable hydrogel-based drug delivery systems for local cancer therapy. *Drug Discov. Today* **2016**, *21*, 1835–1849. [CrossRef]
26. Salagierski, M.; Wojciechowska, A.; Zajac, K.; Klatte, T.; Thompson, R.H.; Cadeddu, J.A.; Kaouk, J.; Autorino, R.; Ahrar, K.; Capitanio, U.; et al. The Role of Ablation and Minimally Invasive Techniques in the Management of Small Renal Masses. *Eur. Urol. Oncol.* **2018**, *1*, 395–402. [CrossRef]
27. Nieuwenhuizen, S.; Dijkstra, M.; Puijk, R.S.; Geboers, B.; Ruarus, A.H.; Schouten, E.A.; Nielsen, K.; de Vries, J.J.J.; Bruynzeel, A.M.E.; Scheffer, H.J.; et al. Microwave Ablation, Radiofrequency Ablation, Irreversible Electroporation, and Stereotactic Ablative Body Radiotherapy for Intermediate Size (3–5 cm) Unresectable Colorectal Liver Metastases: A Systematic Review and Meta-analysis. *Curr. Oncol. Rep.* **2022**, *24*, 793–808. [CrossRef]
28. Han, K.; Kim, J.H.; Yang, S.G.; Park, S.H.; Choi, H.-K.; Chun, S.-Y.; Kim, P.N.; Park, J.; Lee, M. A Single-Center Retrospective Analysis of Periprocedural Variables Affecting Local Tumor Progression after Radiofrequency Ablation of Colorectal Cancer Liver Metastases. *Radiology* **2021**, *298*, 212–218. [CrossRef]
29. Nam, J.; Son, S.; Ochyl, L.J.; Kuai, R.; Schwendeman, A.; Moon, J.J. Chemo-photothermal therapy combination elicits anti-tumor immunity against advanced metastatic cancer. *Nat. Commun.* **2018**, *9*, 1074. [CrossRef]
30. Valerio, M.; Ahmed, H.U.; Emberton, M.; Lawrentschuk, N.; Lazzeri, M.; Montironi, R.; Nguyen, P.L.; Trachtenberg, J.; Polascik, T.J. The Role of Focal Therapy in the Management of Localised Prostate Cancer: A Systematic Review. *Eur. Urol.* **2014**, *66*, 732–751. [CrossRef]
31. Yakkala, C.; Denys, A.; Kandalaft, L.; Duran, R. Cryoablation and immunotherapy of cancer. *Curr. Opin. Biotechnol.* **2020**, *65*, 60–64. [CrossRef]

32. Chen, Q.; Chen, M.C.; Liu, Z. Local biomaterials-assisted cancer immunotherapy to trigger systemic antitumor responses. *Chem. Soc. Rev.* **2019**, *48*, 5506–5526. [CrossRef]
33. Wei, W.F.; Zhang, X.Y.; Zhang, S.; Wei, G.; Su, Z.Q. Biomedical and bioactive engineered nanomaterials for targeted tumor photothermal therapy: A review. *Mat. Sci. Eng. C Mater.* **2019**, *104*, 109891. [CrossRef]
34. Lin, H.; Gao, S.S.; Dai, C.; Chen, Y.; Shi, J.L. A Two-Dimensional Biodegradable Niobium Carbide (MXene) for Photothermal Tumor Eradication in NIR-I and NIR-II Biowindows. *J. Am. Chem. Soc.* **2017**, *139*, 16235–16247. [CrossRef]
35. Cao, Z.Y.; Feng, L.Z.; Zhang, G.B.; Wang, J.X.; Shen, S.; Li, D.D.; Yang, X.Z. Semiconducting polymer-based nanoparticles with strong absorbance in NIR-II window for in vivo photothermal therapy and photoacoustic imaging. *Biomaterials* **2018**, *155*, 103–111. [CrossRef]
36. Guo, B.; Sheng, Z.; Hu, D.; Liu, C.; Zheng, H.; Liu, B. Through Scalp and Skull NIR-II Photothermal Therapy of Deep Orthotopic Brain Tumors with Precise Photoacoustic Imaging Guidance. *Adv. Mater.* **2018**, *30*, 1802591. [CrossRef]
37. Tang, Z.M.; Zhao, P.R.; Ni, D.L.; Liu, Y.Y.; Zhang, M.; Wang, H.; Zhang, H.; Gao, H.B.; Yao, Z.W.; Bu, W.B. Pyroelectric nanoplatform for NIR-II-triggered photothermal therapy with simultaneous pyroelectric dynamic therapy. *Mater. Horizons* **2018**, *5*, 946–952. [CrossRef]
38. Zhu, Y.; Wang, Y.; Williams, G.R.; Fu, L.; Wu, J.; Wang, H.; Liang, R.; Weng, X.; Wei, M. Multicomponent Transition Metal Dichalcogenide Nanosheets for Imaging-Guided Photothermal and Chemodynamic Therapy. *Adv. Sci.* **2020**, *7*, 2000272. [CrossRef]
39. Hu, K.; Xie, L.; Zhang, Y.; Hanyu, M.; Yang, Z.; Nagatsu, K.; Suzuki, H.; Ouyang, J.; Ji, X.; Wei, J.; et al. Marriage of black phosphorus and Cu^{2+} as effective photothermal agents for PET-guided combination cancer therapy. *Nat. Commun.* **2020**, *11*, 2778. [CrossRef]
40. Wu, F.; Lu, Y.; Mu, X.; Chen, Z.; Liu, S.; Zhou, X.; Liu, S.; Li, Z. Intriguing H-Aggregates of Heptamethine Cyanine for Imaging-Guided Photothermal Cancer Therapy. *ACS Appl. Mater. Interfaces* **2020**, *12*, 32388–32396. [CrossRef]
41. Wang, Q.; Dai, Y.; Xu, J.; Cai, J.; Niu, X.; Zhang, L.; Chen, R.; Shen, Q.; Huang, W.; Fan, Q. All-in-One Phototheranostics: Single Laser Triggers NIR-II Fluorescence/Photoacoustic Imaging Guided Photothermal/Photodynamic/Chemo Combination Therapy. *Adv. Funct. Mater.* **2019**, *29*, 1901480. [CrossRef]
42. Zhou, J.; Li, M.; Hou, Y.; Luo, Z.; Chen, Q.; Cao, H.; Huo, R.; Xue, C.; Sutrisno, L.; Hao, L.; et al. Engineering of a Nanosized Biocatalyst for Combined Tumor Starvation and Low-Temperature Photothermal Therapy. *ACS Nano* **2018**, *12*, 2858–2872. [CrossRef]
43. Jiang, Y.; Li, J.; Zhen, X.; Xie, C.; Pu, K. Dual-Peak Absorbing Semiconducting Copolymer Nanoparticles for First and Second Near-Infrared Window Photothermal Therapy: A Comparative Study. *Adv. Mater.* **2018**, *30*, 1705980. [CrossRef]
44. Jung, H.S.; Verwilst, P.; Sharma, A.; Shin, J.; Sessler, J.L.; Kim, J.S. Organic molecule-based photothermal agents: An expanding photothermal therapy universe. *Chem. Soc. Rev.* **2018**, *47*, 2280–2297. [CrossRef]
45. Li, J.; Rao, J.; Pu, K. Recent progress on semiconducting polymer nanoparticles for molecular imaging and cancer photothermy. *Biomaterials* **2018**, *155*, 217–235. [CrossRef]
46. Cheng, Y.; Chang, Y.; Feng, Y.; Jian, H.; Tang, Z.; Zhang, H. Deep-Level Defect Enhanced Photothermal Performance of Bismuth Sulfide-Gold Heterojunction Nanorods for Photothermal Therapy of Cancer Guided by Computed Tomography Imaging. *Angew. Chem. Int. Ed.* **2018**, *57*, 246–251. [CrossRef]
47. Tang, W.; Dong, Z.; Zhang, R.; Yi, X.; Yang, K.; Jin, M.; Yuan, C.; Xiao, Z.; Liu, Z.; Cheng, L. Multifunctional Two-Dimensional Core-Shell MXene@Gold Nanocomposites for Enhanced Photo-Radio Combined Therapy in the Second Biological Window. *ACS Nano* **2019**, *13*, 284–294. [CrossRef]
48. Wang, Q.; Wang, H.; Yang, Y.; Jin, L.; Liu, Y.; Wang, Y.; Yan, X.; Xu, J.; Gao, R.; Lei, P.; et al. Plasmonic Pt Superstructures with Boosted Near-Infrared Absorption and Photothermal Conversion Efficiency in the Second Biowindow for Cancer Therapy. *Adv. Mater.* **2019**, *31*, 1904836. [CrossRef]
49. Gu, Z.; Zhu, S.; Yan, L.; Zhao, F.; Zhao, Y. Graphene-Based Smart Platforms for Combined Cancer Therapy. *Adv. Mater.* **2019**, *31*, 1800662. [CrossRef]
50. Liu, H.; Li, C.; Qian, Y.; Hu, L.; Fang, J.; Tong, W.; Nie, R.; Chen, Q.; Wang, H. Magnetic-induced graphene quantum dots for imaging-guided photothermal therapy in the second near-infrared window. *Biomaterials* **2020**, *232*, 119700. [CrossRef]
51. Liang, X.; Ye, X.; Wang, C.; Xing, C.; Miao, Q.; Xie, Z.; Chen, X.; Zhang, X.; Zhang, H.; Mei, L. Photothermal cancer immunotherapy by erythrocyte membrane-coated black phosphorus formulation. *J. Control. Release* **2019**, *296*, 150–161. [CrossRef] [PubMed]
52. Choi, J.R.; Yong, K.W.; Choi, J.Y.; Nilghaz, A.; Lin, Y.; Xu, J.; Lu, X. Black Phosphorus and its Biomedical Applications. *Theranostics* **2018**, *8*, 1005–1026. [CrossRef]
53. Zhou, Z.; Wang, X.; Zhang, H.; Huang, H.; Sun, L.; Ma, L.; Du, Y.; Pei, C.; Zhang, Q.; Li, H.; et al. Activating Layered Metal Oxide Nanomaterials via Structural Engineering as Biodegradable Nanoagents for Photothermal Cancer Therapy. *Small* **2021**, *17*, 2007486. [CrossRef] [PubMed]
54. Huang, X.; Zhang, W.; Guan, G.; Song, G.; Zou, R.; Hu, J. Design and Functionalization of the NIR-Responsive Photothermal Semiconductor Nanomaterials for Cancer Theranostics. *Acc. Chem. Res.* **2017**, *50*, 2529–2538. [CrossRef]
55. Liu, Y.; Ji, X.; Liu, J.; Tong, W.W.L.; Askhatova, D.; Shi, J. Tantalum Sulfide Nanosheets as a Theranostic Nanoplatform for Computed Tomography Imaging-Guided Combinatorial Chemo-Photothermal Therapy. *Adv. Funct. Mater.* **2017**, *27*, 1703261. [CrossRef] [PubMed]

56. Cheng, H.; Wen, M.; Ma, X.; Kuwahara, Y.; Mori, K.; Dai, Y.; Huang, B.; Yamashita, H. Hydrogen Doped Metal Oxide Semiconductors with Exceptional and Tunable Localized Surface Plasmon Resonances. *J. Am. Chem. Soc.* **2016**, *138*, 9316–9324. [CrossRef]
57. Cheng, L.; Liu, J.; Gu, X.; Gong, H.; Shi, X.; Liu, T.; Wang, C.; Wang, X.; Liu, G.; Xing, H.; et al. PEGylated WS$_2$ Nanosheets as a Multifunctional Theranostic Agent for in vivo Dual-Modal CT/Photoacoustic Imaging Guided Photothermal Therapy. *Adv. Mater.* **2014**, *26*, 1886–1893. [CrossRef]
58. Guan, G.; Wang, X.; Huang, X.; Zhang, W.; Cui, Z.; Zhang, Y.; Lu, X.; Zou, R.; Hu, J. Porous cobalt sulfide hollow nanospheres with tunable optical property for magnetic resonance imaging-guided photothermal therapy. *Nanoscale* **2018**, *10*, 14190–14200. [CrossRef]
59. Liu, J.; Zheng, X.; Yan, L.; Zhou, L.; Tian, G.; Yin, W.; Wang, L.; Liu, Y.; Hu, Z.; Gu, Z.; et al. Bismuth Sulfide Nanorods as a Precision Nanomedicine for in Vivo Multimodal Imaging-Guided Photothermal Therapy of Tumor. *ACS Nano* **2015**, *9*, 696–707. [CrossRef]
60. Yang, K.; Yang, G.; Chen, L.; Cheng, L.; Wang, L.; Ge, C.; Liu, Z. FeS nanoplates as a multifunctional nano-theranostic for magnetic resonance imaging guided photothermal therapy. *Biomaterials* **2015**, *38*, 1–9. [CrossRef]
61. Yang, W.; Guo, W.; Le, W.; Lv, G.; Zhang, F.; Shi, L.; Wang, X.; Wang, J.; Wang, S.; Chang, J.; et al. Albumin-Bioinspired Gd:CuS Nanotheranostic Agent for In Vivo Photoacoustic/Magnetic Resonance Imaging-Guided Tumor-Targeted Photothermal Therapy. *ACS Nano* **2016**, *10*, 10245–10257. [CrossRef] [PubMed]
62. Lei, Z.; Zhang, W.; Li, B.; Guan, G.; Huang, X.; Peng, X.; Zou, R.; Hu, J. A full-spectrum-absorption from nickel sulphide nanoparticles for efficient NIR-II window photothermal therapy. *Nanoscale* **2019**, *11*, 20161–20170. [CrossRef]
63. Zhang, C.; Sun, W.; Wang, Y.; Xu, F.; Qu, J.; Xia, J.; Shen, M.; Shi, X. Gd-/CuS-Loaded Functional Nanogels for MR/PA Imaging-Guided Tumor-Targeted Photothermal Therapy. *ACS Appl. Mater. Interfaces* **2020**, *12*, 9107–9117. [CrossRef] [PubMed]
64. Sun, H.; Zhang, Y.; Chen, S.; Wang, R.; Chen, Q.; Li, J.; Luo, Y.; Wang, X.; Chen, H. Photothermal Fenton Nanocatalysts for Synergetic Cancer Therapy in the Second Near-Infrared Window. *ACS Appl. Mater. Interfaces* **2020**, *12*, 30145–30154. [CrossRef]
65. Liu, S.; Pan, X.T.; Liu, H.Y. Two-Dimensional Nanomaterials for Photothermal Therapy. *Angew. Chem. Int. Ed.* **2020**, *59*, 5890–5900. [CrossRef]
66. Hu, J.-J.; Cheng, Y.-J.; Zhang, X.-Z. Recent advances in nanomaterials for enhanced photothermal therapy of tumors. *Nanoscale* **2018**, *10*, 22657–22672. [CrossRef]
67. Wu, Y.; Liang, Y.; Liu, Y.; Hao, Y.; Tao, N.; Li, J.; Sun, X.; Zhou, M.; Liu, Y.N. A Bi$_2$S$_3$-embedded gellan gum hydrogel for localized tumor photothermal/antiangiogenic therapy. *J. Mater. Chem. B* **2021**, *9*, 3224–3234. [CrossRef]
68. Sun, S.-K.; Wu, J.-C.; Wang, H.; Zhou, L.; Zhang, C.; Cheng, R.; Kan, D.; Zhang, X.; Yu, C. Turning solid into gel for high-efficient persistent luminescence-sensitized photodynamic therapy. *Biomaterials* **2019**, *218*, 119328. [CrossRef]
69. Ding, X.; Wang, Y. Weak Bond-Based Injectable and Stimuli Responsive Hydrogels for Biomedical Applications. *J. Mater. Chem. B* **2017**, *5*, 887–906. [CrossRef]
70. Chandel, A.K.S.; Nutan, B.; Raval, I.H.; Jewrajka, S.K. Self-Assembly of Partially Alkylated Dextran- graft-poly[(2-dimethylamino)ethyl methacrylate] Copolymer Facilitating Hydrophobic/Hydrophilic Drug Delivery and Improving Conetwork Hydrogel Properties. *Biomacromolecules* **2018**, *19*, 1142–1153. [CrossRef]
71. Jiang, K.; Song, X.; Yang, L.; Li, L.; Wan, Z.; Sun, X.; Gong, T.; Lin, Q.; Zhang, Z. Enhanced antitumor and anti-metastasis efficacy against aggressive breast cancer with a fibronectin-targeting liposomal doxorubicin. *J. Control. Release* **2018**, *271*, 21–30. [CrossRef] [PubMed]
72. Henschke, C.I.; Yankelevitz, D.F.; McCauley, D.I.; Rifkin, M.; Fiore, E.S.; Austin, J.H.M.; Pearson, G.D.N.; Shiau, M.C.; Kopel, S.; Klippenstein, D.; et al. CT screening for lung cancer: Diagnoses resulting from the New York Early Lung Cancer Action Project. *Radiology* **2007**, *243*, 239–249.

Article

Quercetin/Hydroxypropyl-β-Cyclodextrin Inclusion Complex-Loaded Hydrogels for Accelerated Wound Healing

Nutsarun Wangsawangrung [1], Chasuda Choipang [1,2], Sonthaya Chaiarwut [1,2], Pongpol Ekabutr [1], Orawan Suwantong [3,4], Piyachat Chuysinuan [5], Supanna Techasakul [5] and Pitt Supaphol [1,2,*]

[1] The Petroleum and Petrochemical College, Chulalongkorn University, Bangkok 10330, Thailand
[2] Research Unit on Herbal Extracts-Infused Advanced Wound Dressing, Chulalongkorn University, Bangkok 10330, Thailand
[3] School of Science, Mae Fah Luang University, Chiang Rai 57100, Thailand
[4] Center of Chemical Innovation for Sustainability (CIS), Mae Fah Luang University, Chiang Rai 57100, Thailand
[5] Laboratory of Organic Synthesis, Chulabhorn Research Institute, Bangkok 10210, Thailand
* Correspondence: pitt.s@chula.ac.th; Tel.: +66-2-2184-117

Abstract: This study concentrated on developing quercetin/cyclodextrin inclusion complex-loaded polyvinyl alcohol (PVA) hydrogel for enhanced stability and solubility. Quercetin was encapsulated in hydroxypropyl-β-cyclodextrin (HP-β-CD) by the solvent evaporation method. The prepared quercetin/HP-β-CD inclusion complex showed 90.50 ± 1.84% encapsulation efficiency (%EE) and 4.67 ± 0.13% loading capacity (%LC), and its successful encapsulation was confirmed by FT-IR and XRD. The quercetin/HP-β-CD inclusion complex was well dispersed in viscous solutions of PVA in various amounts (0.5, 1.0, 1.5. 2.5, and 5.0% w/v ratio), and the drug-loaded polymer solution was physically crosslinked by multiple freeze–thaw cycles to form the hydrogel. The cumulative amount of quercetin released from the prepared hydrogels increased with increasing concentrations of the inclusion complex. The introduction of the inclusion complex into the PVA hydrogels had no influence on their swelling ratio, but gelation and compressive strength reduced with increasing inclusion complex concentration. The potential cytotoxicity of quercetin/HP-β-CD inclusion complex hydrogels was evaluated by MTT assay and expressed as % cell viability. The results show biocompatibility toward NCTC 929 clone cells. The inhibitory efficacy was evaluated with 2, 2-diphenyl-1-picrylhydrazyl (DPPH) free radical scavenging assay, and the results show a higher level of antioxidant activity for quercetin/HP-β-CD inclusion complex hydrogels compared with free quercetin. The findings of our study indicate that the developed quercetin/HP-β-CD inclusion complex hydrogels possess the required properties and can be proposed as a quercetin delivery system for wound-healing applications.

Keywords: quercetin; cyclodextrin; polyvinyl alcohol; inclusion complex; hydrogel

Citation: Wangsawangrung, N.; Choipang, C.; Chaiarwut, S.; Ekabutr, P.; Suwantong, O.; Chuysinuan, P.; Techasakul, S.; Supaphol, P. Quercetin/Hydroxypropyl-β-Cyclodextrin Inclusion Complex-Loaded Hydrogels for Accelerated Wound Healing. *Gels* **2022**, *8*, 573. https://doi.org/10.3390/gels8090573

Academic Editors: Kiat Hwa Chan and Yang Liu

Received: 8 August 2022
Accepted: 6 September 2022
Published: 8 September 2022

Publisher's Note: MDPI stays neutral with regard to jurisdictional claims in published maps and institutional affiliations.

Copyright: © 2022 by the authors. Licensee MDPI, Basel, Switzerland. This article is an open access article distributed under the terms and conditions of the Creative Commons Attribution (CC BY) license (https://creativecommons.org/licenses/by/4.0/).

1. Introduction

At some point, everyone will experience an injury that involves skin or body tissue damage, and it is important to care for it properly. There are many wound care products on the market, and several factors need to be considered when choosing an appropriate dressing for each type of wound, including the color and depth of the wound, degree of infection, amount of exudate, and condition of the peri-wound skin [1]. Normally, an ideal dressing should have the following properties: it should maintain a moist environment, protect from bacterial invasion, accelerate the formation of epithelialization, provide thermal insulation, and be easy to remove without debris after healing [2].

Today, wound dressings have been developed to facilitate biological functions, such as antibacterial, antifungal, and antiviral functions, compared with traditional wound dressings, which just cover wounds. Herbal drugs from medicinal plants are widely incorporated in wound dressing materials to improve the healing properties [3,4]. Quercetin is an interesting component of the flavonoid family that is found in onions, apples, berries, tea, and tomatoes [5,6]. In addition, it has physicochemical features and absorption ability, is a source of dietary nutrition, and principally impacts inflammation and immunological function. Research has shown that ingesting 50 to 800 mg of quercetin per day (quercetin accounts for 75%) may have potential health advantages owing to its easy absorption in the organs responsible for metabolism and elimination [7]. The antioxidant, anti-inflammatory, antibacterial, antiviral, anticancer, antitoxic, and immunomodulatory effects of quercetin prove that it has potential therapeutic value for wound healing.

However, the major problems that limit its biological activity for wound healing are poor solubility, low bioavailability, and hydrophobicity, which means it generally dissolves more readily in organic solution [8,9]. To overcome the disadvantages of quercetin, a drug-delivery system can be used, for example, that forms an inclusion complex with host molecules such as cyclodextrins (CDs), which is one of the most popular techniques [10].

Cyclodextrins, generated by the action of an enzyme on starch, are cyclic oligosaccharides linked by α-1,4 glycosidic bonds that are normally composed of six (α-CD), seven (β-CD), or eight (γ-CD) α-D-glucopyranose units. Because cyclodextrins enhance hydrophobic–hydrophilic interactions between proteins and other molecules, they are employed in the food processing industry to create reduced-cholesterol products and increase the bioavailability of desired molecules [11]. In general, CDs have a cone-shaped structure, with a hydrophilic outer surface and a hydrophobic inner cavity. Due to the hydrophobic cavity, CDs can trap hydrophobic molecules to form host–guest complexes for use in the pharmaceutical industry [12]. CDs are used to encapsulate odors and flavors to avoid food discoloration and minimize off-flavors and can be applied in food packaging [13]. In addition, CDs are used in many medical applications, such as wound dressings. Inclusion complexes containing curcumin and hydroxypropyl–cyclodextrin have been added to bacterial cellulose hydrogels to enhance wound healing [14]. However, mainly hydroxypropyl-β-cyclodextrin (HP-β-CD) is employed as a complexing agent to improve the water solubility and bioavailability of poorly soluble medicines and enhance their stability [15].

Hydrogels are three-dimensional natural or synthetic polymers that contain more than 90% water. The polymer chains are crosslinked to form a network to maintain the hydrogel structure, and the crosslinking process can be either physical or chemical. Hydrogels have a hydrophilic porous structure that can absorb a certain amount of water depending on many factors such as the pore size or the polymer type [16]. Hydrogels are popular for wound dressing because of several benefits including the ability to maintain a moist environment and absorb exudates, good biocompatibility, low adhesion to wound tissue, reduced pressure and shear forces, protection from bacterial invasion, and thermal insulation [16,17]. Hydrogels also have the ability to cool and soothe the skin, which is helpful for pain relief of burns or painful wounds [18]. Polyvinyl alcohol (PVA) is a biodegradable and water-soluble hydrophilic polymer that is produced by hydrolyzing polyvinyl acetate (PVAc). This polymer has good mechanical properties, excellent thermal stability, low toxicity, and high gas barrier properties, so it is widely used in many applications, including biomedical, food packaging, paper coating, and textile sizing and as a thickener and emulsion stabilizer, or it can be blended with other polymers to increase the overall properties. PVA hydrogel wound dressings can be produced by physical crosslinking through the cyclic freeze–thaw process [19]. Due to the PVA structure, they can form H-bonding between −OH groups during freezing, leading to crosslinking of their polymer chains (crystalline zone). However, the freezing conditions, including temperature, time, and number of cycles, could affect the degree of physical crosslinking or crystallinity of the obtained hydrogels. At lower freezing temperature, there are greater interactions between PVA chains and more crystal formation. Small water crystals are formed and sublimated at low freezing temperatures, resulting in a porous structure [20], whereas more freeze–thaw cycles lead to further crystal formation and a higher degree of physical crosslinking [21].

The purpose of this work was to form an inclusion complex by encapsulating quercetin into HP-β-CD to overcome its limitations and to develop new hydrogels incorporating the inclusion complex to accelerate wound healing. Drug-loaded hydrogels were evaluated for their physical properties in terms of gelation, swelling, and compression. The quercetin released from the prepared hydrogels and the cytotoxic effects on an NCTC 929 clone cells were also investigated regarding the biological abilities of the drug-loaded hydrogels.

2. Results and Discussion
2.1. Characterization of Quercetin/HP-β-CD Inclusion Complex

FTIR spectroscopy was used to investigate the successful encapsulation of quercetin into HP-β-CD (Figure 1). The FTIR spectrum of quercetin (Figure 1a) showed –OH groups stretching at 3395 and 3268 cm^{-1}; C=O aryl ketone stretching at 1663 cm^{-1}; C=C aromatic ring stretching at 1605, 1560, and 1518 cm^{-1}; and C–O aryl ether ring stretching, C–O phenol stretching, and C–CO–C ketone stretching and bending at 1256, 1194, and 1164 cm^{-1}, respectively [22]. For HP-β-CD (Figure 1b), the characteristic broad peak of OH group stretching was shown at 3344 cm^{-1}; C–H stretching was detected at 2926 cm^{-1}; and C-O-C glucose unit stretching occurred at 1020 cm^{-1} [23]. The FTIR spectrum of the physical mixture of the two (Figure 1c) showed characteristic peaks of quercetin, including C=O aryl ketone stretching (1663 cm^{-1}), C=C aromatic ring stretching (1609, 1560, and 1520 cm^{-1}), and C–CO–C ketone stretching and bending (1165 cm^{-1}), as well as C–O–C glucose unit stretching of HP-β-CD. When quercetin was encapsulated, its characteristic peaks seemed to be masked (Figure 1d). The C=O aryl ketone stretching (1663 cm^{-1}), C=C aromatic ring stretching (1605, 1560, and 1518 cm^{-1}), and C–CO–C ketone stretching and bending (1164 cm^{-1}) peaks of quercetin were absent in the inclusion complex, which may be because the quercetin was entrapped in the HP-β-CD cavities [24]. Signals of the characteristic bands of quercetin shifted from 3282 cm^{-1} to 3142 cm^{-1} O–H bending signals showing the formation of quercetin/HP-β-CD inclusion complex [25].

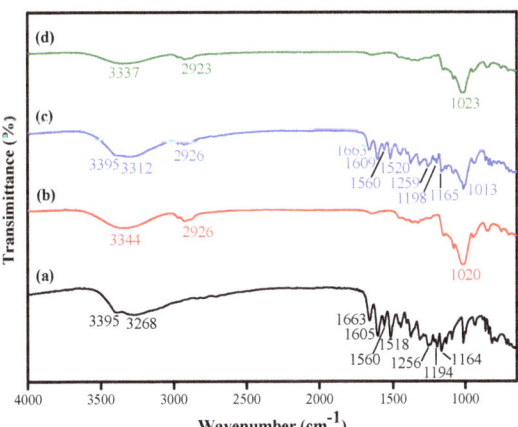

Figure 1. FTIR spectra of (**a**) quercetin, (**b**) HP-β-CD, (**c**) quercetin/HP-β-CD physical mixture, and (**d**) quercetin/HP-β-CD inclusion complex.

The XRD patterns provide further evidence of the successful encapsulation of quercetin in HP-β-CD (Figure 2). Characteristic diffraction peaks (Figure 2a) indicated that quercetin existed in a crystalline form, with peaks at three 2θ positions (10.66°, 12.34°, and 27.26°) [26], while HP-β-CD (Figure 2b) showed broad peaks of amorphous materials caused by random X-ray scattering [27]. After the complexation of quercetin and HP-β-CD, there were no noticeable crystalline peaks of quercetin in the inclusion complex (Figure 2d) because the CD cavities hindered its crystalline structure, unlike the physical mixture, which showed crystalline and amorphous behavior (Figure 2c) of quercetin and HP-β-CD, respectively. These results suggest that quercetin was successfully encapsulated into HP-β-CD cavities, leading to the loss of characteristic peaks. Similarly, Zhu et al. encapsulated l-menthol, which is crystalline (2θ = 8.0°, 14.0°, 20.5°, and 21.6°), into amorphous HP-β-CD and observed no sharp peaks of l-menthol [28]. This result agrees with Kim et al., who reported similar XRD patterns of some selective flavonoids in hydroxypropyl-β-cyclodextrin inclusion complexes, which was indicative of the amorphous nature of the complex [29].

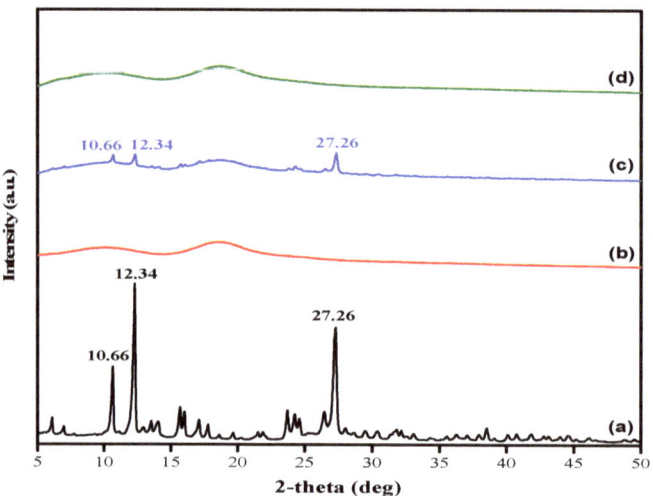

Figure 2. XRD patterns of (**a**) quercetin, (**b**) HP-β-CD, (**c**) quercetin/HP-β-CD physical mixture, and (**d**) quercetin/HP-β-CD inclusion complex.

The morphologies of quercetin, HP-β-CD, and quercetin/HP-β-CD physical mixture and inclusion complex were observed at different magnifications on SEM (Figure 3). Quercetin (Figure 3a) presented a strip-like structure with a smooth surface [30], while HP-β-CD (Figure 3b) appeared as rough spheres [31]. In the physical mixture (Figure 3c), both quercetin and HP-β-CD kept their own structure, with particles of HP-β-CD embedded with quercetin particles [32]. On the other hand, the morphology and shape of quercetin/HP-β-CD inclusion complex obtained from the solvent evaporation method were completely different from those of the original product (Figure 3d). Our results were consistent with Pradhan et al. [33] in that the morphology of Berberis anthocyanin-loaded CD inclusion complex revealed the rod shape. The encapsulation of quercetin with CD demonstrated the electrostatic and hydrogel bond interactions between quercetin and CD.

2.2. Encapsulation Efficiency and Loading Capacity of Quercetin in the Inclusion Complex

Cyclodextrins are nontoxic cyclic oligosaccharides consisting of α1,4-glycosidic bonds. CDs are distinguished by hydrophilic surfaces and hydrophobic cavities. The cavities are suitable for inclusion complexes with hydrophobic drugs (quercetins) to improve the aqueous solubility and yield Van der Waals and hydrophobic forces [34,35]. Quercetin/HP-β-CD inclusion complex was formed with optimum loading and encapsulation efficacy, achieved at a mass ratio of HP-β-CD to quercetin of 4.35:1. Loading capacity (LC) is the ratio of mass of encapsulated compound to mass of the polymer (Equation (1)), and encapsulation efficacy (EE%) refers to the ratio of the mass of the encapsulated compounds to the total mass of the compounds (Equation (2)). Encapsulation is considered to be an excellent carrier of compounds when LC id $\geq$ 5%, and %EE in the range of 70–100% indicates excellent encapsulation capacity.

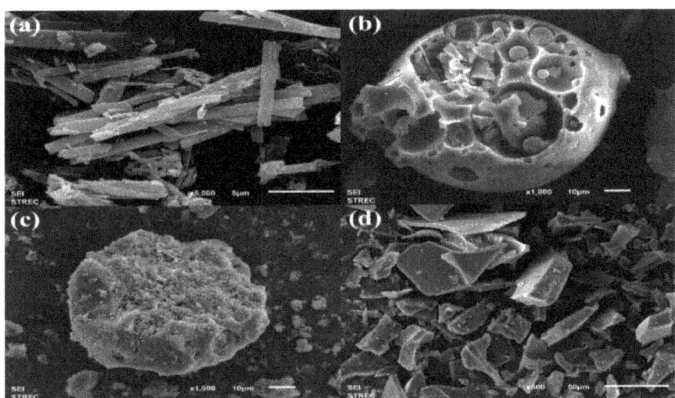

Figure 3. SEM images of (**a**) quercetin, (**b**) HP-β-CD, (**c**) quercetin/HP-β-CD physical mixture, and (**d**) quercetin/HP-β-CD inclusion complex.

The absorbance of ~1 mg/mL inclusion complex in DMSO was 0.68917 ± 0.01715, which can be calculated back to the concentration of quercetin using the predetermined calibration curve (R^2 = 0.99758). The estimated total quercetin content in the inclusion complex was 0.4554 ± 0.0089 g.

The %EE and %LC of quercetin were quantified with UV-Vis spectrophotometry at a wavelength of 379 nm and calculated using Equations (1) and (2), respectively. The %EE of the obtained quercetin/HP-β-CD inclusion complex was 90.50 ± 1.84%, and the %LC was 4.67 ± 0.13%. The %LC of the inclusion complex seemed to be quite low but was not. The molecular weight of HP-β-CD was greater than that of quercetin by around 4.6 times. We also used the excess molar ratio of HP-β-CD to prepare the inclusion complex. That explains why when we calculated %LC, which compares the drug in the complexation (or inclusion complex) in terms of weight, the molar ratio seemed to be low. This was similar to the results of Hadian et al., who prepared four formulations of geraniol/β-CD inclusion complexes by varying the mole ratio of GR:β-CD (0.44:0.13, 0.44:0.2, 0.44:0.4, and 0.44:1), and showed that %LC gradually decreased with increasing mole ratio (by 7.8 ± 0.70, 6.9 ± 0.1 6.6 ± 0.28 and 6.5 ± 0.6, respectively) [36].

2.3. Gelation of PVA Hydrogels Loaded with Quercetin/HP-β-CD Inclusion Complex

The gelation of neat and all prepared drug-loaded hydrogels in PBS (pH = 7.4) at 37 °C for 24 h was calculated using Equation (3). The experiments were repeat five times, and values were averaged. The gelation of neat PVA hydrogel was 97.40 ± 0.62%, while the gelation of drug-loaded hydrogels ranged from 74.52 ± 0.86% to 94.36 ± 0.80%, as shown in Table 1.

Table 1. Gelation value and mass content of PVA hydrogels containing various amounts of quercetin/HP-β-CD inclusion complex. Data shown as mean ± standard deviation (*n* = 5). IC as quercetin/HP-β-CD inclusion complex.

Sample	Gelation (%)	Mass Content (%)	
		IC	PVA
PVAIC0	97.40 ± 0.62	0.00	100.00
PVAIC0.5	94.36 ± 0.80	4.67	95.24
PVAIC1.0	92.29 ± 0.71	9.09	90.91
PVAIC1.5	89.88 ± 0.83	13.04	86.96

Table 1. Cont.

Sample	Gelation (%)	Mass Content (%)	
		IC	PVA
PVAIC2.5	85.59 ± 0.95	20.00	80.00
PVAIC5.0	74.52 ± 0.86	33.33	66.67

The percent gelation of drug-loaded hydrogel clearly gradually decreased with increasing inclusion complex content in the hydrogels. The reduction in hydrogel weight may have occurred for the following reasons. The first is that hydrogel swells when immersed in PBS, leading to a loose structure, so the quercetin can dissolve and be released out of the hydrogels [20]. The second reason is that when loading the quercetin/HP-β-CD inclusion complex into the hydrogels, the inclusion complex may disrupt the crosslink formation of the hydrogel, so some PVA that did not participate in crystallite formation could have dissolved into the solution [21].

2.4. Swelling Ratio of PVA Hydrogels Loaded with Quercetin/HP-β-CD Inclusion Complex

The swelling ratio of neat and drug-loaded hydrogels in PBS (pH = 7.4) at 37 °C at 10, 30, 60, 180, and 360 min was calculated using Equation (4). The experiment was repeated three times at each release time point, and values were averaged. The results show that the swelling ratio of hydrogels with various drug amounts was slightly higher than that of neat PVA hydrogel, but there was not much difference (Figure 4). After the first 60 min of the experiment, the swelling ratio of all hydrogels dramatically increased to around 4.3%, then gradually increased to around 5.8% at 180 min and became relatively stable at around 6.4% after immersion in PBS for 360 min. The incorporation of quercetin/HP-β-CD inclusion complex into the PVA hydrogels had no effect on the swelling ratio, which may be caused by the similar conditions of the freeze–thaw process that we used to prepare the hydrogels. Therefore, the number and temperature of freezing–thawing cycles could be the most important factors controlling the swelling ratio of hydrogels. We can also add other polymers or materials, such as gelatin or chitosan, to increase the swellability of PVA hydrogels [20].

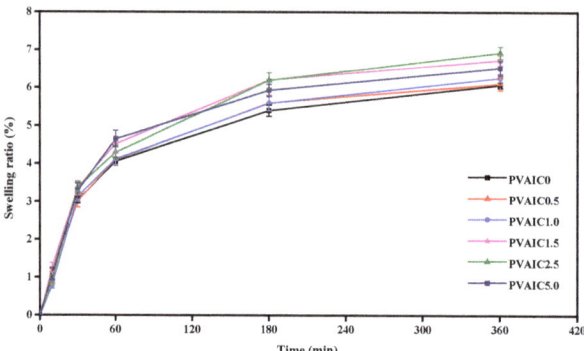

Figure 4. Swelling ratios of PVA hydrogels containing 0% (PVAIC0), 0.5% (PVAIC0.5), 1.0% (PVAIC1.0), 1.5% (PVAIC1.5), 2.5% (PVAIC2.5), and 5.0% (PVAIC5.0) quercetin/HP-β-CD inclusion complex. Data shown as mean ± standard deviation (n = 3).

2.5. Mechanical Test of PVA Hydrogels Loaded with Quercetin/HP-β-CD Inclusion Complex

The compressive strength of neat and all prepared drug-loaded hydrogels was tested, and the experiment was repeated seven times to obtain average values. The stiffness and compressive modulus of neat PVA hydrogel were 27,318 ± 831 N/m and 1.19 ± 0.07 MPa, respectively, while the values for drug-loaded hydrogels were 19,520 ± 1066 to 26,199 ± 1275 N/m and 0.88 ± 0.04 to 1.12 ± 0.08 MPa, respectively (Table 2). As the amount of inclusion complex in the hydrogels increased, stiffness and compressive modulus seemed to gradually decrease. This may have resulted from the molecules of the inclusion complex being distributed in the PVA structure, leading to a lower crystalline zone [20] and decreased crosslink density of the PVA hydrogel [37], so it was not able to withstand the compressive force compared with lower drug content hydrogels. Water content is another factor that can affect the mechanical properties of hydrogels because when the water in hydrogel evaporates, the hydrogel becomes stronger, so it will have stronger resistance against compressive force than softer or lower-water-content hydrogels [38]. Qing et al. prepared PVA hydrogel dressings that incorporated N-succinyl chitosan (NSCS) and lincomycin, and the results revealed that the introduction of 10% NSCS reduced the compression strength of pure PVA hydrogel from 0.33 to 0.25 MPa because NSCS increases the distance between PVA chains and weakens the molecular interaction force. However, when they further increased NSCS content, hydrogel bonding occurred, and the hydrogel became stronger, leading to enhanced compression strength (0.75 MPa). As NSCS content was further increased to 40 and 50%, the compression strength in turn decreased due to the low crosslinking density resulting from the decreased gel fraction [39].

Table 2. Mechanical test of PVA hydrogels containing various amounts of quercetin/HP-β-CD inclusion complex. Data shown as mean ± standard deviation (n = 7).

Sample	Stiffness (N/m)	Compressive Modulus (MPa)
PVAIC0	27,318 ± 831	1.19 ± 0.07
PVAIC0.5	26,199 ± 1275	1.12 ± 0.08
PVAIC1.0	24,878 ± 758	1.09 ± 0.10
PVAIC1.5	24,011 ± 941	0.99 ± 0.06
PVAIC2.5	21,817 ± 1001	0.92 ± 0.04
PVAIC5.0	19,520 ± 1066	0.88 ± 0.04

2.6. In Vitro PVA Hydrogels Loaded with Quercetin/HP-β-CD Inclusion Complex Dissolution and Release Kinetics Study

The characteristics of quercetin release from the prepared hydrogels were investigated using the total immersion method over a period of 48 h in PBS (pH = 7.4) and incubated at 37 °C under continuous stirring. The cumulative release of quercetin from all hydrogels was calculated using Equation (5) and reported in parts per million. As shown in Figure 5, the cumulative amount of quercetin released from the hydrogels increased rapidly with increasing immersion time for the first 5 or 8 h of the experiment, then gradually increased and reached a plateau. After a period of 48 h, the cumulative quercetin quantities released from the quercetin/HP-β-CD inclusion complex PVA hydrogels were 81.66 ± 17.42, 142.07 ± 30.43, 167.64 ± 23.30, 211.74 ± 27.78, and 286.10 ± 27.71 ppm for PVAIC0.5, PVAIC1.0, PVAIC1.5, PVAIC2.5, and PVAIC5.0, respectively. The greatest cumulative amounts of CIP released were observed when the quercetin concentration of PVAIC5.0 was used, and therefore, it can be concluded that when a greater concentration of quercetin is introduced to the nanoparticles, the drug release will increase. In the USA, the mean quercetin intake was approximately 14.90 to 16.39 mg per day [40]. Thus, the release characteristics of quercetin for the PVAIC5.0 hydrogels (close to 28.6 mg/100 g of sample) was found to be about 1.9-times higher than the oral solution of the quercetin intake per day.

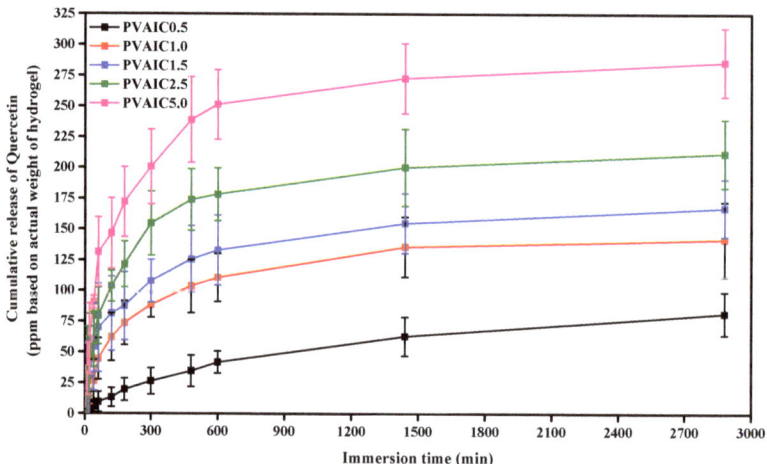

Figure 5. Release profiles of quercetin from PVA hydrogels containing 0.5% (PVAIC0.5), 1.0% (PVAIC1.0), 1.5% (PVAIC1.5), 2.5% (PVAIC2.5), and 5.0% (PVAIC5.0) of quercetin/HP-β-CD inclusion complex. Data shown as mean ± standard deviation (n = 3).

The drug-release kinetics of quercetin from PVA hydrogels loaded with quercetin/HP-β-CD inclusion complex were fitted with Zero-order and Korsmeyer–Peppas models to reveal the kinetics of drug release. Release constant (k) and release exponent (n) were calculated to identify the release mechanism, and the fitted model parameters and correlation coefficients (r^2) for the release profiles are given in Table 3. An r^2 coefficient of determination over 0.95 was considered to reflect the goodness of fit of the release model [41]. The release data from PVA hydrogels loaded with quercetin/HP-β-CD inclusion complex fit with the Korsmeyer–Peppas model. The release exponent (n) values can be used to indicate the release rate mechanism. The PVAIC0.5, PVAIC1.0, PVAIC1.5, PVAIC2.5, and PVAIC5.0 hydrogels had n values of 1.31, 0.77, 0.68, 0.58, and 0.77, respectively. The n values between 0.89 and 1 indicate that the release of quercetin from PVAIC1.0 to PVAIC5.0 were controlled by the mechanism of anomalous (non-Fickian) diffusion. The diffusion exponent for PVAIC0.5 is in the range of 0.89 and 1, which denotes the case II transport mechanism (zero-order kinetics).

Table 3. Modeling results of quercetin released from PVA hydrogels loaded with quercetin/HP-β-CD inclusion complex fitting with Zero-order and Korsmeyer–Peppas models.

Sample	Release Kinetic Models			
	Zero-Order		Korsmeyer–Peppas	
	k	r^2	n	r^2
PVAIC0.5	0.78	0.7953	1.31	0.995
PVAIC1.0	0.67	0.9127	0.77	0.9671
PVAIC1.5	1.10	0.844	0.68	0.9526
PVAIC2.5	1.49	0.7033	0.58	0.9748
PVAIC5.0	2.31	0.6468	0.77	0.9535

2.7. In Vitro Antioxidant Activity: DPPH-Radical Scavenging Ability Assay

The encapsulation of quercetin within cyclodextrin increases the antioxidant activity compared with free molecules [42]. The antioxidant activity of quercetin/HP-β-CD inclusion complex is widely used in 1,1-diphenyl-2-picrylhydrazyl (DPPH) radical scavenging assay in comparison with free counterparts and the results were shown in Figure 6. A concentration of 120 µg/mL quercetin/HP-β-CD inclusion complex was determined to have 74.80% DPPH inhibition compared with free quercetin, with 61.23% of inhibition, and ascorbic acid with 74.80%. HP-β-CD inclusion complex showed significant antioxidant activity compared with quercetin alone (Figure 6) in a dose-dependent manner. The results indicate that HP-β-CD inclusion complex is associated with the sustained release of quercetin and increases the stability of incorporated bioactive compounds such as quercetin, significantly improving its antioxidant activity [43].

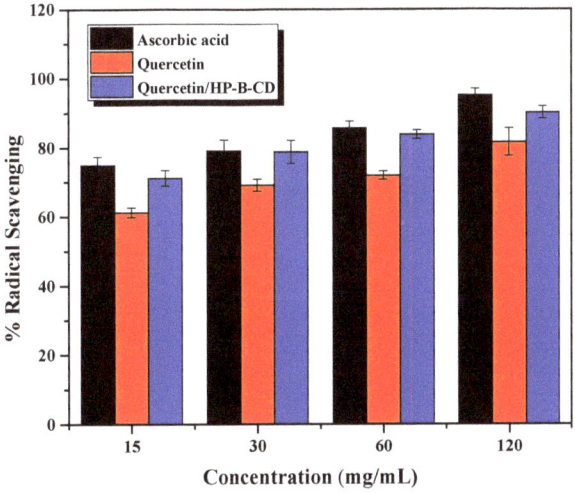

Figure 6. Scavenging activity of ascorbic acid, quercetin and quercetin/ HP-β-CD as determined by DPPH assay. Values expressed as mean ± standard deviation (n = 5).

2.8. Cytotoxicity Evaluation

The indirect cytotoxicity evaluation of all samples toward NCTC 929 clone cells investigated the potential use of drug loaded PVA hydrogels as wound dressing material. Figure 7 shows the viability of cells obtained from MTT assay after being cultured with various concentrations of media extracted from neat PVA hydrogel as blank PVA, and PVA hydrogels loaded with quercetin/HP-β-CD inclusion complex (PVAIC0.5, PVAIC1.0, PVAIC1.5, PVAIC2.5, and PVAIC5.0) compared with that obtained after cells were cultured with fresh SFM. According to the in vitro cytotoxicity standard, a reduction of cell viability by more than 30% is considered a cytotoxic effect [44]. The viability of NCTC 929 clone cells with all extraction medium concentrations, i.e., neat PVA, PVAIC0.5, PVAIC1.0, PVAIC1.5, PVAIC2.5, and PVAIC5.0, were 92.7–96.3, 76.8–88.2, 80.0–97.5, 74.7–92.9, 87.6–96.8, and 78.3–85.3%, respectively. Cell viability was higher than 70% in all prepared hydrogels loaded with quercetin/HP-β-CD inclusion complex. Therefore, all hydrogels loaded with quercetin/HP-β-CD inclusion complex with extraction medium concentrations of 0.5, 5, 10, 25, and 50 mg/mL had cell viability of more than 70%, indicating that these conditions were non-cytotoxic, and the hydrogels have potential in wound dressing applications.

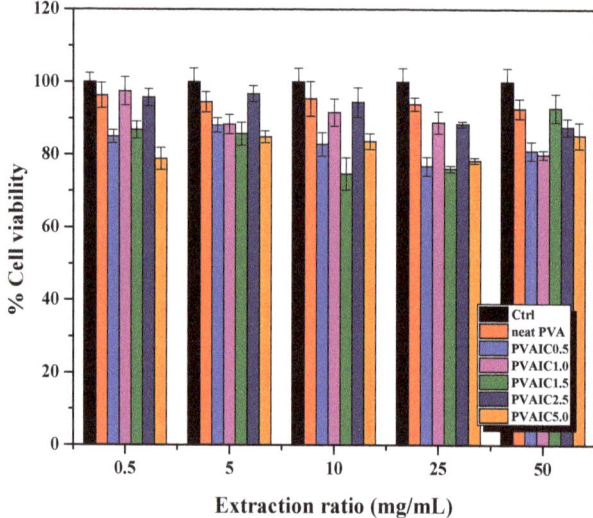

Figure 7. Indirect cytotoxicity of PVA hydrogels loaded with various amounts of quercetin/HP-β-CD inclusion complex with extraction medium: neat PVA, 0.5, 5, 10, 25, and 50 mg/mL cultured with NCTC 929 clone cells.

3. Conclusions

We successfully prepared PVA hydrogels containing quercetin/HP-β-CD inclusion complex that achieved efficient drug loading and effective encapsulation. The quercetin/HP-β-CD inclusion complex was prepared with an excess molar ratio of HP-β-CD via the solvent evaporation method. The successful encapsulation of quercetin in HP-β-CD was confirmed by FT-IR and XRD, in which characteristic peaks of quercetin could not be observed after encapsulation due to the hindering of CD cavities. As observed on SEM, the quercetin/HP-β-CD inclusion complex demonstrated rod-shaped particles. The prepared quercetin/HP-β-CD inclusion complex had 90.50 ± 1.84% encapsulation efficiency and 4.67 ± 0.13% loading capacity. It was incorporated into PVA hydrogels in various amounts (0, 0.5, 1.0, 1.5, 2.5, and 5.0% MIC/$V_{solvent\ ratio}$). The hydrogels were evaluated for their potential use as wound dressing in terms of gelation, swelling, compression, release characteristics, antioxidant properties, and in vitro non-cytotoxicity. The incorporation of quercetin/HP-β-CD inclusion complex into PVA hydrogels had no effect on the swelling ratio of the hydrogels. The gelation and compressive strength of hydrogels decreased with increasing inclusion complex content, whereas the cumulative release of quercetin from hydrogels increased with higher amounts of inclusion complex.

Hydrogels loaded with quercetin/HP-β-CD inclusion complex investigated for their cytotoxicity in vitro by MTT assay were shown to be nontoxic toward mouse fibroblast NCTC 929 clone cells at the test concentrations. The results showed significantly increased antioxidant properties of quercetin/HP-β-CD inclusion complex compared with free quercetin. These findings suggest that the use of quercetin/HP-β-CD inclusion complex ensures non-cytotoxicity and potential antioxidant activity. Hence, hydrogel with quercetin/HP-β-CD inclusion complex is an attractive candidate for the encapsulation of bioactive compounds for biomedical applications.

4. Materials and Methods

4.1. Materials

Quercetin hydrate ($C_{15}H_{10}O_7 \cdot xH_2O$, Mw 302.24, >96% purity) and hydroxypropyl-β-cyclodextrin (HP-β-CD, Mw ~1380–1500, >99.0% purity) were purchased from Tokyo Chemical Industry (Portland, OR, USA). Polyvinyl alcohol (PVA; Mw 89,000–98,000, 99+%

hydrolyzed) was purchased from Sigma Aldrich (Saint Louis, MO, USA). Methanol (MeOH; HPLC grade) and dimethyl sulfoxide (DMSO; AR grade) were purchased from RCI Labscan (Bangkok, Thailand).

4.2. Formation of Quercetin/HP-β-CD Inclusion Complex

Lyophilized formulations of quercetin/HP-β-CD inclusion complex were prepared by freeze-drying using the neutralization method [45]. Freeze-drying is an excellent method for preserving a wide variety of heat-sensitive materials such as proteins, microbes, pharmaceutical agents, tissues, and plasma [46]. The molar ratio of quercetin to HP-β-CD, following Lan et al., was 1:4.35 [24]. First, 10 g of HP-β-CD was dissolved in 50 mL distilled water, and 0.5 g of quercetin was dissolved in 70 mL MeOH at 60 °C to obtain a homogeneous solution. Then, both solutions were mixed and refluxed at 65 °C overnight. The solution was stirred continuously at room temperature for 5 h for complete encapsulation. Next, MeOH was removed by heating the solution at 75 °C for 2 h. The solution was freeze-dried after being filtered (0.45 μm) and frozen at −20 °C. Finally, quercetin/HP-β-CD inclusion complex was obtained after freeze-drying at −50 °C for 48 h. The resulting lyophilized quercetin/HP-β-CD inclusion complex powder was stored in airtight containers and protected from light until use. Freeze-dried drug–cyclodextrin complexes were used for hydrogel preparation.

4.3. Characterization of Quercetin/HP-β-CD Inclusion Complex

Fourier-transform infrared spectrometry (FT-IR; Nicolet iS5 with iD7, Thermo Scientific, Waltham, MA, USA) was performed to reveal the chemical structure and group interaction of inclusion complex and analyzed in ATR mode with a framework region of 650 to 4000 cm^{-1} with 64 scans at resolution of 4 cm^{-1}. Samples were prepared as KBr pellets.

The crystalline structure of quercetin/HP-β-CD inclusion complex was examined by X-ray diffractometry (XRD; SmartLab, Rigaku, Japan) with diffraction angle of 2θ ranging from 5 to 50° at a scanning rate of 0.02°/s.

A scanning electron microscope (SEM; JEOL, JSM-6610LV, Japan) was used to observe the morphology of quercetin/HP-β-CD inclusion complex with 15 kV accelerating voltage and magnification of 5000, 1000, and 500 times after the samples were vacuum-coated with a fine layer of gold.

4.4. Determination of Encapsulation Efficiency and Loading Capacity of Quercetin/HP-β-CD Inclusion Complex

The drug-loading process was carried out through a diffusional mechanism, and the encapsulation efficiency (%EE) was used to illustrate the quantity of added drug encapsulated in the formulation. The amount of encapsulated quercetin was determined using a direct method [18] and recorded on a calibration curve at 379 nm.

Briefly, 20 mg of quercetin/HP-β-CD inclusion complex was dissolved in 20 mL DMSO. The amount of quercetin was quantified using a UV-Vis spectrophotometer at a wavelength that provided the maximum absorbance (379 nm), and the concentration was back-calculated from the predetermined quercetin calibration curve. Encapsulation efficiency (%EE) and loading capacity (%LC) were calculated from the following equations [47]:

$$EE\,(\%) = \frac{W_{Entrapped\ Quercetin}}{W_{Quercetin\ loaded}} \times 100 \quad (1)$$

$$LC\,(\%) = \frac{W_{Entrapped\ Quercetin}}{W_{Sample}} \times 100 \quad (2)$$

4.5. Preparation of Quercetin/HP-β-CD Inclusion Complex-Loaded Hydrogels

Quercetin/HP-β-CD inclusion complex was loaded into 10% *w/v* PVA hydrogels. Briefly, 2 g of PVA was slowly dissolved in 20 mL distilled water at 80–100 °C for 40 min.

Then, an accurate weight of quercetin/HP-β-CD inclusion complex was added to PVA solutions at concentrations of 0, 0.5, 1.0, 1.5, 2.5, and 5.0% w/v, which were labeled as conditions PVAIC0, PVAIC0.5, PVAIC1.0, PVAIC1.5, PVAIC2.5, and PVAIC5.0, respectively. Then, the solutions were stirred continuously for 20 min to obtain homogeneous solutions with uniform distribution. The homogeneous solutions were then sonicated in an 80 °C ultrasonic bath to remove air bubbles. Then, the viscous solutions were poured into 90 mm Petri dishes and covered with aluminum foil. The drug-loaded PVA hydrogels were physically crosslinked by repeated freeze–thaw cycles, with freezing at −20 °C for 20 h then thawing at room temperature for 4 h [48]. This freeze–thaw process was repeated for 4 cycles to obtain quercetin/HP-β-CD inclusion complex-loaded hydrogels. The Petri dishes containing prepared hydrogels were then sealed with parafilm and kept in the refrigerator.

4.6. Gelation of PVA Hydrogels Loaded with Quercetin/HP-β-CD Inclusion Complex

To investigate the gel fractions of the hydrogels used in the study, it was first necessary to ensure that the samples had achieved a constant weight. The prepared hydrogels were cut into 0.5 cm radius round and then dried in an oven at 50 °C for 48 h to obtain the dry weight (W_{dry}). After that, the dried hydrogels were submerged in phosphate-buffered saline (PBS, pH = 7.4) at 37 °C for 24 h with agitation (100 rpm) and then were dried again at 50 °C for 48 h to obtain the hydrogel weight after extraction ($W_{after\ extraction}$):

$$\text{Gelation (\%)} = \frac{W_{after\ extraction}}{W_{dry}} \times 100 \qquad (3)$$

4.7. Swelling Ratio of PVA Hydrogels Loaded with Quercetin/HP-β-CD Inclusion Complex

The swelling ratio was measured by taking the dried hydrogel (m_{dry}) and the wet mass (m_{wet}) of the hydrogel after immersion in a solution for a specified time [49]. First, each hydrogel was cut into circles of 0.5 cm radius that were then immersed in 5 mL of phosphate-buffered saline (PBS, pH = 7.4) at 37 °C with agitation (100 rpm) for 10, 30, 60, 180, and 360 min. Finally, the swelling ratio of the hydrogels was calculated using the following equation:

$$\text{Swelling ratio (\%)} = \frac{m_{wet} - m_{dry}}{m_{dry}} \times 100 \qquad (4)$$

4.8. Mechanical Test of PVA Hydrogels Loaded with Quercetin/HP-β-CD Inclusion Complex

To observe the effect of loading the drug into the hydrogel on its mechanical properties, compression testing was performed. The hydrogels were cut into circles of 0.5 cm radius in all experiments. The compressive strength of neat PVA hydrogel and drug-loaded hydrogels was measured with a universal testing machine (LRX-Plus, AMETEK Lloyd Instruments Ltd., Hampshire, UK) in terms of stiffness and compressive modulus. The loading cell was set at a 500 N, with 0.05 N preload and 1 mm/min speed, and the measurement was stopped when the load reached 10 N [50].

4.9. In Vitro PVA Hydrogels Loaded with Quercetin/HP-β-CD Inclusion Complex Dissolution and Release Kinetics Study

The cumulative amount of released quercetin/HP-β-CD from hydrogels was measured using an immersion method according to Chuysinuan et al. [51] with minor modifications. The disc-shaped specimens (2.8 cm in diameter) were immersed in phosphate-buffered saline (PBS, pH = 7.4) and incubated at 37 °C under continuous stirring. The sample solution (1 mL) was withdrawn at specific time intervals, and an equal amount of fresh release medium was added. The absorbance of released quercetin was measured with a UV-Vis spectrophotometer (LAMBDA 850+, PerkinElmer, Waltham, MA, USA) at a wavelength of 370 nm. The obtained data were back calculated from the predetermined quercetin calibration curve ($R^2 = 0.9944$) to determine the amounts of quercetin released from the hydrogel samples. The cumulative release of quercetin from each hydrogel at different time points (with measurement carried out in triplicate at each time point) was reported as parts per million of the amounts of quercetin released ($W_{\text{Quercetin released}}$) divided by the weight of hydrogels (W_{hydrogel}):

$$\text{Cumulative release (\%)} = \frac{W_{\text{Quercetin released}}}{W_{\text{hydrogel}}} \times 10^6 \tag{5}$$

A release kinetics study was used to ascertain the release mechanism, and the quercetin release data were fitted to Korsmeyer–Peppas and zero-order models. The first 60% of the drug release data was used to fit the models.

The Korsmeyer–Peppas model is a simple model that describes drug release from polymer nanoparticle systems [52,53]:

$$M_t/M_\infty = kt^n$$

where M_t/M_∞ is the quercetin release fraction at time t, k is a constant incorporating geometric structural features, and n is the release exponent indicating the release rate mechanism. The value of n indicates the mechanism of the release; a value around 0.45 indicates case I (Fickian) diffusion, between 0.45 and 0.89 indicates anomalous (non-Fickian) diffusion, and between 0.89 and 1 indicates case II transport (zero-order kinetics).

A zero-order kinetic model is used to describe drugs that are released slowly with a constant concentration and can be characterized as ideal kinetic model in that it maintains constant drug levels during the delivery process [54,55]:

$$M_t/M_\infty = kt$$

where M_t/M_∞ is the quercetin release fraction at time t, with zero-order rate constant, and t is investigation time.

4.10. In Vitro Antioxidant Characterization

The antioxidant activity of free quercetin and quercetin/HP-β-CD inclusion complex was evaluated using 2, 2-diphenyl-1-picrylhydrazyl (DPPH) free radical scavenging assay following [3]. First, quercetin and quercetin/HP β CD in various concentrations (15, 30, 60, and 120 µg/mL) were dissolved in methanol. Then, the samples (1 mL) were mixed with 3 mL DPPH solution (0.1 mM in methanol). The mixture was incubated for 30 min in the dark. Finally, the absorbance of the reaction mixture was measured using a microplate reader at 517 nm. Radical scavenging activity (%) was calculated using the following equation. Ascorbic acid was used as a standard and all analyses were performed in triplicate:

$$\text{DPPH radical scavenging activity (\%)} = \frac{(\text{Absorbance of control} - \text{Absorbance of tested sample})}{\text{Absorbance of control}} \times 100 \tag{6}$$

4.11. Cytotoxicity Evaluation

While CDs are chosen for pharmaceutical formulations to increase the solubility, bioavailability, and stability of many medications, their structure and cytotoxic capability are crucial for more effective drug delivery [56]. To evaluate the potential biomedical applications of PVA hydrogels loaded with various amounts of quercetin/HP-β-CD inclusion complex, their biocompatibility in terms of indirect cytotoxicity toward NCTC 929 clone cells (ATCC-CCL-1, Rockville, MD, USA (17th passage) was evaluated in accordance with the ISO10993-5 standard test method [57]. First, cells were cultured in Dulbecco's Modified Eagle Medium (DMEM (1×); GIBCO, Waltham, MA, USA) containing 10% fetal bovine serum (FBS; GIBCO, USA) and 1% antibiotic–antimycotic agent (GIBCO, Waltham, MA, USA). Then, the cells were seeded into 96-well tissue-culture polystyrene (TCPS) plates (SPL Lifescience, Pochon, Korea) at 8000 cells/well and then incubated at 37 °C in a humidified atmosphere containing 95% air and 5% CO_2. Next, the hydrogel samples were sterilized by exposure to UV radiation for 30 min/side and were immersed in serum-free medium (SFM, containing DMEM and 1% antibiotic–antimycotic agent) in a 96-well TCPS plate. The samples were incubated for 24 h to produce sample extraction at five ratios of extraction medium (0.5, 5, 10, 25, and 50 mg/mL). NCTC 929 clone cells were cultured separately in culture medium in wells of TCPS plates at 8000 cells/well for 24 h to allow cells to attach to the well surface. After that, the medium was replaced with an extraction medium, and mouse fibroblast L929 cells were incubated for a further 24 h.

The viability of cells cultured in each extraction medium was determined with 3-(4,5-dimethylthiazol-2-yl)-2,5-diphenyltetrazolium bromide (MTT) assay. After treatment, the medium was removed, and the samples were washed with PBS; then, MTT solution (0.5 mg/mL) was added, and samples were incubated for 3 h. After that, the MTT solution was removed from the well and replaced by DMSO (Labscan, Bangkok, Thailand) to dissolve the formazan crystals. Finally, the absorbance of the solutions was measured at 570 nm using a microplate reader (BioTek Instruments, Winooski, VT, USA) to investigate cell viability. The viability of cells cultured with fresh SFM was used as a control.

Author Contributions: Conceptualization, supervision, and funding acquisition, P.S.; Data curation, formal analysis, investigation, and methodology, N.W., C.C., S.C., O.S. and P.C.; Project administration, P.E.; Data curation, validation, and manuscript review, O.S., P.C. and S.T.; Visualization, N.W., C.C. and S.C.; Writing—original draft, N.W., C.C. and S.C. All authors have read and agreed to the published version of the manuscript.

Funding: This research was funded by the 90th Anniversary of Chulalongkorn University Scholarship Batch 50: GCUGR1125643048M No. 2–14 and Fundamental Fund 2565: FF65, Chulalongkorn University, Thailand.

Institutional Review Board Statement: Not applicable.

Informed Consent Statement: Not applicable.

Data Availability Statement: Not applicable.

Acknowledgments: This work was conducted with the support of the Herbal Extract-Infused Advanced Wound Dressing Research Unit, Rachadaphiseksomphot Endowment Fund, Chulalongkorn University, Thailand.

Conflicts of Interest: The authors declare no conflict of interest.

Abbreviations

CDs	Cyclodextrins
DMSO	Dimethyl sulfoxide
DPPH2	2-diphenyl-1-picrylhydrazyl
EE	Encapsulation efficiency
FT-IR	Fourier-transform infrared spectrometry
GR	Geraniol
HP-β-CD	Hydroxypropyl-β-cyclodextrin
IC	Quercetin/HP-β-CD inclusion complex
LC	Loading capacity
m	Mass
MeOH	Methanol
MTT	3-[4,5-dimethylthiazol-2-yl]-2,5 diphenyl tetrazolium bromide
NCTC 929	Connective mouse tissue (L cell, L-929, derivative of Strain L)
NSCS	N–succinyl chitosan
PBS	Phosphate-buffered saline

PVA	Polyvinyl alcohol
PVAc	Polyvinyl acetate
SEM	Scanning electron microscope
TCPS	Tissue-culture polystyrene
W	Weight
XRD	X-ray powder diffraction

References

1. Cuzzell, J. Choosing a wound dressing. *Geriatr. Nurs.* **1997**, *18*, 260–265. [CrossRef]
2. Dhivya, S.; Padma, V.V.; Santhini, E. Wound dressings—A review. *BioMedicine* **2015**, *5*, 22. [CrossRef] [PubMed]
3. Maver, T.; Kurecic, M.; Smrke, D.; Stana-Kleinschek, K.; Maver, U. Plant-derived medicines with potential use in wound treatment. In *Herbal Medicine*, 1st ed.; Builders, P., Ed.; IntechOpen: London, UK, 2019; pp. 121–150. [CrossRef]
4. Shedoeva, A.; Leavesley, D.; Upton, Z.; Fan, C. Wound Healing and the Use of Medicinal Plants. *J. Evid. Based Complement. Altern. Med.* **2019**, *2019*, 2684108. [CrossRef]
5. Bhatt, K.; Flora, S.J.S. Oral co-administration of α-lipoic acid, quercetin and captopril prevents gallium arsenide toxicity in rats. *Environ. Toxicol. Pharmacol.* **2009**, *28*, 140–146. [CrossRef] [PubMed]
6. Heinonen, I.M.; Meyer, A.S. Antioxidants in fruits, berries and vegetables. In *Fruit and Vegetable Processing*, 1st ed.; Jongen, W., Ed.; Woodhead Publishing Limited: Cambridge, OH, USA, 2002; pp. 23–51. [CrossRef]
7. Li, Y.; Yao, J.; Han, C.; Yang, J.; Chaudhry, M.T.; Wang, S.; Liu, H.; Yin, Y. Quercetin, Inflammation and Immunity. *Nutrients* **2016**, *8*, 167. [CrossRef] [PubMed]
8. Nathiya, S.; Durga, M.; Devasena, T. Quercetin, encapsulated quercetin and its application—A review. *Int. J. Pharm. Pharm. Sci.* **2014**, *6*, 20–26.
9. Razmara, R.S.; Daneshfar, A.; Sahraei, R. Solubility of Quercetin in Water + Methanol and Water + Ethanol from (292.8 to 333.8) K. *J. Chem. Eng. Data* **2010**, *55*, 3934–3936. [CrossRef]
10. Liu, M.; Dong, L.; Chen, A.; Zheng, Y.; Sun, D.; Wang, X.; Wang, B. Inclusion complexes of quercetin with three β-cyclodextrins derivatives at physiological pH: Spectroscopic study and antioxidant activity. *Spectrochim. Acta A Mol. Biomol. Spectrosc.* **2013**, *115*, 854–860. [CrossRef]
11. Singh, P.; Kumar, S. Chapter 2—Microbial enzyme in food biotechnology. In *Enzymes in Food Biotechnology*, 1st ed.; Kuddus, M., Ed.; Academic Press: London, UK, 2019; pp. 19–28.
12. Del Valle, E.M.M. Cyclodextrins and their uses: A review. *Process Biochem.* **2004**, *39*, 1033–1046. [CrossRef]
13. Kfoury, M.; Hădărugă, N.G.; Hădărugă, D.I.; Fourmentin, S. 4—Cyclodextrins as encapsulation material for flavors and aroma. In *Encapsulations*; Grumezescu, A.M., Ed.; Academic Press: Cambridge, MA, USA, 2016; Volume 2, pp. 127–192.
14. Gupta, A.; Keddie, D.J.; Kannappan, V.; Gibson, H.; Khalil, I.R.; Kowalczuk, M.; Martin, C.; Shuai, X.; Radecka, I. Production and characterisation of bacterial cellulose hydrogels loaded with curcumin encapsulated in cyclodextrins as wound dressings. *Eur. Polym. J.* **2019**, *118*, 437–450. [CrossRef]
15. Esmaeilpour, D.; Hussein, A.A.; Almalki, F.A.; Shityakov, S.; Bordbar, A.K. Probing inclusion complexes of ?-hydroxypropyl-β-cyclodextrin with mono-amino mono-carboxylic acids: Physicochemical specification, characterization and molecular modeling. *Heliyon* **2020**, *6*, e03360. [CrossRef] [PubMed]
16. Zhang, M.; Zhao, X. Alginate hydrogel dressings for advanced wound management. *Int. J. Biol. Macromol.* **2020**, *162*, 1414–1428. [CrossRef]
17. Boonkaew, B.; Barber, P.M.; Rengpipat, S.; Supaphol, P.; Kempf, M.; He, J.; John, V.T.; Cuttle, L. Development and Characterization of a Novel, Antimicrobial, Sterile Hydrogel Dressing for Burn Wounds: Single–Step Production with Gamma Irradiation Creates Silver Nanoparticles and Radical Polymerization. *J. Pharm. Sci.* **2014**, *103*, 3244–3253. [CrossRef]
18. Weller, C.; Weller, C.; Team, V. 4—Interactive dressings and their role in moist wound management. In *Advanced Textiles for Wound Care*, 2nd ed.; Rajendran, S., Ed.; Woodhead Publishing: Cambridge, UK, 2019; pp. 105–134.
19. Tsou, Y.-H.; Khoneisser, J.; Huang, P.-C.; Xu, X. Hydrogel as a bioactive material to regulate stem cell fate. *Bioact. Mater.* **2016**, *1*, 39–55. [CrossRef] [PubMed]
20. Figueroa-Pizano, M.D.; Vélaz, I.; Martínez-Barbosa, M.E. A Freeze-Thawing Method to Prepare Chitosan-Poly(vinyl alcohol) Hydrogels without Crosslinking Agents and Diflunisal Release Studies. *J. Vis. Exp.* **2020**, *155*, e59636. [CrossRef] [PubMed]
21. Hassan, C.; Peppas, N. Structure and Morphology of Freeze/Thawed PVA Hydrogels. *Macromolecules* **2000**, *33*, 2472–2479. [CrossRef]
22. Catauro, M.; Papale, F.; Bollino, F.; Piccolella, S.; Marciano, S.; Nocera, P.; Pacifico, S. Silica/quercetin sol-gel hybrids as antioxidant dental implant materials. *Sci. Technol. Adv. Mater.* **2015**, *16*, 035001. [CrossRef]
23. Da Silva Júnior, W.F.; Bezerra de Menezes, D.L.; de Oliveira, L.C.; Koester, L.S.; Oliveira de Almeida, P.D.; Lima, E.S.; de Azevedo, E.P.; da Veiga Júnior, V.F.; Neves de Lima, Á.A. Inclusion Complexes of β and HPβ-Cyclodextrin with α, β Amyrin and In Vitro Anti-Inflammatory Activity. *Biomolecules* **2019**, *9*, 241. [CrossRef]
24. Lan, Q.; Di, D.; Wang, S.; Zhao, Q.; Gao, Y.; Chang, D.; Jiang, T. Chitosan-N-acetylcysteine modified HP-β-CD inclusion complex as a potential ocular delivery system for anti-cataract drug: Quercetin. *J. Drug Deliv. Sci. Technol.* **2020**, *55*, 101407. [CrossRef]

25. Pereira, B.A.; Silva, M.A.; Barroca, J.M.; Marques, M.P.M.; Braga, S.S. Physicochemical Properties, Antioxidant Action and Practical Application in Fresh Cheese of the Solid Inclusion Compound γ-Cyclodextrin•quercetin, in Comparison with β-Cyclodextrin•quercetin. *Arab. J. Chem.* **2020**, *13*, 205–215. [CrossRef]
26. Aytac, Z.; Ipek, S.; Durgun, E.; Uyar, T. Antioxidant electrospun zein nanofibrous web encapsulating quercetin/cyclodextrin inclusion complex. *J. Mater. Sci.* **2018**, *53*, 1527–1539. [CrossRef]
27. Han, D.; Han, Z.; Liu, L.; Wang, Y.; Xin, S.; Zhang, H.; Yu, Z. Solubility Enhancement of Myricetin by Inclusion Complexation with Heptakis-O-(2-Hydroxypropyl)-β-Cyclodextrin: A Joint Experimental and Theoretical Study. *Int. J. Mol. Sci.* **2020**, *21*, 766. [CrossRef] [PubMed]
28. Zhu, G.; Xiao, Z.; Zhu, G.; Zhou, R.; Niu, Y. Encapsulation of l-menthol in hydroxypropyl-β-cyclodextrin and release characteristics of the inclusion complex. *Pol. J. Chem. Technol.* **2016**, *18*, 110–116. [CrossRef]
29. Kim, S.J. Study of flavonoid/hydroxypropyl-β-cyclodextrin inclusion complexes by UV-Vis, FT-IR, DSC, and X-Ray diffraction analysis. *Prev. Nutr. Food Sci.* **2020**, *25*, 449–456. [CrossRef] [PubMed]
30. Lv, R.; Qi, L.; Zou, Y.; Zou, J.; Luo, Z.; Shao, P.; Tamer, T.M. Preparation and structural properties of amylose complexes with quercetin and their preliminary evaluation in delivery application. *Int. J. Food Prop.* **2019**, *22*, 1445–1462. [CrossRef]
31. Ding, Y.; Pang, Y.; Vara Prasad, C.; Wang, B. Formation of inclusion complex of enrofloxacin with 2-hydroxypropyl-β-cyclodextrin. *Drug Deliv.* **2020**, *27*, 334–343. [CrossRef]
32. Pralhad, T.; Rajendrakumar, K. Study of freeze-dried quercetin–cyclodextrin binary systems by DSC, FT-IR, X-ray diffraction and SEM analysis. *J. Pharm. Biomed.* **2004**, *34*, 333–339. [CrossRef]
33. Pradhan, C.P.; Mandal, A.; Dutta, A.; Sarkar, R.; Kundu, A.; Saha, S. Delineating the behavior of Berberis anthocyanin/β-cyclodextrin inclusion complex in vitro: A molecular dynamics approach. *LWT* **2022**, *157*, 113090. [CrossRef]
34. Choipang, C.; Buntum, T.; Chuysinuan, P.; Techasakul, S.; Supaphol, P.; Suwantong, O. Gelatin scaffolds loaded with asiaticoside/2-hydroxypropyl-β-cyclodextrin complex for use as wound dressings. *Polym. Adv. Technol.* **2021**, *32*, 1187–1193. [CrossRef]
35. Al-Qubaisi, M.S.; Rasedee, A.; Flaifel, M.H.; Eid, E.E.M.; Hussein-Al-Ali, S.; Alhassan, F.H.; Salih, A.M.; Hussein, M.Z.; Zainal, Z.; Sani, D.; et al. Characterization of thymoquinone/hydroxypropyl-β-cyclodextrin inclusion complex: Application to anti-allergy properties. *Eur. J. Pharm. Sci.* **2019**, *133*, 167–182. [CrossRef]
36. Hadian, Z.; Maleki, M.; Abdi, K.; Atyabi, F.; Mohammadi, A.; Khaksar, R. Preparation and Characterization of Nanoparticle β-Cyclodextrin:Geraniol Inclusion Complexes. *Iran J. Pharm. Res.* **2018**, *17*, 39–51. [PubMed]
37. Afshar, M.; Dini, G.; Vaezifar, S.; Mehdikhani, M.; Movahedi, B. Preparation and characterization of sodium alginate/polyvinyl alcohol hydrogel containing drug-loaded chitosan nanoparticles as a drug delivery system. *J. Drug Deliv. Sci. Technol.* **2020**, *56*, 101530. [CrossRef]
38. Chou, S.-F.; Woodrow, K.A. Relationships between mechanical properties and drug release from electrospun fibers of PCL and PLGA blends. *J. Mech. Behav. Biomed. Mater.* **2017**, *65*, 724–733. [CrossRef] [PubMed]
39. Qing, X.; He, G.; Liu, Z.; Yin, Y.; Cai, W.; Fan, L.; Fardim, P. Preparation and properties of polyvinyl alcohol/N–succinyl chitosan/lincomycin composite antibacterial hydrogels for wound dressing. *Carbohydr. Polym.* **2021**, *261*, 117875. [CrossRef] [PubMed]
40. Sampson, L.; Rimm, E.; Hollman, P.C.; de Vries, J.H.; Katan, M.B. Flavonol and flavone intakes in US health professionals. *J. Am. Diet. Assoc.* **2002**, *102*, 1414–1420. [CrossRef]
41. Cai, B.; Engqvist, H.; Bredenberg, S. Development and evaluation of a tampering resistant transdermal fentanyl patch. *Int. J. Pharm.* **2015**, *488*, 102–107. [CrossRef] [PubMed]
42. Thanyacharoen, T.; Chuysinuan, P.; Techasakul, S.; Nooeaid, P.; Ummartyotin, S. Development of a gallic acid-loaded chitosan and polyvinyl alcohol hydrogel composite: Release characteristics and antioxidant activity. *Int. J. Biol. Macromol.* **2018**, *107*, 363–370. [CrossRef] [PubMed]
43. Nalini, T.; Basha, S.K.; Sadiq, A.M.; Kumari, V.S. In vitro cytocompatibility assessment and antibacterial effects of quercetin encapsulated alginate/chitosan nanoparticle. *Int. J. Biol. Macromol.* **2022**, *219*, 304–311. [CrossRef]
44. Srivastava, G.K.; Alonso-Alonso, M.L.; Fernandez-Bueno, I.; Garcia-Gutierrez, M.T.; Rull, F.; Medina, J.; Coco, R.M.; Pastor, J.C. Comparison between direct contact and extract exposure methods for PFO cytotoxicity evaluation. *Sci. Rep.* **2018**, *8*, 1425. [CrossRef]
45. Figueiras, A.; Carvalho, R.A.; Ribeiro, L.; Torres-Labandeira, J.J.; Veiga, F.J.B. Solid-state characterization and dissolution profiles of the inclusion complexes of omeprazole with native and chemically modified β-cyclodextrin. *Eur. J. Pharm. Biopharm.* **2007**, *67*, 531–539. [CrossRef]
46. Kulkarni, A.; Dias, R.; Ghorpade, V. Freeze dried multicomponent inclusion complexes of quercetin: Physicochemical evaluation and pharmacodynamic study. *Marmara Pharm. J.* **2019**, *23*, 403–414. [CrossRef]
47. Niu, Y.; Deng, J.; Xiao, Z.; Kou, X.; Zhu, G.; Liu, M.; Liu, S. Preparation and slow release kinetics of apple fragrance/β-cyclodextrin inclusion complex. *J. Therm. Anal. Calorim.* **2020**, *143*, 3775–3781. [CrossRef]
48. Ahsan, A.; Farooq, M.A. Therapeutic potential of green synthesized silver nanoparticles loaded PVA hydrogel patches for wound healing. *J. Drug Deliv. Sci. Technol.* **2019**, *54*, 101308. [CrossRef]
49. Raghuwanshi, V.S.; Garnier, G. Characterisation of hydrogels: Linking the nano to the microscale. *Adv. Colloid Interface Sci.* **2019**, *274*, 102044. [CrossRef]

50. Fan, W.; Zhang, Z.; Liu, Y.; Wang, J.; Li, Z.; Wang, M. Shape memory polyacrylamide/gelatin hydrogel with controllable mechanical and drug release properties potential for wound dressing application. *Polymer* **2021**, *226*, 123786. [CrossRef]
51. Chuysinuan, P.; Thanyacharoen, T.; Thongchai, K.; Techasakul, S.; Ummartyotin, S. Preparation of chitosan/hydrolyzed collagen/hyaluronic acid based hydrogel composite with caffeic acid addition. *Int. J. Biol. Macromol.* **2020**, *162*, 1937–1943. [CrossRef]
52. Korsmeyer, R.W.; Gurny, R.; Doelker, E.; Buri, P.; Peppas, N.A. Mechanisms of solute release from porous hydrophilic polymers. *Int. J. Pharm.* **1983**, *15*, 25–35. [CrossRef]
53. Brunner, E. Reaktionsgeschwindigkeit in heterogenen Systemen. *Z. Phys. Chem.* **1904**, *47*, 56–102. [CrossRef]
54. Rasetti-Escargueil, C.; Grangé, V. Pharmacokinetic profiles of two tablet formulations of piroxicam. *Int. J. Pharm.* **2005**, *295*, 129–134. [CrossRef]
55. Freitas, M.N.; Marchetti, J.M. Nimesulide PLA microspheres as a potential sustained release system for the treatment of inflammatory diseases. *Int. J. Pharm.* **2005**, *295*, 201–211. [CrossRef]
56. Staedler, D.; Idrizi, E.; Kenzaoui, B.H.; Juillerat-Jeanneret, L. Drug combinations with quercetin: Doxorubicin plus quercetin in human breast cancer cells. *Cancer Chemother. Pharmacol.* **2011**, *68*, 1161–1172. [CrossRef] [PubMed]
57. Kudłacik-Kramarczyk, S.; Drabczyk, A.; Głąb, M.; Alves-Lima, D.; Lin, H.; Douglas, T.E.L.; Kuciel, S.; Zagórska, A.; Tyliszczak, B. Investigations on the impact of the introduction of the Aloe vera into the hydrogel matrix on cytotoxic and hydrophilic properties of these systems considered as potential wound dressings. *Mater. Sci. Eng. C.* **2021**, *123*, 111977. [CrossRef] [PubMed]

Article

The Preparation of Novel P(OEGMA-co-MEO₂MA) Microgels-Based Thermosensitive Hydrogel and Its Application in Three-Dimensional Cell Scaffold

Yang Liu *, Yu-Ning Luo, Pei Zhang, Wen-Fei Yang, Cai-Yao Zhang and Yu-Li Yin

Hunan Provincial Key Laboratory of Tumor Microenvironment Responsive Drug Research,
Hunan Province Cooperative Innovation Center for Molecular Target New Drug Study,
School of Pharmacy, Hengyang Medical School, University of South China, Hengyang 421001, China;
luo1748058392@163.com (Y.-N.L.); zp1832711702@163.com (P.Z.); yang18138800314@163.com (W.-F.Y.);
caiyao.zhang@nuclover.com (C.-Y.Z.); yinyulimy@163.com (Y.-L.Y.)
* Correspondence: liuyanghxl@126.com; Tel.: +86-0734-8281296

Abstract: Thermosensitive hydrogel scaffolds have attracted particular attention in three-dimensional (3D) cell culture. It is very necessary to develop a type of thermosensitive hydrogel material with low shrinkage, and excellent biocompatibility and biodegradability. Here, five types of thermosensitive microgels with different volume phase transition temperature (VPTT) or particle sizes were first synthesized using 2-methyl-2-propenoic acid-2-(2-methoxyethoxy) ethyl ester (MEO₂MA) and oligoethylene glycol methyl ether methacrylate (OEGMA) as thermosensitive monomers by free radical polymerization. Their VPTT and particle sizes were investigated by a nanometer particle size meter and an ultraviolet spectrophotometer. The feasibility of using these P(OEGMA-co-MEO₂MA) microgels to construct thermosensitive hydrogel by means of the thermal induction method is discussed for the first time. The prepared thermosensitive hydrogel with the optimum performance was screened for in situ embedding and three-dimensional (3D) culture of MCF-7 breast cancer cells. The experimental results of AO/EB and MTT methods indicate that the pioneering scaffold material has prominent biocompatibility, and cells grow rapidly in the 3D scaffold and maintain high proliferative capacity. At the same time, there is also a tendency to aggregate to form multicellular spheres. Therefore, this original P(OEGMA-co-MEO₂MA) thermosensitive hydrogel can serve as a highly biocompatible and easily functionalized 3D cell culture platform with great potential in the biomedical area.

Keywords: thermosensitive; hydrogel; P(OEGMA-co-MEO₂MA); microgel; scaffold; 3D culture

1. Introduction

In recent years, the use of tissue engineering technology for in vitro three-dimensional (3D) cell culture to construct 3D cell models for tumor research, drug screening, and tissue repair has aroused great interest in biomedical researchers and clinical workers [1–3]. Among them, scaffold materials play a key role. As the carrier of seed cells and bioactive growth factors, scaffold materials can provide a favorable microenvironment for cell growth, reproduction, metabolism, and other physiological activities, and induce cells to form a 3D model [4–7]. Therefore, an excellent scaffold material should have a 3D porous network structure to facilitate cell insertion and transport of nutrients and metabolic waste. At the same time, it should have ideal biocompatibility, degradability and mechanical strength matching the properties of the extracellular matrix.

Hydrogel is a kind of soft material with a 3D network structure, formed through various physical or chemical crosslinking methods [8–11]. Hydrogels have extremely similar physical properties to extracellular matrices, which have been widely used as scaffold materials for 3D cell culture [12–14]. Among them, thermosensitive polymers-based hydrogel scaffolds have attracted particular attention [15–17]. Thermosensitive polymers generally appear liquid at room temperature, and can be evenly mixed with cells, drugs, growth factors, etc. When the temperature rises to its low critical dissolution temperature (LCST), or volume phase transition temperature (VPTT), and appropriate ions are present, the thermosensitive polymer undergoes in situ gelation to form injectable 3D hydrogels. Thus, it can nondestructively embed cells in situ and promote further cell growth. By cooling it to liquefy it, cells, or complex cell aggregates, can also be easily released.

Poly(N-isopropylacrylamide) (PNIPAM) is the most studied thermosensitive material with LCST or VPTT of 32 °C, and is often used as a scaffold material [18–20]. For instance, Ekerdt and colleagues successfully constructed a type of thermosensitive hyaluronic acid (HA)-PNIPAM brush polymer through thiol-ene click chemistry, which could mix with human embryonic stem cells (hESCs), or human induced pluripotent stem cells (hiPSCs), at 4 °C, and gel rapidly to form a hydrogel in situ at 37 °C [21]. After five days of culture, cells were able to grow rapidly and aggregate to form regular multicellular spheroids, while maintaining cell pluripotency after multiple passages. Moreover, this type of hydrogel was able to release the cells after cooling, liquefaction and centrifugation [22]. Liang et al. also synthesized a type of dendritic thermosensitive polymer from PNIPAM, dimeric glycerol, and polyethylene glycol (PEG). This polymer could also be used as a scaffolding material for the culture of hiPSCs, which were similarly able to release cells after simple cooling, liquefaction and centrifugation [23].

To further adjust and improve the properties of formed thermosensitive hydrogel, spherical thermosensitive polymers, namely microgels, were also used for the fabrication of hydrogel scaffolds. Microgels are spherical hydrogel nanoparticles with particle sizes from 10 nm to a few microns, which have the advantages of simple preparation, fast response and easy modification and are of considerable interest in the biomedical field [24–27]. Gan et al. successfully constructed this type of thermosensitive hydrogel scaffold using PNIPAM-based thermosensitive microgels for the first time. The hydrogel thus formed has an interconnected porous structure, which is highly conducive to the transport of oxygen, nutrients and cellular metabolites. Human embryonic kidney (HEK) 293T cells could be encapsulated in this hydrogel and grew well [28]. We also used PNIPAM-based microgels as scaffold materials to encapsulate aggregates of Human hepatocellular carcinomas HepG2 cells, in which multicellular spheroids of HepG2 cells could be quickly obtained [29]. In addition, Shen et al. also used the more hydrophilic NIPAM and acrylamide (AAm) copolymer microgels as scaffold materials for the culture of mouse melanoma cells [30]. It was found that the cells gradually grew in a single dispersed state. The larger the particle size of the microgels used, the better the activity of the cells. The Dai group also used PNIPAM-based microgels with a small negative charge and a small positive charge as cell scaffolds to study the culture and behavior of mouse embryonic mesenchymal stem cells (MSCS), respectively [31,32].

Despite the numerous advantages of PNIPAM thermosensitive polymers, the thermosensitive hydrogels formed from them suffer from a certain degree of shrinkage over time, which severely limits cell growth and activity. Although it is possible to reduce the shrinkage of the hydrogels by introducing acrylic acid (AA) with negative charge [33], blending PEG [29,34] and adjusting the particle size of the microgels [35], the effect is still not satisfactory. In addition, PNIPAM polymers are toxic to some extent and difficult to degrade. Therefore, it is an urgent problem to find a type of thermosensitive gel material with low shrinkage, and excellent biocompatibility and biodegradability. In recent years, novel thermosensitive polymers based on oligo(ethylene glycol) methacrylate (OEGMA) and 2-(2-methoxyethoxy) ethyl acrylate (MEO_2MA) have aroused great interest from researchers because of their excellent biocompatibility and biodegradability [36,37]. OEGMA and MEO_2MA are two types of PEG-based macromolecular monomers with similar structures, in which the LCST of POEGMA homopolymer is about 90 °C and often acts as a hydrophilic chain segment, while $PMEO_2MA$ homopolymer has an LCST of about 19 °C and often acts as a hydrophobic chain segment. The copolymers formed from them also have outstanding thermosensitive behavior. The LCST or VPTT could vary from 19 to 90 °C by simply adjusting the relative proportions of OEGMA and MEO$_2$MA in the polymerization process. Several studies have reported using these novel thermosensitive polymers for the culture of various types of cells. For instance, Anderson et al. prepared a brush-like P(OEGMA-co-MEO$_2$MA) thermosensitive polymer by adjusting the ratio of OEGMA and MEO$_2$MA. It was used as a matrix material to study the effect on adhesion and morphology of L-929 mouse fibroblasts [38,39]. It was found that the combination of thermosensitive polymers with a smaller relative proportion of hydrophilic OEGMA was more favorable for cell growth and adhesion. In addition, Bakaic et al. [40] constructed an injectable hydrogel scaffold with temperature and pH dual responsiveness using chemical crosslinking between neutral or charged aldehyde functionalized POEGMA copolymers and hydrazide functionalized POEGMA copolymer precursors. They found that the charged hydrogel was more supportive of the two-dimensional (2D) cell adhesive of 3T3 mouse fibroblasts and the 3D stabilization and proliferative of ARPE-19 human retinal epithelial cells.

In this manuscript, we discuss the feasibility of using P(OEGMA-co-MEO$_2$MA) spherical microgels to construct thermosensitive hydrogel 3D cell scaffolds for the first time. We first used OEGMA and MEO$_2$MA to prepare a series of P(OEGMA-co-MEO$_2$MA) microgels with different VPTT and sizes through free radical polymerization. Then, novel P(OEGMA-co-MEO$_2$MA) microgels-based thermosensitive hydrogels were prepared by the thermal induction method under conditions of heating and the presence of salts. MCF-7 human breast cancer cells were used as the model cells for in-situ embedding and 3D co-culture in this novel thermosensitive hydrogel. To the best of our knowledge, there are no reports related to P(OEGMA-co-MEO$_2$MA)-based microgels as scaffolds for 3D cell culture. Therefore, our study may provide new ideas for the development of thermosensitive hydrogel scaffolds.

2. Results and Discussion
2.1. The Synthesis of P(OEGMA-co-MEO$_2$MA) Thermosensitive Microgels

In this work, to investigate the influence of the properties of microgels on subsequent hydrogel formation, five types of P(OEGMA-co-MEO$_2$MA) microgel samples, with different thermosensitivities or sizes, were synthesized by a free radical polymerization reaction [26].

As shown in Table 1, we first synthesized M1-M3 samples through changing the mass ratio of thermosensitive monomers OEGMA and MEO$_2$MA. We measured the particle sizes of microgel aqueous solutions at different temperatures using a nanometer particle size meter. It can be seen from Figure 1A that the particle sizes of all the three P(OEGMA-co-MEO$_2$MA) microgels decreased gradually with increase in temperature, which indicated that all of them had great thermosensitivity, similar to linear and branched P(OEGMA-co-MEO$_2$MA) polymers. At lower temperatures, due to the strong hydrogen bonding of oxygen atoms in the microgel, the microgel particle has strong hydrophilicity and high water content, so the particle size is larger, which is manifested as volume swelling. When the temperature gradually increases, the hydrogen bond is gradually destroyed, and the microgel gradually shows strong hydrophobic effect derived from alkyl groups, so the internal water content decreases, the volume shrinks, and the particle size decreases. The temperature at which the slope of the curve changes the most is defined as the VPTT. As shown in the figure, the VPTT of M1, M2, and M3 were about 21 °C, 2 °C, and 2 °C, respectively, which means that with the increase of the relative proportion of OEGMA in the monomers, the final microgel has a higher VPTT. This is because, compared with MEO$_2$MA, OEGMA has more ether oxygen bonds in the molecular structure, which can form more hydrogen bonds with water molecules; so, the more OEGMA content, the stronger the hydrophilicity of the formed P(OEGMA-co-MEO$_2$MA) microgel. In addition, we found that the sizes of M3 were larger than others at most temperatures, which may also be due to the higher content of OEGMA increasing the hydrophilicity and swelling of this microgel. This is in agreement with what was reported in the literature [41].

Table 1. The feeding amount of each sample during the preparation process.

Sample	n (MEO$_2$MA): n (OEGMA)	m/g				
		MEO$_2$MA	OEGMA	SDS	BIS	KPS
M1	95% : 3%	2.5002	0.1259	0.0284	0.0431	0.0812
M2	90% : 8%	2.3664	0.3368	0.0284	0.0431	0.0812
M3	85% : 13%	2.2371	0.5462	0.0284	0.0431	0.0812
M4	85% : 13%	2.2371	0.5462	0.0284	0.0431	0.0812
M5	85% : 13%	2.2371	0.5462	0.0284	0.0431	0.0812

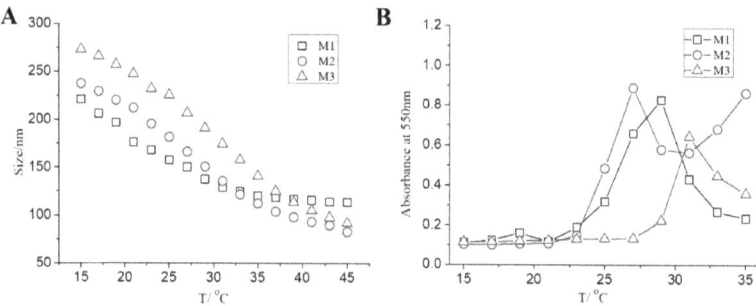

Figure 1. Hydrodynamic radius (A) and absorbance (B) of P(OEGMA-co-MEO$_2$MA) microgel samples M1-M3 as a function of temperature.

In order to further study the thermosensitivity of the above prepared P(OEGMA-co-MEO$_2$MA) microgels, we used phosphate buffer solution (PBS) to configure a certain concentrated dilute solution of microgel samples, and measured the change of absorbance at different temperatures by using an UV-Vis spectrophotometer. As shown in Figure 1B, when the temperature was lower, the absorbance of all the three microgel solutions was basically the same, but when the temperature exceeded a certain higher critical value, it mutated suddenly. The reason for this phenomenon is that with increase of temperature, hydrophilic microgels gradually become hydrophobic. When a critical temperature is reached, microgels gather together to form small aggregates, resulting in increased light refraction, thus decreasing light transmittance and increasing absorbance. Actually, the critical temperature is the VPTT. Similarly, we found that the VPTT of three samples was 21 °C, 23 °C and 27 °C, respectively. With increase of the relative proportion of OEGMA, the VPTTs of the samples M1, M2, M3 increased gradually, which was consistent with previous results.

In addition, the effect of the dosage of surfactant sodium dodecyl sulfate (SDS) on P(OEGMA-co-MEO$_2$MA) microgel properties was also investigated. We synthesized microgel samples M4 and M5, and compared them with sample M3. The SDS dosages in their synthesis processes were reduced (Table 1). We first also measured the particle sizes of microgel solutions at different temperatures. As shown in Figure 2A, the particle sizes of M4 and M5 also decreased gradually with increase of temperature, which indicated the presence of excellent thermosensitivity. From M3, M4 to M5, with reduction of the SDS dosages, the sizes of the obtained microgels increased at measured temperatures, especially at lower temperatures. For instance, at 15 °C, the sizes of M3, M4 and M5 were 273.3 nm, 363.5 nm, and 406.7 nm, respectively. Moreover, we found that all these three types of microgel samples had the same VPTT, around 27 °C. The results of absorbance measurements at different temperatures, shown in Figure 2B, also indicates that the VPTT remains constant regardless of the SDS dosage. These phenomena indicated that with the decrease of SDS dosage, the relative particle size of the microgel generally increased, but the VPTT remained basically unchanged. This is because SDS in the reaction process does not directly participate in the polymerization reaction, but plays the role of stabilizing the parent particles of microgel by forming micelles. So, the SDS dosage will not affect the thermosensitivities and VPTTs of obtained microgels. The lower the SDS content, the larger the size of the micelle formed from the SDS, so the size of the final P(OEGMA-co-MEO$_2$MA) microgel particle is larger.

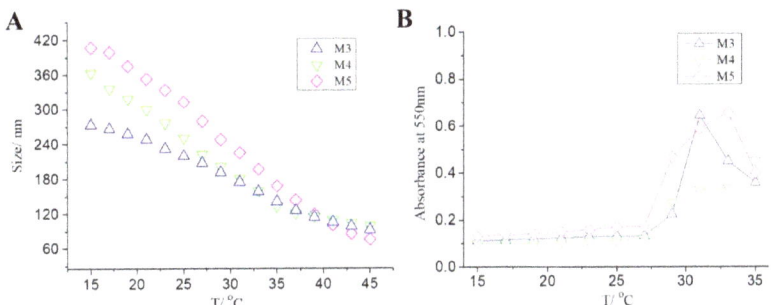

Figure 2. Hydrodynamic radius (**A**) and absorbance (**B**) of P(OEGMA-co-MEO$_2$MA) microgel samples M3-M5 as a function of temperature.

In summary, three types of thermosensitive P(OEGMA-co-MEO$_2$MA) microgels, with different VPTTs, were synthesized by free radical emulsion polymerization through change in the mass ratio of OEGMA and MEO$_2$MA. With increase of the content of OEGMA, the VPTT of the obtained microgels also increased. In addition, two types of thermosensitive P(OEGMA-co-MEO$_2$MA) microgels, with different sizes and the same VPTT, were also synthesized by polymerization through change of the SDS dosage. With the decrease of the dosages of SDS, the sizes of the obtained microgels increased.

2.2. The Preparation of P(OEGMA-co-MEO$_2$MA) Thermosensitive Microgels-Based Hydrogel

Next, we studied the feasibility of using P(OEGMA-co-MEO$_2$MA) microgels to construct thermosensitive hydrogels through the thermal induction method. Referring to our previously reported method [29], we configured the PBS solutions of microgel samples with a concentration of 3.0 wt%, and placed them at 37 °C, which was higher than their VPTT. Then, we observed the formation of hydrogels by the inverted method and photographed the changes of these hydrogels.

As shown in Figure 3, after maintaining for 10 min at 37 °C, all three types of P(OEGMA-co-MEO$_2$MA) microgels could gel and form white hydrogels. This result was consistent with that of the traditional PNIPAM thermosensitive polymer. When the temperature was higher than their VPTT, the microgels, in their hydrophobic contraction state, would aggregate under the synergistic influence of hydrophobicity and the shielding effect of ions from PBS, thus forming hydrogel macroscopically. The gelation times were about 5 min to 10 min. Meanwhile, we found that the hydrogels formed from M1 and M2 showed obvious shrinkage after 2 h, and the shrinkage degree increased with time. After 48 h, the volume of the hydrogels formed from M1 and M2 was less than 1/2 of the initial value. This phenomenon of volume shrinkage has been reported several times regarding traditional PNIPAMs and other thermosensitive hydrogels, because of the dehydration of thermosensitive materials at higher temperature [19,20]. As an ideal scaffold material, the volume contraction should be avoided as far as possible in order to maintain sufficient cell growth space and ensure the circulation of nutrients and metabolic waste [4,6].

Encouragingly, we found that the volume change of the hydrogel formed by the prepared sample M3 with a higher VPTT was less than 5% and no significant shrinkage was observed. So, we also observed and recorded the formation of hydrogels from M4 and M5 microgels with the same VPTT and larger sizes. It can be seen from Figure 4 that the hydrogels formed from M4 and M5 also showed obvious shrinkage after 2 h, and the shrinkage degree increased with time, which was hardly surprising.

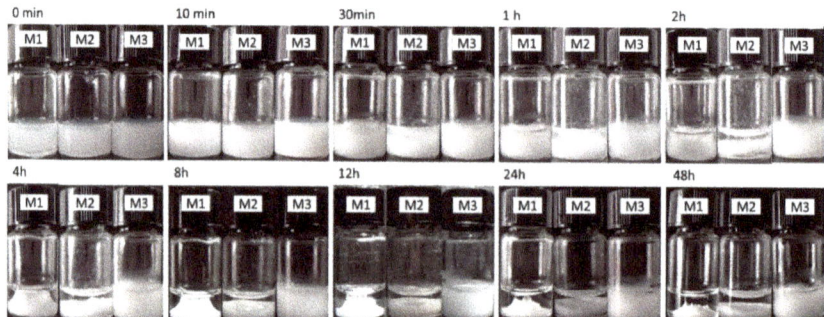

Figure 3. Photographs of the P(OEGMA-co-MEO$_2$MA) hydrogels formed from M1–M3 microgels at different times.

Figure 4. Photographs of the P(OEGMA-co-MEO$_2$MA) hydrogels formed from M3–M5 microgels at different times.

We further investigated the quantitative relation between the volume shrinkage degree V_t/V_0 of the hydrogel (the ratio of the volume of the formed hydrogel at a certain time point to the initial volume) and time (Figure 5). We found that the hydrogels formed from M1 or M2 contracted rapidly in the first 8 h, when the volume was less than half of the initial volume. After 48 h, the volumes of the samples were about 20% and 30% of the initial volume, respectively. The shrinkage degree was very large. However, the volume of hydrogel formed from M3 still had a volume of more than 98% of the initial volume after 48 h without significant shrinkage. We speculated that the reason why the hydrogel formed by M3 did not shrink significantly may be related to its containing longer OEGMA chain fragments with excellent hydrophilic properties, which could hinder the shrinkage of the hydrogel. In our previous report [29,34], we found that adding a small amount of PEG to the prepared PNIPAM microgel significantly reduced the shrinkage of the resulting hydrogel. Due to the presence of a large number of PEG fragments in OEGMA, we thought that OEGMA may have a similar effect to PEG in reducing hydrogel shrinkage.

We also compared the change of V_t/V_0 of the hydrogels formed from M3, M4, and M5 microgels. It was shown that, as the size of P(OEGMA-co-MEO$_2$MA) microgel increased, the shrinkage of hydrogel also increased. For instance, after 48 h, the volume of the hydrogel samples formed from M3, M4, and M5 were about 98%, 19% and 17% of the initial volume, respectively. We think the reason was that the larger the particle size of the microgel particles with the same VPTT, the higher the water content. Therefore, when at a certain temperature higher than their VPTT, the shrinkage degree of microgels is greater, resulting in more solvents being extruded from the microgels, which would also cause more volume shrinkage of the macroscopic hydrogels.

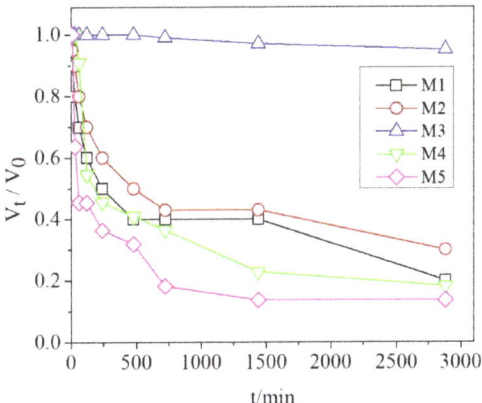

Figure 5. The change of V_t/V_0 (shrinkage degree) of P(OEGMA-co-MEO$_2$MA) hydrogels formed from microgels as a function of time.

These results indicated that, by adjusting the thermosensitivity (VPTT) and particle size of these synthesized P(OEGMA-co-MEO$_2$MA) microgels, the properties of the obtained thermosensitive hydrogels can be adjusted, especially volume shrinkage. We also found that the obtained P(OEGMA-co-MEO$_2$MA) thermosensitive hydrogels showed a small shrinkage degree only by controlling the synthesis conditions of P(OEGMA-co-MEO$_2$MA) microgels; while for the commonly used PNIPAM hydrogels, complex operations, such as physical blending or copolymerization, were required to achieve the same effect. Therefore, the P(OEGMA-co-MEO$_2$MA) thermosensitive hydrogel showed better application prospects in the biomedical field. In following work, we will continue to conduct in-depth studies on P(OEGMA-co-MEO$_2$MA) microgel-based hydrogels, especially the gelation mechanism, gelation conditions and rheological behavior.

2.3. P(OEGMA-co-MEO$_2$MA) Thermosensitive Hydrogels for 3D Cell Culture

A hydrogel without shrinkage is more suitable for cell embedding and 3D growth. Therefore, we selected microgel sample M3 as the scaffold material for subsequent 3D cell culture. A type of tumor cell, MCF-7 human breast cancer cell, was chosen as the model cell to explore the feasibility of using this hydrogel to construct a 3D cell model [29,30,33–35]. After a certain concentration of microgel dispersion, cells were evenly mixed at room temperature, and then transferred to 48-well culture plates and placed in a cell incubator at 37 °C for culturing. About 1 h later, a stable hydrogel was formed and the cells were encapsulated in situ in it. Part of the DMEM cell culture medium was added to continue this culture process.

Firstly, we investigated the viability of cells in the P(OEGMA-co-MEO$_2$MA) hydrogel scaffold using MTT assay. As shown in Figure 6, the absorbance values measured by the MTT method increased significantly in the first two days of cell culture, which may be related to a rapid increase in the number of living cells in this hydrogel scaffold. This result also indirectly reflected the favorable biocompatibility of this hydrogel scaffold material. The increase of cell viability slowed down after the third day, which may be related to the accumulation of metabolic waste, and the reduction of nutrients and relative living space caused by the increase in the number of cells.

We further observed the growth and morphology of cells in P(OEGMA-co-MEO$_2$MA) thermosensitive hydrogel scaffolds by AO/EB staining, as shown in Figure 7. On the day of cell embedding (day 0), we found that dispersed MCF-7 cells did not adhere to the hydrogel scaffold, but grew uniformly in a single spherical shape in the scaffold. It is well known that cells can only stick to surfaces that are hydrophobic enough, so this cell behavior in the scaffold may be caused by the hydrophilic environment inside the scaffold. After 1 day of culture, some smaller cell clusters could be observed, which may have formed by cell division or cell aggregation. As the culture time was further extended, cells grew faster and the number of cells increased, and most of them showed a green color in the observed field of view, indicating that the cells were still alive and the hydrogel scaffold could maintain the cell growth. When cultured for 7 days, we observed that most of the cells were able to form small multicellular spheres, which may be due to interaction between cells in adjacent cell clusters as the cells continued to spread and grow within the hydrogel scaffold, thus aggregating to form multicellular spheres. Therefore, we can speculate that this type of thermosensitive P(OEGMA-co-MEO$_2$MA) hydrogel scaffold can be used to construct tumor multicellular sphere models and may have wide application prospects in drug screening and tumor research. Besides, based on our previous research experience [29], this type of hydrogel scaffold should also be able to be used for 3D culture of many other types of cells, which may extend its range of applications.

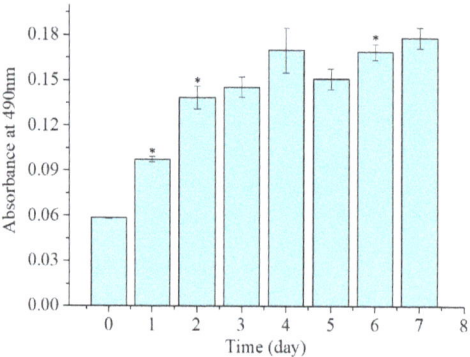

Figure 6. Viability of MCF-7 cells in formed thermosensitive P(OEGMA-co-MEO$_2$MA) hydrogel as assessed by MTT assay. The data shown are the mean of three independent experiments. Error bars indicate the standard deviations. An asterisk (*) indicates a significant difference between this group of data and the previous group ($p < 0.05$).

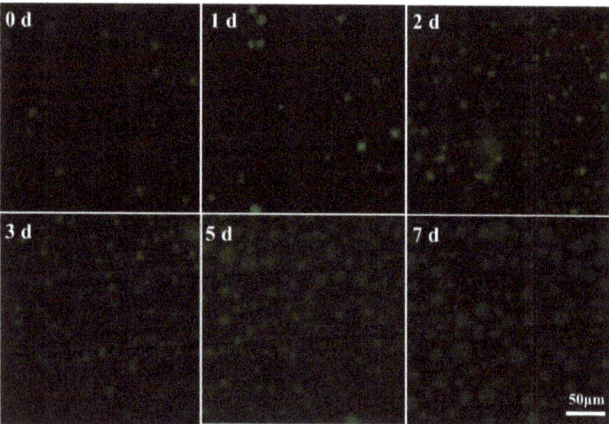

Figure 7. Fluorescence images of MCF-7 cells cultured in the formed thermosensitive P(OEGMA-co-MEO$_2$MA) hydrogel. The cells were stained with acridine orange (AO) and ethidium bromide (EB) beforehand.

3. Conclusions

Here, by using MEO$_2$MA and OEGMA, having excellent biocompatibility and biodegradability as monomers, we synthesized five types of P(OEGMA-co-MEO$_2$MA) thermosensitive microgel samples with different VPTTs or particle sizes by free radical polymerization. Their VPTTs and particle sizes were investigated by means of a nanometer particle size meter and ultraviolet spectrophotometer. With increase of the content of OEGMA, the VPTT of the obtained microgels increased. With decrease of the dosage of SDS, the sizes of the obtained microgels increased. All the types of prepared microgels could aggregate to form 3D hydrogels by thermal induction. All these hydrogels still showed characteristic shrinkage of thermosensitive hydrogels. However, we found that with the increase of the VPTT, or the decrease of the particle size, the less obvious the contraction. Among them, the volume of hydrogel formed from M3 still had a volume of more than 98% of the initial volume after 48 h without significant shrinkage. These hydrogels could be used as a scaffold for 3D culture of MCF-7 model cells and showed excellent biocompatibility. Cells can grow in this scaffold and tend to form multicellular sphere models. This is a simple method for the preparation of thermosensitive hydrogel 3D cell scaffolds, which shows enormous application prospects in the biomedical field, such as tumor research, drug screening, and tissue engineering.

4. Materials and Methods

4.1. Materials

Oligo(ethylene glycol) methacrylate (OEGMA, Mn = 300), 2-(2-methoxyethoxy) ethyl acrylate (MEO$_2$MA), N,N′-methylene diacrylamide (BIS), sodium dodecyl sulfate (SDS), were obtained from Tianjin Heowns Biochemical Technology Co, LTD; potassium persulfate (KPS) was obtained from Tianjin Damao Reagent Factory; Phosphate buffer (PBS), DMEM medium, fetal bovine serum, streptomycin mixture, trypsin, thiazole blue (MTT), acridine orange (AO), ethidium bromide (EB) were obtained from Wuhan Rutgers Biotechnology Co; MCF-7 breast cancer cells were purchased from the Stem Cell Bank of the Chinese Academy of Sciences.

4.2. Synthesis and Characterization of P(OEGMA-co-MEO$_2$MA) Thermosensitive Microgel

In this manuscript, we first used MEO$_2$MA and OEGMA as polymerization monomers, BIS as crosslinker, SDS as surfactant, and KPS as initiator to generate P(OEGMA-co-MEO$_2$MA) thermosensitive microgels by free radical polymerization at 70 °C [26] (Scheme 1). The content of BIS was 2%, the mass of KPS were 0.0812 g, respectively. Three types of microgel samples M1, M2 and M3 were synthesized by changing the molar ratios of MEO$_2$MA and OEGMA as 95 : 3, 90 : 8, and 85 : 13, respectively. In contrast to the synthesis of M3, another two types of microgel samples, M4 and M5, were synthesized by reducing the SDS dosage, while keeping the other substances constant. The detail feeding amount of each sample is in Table 1.

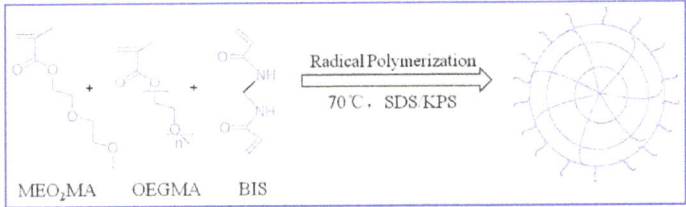

Scheme 1. The synthesis of P(OEGMA-co-MEO$_2$MA) thermosensitive microgel.

The typical preparation process for sample M1 was as follows: 2.5002 g MEO$_2$MA, 0.1259 g OEGMA, 0.0431 g BIS and 0.0284 g SDS were weighed accurately, dissolved in 100 mL deionized water, and transferred to a three-necked flask. After being passed into N$_2$

for 1 h, KPS (dissolved in 5 mL of deionized water) was added to initiate polymerization under magnetic stirring. After 5 h, the reaction was terminated and transferred to a dialysis belt and dialysis for 7 days. Then, the sample was freeze-dried and preserved.

In order to test the thermosensitivities of these samples, the changes of particle size and absorbance (turbidity) with temperature were measured by a nanometer particle size meter and an ultraviolet spectrophotometer, respectively.

4.3. Preparation and Characterization of P(OEGMA-co-MEO$_2$MA) Microgels-Based Thermosensitive Hydrogel

The thermosensitive P(OEGMA-co-MEO$_2$MA) hydrogels were prepared through the thermal induction method [29].

A certain amount of P(OEGMA-co-MEO$_2$MA) microgel samples were weighed and dissolved in phosphate buffered saline (PBS) to obtain a final solution with mass fraction of 3.0 wt%. Then, they were placed in a physiological temperature environment, which was higher than their VPTTs. The formation process of thermosensitive hydrogel from microgels and its stability and shrinkage were observed by the inverted method and recorded by camera. The gelation time was defined as the time point at which the inverted hydrogel no longer flowed visually.

To quantitatively analyze the shrinkage degree of the obtained hydrogels, we calculated the ratio of the volume of the obtained hydrogel at certain time point (V_t) to the initial volume (V_0) by measuring with a ruler. The thermosensitive hydrogel sample with minimum shrinkage was screened out for the next cell experiments.

4.4. P(OEGMA-co-MEO$_2$MA) Thermosensitive Hydrogels for 3D Cell Culture

4.4.1. Cell Culture

The model cells used in this experiment were human breast cancer cells (MCF-7), which were cultured in a cell culture incubator using DMEM medium. The main components of the culture medium were 90% DMEM high glucose medium, 10% fetal bovine serum, and 100 unit/mL penicillin/streptomycin double antibodies. The culture temperature was 37 °C and the carbon dioxide concentration was 5%.

4.4.2. Cells Embedded and Cultured in Hydrogel Scaffold

A certain concentration of P(OEGMA-co-MEO$_2$MA) microgel solution was prepared and mixed well with an equal volume of the uniform cell suspension to form a mixed solution, in which the cell concentration was about 1×10^5 cells/mL and the microgel concentration was 3.0 wt%. To a 48-well culture plate, 0.5 mL of the above cell/gel mixture was added and placed in a cell culture incubator. With increasing temperature, the solution gelled gradually. After the morphology of the gel in the culture plate was fixed, 0.5 mL of the culture solution was added to each well and recorded as day 0. The culture solution was changed regularly to ensure a nutritious environment for cell growth. The whole process is shown in Scheme 2.

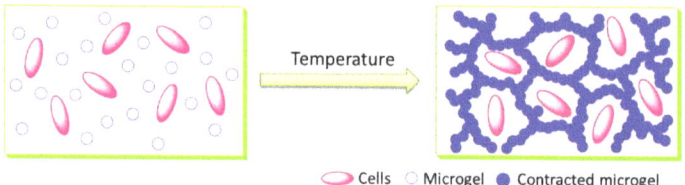

Scheme 2. The preparation of the P(OEGMA-co-MEO$_2$MA) thermosensitive hydrogel scaffold.

4.4.3. Cell Viability (MTT Assay)

MTT assay was used to detect the viability of cells in the scaffolds. From day 0 of cell culture, three wells were selected each day and 100.0 µL of MTT solution was added to the wells. After continuing to incubate for 4 h, the gel pieces were removed, and 10.0 mL of DMSO was added. Then, the mixture was shaken for 10 min in the dark to dissolve the formed formazan. The absorbance of the solution at 490 nm was measured using a UV-Vis spectrophotometer, and the MTT curve of cells was plotted.

4.4.4. Cell Activity and Morphology (AO/EB Staining)

From day 0 of cell culture, three wells were selected out after culturing for certain days, and one drop of AO/EB double staining solution was added to the selected wells. After further incubation for 1h, the cells were removed, and the morphology and growth state of the cells in the scaffold were observed by an inverted fluorescence microscope.

Author Contributions: Conceptualization, Y.L.; methodology, Y.-N.L., P.Z., W.-F.Y., C.-Y.Z. and Y.-L.Y.; validation, Y.L., Y.-N.L. and Y.-L.Y.; formal analysis, Y.-N.L., P.Z., W.-F.Y., C.-Y.Z. and Y.-L.Y.; investigation, P.Z., W.-F.Y. and C.-Y.Z.; resources, Y.L.; writing—original draft preparation, Y.-N.L., P.Z., W.-F.Y., C.-Y.Z. and Y.-L.Y.; writing—review and editing, Y.L., Y.-N.L. and Y.-L.Y.; visualization, Y.-N.L., P.Z., W.-F.Y., C.-Y.Z. and Y.-L.Y.; supervision, Y.L.; project administration, Y.L.; funding acquisition, Y.L. All authors have read and agreed to the published version of the manuscript.

Funding: This research was funded by the Natural Science Foundation of Hunan Province (No. 2021JJ30597), Scientific Research Projects of Health Commission of Hunan Province (No. 202113022002), and Hengyang Guided Science and Technology Project (No. 2020jh042809).

Institutional Review Board Statement: Not applicable.

Informed Consent Statement: Not applicable.

Data Availability Statement: Not applicable.

Conflicts of Interest: The authors declare no conflict of interest.

References

1. Khademhosseini, A.; Langer, R. A decade of progress in tissue engineering. *Nat. Protoc.* **2016**, *11*, 1775–1781. [CrossRef] [PubMed]
2. Esdaille, C.J.; Washington, K.S.; Laurencin, C.T. Regenerative engineering: A review of recent advances and future directions. *Regen. Med.* **2021**, *16*, 495–512. [CrossRef] [PubMed]
3. Atala, A.; Kasper, F.K.; Mikos, A.G. Engineering complex tissues. *Sci. Transl. Med.* **2012**, *4*, 160rv12. [CrossRef] [PubMed]
4. Gaharwar, A.K.; Singh, I.; Khademhosseini, A. Engineered biomaterials for in situ tissue regeneration. *Nat. Rev. Mater.* **2020**, *5*, 686–705. [CrossRef]
5. Reddy, M.S.B.; Ponnamma, D.; Choudhary, R.; Sadasivuni, K.K. A comparative review of natural and synthetic biopolymer composite scaffolds. *Polymers* **2021**, *13*, 1105. [CrossRef]
6. Koons, G.L.; Diba, M.; Mikos, A.G. Materials design for bone-tissue engineering. *Nat. Rev. Mater.* **2020**, *5*, 584–603. [CrossRef]
7. Ebhodaghe, S.O. Natural polymeric scaffolds for tissue engineering applications. *J. Biomater. Sci. Polym. Ed.* **2021**, *32*, 2144–2194. [CrossRef]
8. Bashir, S.; Hina, M.; Iqbal, J.; Rajpar, A.H.; Mujtaba, M.A.; Alghamdi, N.A.; Wageh, S.; Ramesh, K.; Ramesh, S. Fundamental concepts of hydrogels: Synthesis, properties, and their applications. *Polymers* **2020**, *12*, 2702. [CrossRef]
9. Jian, X.; Feng, X.; Luo, Y.; Li, F.; Tan, J.; Yin, Y.; Liu, Y. Development, preparation, and biomedical applications of DNA-based hydrogels. *Front. Bioeng. Biotechnol.* **2021**, *9*, 661409. [CrossRef]
10. Ahmed, E.M. Hydrogel: Preparation, characterization, and applications: A review. *J. Adv. Res.* **2015**, *6*, 105–121. [CrossRef]
11. Feng, X.; Luo, Y.; Li, F.; Jian, X.; Liu, Y. Development of natural-drugs-based low-molecular-weight supramolecular gels. *Gels* **2021**, *7*, 105. [CrossRef] [PubMed]
12. Spicer, C.D. Hydrogel scaffolds for tissue engineering: The importance of polymer choice. *Polym. Chem.* **2020**, *11*, 184–219. [CrossRef]
13. Portnov, T.; Shulimzon, T.R.; Zilberman, M. Injectable hydrogel-based scaffolds for tissue engineering applications. *Rev. Chem. Eng.* **2017**, *33*, 91–107. [CrossRef]
14. Radulescu, D.-M.; Neacsu, I.A.; Grumezescu, A.-M.; Andronescu, E. New insights of scaffolds based on hydrogels in tissue engineering. *Polymers* **2022**, *14*, 799. [CrossRef] [PubMed]
15. Zhang, Y.; Yu, J.-K.; Ren, K.; Zuo, J.; Ding, J.; Chen, X. Thermosensitive hydrogels as scaffolds for cartilage tissue engineering. *Biomacromolecules* **2019**, *20*, 1478–1492. [CrossRef]
16. Suntornnond, R.; An, J.; Chua, C.K. Bioprinting of thermoresponsive hydrogels for next generation tissue engineering: A review. *Macromol. Mater. Eng.* **2017**, *302*, 1600266. [CrossRef]
17. Doberenz, F.; Zeng, K.; Willems, C.; Zhang, K.; Groth, T. Thermoresponsive polymers and their biomedical application in tissue engineering–A review. *J. Mater. Chem. B* **2020**, *8*, 607–628. [CrossRef]

18. Dou, Q.Q.; Liow, S.S.; Ye, E.Y.; Lakshminarayanan, R.; Loh, X.J. Biodegradable thermogelling polymers: Working towards clinical applications. *Adv. Healthc. Mater.* **2014**, *3*, 977. [CrossRef]
19. Nagase, K.; Yamato, M.; Kanazawa, H.; Okano, T. Poly(N-isopropylacrylamide)-based thermoresponsive surfaces provide new types of biomedical applications. *Biomaterials* **2018**, *153*, 27–48. [CrossRef]
20. Tang, L.; Wang, L.; Yang, X.; Feng, Y.; Li, Y.; Feng, W. Poly(N-isopropylacrylamide)-based smart hydrogels: Design, properties and applications. *Prog. Mater. Sci.* **2020**, *115*, 100702. [CrossRef]
21. Ekerdt, B.; Fuentes, C.M.; Lei, Y.; Adil, M.M.; Ramasubramanian, A.; Segalman, R.A.; Schaffer, D.V. Thermoreversible hyaluronic acid-PNIPAAm hydrogel systems for 3D stem cell culture. *Adv. Health Mater.* **2018**, *7*, e1800225. [CrossRef] [PubMed]
22. Kim, G.; Jung, Y.; Cho, K.; Lee, H.J.; Koh, W.-G. Thermoresponsive poly(N-isopropylacrylamide) hydrogel substrates micropatterned with poly(ethylene glycol) hydrogel for adipose mesenchymal stem cell spheroid formation and retrieval. *Mater. Sci. Eng. C* **2020**, *115*, 111128. [CrossRef] [PubMed]
23. Liang, W.; Bhatia, S.; Reisbeck, F.; Zhong, Y.; Singh, A.K.; Li, W.; Haag, R. Thermoresponsive hydrogels as microniches for growth and controlled release of induced pluripotent stem cells. *Adv. Funct. Mater.* **2021**, *31*, 2010630. [CrossRef]
24. Li, F.; Luo, Y.; Feng, X.; Guo, Y.; Zhou, Y.; He, D.; Xie, Z.; Zhang, H.; Liu, Y. Two-dimensional colloidal crystal of soft microgel spheres: Development, preparation and applications. *Colloids Surfaces B Biointerfaces* **2022**, *212*, 112358. [CrossRef] [PubMed]
25. Liu, Y.; Guan, Y.; Zhang, Y. Facile assembly of 3D binary colloidal crystals from soft microgel spheres. *Macromol. Rapid Commun.* **2014**, *35*, 630–634. [CrossRef] [PubMed]
26. Guan, Y.; Zhang, Y.J. PNIPAM microgels for biomedical applications: From dispersed particles to 3D assemblies. *Soft Matter* **2011**, *7*, 6375–6384. [CrossRef]
27. Gan, T.; Zhang, Y.; Guan, Y. In situ gelation of P(NIPAM-HEMA) microgel dispersion and its applications as injectable 3D cell scaffold. *Biomacromolecules* **2009**, *10*, 1410–1415. [CrossRef]
28. Liu, Y.; Guan, Y.; Zhang, Y. Chitosan as inter-cellular linker to accelerate multicellular spheroid generation in hydrogel scaffold. *Polymer* **2015**, *77*, 366–376. [CrossRef]
29. Shen, J.; Ye, T.; Chang, A.; Wu, W.; Zhou, S. A colloidal supra-structure of responsive microgels as a potential cell scaffold. *Soft Matter* **2012**, *8*, 12034–12042. [CrossRef]
30. Shen, Z.; Bi, J.; Shi, B.; Nguyen, D.; Xian, C.J.; Zhang, H.; Dai, S. Exploring thermal reversible hydrogels for stem cell expansion in three-dimensions. *Soft Matter* **2012**, *8*, 7250–7257. [CrossRef]
31. Shen, Z.; Mellati, A.; Bi, J.; Zhang, H.; Dai, S. A thermally responsive cationic nanogel-based platform for three-dimensional cell culture and recovery. *RSC Adv.* **2014**, *4*, 29146–29156. [CrossRef]
32. Gan, T.; Guan, Y.; Zhang, Y. Thermogelable PNIPAM microgel dispersion as 3D cell scaffold: Effect of syneresis. *J. Mater. Chem.* **2010**, *20*, 5937–5944. [CrossRef]
33. Cheng, D.; Wu, Y.; Guan, Y.; Zhang, Y. Tuning properties of injectable hydrogel scaffold by PEG blending. *Polymer* **2012**, *53*, 5124–5131. [CrossRef]
34. Gu, J.; Zhao, Y.; Guan, Y.; Zhang, Y. Effect of particle size in a colloidal hydrogel scaffold for 3D cell culture. *Colloids Surfaces B Biointerfaces* **2015**, *136*, 1139–1147. [CrossRef]
35. Vancoillie, G.; Frank, D.; Hoogenboom, R. Thermoresponsive poly(oligo ethylene glycol acrylates). *Prog. Polym. Sci.* **2014**, *39*, 1074–1095. [CrossRef]
36. Badi, N. Non-linear PEG-based thermoresponsive polymer systems. *Prog. Polym. Sci.* **2016**, *66*, 54–79. [CrossRef]
37. Anderson, C.R.; Abecunas, C.; Warrener, M.; Laschewsky, A.; Wischerhoff, E. Effects of methacrylate-based thermoresponsive polymer brush composition on fibroblast adhesion and morphology. *Cell. Mol. Bioeng.* **2016**, *10*, 75–88. [CrossRef]
38. Anderson, C.R.; Gambinossi, F.; DiLillo, K.M.; Laschewsky, A.; Wischerhoff, E.; Ferri, J.K.; Sefcik, L.S. Tuning reversible cell adhesion to methacrylate-based thermoresponsive polymers: Effects of composition on substrate hydrophobicity and cellular responses. *J. Biomed. Mater. Res. Part A* **2017**, *105*, 2416–2428. [CrossRef]
39. Bakaic, E.; Smeets, N.M.; Badv, M.; Dodd, M.; Barrigar, O.; Siebers, E.; Lawlor, M.; Sheardown, H.; Hoare, T. Injectable and degradable poly(oligoethylene glycol methacrylate) hydrogels with tunable charge densities as adhesive peptide-free cell scaffolds. *ACS Biomater. Sci. Eng.* **2017**, *4*, 3713–3725. [CrossRef]
40. Tatry, M.-C.; Galanopoulo, P.; Waldmann, L.; Lapeyre, V.; Garrigue, P.; Schmitt, V.; Ravaine, V. Pickering emulsions stabilized by thermoresponsive oligo(ethylene glycol)-based microgels: Effect of temperature-sensitivity on emulsion stability. *J. Colloid Interface Sci.* **2020**, *589*, 96–109. [CrossRef]
41. Keerl, M.; Pedersen, J.S.; Richtering, W. Temperature sensitive copolymer microgels with nanophase separated structure. *J. Am. Chem. Soc.* **2009**, *131*, 3093–3097. [CrossRef] [PubMed]

Article

Typical Fluorescent Sensors Exploiting Molecularly Imprinted Hydrogels for Environmentally and Medicinally Important Analytes Detection

Lihua Zou [1,†], Rong Ding [2,†], Xiaolei Li [2], Haohan Miao [3], Jingjing Xu [1,2,*] and Guoqing Pan [3,*]

[1] Center for Molecular Recognition and Biosensing, School of Life Sciences, Shanghai University, Shanghai 200444, China; 18201738579@163.com
[2] Sino-European School of Technology of Shanghai University, Shanghai University, Shanghai 200444, China; Rongding_2021@163.com (R.D.); xiaolei_li2020@163.com (X.L.)
[3] Institute for Advanced Materials, School of Materials Science and Engineering, Jiangsu University, Zhenjiang 212013, China; haohan_miao@163.com
* Correspondence: jingjing_xu@shu.edu.cn (J.X.); panguoqing@ujs.edu.cn (G.P.)
† These authors contributed equally to this work.

Abstract: In this work, two typical fluorescent sensors were generated by exploiting molecularly imprinted polymeric hydrogels (MIPGs) for zearalenone (ZON) and glucuronic acid (GA) detection, via the analyte's self-fluorescence property and receptor's fluorescence effect, respectively. Though significant advances have been achieved on MIPG-fluorescent sensors endowed with superior stability over natural receptor-sensors, there is an increasing demand for developing sensing devices with cost-effective, easy-to-use, portable advantages in terms of commercialization. Zooming in on the commercial potential of MIPG-fluorescent sensors, the MIPG_ZON is synthesized using zearalanone (an analogue of ZON) as template, which exhibits good detection performance even in corn samples with a limit of detection of 1.6 μM. In parallel, fluorescein-incorporated MIPG_GA is obtained and directly used for cancer cell imaging, with significant specificity and selectivity. Last but not least, our consolidated application results unfold new opportunities for MIPG-fluorescent sensors for environmentally and medicinally important analytes detection.

Keywords: molecularly imprinted polymeric hydrogels; synthetic receptors; zearalenone; glucuronic acid; fluorescent sensors

1. Introduction

Due to the tight correlation between human health and environmental pollution that results in various diseases, environmental monitoring of pollutants and biomedical diagnosis of biomarkers are global concerns [1,2]. Hence, there is an increasing demand for developing sensors for the selective detection of various targets, such as abused pesticides and overproduced mycotoxins in food samples [3–6], as well as biomarkers and tumor cells that represent related diseases [7–10]. Moreover, the obtained sensors should be fabricated cost-effectively and capable of rapid, sensitive, portable applications. Traditionally, a natural receptor, such as enzymes, antibodies, DNA, and cells are used to react with the analyte of interest. A significant advancement has been achieved in sensors development, especially pertaining to routine procedures based on these biomolecules [11]. Despite their excellent specificity, these biosensors suffer from reagent stability and availability, high cost and cumbersome analysis procedures, etc. [12]. In this context, researchers eagerly seek synthetic tailor-made receptors capable of selectively recognizing and binding target molecules with good affinity, while being chemically and thermally stable, easy to prepare, and reusable [13]. One of the most promising strategies to create such synthetic receptors is molecular imprinting technology (MIT), which is based on the co-polymerization of functional monomers with the crosslinking agent in the presence of the template (the target molecule or a derivative thereof). After removal of the template which leaves cavities with a size, shape, and functionality complementary to the template, a tailor-made receptor called molecularly imprinted polymer (MIP) is synthesized [14,15].

Owning to the manufacture advances, the MIT has been used for a wide range of target molecules, from small molecules to macromolecules and even larger entities [16,17]. The synthesized MIP can be widely used in analyte purification, biosensing, and bioimaging, as well as drug delivery and cancer therapy [18–22], due to its antibody-like affinity, significant selectivity, and excellent biocompatibility [17]. However, the MIP advances towards industrial use very slowly, and even the most promising MIP-based sensors are not widely used in the industry, mainly due to the inconvenience in manufacturing and real-world use [23]. Therefore, the commercialization of MIP-based sensors relies on the simplified synthesis process and portable property for the improvement of consolidated applications in various scenarios [24].

Aiming at developing convenient MIP-based sensors towards commercial applications in the future, two typical fluorescent sensors were generated, possessing the sensing mechanism either via the analyte's self-fluorescence property or receptor's fluorescence effect, respectively (Scheme 1). Specifically, a molecularly imprinted polymeric hydrogel (MIPG) was synthesized for zearalenone (ZON), a mycotoxin produced by several species of Fusarium molds to contaminate cereals, in particular corn, barley, oats, wheat, and sorghum [25], causing toxic effects in humans and animals, including nephrotoxic, neurotoxic, carcinogenic, estrogenic, and immunosuppressive effects [26]. Specifically, MIPG_ZON was developed using an optimized amount of 4-vinylpyridine (4-VPY) as a functional monomer and ethylene glycol dimethacrylate (EGDMA) as a cross-linking monomer against the zearalanone (ZAN, an analogue of ZON but without a double bond, which may interfere with the polymerization process). Afterwards, the synthesized MIPG_ZON was grafted on the glass plate to detect the ZON in corn juice. In parallel, a fluorescein-incorporated MIPG was generated for the direct recognition of glucuronic acid (GA), a small monosaccharide as the epitope of hyaluronic acid, the biomarker of cancerous or infectious cells on the cell surface [27–29]. The obtained MIPG_GA was fluorescein dimethacrylate (PolyFluor® 511) cross-linked using two functional monomers, (4-acrylamidophenyl)(amino)methaniminium acetate (AAB) that interacts with the carboxylic group of GA with high affinity, and methacrylamide (MAM) which provides hydrogen bonds to GA. The final application of MIP_GA in cancer cell imaging showed a great potential of our MIP as a fluorescent probe for biomedical analysis.

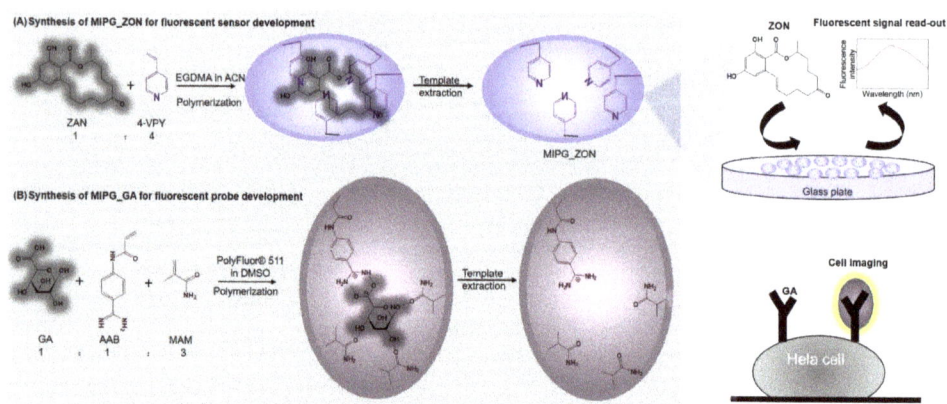

Scheme 1. Schematic illustration of molecular imprinting technology for two typical fluorescent sensors development. (A) The molecular imprinting process for the synthesis of MIPG_ZON, which can be pasted on the glass substrate in order to work as a sensor for ZON detection by reading out the fluorescence signal of ZON. (B) The molecular imprinting process for the synthesis of fluoresce-incorporated MIPG_GA as a fluorescent probe which can be directly used for cancer cell imaging.

2. Results and Discussion

2.1. Characterization of Synthesized MIPGs_ZON

Three MIPGs_ZON using the functional monomer 4-VPY at different proportions were synthesized (Table 1), with the expectation that the "π–π" interaction between ZON and the monomer would lead to strong affinity. According to our preliminary study (Figure S3), the double bond in the chemical structure of ZON interferes with the imprinting process, which brought no difference between MIPG and NIPG. To address this problem, an analogue ZAN was used as a template for MIPGs_ZON synthesis. The hydrodynamic diameter of the ZAN imprinted polymers was analyzed by the dynamic light scattering (DLS) on a Zeta-sizer NanoZS (results are summarized in Table 1), which shows a very homogenous size for each MIPG_ZON.

Table 1. Composition of ZAN imprinted polymers synthesized in 400 μL anhydrous ACN and their corresponding sizes.

Polymer	ZAN (mmol)	4-VPY (mmol)	EGDMA (mmol)	ABDV (mg)	Size (nm)	PDI
MIPG1	0.03	0.06	0.6	3.1	345 ± 5	0.163
NIPG1	-	0.06	0.6	3.1	455 ± 16	0.160
MIPG2	0.03	0.12	0.6	3.2	443 ± 11	0.137
NIPG2	-	0.12	0.6	3.2	483 ± 6	0.199
MIPG3	0.03	0.18	0.6	3.4	468 ± 12	0.156
NIPG3	-	0.18	0.6	3.4	503 ± 9	0.085

2.2. Equilibrium Binding Studies on MIPGs_ZON

In Figure 1A, our hydrogel is a dense three-dimensional fiber network as observed under the scanning electron microscope (SEM). The relevant component analysis was carried out by a Fourier transform infrared spectrometer (FTIR, Figure S4). By magnifying the SEM image, the thickness of the fibers is found to be at the nanometer level. Figure 1B represents the binding behavior of the three MIPGs_ZON ranging from 0.1 to 5 mg/mL towards 2.5 μM ZON in ACN, indicating that the ratio ZAN: 4-VPY: EGDMA of 1:4:20 corresponding to MIPG2 gave the best binding specificity, especially for low polymer concentrations where 0.1 mg of MIP can bind to 0.5 nmol of ZON, leading to an imprinting factor of 3.0. Afterwards, 1 mg of MIPG2 and NIPG2 were incubated overnight with ZON at concentrations varying from 0.5 to 10 μM in 1 mL ACN. As shown in Figure 1C, the Langmuir-type isotherm equation was used for plotting ZON bound versus free in case of MIPG2 for the specific binding analysis, in which the binding capacity B_{max} was found to be 5.8 nmol/mg, and the dissociation constant K_d was 7.9 μM with $R^2 = 0.99$. At the same time, the Freundlich model was used in NIPG2 cases for the non-specific adsorption analysis, it was found that $B_{max} = 4.3$ nmol/mg and $K_d = 7.1$ μM with $R^2 = 0.98$. Therefore, MIPG2 presents higher binding capacity while a similar affinity towards ZON, with respect to NIPG2. As shown in Figure 1D, the corresponding Scatchard plot, ZON bound/free versus ZON bound is represented by a straight line, indicating a homogeneous MIPG2. The variables N and K_a in the Scatchard equation ($B/F = K_aN - K_aB$) correspond to the total number of binding sites and the association constant, which were found to be $\sim 5.1 \times 10^5$ M^{-1} and 4.0 nmol/mg, respectively.

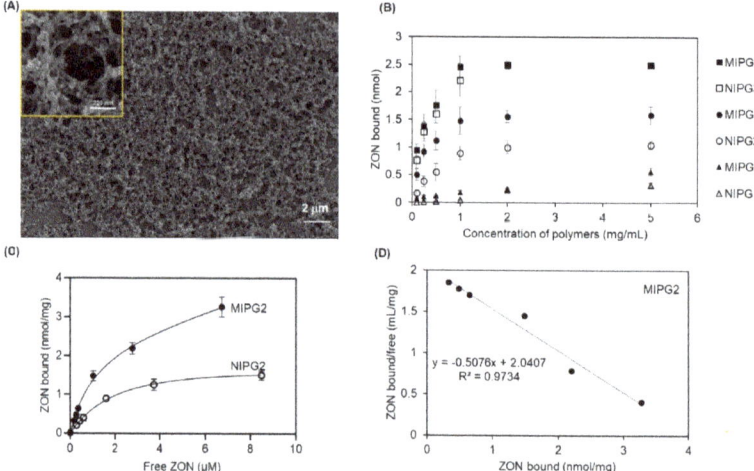

Figure 1. (**A**) SEM image of the MIPGs_ZON. (**B**) Equilibrium binding isotherms for 2.5 µM ZON on MIPGs_ZON at concentrations ranging from 0.1 to 5 mg/mL in 1 mL ACN. Data are means from three independent experiments. The error bars represent standard deviations. (**C**) Determination of the binding capacity of 1 mg MIPG2 and NIPG2 towards ZON varying from 0.5 to 10 µM in 1 mL ACN, respectively. Data are means from three independent experiments. The error bars represent standard deviations. (**D**) Corresponding Scatchard plot of MIPG2.

2.3. Application of MIPGs_ZON-Based Fluorescent Sensor for Real Sample Tests

In order to study the anti-interference ability of the MIPG2, the structurally similar reagents tetracycline (TC), amoxicillin (AMO), and levofloxacin (LEV) were used for the selectivity test. As shown in Figure 2A, MIPG2 exhibited lower binding towards other structural analogs with respect to ZON, demonstrating a great potential in anti-interference detecting. On the other hand, the NIPG2 shows a high cross-reactivity towards TC and AMO. Therefore, our MIPG2 is promising to detect ZON in real samples. Furthermore, the colloidal stability and batch-to-batch reproducibility were investigated prior to the real sample tests. As shown in Figure 2B, when 1 mg MIPG2 (of the same batch) was used to detect 2.5 µM ZON repeatedly after each template extraction, the binding amount was almost constant during one month, indicating its reversibility property. Meanwhile, 1 mg MIPG2 (of a different batch) was employed to recognize 2.5 µM ZON in 1 and 2 years, the amount of ZON bound remained the same, showing its excellent stability and reproducibility for sensor development.

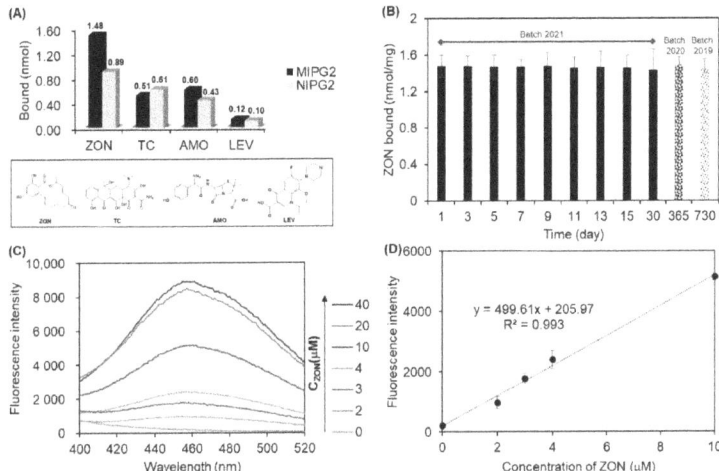

Figure 2. (**A**) Selectivity test of 1 mg MIPG2 and NIPG2 towards ZON and three structurally similar reagents tetracycline (TC), amoxicillin (AMO), and levofloxacin (LEV) at 2.5 μM, respectively. The chemical structure of three structurally similar reagents are given below. (**B**) Detection record of 1 mg MIPG2 (of the same batch) towards 2.5 μM ZON during 30 days after each template extraction, as well as 1 mg MIPG2 (of a different batch) towards 2.5 μM ZON in 1 and 2 years, respectively. (**C**) Analyte-dependent fluorescence emission spectra of MIPG2 with respect to ZON ranging from 0 to 40 μM in corn juice. (**D**) Lear fitting plot of fluorescence intensity of MIPG2 towards ZON in corn juice. Corn juice tests were performed on glass substrates coated with MIPG2.

It is well-known that ZON is a mycotoxin produced by several species of Fusarium that may contaminate cereals, in particular corn, barley, oats, wheat, and sorghum. Thus, ZON contamination can cause toxic effects in humans and animals, including nephrotoxic, neurotoxic, carcinogenic, estrogenic, and immunosuppressive effects [25,26]. Since MIPG2 exhibits good specificity and selectivity for ZON, they were then adhered on a glass substrate in order to work as a fluorescent sensor for ZON detection in real samples. Herein, we bought some corn juice as ZON-containing practical samples, and diluted it 4 times with ACN for testing. Using the standard recovery method, ZON at concentrations ranging from 0 to 40 μM were spiked in the corn juice. According to the results of ZON detection in real samples (Figure 2C,D), our MIPG2-based optical sensor may detect ZON at the linear concentration range of 0–10 μM with a limit of detection (LOD) of 1.6 μM (calculated according to the equation: $D = 3\sigma/k$, where σ is the relative standard deviation of the blank sample, k is the slope of the calibration line), showing great potential for mycotoxin detection in food testing. Although, biological molecular recognition elements exhibit lower LOD, such as the LOD of antibody-based sensors for ZON is at the nM level [30,31], and aptamer-based sensors can detect ZON at the pM level [32,33], such biosensors are expensive and have a short service life. Compared with the chromatography method which exhibited a LOD in the nM range [5], our MIPG_ZON-based fluorescent sensor is more convenient for commercial applications. Inspired by recent quantum dots-incorporated MIPG [4], our MIPG_ZON promises to be smaller and effective, and the optimization could be established on this protocol.

2.4. Synthesis and Binding Characterization of MIPG_GA

The aim of this part is to synthesize a fluorescent MIPG_GA, which is capable of specifically binding towards GA (an epitope of hyaluronic acid which is usually overexpressed on cancer cells) for cell imaging. To achieve this goal, the particle size of MIPG_GA must be controlled at the nanometer scale, and the MIPG must exhibit good compatibility with water. Hence, our MIPG_GA was generated using AAB and MAM as functional monomers to create specific binding sites (Scheme 1). According to our intensive research [28,29,34], AAB is able to strongly interact with the carboxylic group and MAM is a co-monomer that establishes a hydrogen interaction with the target molecule. As shown in Figure 3A,B, the hydrodynamic size of MIPG_GA and NIPG_GA was found to be 520 and 585 nm, respectively. The SEM image (Figure 3C) shows the three-dimensional nanofiber structure of the hydrogel (FTIR analysis in Figure S5). Using these polymers to evaluate the binding performance towards GA, different concentrations of the polymers ranging from 0.1 to 2 mg/mL were incubated with 1 mM GA. After the subtraction of unbound GA from the total amount of the analyte added by the well-established Dubois method, the amount of bound GA was determined for each concentration of polymers (Figure 3D). Although the NIPG_GA bound significant amount of analyte was due to the presence of AB, which might be largely distributed on the polymer surface and very accessible to GA. It was found that MIPG_GA bound specifically to GA as the binding to NIPG_GA was lower.

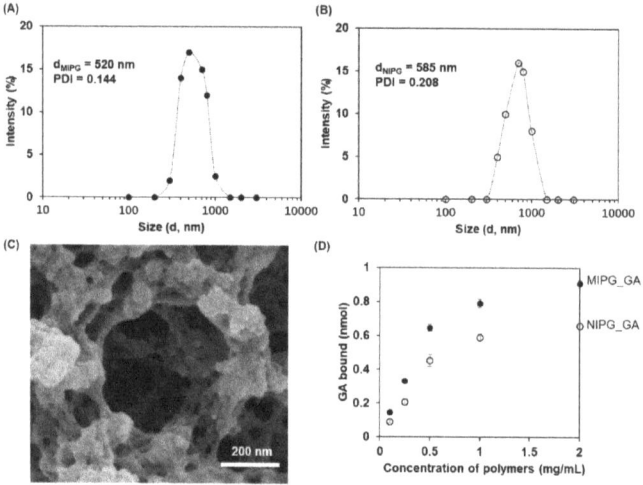

Figure 3. Dynamic light scattering measurements of 0.5 mg/mL (**A**) MIPG_GA and (**B**) NIPG_GA in H$_2$O (mean value of two experiments). (**C**) SEM image of MIPG_GA. (**D**) Equilibrium binding isotherm for 1 mM GA on MIPG_GA and NIPG_GA at the concentrations ranging from 0.1 to 2 mg/mL in water. Data are means of three independent experiments. The error bars represent standard deviations.

2.5. Application of MIPG_GA-Based Fluorescent Probe for Cell Imaging

The objective of this part is to observe the interaction between fluorescent polymers and Hela cells before or after the enzyme treatment, in order to demonstrate the potential of our MIPG_GA to detect cancer cells in clinical uses. The images shown in Figure 4A,B correspond to the fluorescence intensity analysis for Hela cells incubated with MIPG_GA or NIPG_GA, respectively, indicating that the MIPG_GA specifically bound towards the epitope of hyaluronic acid on Hela cells. Our fluorescent probe is more convenient for commercializing due to its excellent stability, though less sensitive with respect to the hyaluronic acid-binding protein (HABP, Figure S6) [35–37]. Moreover, compared with nano-sized MIP containing quantum dots [28,29], our MIPG_GA is easier to fabricate using a broad-spectrum cross-linker FAM-based monomer for signaling, showing the universal utility of this strategy.

Furthermore, according to Figure 4C, the MIPG_GA obtained from fluorescence microscope images exhibited more binding towards Hela cells with respect to the NIPG_GA. Particularly, hyaluronidase effectively removes GA on the surface of Hela cells by breaking down hyaluronic acids. This is why the fluorescence intensity for hyaluronidase treated Hela cells became lower than the untreated cells. On the other hand, N-acetylglucosaminidase plays a role to hydrolyze the non-reducing terminal of N-acetyl-D-glucosamine, whereby more GA is obtained. Thus, the fluorescence intensity of N-acetylglucosaminidase treated Hela cells increased. In addition, the MIPG_GA exhibited almost the same cell imaging fluorescence intensity in 2 years, indicating its excellent stability.

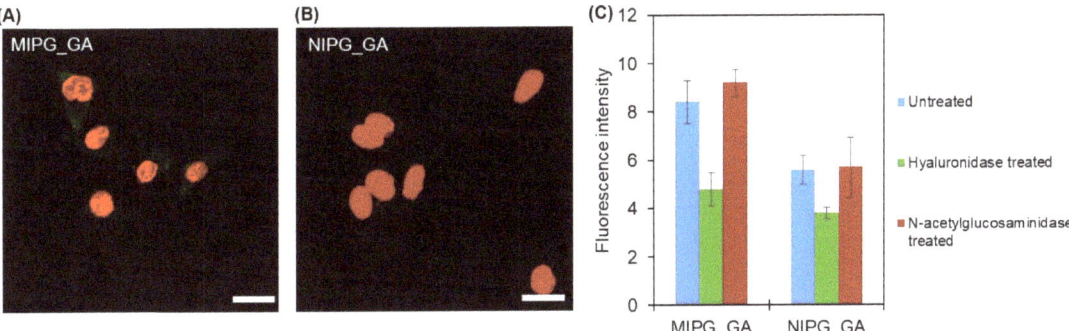

Figure 4. Confocal microscope images of fixed Hela cells incubated MIPG_GA (**A**) and NIPG_GA (**B**), that exhibit green fluorescence under proper filter sets of confocal microscopy. Nuclei stained with PI (red). Scale bar: 50 μm. (**C**) Summarized histogram results showing the binding performance of polymers towards untreated, hyaluronidase treated, and N-acetylglucosaminidase treated Hela cells. Data are means of three independent experiments. The error bars represent standard deviations.

3. Conclusions

In summary, a MIP-based fluorescent sensor for ZON detection was successfully prepared and applied for tracing ZON in real samples. In parallel, a MIP-based fluorescent probe was fabricated and directly used for cell imaging. These two examples unfold the beauty of molecular imprinting technology in front of our eyes, indicating the perspectives and versatility of this technique for designing and fabricating fluorescent sensors towards commercial applications. The biggest gain in this work is the discovery of the road towards simple and real-world applications for the MIPGs, showing a great potential of this kind of fluorescent sensors in the detection of analyte of interest in food and biological samples.

4. Materials and Methods

4.1. Reagents

The (4-acrylamidophenyl)(amino)methaniminium acetate (AAB) was synthesized as previously described [34]. The 4-vinylpyridine (4-VPY), methacrylamide (MAM), ethylene glycol dimethacrylate (EGDMA), azo-bis-dimethylvaleronitrile (ABDV), azo-bisisobutyronitrile (AIBN), zearalanone (ZAN), zearalenone (ZON), hyaluronidase, N-acetylglucosaminidase, hyaluronic acid-binding protein (HABP), biotin, streptavidin, and fluorescein-5-isothiocyanate (FITC) was purchased from Sigma-Aldrich (Shanghai, China). Fluorescein dimethacrylate (PolyFluor® 511) was obtained from Polysciences (Hirschberg an der Bergstrasse, Germany). Glucuronic acid (GA), tetracycline (TC), amoxicillin (AMO), and levofloxacin (LEV) came from Sinopharm Chemical Reagent Co., Ltd. (Shanghai, China). All the solvents used in the experiments as well as acids and bases are of analytical grade and anhydrous for acetonitrile (ACN) and dimethylsulfoxide (DMSO). The materials for cells culturing are: Phosphate buffer saline (PBS), penicillin/streptomycin, high glucose Dulbecco's Modified Eagle's Medium (DMEM) supplied with L-glutamine, fetal bovine serum (FBS), trypsin, ethylenediaminetetraacetic acid, and were obtained from Thermo Fisher Scientific (Shanghai, China).

4.2. Synthesis of MIPGs_ZON and MIPG_GA

MIPGs_ZON were synthesized in the presence of the ZAN using different functional monomer ratios. The 4-VPY was distilled under reduced pressure before use. First of all, 10 mg (0.03 mmol) ZAN, 6.4 or 12.7 or 19.2 μL (0.06, 0.12, 0.18 mmol) 4-VPY, and 113.1 μL (0.60 mmol) EGDMA was mixed in 400 μL of ACN. After sonication for 5 min, the initiator ABDV (1 mol% with respect to the number of double bonds) was added. Then, the prepolymerization mixture was purged with nitrogen for 5 min and placed in a water bath maintained at 40 °C overnight. The non-imprinted polymeric hydrogels (NIPGs) are synthesized in the same way but without the template. For template extraction, the polymers were washed 3 times with methanol/acetic acid (7/3), once with methanol, 3 times methanol containing 2.8% ammonia, and 3 times with methanol for 1 h at room temperature. The final polymers were obtained after drying in a vacuum overnight.

MIPG_GA was synthesized by mixing 4.85 mg (0.025 mmol) GA, 4.73 mg (0.025 mmol) AAB (1 h pre-incubation with GA), 6.38 mg (0.075 mmol) MAM, and 234 mg (0.5 mmol) PolyFluor® 511 in 1903 μL DMSO. Afterwards, 1.01 mg (0.0062 mmol) AIBN was added and before nitrogen purging for 5 min. The mixture was then heated overnight at 50 °C. The next day, a precipitated polymer was obtained. The NIPGs were synthesized in the same way but without the template. For template extraction, the polymers are washed 3 times with 1 M hydrochloric acid, once with methanol, 3 times methanol containing 2.8% ammonia, and 3 times with methanol for 1 h at room temperature. The final polymers were obtained after drying in a vacuum overnight.

The sizes of synthesized polymers were measured by dynamic light scattering (DLS) analysis on a Zetasizer NanoZS at 25 °C, in ACN for MIPs_ZON and in H_2O for MIP_GA. Scanning electron microscopy (SEM) imaging was carried out on a Quanta FEG 250 scanning electron microscope (FEI Europe), by spraying gold. Prior to the binding studies, the dried MIPGs were added into a dialysis bag for checking whether there was any leakage of the template one day later. If there is any leakage, a second round of washing will be carried out. Afterwards, the MIPG2_ZON and MIPG_GA were analyzed by Fourier transform infrared spectrometer (FTIR), respectively. In addition, 5 mg/mL MIPG2_ZON in ACN and 5 mg/mL MIPG_GA in DMSO images were taken, in order to prove the gel format.

4.3. Equilibrium Binding Studies

For **MIPGs_ZON**, the equilibrium binding experiments with ZON were performed in anhydrous acetonitrile. The particles of 10 mg MIPGs_ZON and NIPGs_ZON were suspended in 1 mL ACN with intensive sonication (ultrasonic power: 80 W, frequency: 40 KHz, 20 min), in order to work as stock solutions, respectively. Then, polymer concentrations at 0.1, 0.25, 0.5, 1, 2, 5 mg/mL were incubated with 2.5 μM ZON in 1 mL ACN overnight. The samples were then centrifuged at 17,500 rpm for 20 min and a 700 μL aliquot of the supernatant was taken for unbound ZON determination. Then, the amount of bound ZON was calculated by subtracting the amount of unbound from the total. A calibration curve of ZON (Ex = 280 nm, Em = 455 nm, slit: 5) was obtained on F-7000 spectrofluorometer (Hitachi High-Technologies, Tokyo, Japan), as shown in Figure S1. To investigate the binding capacity, 1 mg of polymers were incubated with ZON varying from 0.5 to 10 μM in 1 mL ACN, with results plotted with the Langmuir-type isotherm equation. The selectivity study was performed by testing the binding of 1 mg polymers towards 2.5 μM ZON, and three structurally similar reagents: Tetracycline (TC), amoxicillin (AMO), and levofloxacin (LEV).

For **MIPG_GA,** the recognition properties were evaluated using the method of Dubois [28]. Dubois's method is a colorimetric method widely used to determine monosaccharide based on the preconversion of sugars into furfural derivatives upon heating with strong acids, followed by the formation of a colored complex with phenol. Prior to the binding assay, a calibration curve was generated with standard solutions of GA at concentrations ranging from 0.1 to 5 mM in water (Figure S2). Afterwards, the polymers at concentrations varying from 0.1 to 2 mg/mL were incubated with 1 mM GA in 1 mL water overnight. The samples were then centrifuged at 17,500 rpm for 1 h, in order to provide the supernatant for GA determination by Dubois's method.

4.4. Application of MIPG_ZON-Based Fluorescent Sensor for Real Sample Tests

Initially, 100 mg MIPG2_GA and NIPG2_GA were mixed with 480 mg poly (vinyl alcohol) in water and smeared on the glass plates, respectively. Herein, poly (vinyl alcohol) were used as an adhesion agent due to its excellent transparent feature, exhibiting no interference to ZON excitation light. After 30 min of shaking in an oven at 90 °C, the glass plates were left for 1 h at room temperature to cool down. The glass plates coated with MIPG2_ZON and NIPG2_ZON were stored in the fridge at 4 °C until use. For the real sample tests, the spiked samples were prepared using ZON at concentrations from 0–40 μM dissolved in DMSO. The prepared ZON samples were then mixed with corn juice at a ratio of 1:9 for the determination of recovery. For fluorescence intensity measurements after 1 h incubation, the Cary Eclipse fluorescence spectrophotometer (Varian, Palo Alto, CA, USA) was used.

4.5. Application of MIPG_GA-Based Fluorescent Probe for Cell Imaging

Cells were grown in a 5% CO_2 incubator at 37 °C using DMEM supplemented with 10% premium FBS and antibiotics (100 units/mL penicillin and 100 μg/mL streptomycin) in cell culture dishes. For imaging studies, Hela cells were grown on cover slips and then incubated with 0.5 mg/mL polymers or 50 ng/mL FAM-HABP (synthesized in two steps, first with a biotinylated HABP, followed by incubation with streptavidin-FITC [37]) for 1 h. The cell monolayers were then washed 3 times with PBS (pH 7.4) and immediately measured by a Leica TSC SP5 confocal scanning laser microscopy (CSLM) system (Ex = 480, Em = 515).

Supplementary Materials: The following are available online at https://www.mdpi.com/article/10.3390/gels7020067/s1. Figure S1: Fluorescence calibration curve of ZON in ACN (Ex = 280 nm, Em = 455 nm, slit: 5 nm); Figure S2: Calibration curve of glucuronic acid in water (quantified by Dubois's method with absorbance read at 490 nm). Figure S3: Equilibrium binding isotherms for 2.5 μM ZON on ZON-imprinted polymers (MIP) and NIP at concentrations ranging from 0.25 to 2 mg/mL in 1 mL ACN. MIP here were synthesized using the similar protocol for MIPG2, in the presence of ZON. Data are means from three independent experiments. The error bars represent standard deviations. Figure S4: FT-IR spectra of MIPG2_ZON with solution image insert, where some special peaks were highlighted. Figure S5: FT-IR spectra of MIPG_GA with solution image insert, where some special peaks were highlighted. Figure S6: Confocal microscope images of fixed Hela cells incubated MIPG_GA (A) and FAM-HABP (B), that exhibit green fluorescence under proper filter sets of confocal microscopy. Nuclei stained with PI (red). Scale bar: 50 μm.

Author Contributions: Conceptualization, writing original draft, investigation, validation, methodology, L.Z. and R.D.; assistant for methodology, X.L. and H.M.; conceptualization, investigation, writing—review and editing, supervision, funding acquisition, J.X. and G.P. All authors have read and agreed to the published version of the manuscript.

Funding: This research was funded by the National Natural Science Foundation of China (22001162, 21875092) and the Shanghai Sailing Program (20YF1414200).

Institutional Review Board Statement: Not applicable.

Informed Consent Statement: Not applicable.

Acknowledgments: The authors acknowledge the financial support from the National Natural Science Foundation of China (22001162, 21875092) and the Shanghai Sailing Program (20YF1414200).

Conflicts of Interest: The authors declare that they have no known competing financial interests or personal relationships that could have appeared to influence the work reported in this paper.

References

1. Patel, B.R.; Noroozifar, M.; Kerman, K. Review-Nanocomposite-Based Sensors for Voltammetric Detection of Hazardous Phenolic Pollutants in Water. *J. Electrochem. Soc.* **2020**, *167*, 037568. [CrossRef]
2. Sow, W.T.; Ye, F.; Zhang, C.; Li, H. Smart materials for point-of-care testing: From sample extraction to analyte sensing and readout signal generator. *Biosens. Bioelectron.* **2020**, *170*, 112682. [CrossRef] [PubMed]
3. Beloglazova, N.; Lenain, P.; Tessier, M.; Goryacheva, I.; Hens, Z.; De Saeger, S. Bioimprinting for multiplex luminescent detection of deoxynivalenol and zearalenone. *Talanta* **2019**, *192*, 169–174. [CrossRef] [PubMed]
4. Shao, M.; Yao, M.; Saeger, S.D.; Yan, L.; Song, S. Carbon Quantum Dots Encapsulated Molecularly Imprinted Fluorescence Quenching Particles for Sensitive Detection of Zearalenone in Corn Sample. *Toxins* **2018**, *10*, 438. [CrossRef] [PubMed]
5. Lhotska, I.; Gajdosova, B.; Solich, P.; Satinsky, D. Molecularly imprinted vs. reversed-phase extraction for the determination of zearalenone: A method development and critical comparison of sample clean-up efficiency achieved in an on-line coupled SPE chromatography system. *Anal. Bioanal. Chem.* **2018**, *410*, 3265–3273. [CrossRef] [PubMed]
6. Li, G.; Zhang, K.; Fizir, M.; Niu, M.; Sun, C.; Xi, S.; Hui, X.; Shi, J.; He, H. Rational design, preparation and adsorption study of a magnetic molecularly imprinted polymer using a dummy template and a bifunctional monomer. *New J. Chem.* **2017**, *41*, 7092–7101. [CrossRef]
7. Li, Q.; Shinde, S.; Grasso, G.; Caroli, A.; Abouhany, R.; Lanzillotta, M.; Pan, G.; Wan, W.; Rurack, K.; Sellergren, B. Selective detection of phospholipids using molecularly imprinted fluorescent sensory core-shell particles. *Sci. Rep.* **2020**, *10*, 9924. [CrossRef] [PubMed]
8. Li, R.; Feng, Y.; Pan, G.; Liu, L. Advances in Molecularly Imprinting Technology for Bioanalytical Applications. *Sensors* **2019**, *19*, 177. [CrossRef]
9. Ma, Y.; Yin, Y.; Ni, L.; Miao, H.; Wang, Y.; Pan, C.; Tian, X.; Pan, J.; You, T.; Li, B.; et al. Thermo-responsive imprinted hydrogel with switchable sialic acid recognition for selective cancer cell isolation from blood. *Bioact. Mater.* **2021**, *6*, 1308–1317. [CrossRef] [PubMed]
10. Wang, J.; Dai, J.; Xu, Y.; Dai, X.; Zhang, Y.; Shi, W.; Sellergren, B.; Pan, G. Molecularly Imprinted Fluorescent Test Strip for Direct, Rapid, and Visual Dopamine Detection in Tiny Amount of Biofluid. *Small* **2019**, *15*. [CrossRef] [PubMed]
11. Bhalla, N.; Jolly, P.; Formisano, N.; Estrela, P. Introduction to biosensors. *Biosens. Technol. Detect. Biomol.* **2016**, *60*, 1–8.
12. Morales, M.A.; Halpern, J.M. Guide to Selecting a Biorecognition Element for Biosensors. *Bioconjugate Chem.* **2018**, *29*, 3231–3239. [CrossRef] [PubMed]
13. Fuchs, Y.; Soppera, O.; Haupt, K. Photopolymerization and photostructuring of molecularly imprinted polymers for sensor applications-A review. *Anal. Chim. Acta* **2012**, *717*, 7–20. [CrossRef] [PubMed]
14. Haupt, K.; Rangel, P.X.M.; Bui, B.T.S. Molecularly Imprinted Polymers: Antibody Mimics for Bioimaging and Therapy. *Chem. Rev.* **2020**, *120*, 9554–9582. [CrossRef]
15. Xu, J.; Miao, H.; Wang, J.; Pan, G. Molecularly Imprinted Synthetic Antibodies: From Chemical Design to Biomedical Applications. *Small* **2020**, *16*. [CrossRef] [PubMed]
16. Poma, A.; Turner, A.P.F.; Piletsky, S.A. Advances in the manufacture of MIP nanoparticles. *Trends Biotechnol.* **2010**, *28*, 629–637. [CrossRef] [PubMed]
17. Xu, J.; Ambrosini, S.; Tamahkar, E.; Rossi, C.; Haupt, K.; Bui, B.T.S. Toward a Universal Method for Preparing Molecularly Imprinted Polymer Nanoparticles with Antibody-like Affinity for Proteins. *Biomacromolecules* **2016**, *17*, 345–353. [CrossRef] [PubMed]
18. Xu, S.; Wang, L.; Liu, Z. Molecularly Imprinted Polymer Nanoparticles: An Emerging Versatile Platform for Cancer Therapy. *Angew. Chem. Int. Ed.* **2021**, *60*, 3858–3869. [CrossRef] [PubMed]
19. Cheong, W.J.; Yang, S.H.; Ali, F. Molecular imprinted polymers for separation science: A review of reviews. *J. Sep. Sci.* **2013**, *36*, 609–628. [CrossRef] [PubMed]
20. Saylan, Y.; Yilmaz, F.; Ozgur, E.; Derazshamshir, A.; Yavuz, H.; Denizli, A. Molecular Imprinting of Macromolecules for Sensor Applications. *Sensors* **2017**, *17*, 898. [CrossRef]
21. Vaneckova, T.; Bezdekova, J.; Han, G.; Adam, V.; Vaculovicova, M. Application of molecularly imprinted polymers as artificial receptors for imaging. *Acta Biomater.* **2020**, *101*, 444–458. [CrossRef]
22. Korde, B.A.; Mankar, J.S.; Phule, S.; Krupadam, R.J. Nanoporous imprinted polymers (nanoMIPs) for controlled release of cancer drug. *Mater. Sci. Eng. C* **2019**, *99*, 222–230.
23. Wackerlig, J.; Schirhagl, R. Applications of Molecularly Imprinted Polymer Nanoparticles and Their Advances toward Industrial Use: A Review. *Anal. Chem.* **2016**, *88*, 250–261. [CrossRef]
24. Lowdon, J.W.; Dilien, H.; Singla, P.; Peeters, M.; Cleij, T.J.; van Grinsven, B.; Eersels, K. MIPs for commercial application in low-cost sensors and assays—An overview of the current status quo. *Sens. Actuators B* **2020**, *325*, 128973.
25. Urraca, J.L.; Marazuela, M.D.; Merino, E.R.; Orellana, G.; Moreno-Bondi, M.C. Molecularly imprinted polymers with a streamlined mimic for zearalenone analysis. *J. Chromatogr. A* **2006**, *1116*, 127–134. [CrossRef]
26. Lucci, P.; Derrien, D.; Alix, F.; Perollier, C.; Bayoudh, S. Molecularly imprinted polymer solid-phase extraction for detection of zearalenone in cereal sample extracts. *Anal. Chim. Acta* **2010**, *672*, 15–19. [CrossRef]

27. Xuan-Anh, T.; Acha, V.; Haupt, K.; Bernadette Tse Sum, B. Direct fluorimetric sensing of UV-excited analytes in biological and environmental samples using molecularly imprinted polymer nanoparticles and fluorescence polarization. *Biosens. Bioelectron.* **2012**, *36*, 22–28.
28. Demir, B.; Lemberger, M.M.; Panagiotopoulou, M.; Rangel, P.X.M.; Timur, S.; Hirsch, T.; Bui, B.T.S.; Wegener, J.; Haupt, K. Tracking Hyaluronan: Molecularly Imprinted Polymer Coated Carbon Dots for Cancer Cell Targeting and Imaging. *ACS Appl. Mater. Interfaces* **2018**, *10*, 3305–3313. [CrossRef]
29. Panagiotopoulou, M.; Salinas, Y.; Beyazit, S.; Kunath, S.; Duma, L.; Prost, E.; Mayes, A.G.; Resmini, M.; Bui, B.T.S.; Haupt, K. Molecularly Imprinted Polymer Coated Quantum Dots for Multiplexed Cell Targeting and Imaging. *Angew. Chem. Int. Ed.* **2016**, *55*, 8244–8248.
30. Thongrussamee, T.; Kuzmina, N.S.; Shim, W.B.; Jiratpong, T.; Eremin, S.A.; Intrasook, J.; Chung, D.H. Monoclonal-based enzyme-linked immunosorbent assay for the detection of zearalenone in cereals. *Food Addit. Contam. Part A* **2008**, *25*, 997–1006. [CrossRef]
31. Wang, D.; Zhang, Z.; Zhang, Q.; Wang, Z.; Zhang, W.; Yu, L.; Li, H.; Jiang, J.; Li, P. Rapid and sensitive double-label based immunochromatographic assay for zearalenone detection in cereals. *Electrophoresis* **2018**, *39*, 2125–2130. [CrossRef] [PubMed]
32. Caglayan, M.O.; Ustundag, Z. Detection of zearalenone in an aptamer assay using attenuated internal reflection ellipsometry and it's cereal sample applications. *Food Chem. Toxicol.* **2020**, *136*, 111081. [CrossRef]
33. Sun, S.; Xie, Y. An enhanced enzyme-linked aptamer assay for the detection of zearalenone based on gold nanoparticles. *Anal. Methods* **2021**, *13*, 1255–1260. [CrossRef]
34. Nestora, S.; Merlier, F.; Beyazit, S.; Prost, E.; Duma, L.; Baril, B.; Greaves, A.; Haupt, K.; Bui, B.T.S. Plastic Antibodies for Cosmetics: Molecularly Imprinted Polymers Scavenge Precursors of Malodors. *Angew. Chem. Int. Ed.* **2016**, *55*, 6252–6256. [CrossRef]
35. Kolapalli, S.P.; Kumaraswamy, S.B.; Mortha, K.K.; Thomas, A.; Das Banerjee, S. UNIVmAb reactive albumin associated hyaladherin as a potential biomarker for colorectal cancer. *Cancer Biomark.* **2021**, *30*, 55–62. [CrossRef] [PubMed]
36. Li, H.; Guo, L.; Li, J.W.; Liu, N.; Qi, R.; Liu, J. Expression of hyaluronan receptors CD44 and RHAMM in stomach cancers: Relevance with tumor progression. *Int. J. Oncol.* **2000**, *17*, 927–932. [CrossRef]
37. Rangel, P.X.M.; Lacief, S.; Xu, J.; Panagiotopoulou, M.; Kovensky, J.; Bui, B.T.S.; Haupt, K. Solid-phase synthesis of molecularly imprinted polymer nanolabels: Affinity tools for cellular bioimaging of glycans. *Sci. Rep.* **2019**, *9*, 3923. [CrossRef]

Article

Effects of Cryoconcentrated Blueberry Juice as Functional Ingredient for Preparation of Commercial Confectionary Hydrogels

Nidia Casas-Forero [1,2,*], Igor Trujillo-Mayol [1], Rommy N. Zúñiga [3,4], Guillermo Petzold [2,5] and Patricio Orellana-Palma [6,*]

[1] Programa de Doctorado en Ingeniería de Alimentos, Facultad de Ciencias de la Salud y de los Alimentos, Campus Fernando May, Universidad del Bío-Bío, Av. Andrés Bello 720, Chillán 3780000, Chile; igor.trujillo@gmail.com

[2] Laboratorio de Crioconcentración, Departamento de Ingeniería en Alimentos, Facultad de Ciencias de la Salud y de los Alimentos, Campus Fernando May, Universidad del Bío-Bío, Av. Andrés Bello 720, Chillán 3780000, Chile; gpetzold@ubiobio.cl

[3] Departamento de Biotecnología, Facultad de Ciencias Naturales, Matemática y del Medio Ambiente, Campus Macul, Universidad Tecnológica Metropolitana, Las Palmeras 3360, Ñuñoa, Santiago 7800003, Chile; rommy.zuniga@utem.cl

[4] Programa Institucional de Fomento a la Investigación, Desarrollo e Innovación, Universidad Tecnológica Metropolitana, Ignacio Valdivieso 2409, San Joaquín, Santiago 8940577, Chile

[5] Grupo de Crioconcentración de Alimentos y Procesos Relacionados, Universidad del Bío-Bío, Av. Andrés Bello 720, Chillán 3780000, Chile

[6] Departamento de Ingeniería en Alimentos, Facultad de Ingeniería, Campus Andrés Bello, Universidad de La Serena, Av. Raúl Bitrán 1305, La Serena 1720010, Chile

* Correspondence: nidiacf@gmail.com (N.C.-F.); patricio.orellanap@userena.cl (P.O.-P.); Tel.: +56-51-2204000 (P.O.-P.)

Abstract: Hydrogels can absorb and/or retain components in the interstitial spaces due to the 3D cross-linked polymer network, and thus, these matrices can be used in different engineering applications. This study focuses on the physicochemical and textural properties, as well as bioactive compounds and their antioxidant activity stability of commercial hydrogels fortified with cryoconcentrated blueberry juice (CBJ) stored for 35 days. CBJ was added to commercial hydrogels (gelatin gel (GG), aerated gelatin gel (AGG), gummy (GM), and aerated gummy (AGM)). The samples showed a total polyphenol, anthocyanin, and flavonoid content ranging from 230 to 250 mg GAE/100 g, 3.5 to 3.9 mg C3G/100 g, and 120 to 136 mg CEQ/100 g, respectively, and GG and GM showed the lowest bioactive component degradation rate, while AGM presented the highest degradation. GG and GM samples could be stored for up to 21 days without significant changes, while the results indicated ≈15 days for the AGG and AGM samples. Thereby, CBJ offers enormous possibilities to be used as a functional ingredient due to the high nutritional values, and it allows enriching different hydrogel samples, and in turn, the structures of hydrogels protected components during in vitro digestion, enhancing the bioaccessibility after the digestion process.

Keywords: cryoconcentrated blueberry juice; hydrogels; stability; storage; bioactive compounds content; antioxidant activity

1. Introduction

Confectionery hydrogel products have a high popularity and demand among consumers (mainly children and the elderly) due to their visual attractiveness, texture, and mouthfeel [1]. Specifically, the confectionery hydrogel products are manufactured with gelling agents, sucrose, glucose syrup, acids, flavorings, and colorants [2]. Thus, depending on the type of processing and mixture of ingredients, it is possible to obtain hydrogel products with different textures, including gelatin gels with soft texture [3], gummy characterized by a firm, soft and chewy texture structure [4], and marshmallows or aerated gummy (conferred by air bubbles) with a soft and elastic texture [5].

On the other hand, the growing awareness and knowledge (individual and community) among consumers have encouraged food industries to replace ingredients harmful to human health used in many food products. Whereby confectionery hydrogel products have been questioned due to their high sugar content and low nutritional value. This combination has been associated with obesity, diabetes, cardiovascular diseases, and hypertension [6]. Thus, consumers demand healthier confectionery hydrogels foods with lower sugar content and higher natural antioxidants than current commercial products [7]. Therefore, researchers are endeavoring to improve the nutritional value and consumer appreciation of these products. The most common strategies are based on sugar substitution through low-caloric sweeteners, such as sorbitol, isomaltulose, and stevia [6,8,9]. Hence, various bioactive compounds have been incorporated into hydrogels products, such tea extract [7,10], hibiscus extract [11], betalain-rich extract [12], watermelon juice [13], pomegranate juice [14], and cryoconcentrated blueberry juice [15,16].

In particular, cryoconcentrated juices retain their natural sugars, bioactive compounds, and antioxidant activity due to low-temperature processing through cryoconcentration technology [17], display attractive colors, and have interesting health benefits [18–20]. However, it is essential to understand that these bioactive compounds could lose their beneficial bioactivity during storage due to their instability under various ambient factors [21,22]. Therefore, it is important to establish their stability through different storage conditions. In this sense, kinetic models have been used to predict the influence of storage on the stability of bioactive compounds [21]. Tutunchi et al. [22] and Rodríguez-Sánchez et al. [23] reported that betanin and betaxanthins' degradation in gummy candy considers first-order kinetics. However, there are few studies on the degradation kinetics of bioactive compounds in commercial hydrogels based on confectionery products during storage conditions, which is important for the food industry due to their interest in the development of confectionery products with health-promoting effects.

Therefore, the aim of this study was to evaluate physicochemical and textural properties, stability of the bioactive compound, and antioxidant activity in commercial hydrogel products based on confectionery foods fortified with cryoconcentrated blueberry juice during storage conditions (4 °C and 25 °C) over 35 days.

2. Results and Discussion

2.1. Characterization of Hydrogels Samples

2.1.1. Physicochemical Properties

Table 1 summarizes the physicochemical properties of the samples.

Table 1. Physicochemical parameters of gelatin-based confectionery.

Physicochemical Parameter	GG	AGG	GM	AGM
TSS (°Brix)	16.8 ± 0.0 [a]	21.0 ± 0.1 [b]	61.0 ± 1.0 [cd]	60.1 ± 0.2 [c]
pH	5.0 ± 0.0 [a]	5.2 ± 0.0 [c]	5.1 ± 0.0 [b]	5.1 ± 0.0 [b]
Moisture (%)	83.1 ± 0.1 [d]	78.9 ± 0.2 [c]	38.7 ± 0.4 [a]	38.3 ± 2.6 [ab]
Water activity	0.988 ± 0.001 [d]	0.981 ± 0.003 [c]	0.889 ± 0.009 [ab]	0.877 ± 0.008 [a]
Density (kg/m^3)	1083.2 ± 15.9 [b]	408.6 ± 5.3 [a]	1202.1 ± 33.1 [c]	406.5 ± 14.2 [a]
Gas hold-up (ε, %)	ND	62.3 ± 0.9 [a]	ND	66.2 ± 1.5 [b]

[a–d]: Different superscripts within the same row indicate significant differences at $p \leq 0.05$. GG: gelatin gel; AGG: aerated gelatin gel; GM: gummy; AGM: aerated gummy; TSS: total soluble solid; ND: not determined.

Initially, GG and AGG reached a TSS content of around 17 °Brix and 21 °Brix, respectively. Rubio-Arraez et al. [3] reported a similar TSS value (20 °Brix) in gels with citrus juice and non-cariogenic sweeteners. Thus, this result means that CBJ provides the TSS necessary to obtain gelatin gel without sweeteners. While GM and AGM exhibited lower TSS values (≈60 °Brix) than that indicated by studies in gummies and marshmallows [10,11,24,25], which reported values between 70 °Brix and 80 °Brix in the hydrogel samples. This difference may be due to the partial substitution of the syrup for CBJ, resulting in GM and AGM with a lower amount of added sugar. For pH, the values ranged from 5.0 to 5.2. These values are lower than those previously reported for similar products, ranging from 6.0 and 6.9 [5,24,25]. This decrease in pH value can be related to the CBJ added in the samples, which provides organic acids, such as malic, citric, and shikimic acids [26]. In the same way, the moisture content and a_w in the GG and AGG samples were in agreement with the values reported by Rubio-Arraez et al. [3]. For GM and AGM hydrogels, the results were significantly higher than those stated by Mardani et al. [25] and Periche et al. [5,24], who observed a moisture content value close to 16% to 24% and a_w of approximately 0.75 to 0.85. In contrast, Mandura et al. [10] and Šeremet et al. [7] reported high moisture content (34%) for gummy with a white tea infusion as an ingredient. Therefore, these differences could be attributed to TSS and phenolic compounds since the phenolic groups can hold more water due to interactions between the hydroxyl groups of polyphenols, proteins, sugars, and water [27].

For density, the aerated samples (AGG and AGM) showed a reduction of 35% compared to the non-aerated samples (GG and GM) due to the incorporation of air bubbles [15]. This incorporated air (ε) corresponds to 62% and 66% for AGG and AGM, respectively. These results are in line with the values reported by Casas-Forero et al. [15] and Mardani et al. [25] for aerated gelatin gel and marshmallows, respectively. Furthermore, air is the "main ingredient" of aerated confectionery, and it provokes a significant decrease in density [28].

2.1.2. Rheological and Mechanical Properties

Figure 1 shows the rheological behavior of the hydrogel samples.

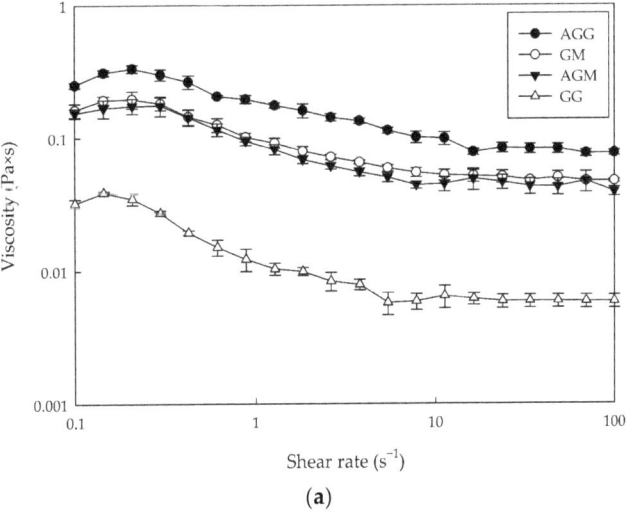

(a)

Figure 1. *Cont.*

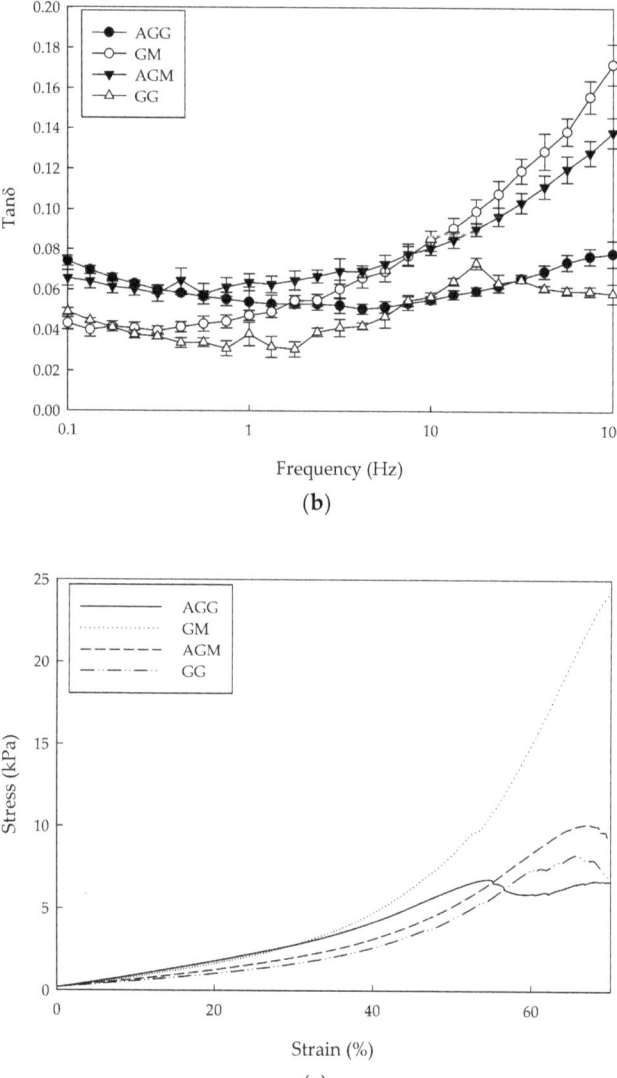

Figure 1. Rheological and mechanical properties of hydrogel samples: (**a**) The changes in viscosity during shear rate; (**b**) The changes in tan δ (G″/G′) during frequency scanning; (**c**) Stress and strain curves in uniaxial compression.

First, all samples exhibited similar behavior. Thus, at low shear rates (<10 s^{-1}), the viscosity decreased with an increasing shear rate, indicating that the samples had a pseudoplastic fluid and shear thinning behavior. Then, at high shear rates (>10 s^{-1}), the viscosity no longer changed with the increase in shear rate (Newtonian fluid) [6]. Similarly, Figure 1a (relationship between apparent viscosity and shear rate) shows that GG had the lowest viscosity (0.006 Pa·s). In contrast, AGG significantly increased its viscosity (0.083 Pa·s), showing values close to those found for GM and AGM (≈0.05 Pa·s). This increase in viscosity is mainly related to an increase in the amount of gelatin added in the AGG, GM, and AGM samples, which was significantly higher than GG hydrogel, with an increase of 2.6, 2.0, and 2.0 times, respectively. This behavior agrees with that reported by Casas-Forero et al. [15]. Furthermore, as described above, aeration in confectionery changes its physical properties, in particular, an increase in viscosity and a decrease in fluidity [28].

In the same way, the value of tan δ (G''/G') is a parameter of the viscoelastic behavior of the materials [29]. As shown in Figure 1b, the tan δ of all samples was less than 0.5 in the entire frequency range, indicating that an elastic behavior predominated over the viscous, which is characteristic of these types of hydrogel products [6].

Additionally, stress-deformation curves of the samples during uniaxial compression are depicted in Figure 1c. All samples exhibited a sigmoid stress–strain behavior, characteristic of foam materials [30]. As can be seen, GG had fracture stress and strain of 8 kPa and 65%, respectively, exhibiting the characteristics of a soft material. The results are consistent with the literature, which reports 4 kPa to 8 kPa and 60% values for fracture stress and strain, respectively [31,32]. AGG was characterized by a weaker and less ductile texture with 6 kPa for fracture stress and 55% for fracture strain. This difference between GG and AGG is attributed to the air bubbles, which reduce the matrix content per unit of sectional [32]. This is in line, with Hartel et al. [28] who indicated that the texture properties of aerated confectionery are largely dependent on the air phase. For GM and AGM, fracture stress and strain values in GM (23 kPa and 69%, respectively) were higher than achieved by AGM (10 kPa and 67%, respectively). This means that GM had a firm and elastic texture, while AGM was a product with a soft elastic texture. Likewise, AGM showed fracture stress close to those obtained in a commercial marshmallow (9 kPa) [33], indicating that it is possible to develop hydrogel foods with CBJ as a source of natural sugar and bioactive compounds without affecting its texture, which would be a potential commercial advantage.

2.1.3. Microstructural Features

Micrographs of the confectionery are shown in Figure 2.

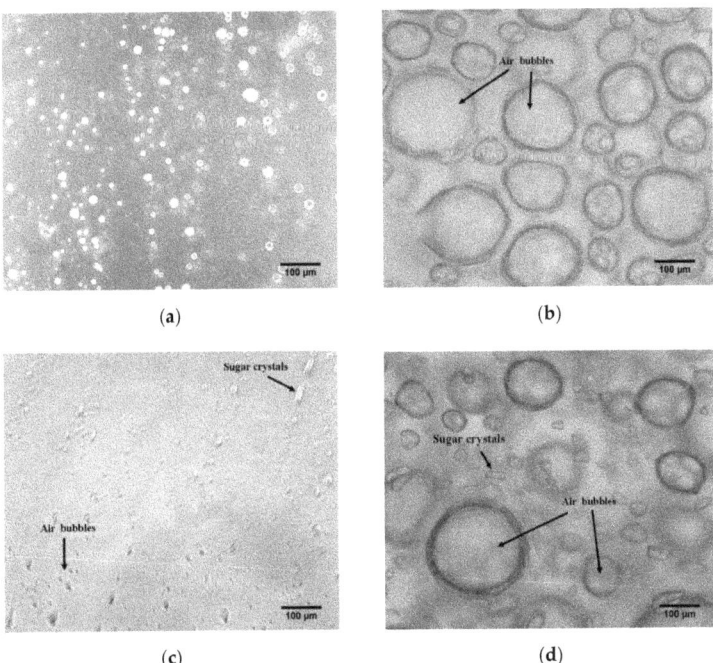

Figure 2. Optical micrograph images of gelatin-based confectionery at 10X: (**a**) GG; (**b**) AGG; (**c**) GM; (**d**) AGM.

GG exhibited a homogeneous structure with a small strand formation (Figure 2a), which is characteristic of the microstructure of bovine gelatin [12]. GM preserved the uniform structure of GG with the formation of some air bubbles and any tiny sugar crystals (Figure 2c). A similar microstructure was reported by de Moura et al. [11] and Kumar et al. [34] in jelly candies with hibiscus extract-encapsulated and vegan gummy candy, respectively. In the case of the aerated samples, AGG and AGM showed small and large air bubbles. Large bubbles with a marked spherical shape were characterized in AGG hydrogels (Figure 2b), while AG showed small bubbles with non-spherical shapes and the formation of tiny sugar crystals (Figure 2d). The difference in bubble size between AGG and AGM can be attributed to the gelatin content, as evidenced by Casas-Forero et al. [15] in the study of aerated gelatin gels that increased small bubbles as the amount of gelatin decreased.

2.2. Stability of Hydrogel Samples during Storage

2.2.1. Stability of TBC Content and Antioxidant Activity

Total bioactive compounds (TBC) and antioxidant activity (AA) values in the initial sample and hydrogel samples are shown in Table 2.

Table 2. Total bioactive compounds (TBC) and antioxidant activity (AA) of the hydrogel samples.

Sample	TBC			AA	
	TPC	TAC	TFC	DPPH	FRAP
CBJ	773.3 ± 8.7 [e]	22.3 ± 0.7 [e]	566.8 ± 7.6 [e]	4585.4 ± 8.5 [e]	4442.0 ± 61.9 [d]
GG	233.4 ± 1.6 [a]	3.9 ± 0.1 [cd]	136.1 ± 2.9 [cd]	782.9 ± 3.0 [b]	1067.8 ± 13.5 [a]
AGG	251.0 ± 3.6 [cd]	3.5 ± 0.2 [a]	133.7 ± 3.7 [c]	759.1 ± 1.1 [a]	1057.3 ± 45.9 [a]
GM	247.7 ± 0.9 [c]	3.7 ± 0.2 [abc]	119.6 ± 3.9 [a]	911.3 ± 2.2 [d]	1075.1 ± 12.6 [ab]
AGM	238.5 ± 3.9 [ab]	3.7 ± 0.1 [ab]	123.9 ± 4.7 [ab]	870.5 ± 4.8 [c]	1083.3 ± 31.8 [ab]

[a–e]: Different superscripts within the same column indicate significant differences at $p \leq 0.05$. CBJ: cryoconcentrated blueberry juice; GG: gelatin gel; AGG: aerated gelatin gel; GM: gummy; AGM: aerated gummy.

The fortification of food with juice rich in bioactive compounds can improve the nutritional and functional values of these products [35]. First, CBJ had a TBC content close to 770 mg GAE/100 g for TPC, 22 mg C3G/100 g for TAC, and 560 mg CEQ/100 g for TFC, and the AA values were approximately 4585 μmol TE/100 g for DPPH and 4440 μmol TE/100 g for FRAP.

On day 0, hydrogel samples displayed a TBC content ranging from 230 to 250 mg GAE/100 g for TPC, 3.5 to 3.9 mg C3G/100 g for TAC, and 120 to 136 mg CEQ/100 g for TFC. These slight variations in TBC content between the samples could be due to processing conditions since all samples were fortified with CBJ at 30% (w/w). Compared to other studies, the TPC obtained was higher than those reported in white tea-based candies (170–180 mg GAE/100 g) [7], pomegranate juice-based candies (72–159 mg GAE/100 g) [14], and jelly candies with rosemary extract (227 mg GAE/100 g) [8]. In contrast, Rivero et al. [9] reported polyphenol content of over 50% in candies with raspberry juice powder (490–550 mg GAE/100 g). These differences between studies can be related to the formulation, and the process conditions used.

In terms of AA, all the samples exhibited similar FRAP values (1050–1085 μmol TE/100 g), while DPPH values varied between samples. Thus, in DPPH, GG and AGG had values of 780 and 760 μmol TE/100 g, respectively, while GM and AGM increased the DPPH results, reaching values of 910 and 870 μmol TE/100 g, respectively. The values are comparable to those reported by Rivero et al. [9], Mandura et al. [10], and Hani et al. [36], who studied the AA of gummy with red pitaya fruit puree, white tea-based candies, and jellies containing honey and propolis, respectively.

Additionally, we observed a proportional trend between CBJ and hydrogel samples. A decrease varying from ≈70% to ≈85% for TBC and DPPH values were observed, while FRAP decreased in all the samples (44.7% for GG, 24.5% for AGG, 39.5% for GM, and 19.1% for AGM). These variations are in agreement with the results reported by Cedeño-Pinos et al. [8], who suggested that the reduction in the AA level in gelatin-based candies may be due to interactions between ingredients, the presence of amino acids, or the interaction of antioxidants with the gelling matrix.

Figure 3 shows the degradation of total bioactive compounds of the hydrogel samples and CBJ during storage.

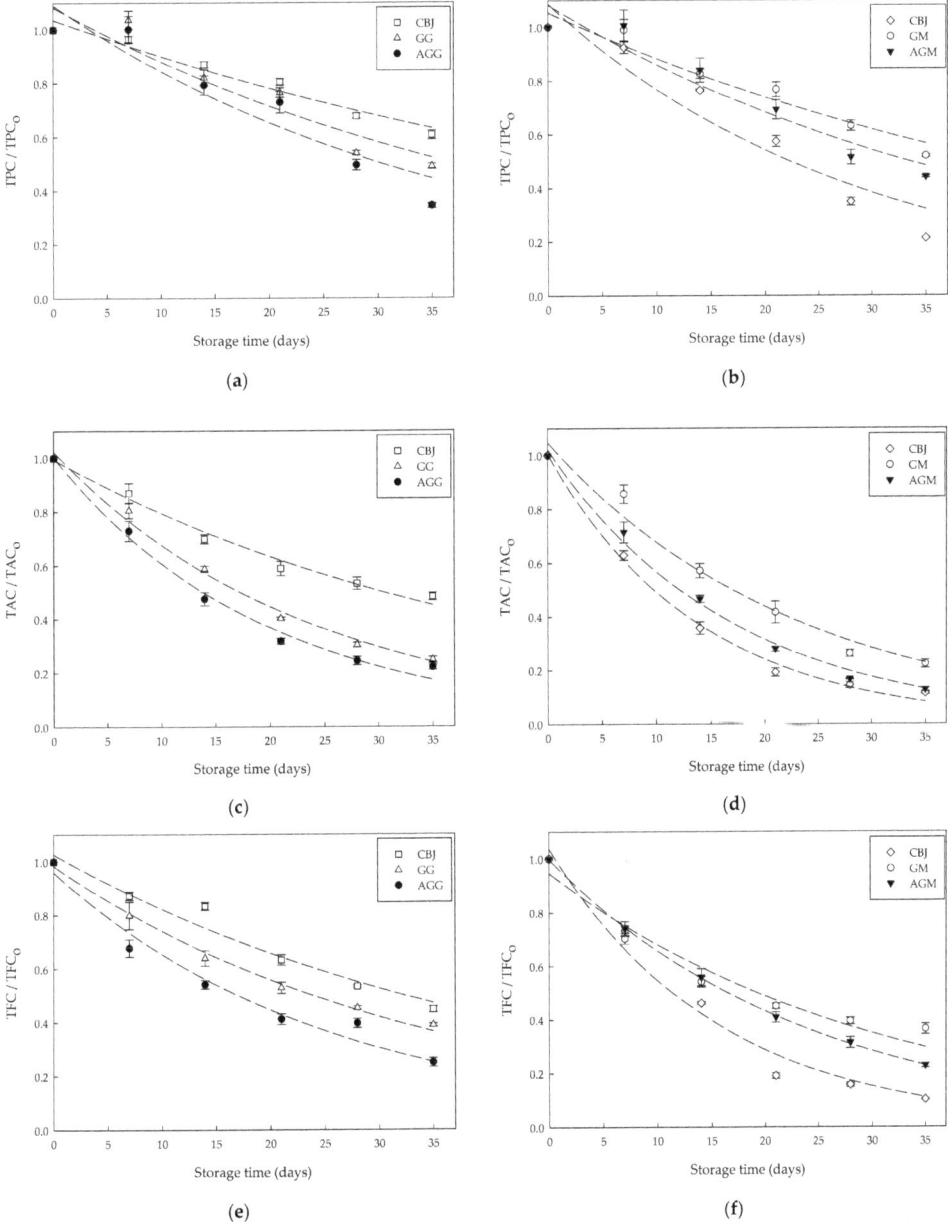

Figure 3. Degradation of total bioactive compounds of the gelatin-based confectionery and CBJ during storage: (**a**) TPC at 4 °C; (**b**) TPC at 25 °C; (**c**) TAC at 4 °C; (**d**) TPC at 25 °C; (**e**) TFC at 4 °C, (**f**) TPC at 25 °C.

The CBJ showed a significant reduction in TBC by increasing the temperature from 4 to 25 °C during storage. This decrease in TBC was between 40% and 50% for CBJ stored at 4 °C and about 80% for CBJ stored at 25 °C. This result was higher than the previously reported value by Orellana-Palma et al. [19] for cryoconcentrated Calafate juice, who reported a loss between 4% and 25% in the TBC after 35 days of storage at 4 °C. Regarding anthocyanin stability studies in blueberry juice, Barba et al. [37] indicated a decrease of about 9% after 7 days of storage at 4 °C in juice treated with PEF. Furthermore, Cortellino and Rizzolo [38] and Zhang et al. [39] observed in thermally treated juice a loss of 60% after 150 days at 20 °C and 80% after ten days at 4 °C, respectively. These differences in the TBC degradation rate during storage can be related to pretreatments before juice storage that could inactivate enzymes, such as polyphenol oxidase and peroxidase, which are considered responsible for the decay of phenols in berry-derived foods [38].

Figure 3a,b show that TPC gradually decreased as the storage time increased. After 35 days, the polyphenols loss was 50% and 65% for GG and AGG, respectively, whereas GM and AGM showed a lower loss with 48% and 55% values. This difference between samples could be related to the moisture content and water activity, since a more significant volume of water favors reactant mobility, contributing to phenolic compounds degradation [40]. Maier et al. [41] also observed a decrease in the phenolic content from 243.6 mg/kg to 82.6 mg/kg in gelatin gels enriched with grape pomace extract during storage. Furthermore, Tutunchi et al. [22] indicated a loss of TPC of 25% to 35% for gummy candy with red beet extract powder at 28 days. Meanwhile, Šeremet et al. [7] reported fluctuations in the TPC during the storage of white tea-based candies.

On the other hand, TAC decreased markedly over 35 days, with a total decrease of 75%, 78%, 77%, and 87% for GG, AGG, GM, and AGM, respectively (Figure 3c,d). These results were consistent with the findings of de Moura et al. [11] and Maier et al. [41]. They reported a reduction of close to 70% in anthocyanin content in jelly candy with hibiscus extract and gelatin gels enriched with grape pomace extract, respectively. Tavares et al. [42] mentioned that anthocyanins are unstable and highly susceptible to degradation under conditions such as pH, temperature, light, oxygen, enzymes, ascorbic acid, and copigments. Likewise, Chen et al. [21] and Teribia et al. [43] suggested that anthocyanin degradation could result from condensation and polymerization reactions with other phenolic compounds present in the sample. Hence, this anthocyanin decrease can lead to undesirable color changes and a reduction in antioxidant activity.

Like the results of the TPC and TAC, the storage also had an unfavorable influence on the TFC (Figure 3e,f). At the end of 35 days of storage, the flavonoids content in GG and AGG stored at 4 °C decreased by approximately 60% and 75%, respectively, and for the GM and AGM samples stored at 25 °C, this reduction was up to 62% and 77%, respectively. In general terms, the stability of TBC in gelatin-based foods was slightly higher than CBJ mainly when stored at 25 °C, possibly due to the interaction of polyphenols with gelatin that led to lower availability of polyphenols for hydrolysis and oxidation reactions [22,23]. Similarly, Kia [44] indicated that the gelatin triple helix could retain antioxidant compounds due to crosslinking and the formation of three-dimensional networks. On the other hand, the TBC stability in samples containing air bubbles inside the food matrix (AGG and AGM) was significantly ($p \leq 0.05$) lower than those samples non-aerated (GG and GM). This can be related to oxygen reacting with antioxidant compounds through hydrogen atom donation of the hydroxyl group to a free radical, leading to a decrease in the bioactive compounds [45].

Figure 4 shows a significant reduction ($p \leq 0.05$) in the values of DPPH and FRAP during the storage period at 4 and 25 °C.

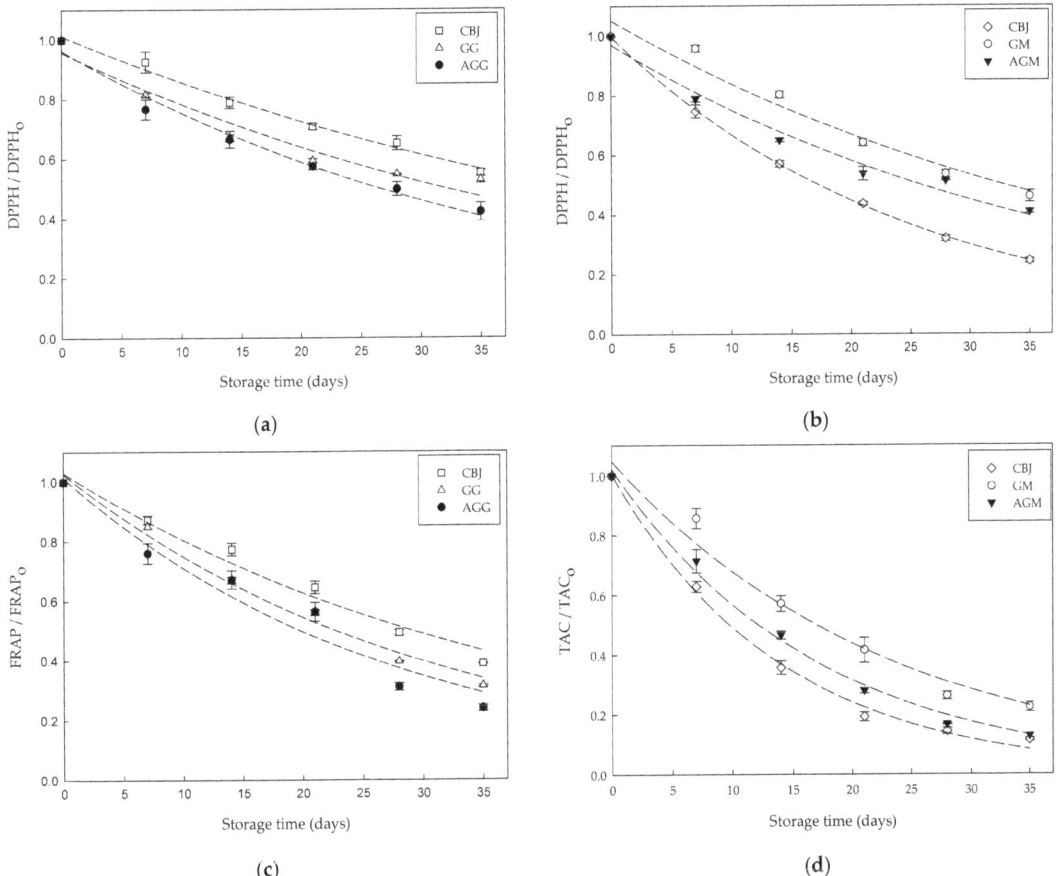

Figure 4. Antioxidant activity stability of hydrogel samples and CBJ during storage: (**a**) DPPH at 4 °C; (**b**) DPPH at 25 °C; (**c**) FRAP at 4 °C; (**d**) FRAP at 25 °C.

DPPH values of hydrogel samples decreased by approximately 54% by day 35. Meanwhile, a higher loss was observed in the FRAP assay, which decreased between 65% and 75% by day 35. This reduction might be due to the degradation of bioactive compounds, mainly anthocyanins [21]. The decrease in AA during storage has been reported in other studies, including a decrease in AA of 64% in jelly prepared with citrus juice [3], 51% in white tea-based candies [10], 21% to 37% in gummy candy with red beet extract powder [22], 15% in vegan gummy candies with betalains nanoliposomes [34], and 55% to 70% in gummy candy with red beet extract [46].

On the other hand, k and $t_{1/2}$ are summarized in Table 3.

Table 3. Kinetic parameters of total bioactive compounds and antioxidant activity in the CBJ and hydrogel samples stored under different conditions.

Assay	Kinetic Parameters	4 °C			25 °C		
		CBJ	GG	AGG	CBJ	GM	AGM
TPC	k (10^{-3} day^{-1})	13.0 ± 0.6 [a]	18.6 ± 0.1 [c]	24.9 ± 1.4 [e]	36.8 ± 0.7 [f]	16.4 ± 0.6 [b]	21.3 ± 1.3 [d]
	$t_{1/2}$ (day)	53.3 ± 2.7 [f]	37.2 ± 0.3 [d]	27.9 ± 1.6 [b]	18.8 ± 0.4 [a]	42.4 ± 1.5 [e]	32.6 ± 1.9 [c]
	R^2	0.94	0.87	0.86	0.89	0.92	0.90
TAC	k (10^{-3} day^{-1})	22.3 ± 1.2 [a]	40.9 ± 1.1 [b]	47.9 ± 1.7 [c]	67.7 ± 2.2 [ed]	43.6 ± 2.4 [b]	60.1 ± 7.2 [d]
	$t_{1/2}$ (day)	31.1 ± 1.8 [e]	16.9 ± 0.4 [d]	14.5 ± 0.5 [c]	10.3 ± 0.3 [a]	15.9 ± 0.9 [cd]	11.5 ± 0.1 [b]
	R^2	0.97	0.99	0.97	0.97	0.98	0.99
TFC	k (10^{-3} day^{-1})	21.8 ± 0.5 [a]	28.4 ± 0.5 [b]	38.8 ± 1.6 [d]	66.7 ± 0.9 [f]	33.1 ± 0.1 [c]	41.9 ± 1.7 [e]
	$t_{1/2}$ (day)	31.8 ± 0.7 [e]	24.4 ± 0.4 [d]	17.9 ± 0.7 [b]	10.4 ± 0.2 [a]	20.9 ± 0.8 [c]	16.5 ± 0.6 [b]
	R^2	0.96	0.98	0.95	0.97	0.90	0.99
DPPH	k (10^{-3} day^{-1})	16.3 ± 0.6 [a]	21.2 ± 0.4 [b]	25.6 ± 0.6 [c]	40.2 ± 0.7 [d]	21.3 ± 0.6 [b]	26.2 ± 0.6 [c]
	$t_{1/2}$ (day)	42.5 ± 1.7 [d]	32.8 ± 0.6 [c]	27.1 ± 0.6 [b]	17.2 ± 0.3 [a]	32.6 ± 0.9 [c]	26.5 ± 0.6 [b]
	R^2	0.98	0.90	0.96	0.98	0.96	0.97
FRAP	k (10^{-3} day^{-1})	24.7 ± 0.6 [a]	31.7 ± 0.2 [b]	37.9 ± 1.7 [d]	54.2 ± 1.5 [f]	34.1 ± 0.3 [c]	41.7 ± 0.6 [e]
	$t_{1/2}$ (day)	28.0 ± 0.7 [e]	21.9 ± 0.2 [d]	18.2 ± 0.8 [c]	12.8 ± 0.3 [a]	20.3 ± 0.2 [c]	16.6 ± 0.2 [b]
	R^2	0.96	0.98	0.94	0.97	0.97	0.99

[a–f]: Different superscripts within the same column indicate significant differences at $p \leq 0.05$. CBJ: cryoconcentrated blueberry juice; GG: gelatin gel; AGG: aerated gelatin gel; GM: gummy; AGM: aerated gummy. TPC: Total polyphenol content, TAC: Total anthocyanin content, TFC: Total flavonoid content, k: kinetic degradation rate, $t_{1/2}$: half-life time, R^2: correlation index.

The TBC and AA degradation was appropriately fitted to a first-order kinetic model ($R^2 > 0.9$), which is similar to the previous studies on the degradation of bioactive compounds in blueberries [41,47]. As expected, in CBJ samples, the storage temperature significantly affected the k values of TBC and AA. The results indicate that CBJ stored at 25 °C degrades three times faster than CBJ stored at 4 °C. Anthocyanins showed the most remarkable changes, decreasing $t_{1/2}$ from 31 days to 10 days. As reported by Tavares et al. [42], storage temperature influences anthocyanin degradation due to their high thermosensitivity.

By comparing samples stored at 4 °C, we observed that the matrix did not have a protective effect against the degradation of TBC and AA, as the value of k was approximately 1.3 and 1.6 times higher for GG and AGG compared to CBJ, respectively. These differences in the value of k could be related to the available water content since CBJ had a high TSS (45 °Brix), exhibited a lower aw (0.911) than the GG and AGG samples (0.988 and 0.981, respectively). Thus, these conditions reduce the degradation of phenolic compounds in CBJ by decreasing the mobility of the reactants within the food [22,40]. In samples stored at 25 °C, a protective effect of gelatin was noticeable on the degradation of TBC and AA. A 52% and 38% lower k were observed in GM and AGM than in CBJ, respectively. Similar results were reported by Tutunchi et al. [22], who indicated that betanin was more stable in gummy candy than a beverage. In this regard, Rodriguez-Sánchez et al. [23] suggested that interactions with proteins may protect some pigments such as betaxanthins from hydrolysis and oxidation. Likewise, Muhamad et al. [48] indicated that sugar has a protective effect on anthocyanins, which could be assigned to its steric interference with reaction products of anthocyanin–phenolic polymerization [49]. The anthocyanins stability results showed a $t_{1/2}$ of 16 and 12 days for GM and AGM, respectively. Tutunchi et al. [22] also observed a similar $t_{1/2}$ in gummy candies enriched with red beet extract powder.

2.2.2. Stability of Color

The CIELAB parameters (L*, a*, b*, and ΔE) of gelatin-based foods and CBJ stored for 35 days are shown in Table 4.

Table 4. Color parameters of CBJ and gelatin-based confectionery during storage.

Sample	Time (Days)	L*	a*	b*	ΔE
4 °C					
CBJ	0	8.0 ± 0.2 d	24.6 ± 0.9 f	4.9 ± 0.2 c	-
	7	8.0 ± 0.0 d	22.5 ± 0.0 e	4.6 ± 0.1 c	2.1 ± 1.0 a
	14	7.1 ± 0.2 c	21.6 ± 0.0 d	4.3 ± 0.0 b	3.1 ± 1.0 b
	21	6.9 ± 0.2 bc	21.0 ± 0.0 c	4.3 ± 0.0 b	3.8 ± 0.9 c
	28	6.7 ± 0.2 b	19.3 ± 0.1 b	3.2 ± 0.1 a	5.7 ± 0.9 d
	35	6.3 ± 0.2 a	17.4 ± 0.6 a	3.2 ± 0.0 a	7.6 ± 1.5 e
GG	0	18.1 ± 0.6 d	7.6 ± 0.4 f	1.1 ± 0.0 d	-
	7	17.2 ± 0.3 cd	7.1 ± 0.2 e	1.1 ± 0.1 d	1.2 ± 0.4 a
	14	16.6 ± 0.3 bc	5.8 ± 0.2 d	0.8 ± 0.1 c	2.4 ± 0.2 b
	21	16.2 ± 0.5 b	4.5 ± 0.1 c	0.7 ± 0.0 c	3.8 ± 0.3 c
	28	15.9 ± 0.6 ab	4.1 ± 0.1 b	0.6 ± 0.0 b	4.3 ± 0.1 d
	35	15.3 ± 0.2 a	3.8 ± 0.1 a	0.5 ± 0.0 a	4.8 ± 0.4 e
AGG	0	60.2 ± 0.5 e	10.2 ± 0.2 f	10.0 ± 0.1 d	-
	7	58.7 ± 0.7 d	9.6 ± 0.1 e	9.9 ± 0.3 d	1.7 ± 0.3 a
	14	56.8 ± 0.3 c	8.4 ± 0.2 d	9.2 ± 0.0 c	4.0 ± 0.0 b
	21	54.8 ± 0.1 b	7.9 ± 0.0 c	8.6 ± 0.1 b	6.0 ± 0.5 c
	28	53.4 ± 0.1 a	5.2 ± 0.0 b	8.5 ± 0.0 b	8.6 ± 0.2 d
	35	51.9 ± 1.4 a	4.6 ± 0.0 a	7.7 ± 0.0 a	10.3 ± 0.9 e
25 °C					
CBJ	0	8.0 ± 0.2 f	24.6 ± 0.9 f	4.9 ± 0.2 e	-
	7	6.8 ± 0.1 e	18.4 ± 0.1 e	3.9 ± 0.1 d	6.4 ± 1.0 a
	14	5.6 ± 0.0 d	13.5 ± 0.5 d	1.3 ± 0.0 c	11.9 ± 1.3 b
	21	4.6 ± 0.0 c	10.7 ± 0.1 c	1.0 ± 0.0 b	14.8 ± 1.0 c
	28	3.6 ± 0.0 b	9.0 ± 0.0 b	0.9 ± 0.0 a	16.7 ± 1.0 d
	35	3.0 ± 0.0 a	7.6 ± 0.0 a	0.9 ± 0.0 a	18.1 ± 1.0 e
GM	0	22.1 ± 0.6 e	4.0 ± 0.1 f	9.4 ± 0.4 d	-
	7	21.0 ± 0.4 d	2.8 ± 0.1 e	9.7 ± 0.1 c	1.7 ± 0.3 a
	14	19.3 ± 0.1 c	2.5 ± 0.1 d	8.7 ± 0.3 b	3.3 ± 0.5 b
	21	18.6 ± 0 4 b	2.4 ± 0.1 c	8.8 ± 0.1 b	3.9 ± 0.3 c
	28	17.9 ± 0.2 a	2.3 ± 0.0 b	8.7 ± 0.3 b	4.6 ± 0.4 d
	35	17.3 ± 0.5 a	1.7 ± 0.0 a	8.1 ± 0.2 a	5.5 ± 0.3 e
AGM	0	56.4 ± 1.6 f	11.3 ± 0.5 f	9.4 ± 0.4 d	-
	7	53.4 ± 0.8 e	11.2 ± 0.1 e	8.9 ± 0.1 c	3.0 ± 1.0 a
	14	47.8 ± 0.6 d	10.9 ± 0.2 d	8.8 ± 0.1 c	8.6 ± 2.2 b
	21	46.0 ± 1.0 c	9.9 ± 0.3 c	8.7 ± 0.3 bc	10.5 ± 0.9 c
	28	44.2 ± 0.5 b	9.1 ± 0.0 b	8.4 ± 0.1 b	12.4 ± 1.6 d
	35	40.7 ± 0.3 a	8.8 ± 0.1 a	7.9 ± 0.1 a	15.9 ± 1.7 e

a–f: Different superscripts within the same column indicate significant differences at $p \leq 0.05$. CBJ: cryoconcentrated blueberry juice; GG: gelatin gel; AGG: aerated gelatin gel; GM: gummy; AGM: aerated gummy. TPC: Total polyphenol content, TAC: Total anthocyanin content, TFC: Total flavonoid content, k: kinetic degradation rate, $t_{1/2}$: half-life time, R^2: correlation index.

On day 0, CBJ had an intense reddish color mainly due to anthocyanins (L* ≈ 7.9, a* ≈ 24.56, and b* ≈ 4.91), and its addition explains the marked color of the samples, as can be seen in Figure 5.

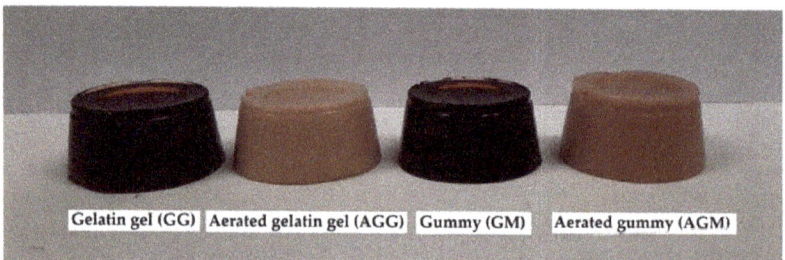

Figure 5. Photographs of the gelatin-based confectionery at day 0.

GG had low L* and higher a* compared to GM, indicating that GG presents a darker reddish color and may be related to a lower gelatin content in the formulation [24], highlighting the intense color of the CBJ. Meanwhile, the aerated samples (AGG and AGM) exhibited an opaque appearance with higher values of L* ($p \leq 0.05$) compared to the non-aerated samples (GG and GM), which is mainly attributed to the air bubbles as reported by Zúñiga et al. [32] and Casas-Forero et al. [15]. In addition, the aeration produces an evident decrease in the darkness of the samples.

In terms of storage, a general decrease in L* values was observed, indicating a progressive darkening. On day 35, CBJ, GG, and AGG showed a decrease of 21%, 16%, and 14%, respectively, stored at 4 °C, while CBJ, GM, and AGM, which were kept at 25 °C, exhibited a reduction of 62%, 22%, and 28%, respectively. Thus, these results denote a protective effect of gelatin on color stability, as indicated by Otálora et al. [12] with gummy candies. Likewise, several authors have reported higher color stability in confectionery, such as gels and candies, made with gelatin as a gelling agent, compared to pectin [7,36,41]. Similarly, the coordinates of a* and b* decreased significantly with time storage ($p \leq 0.05$), indicating that the samples lost their characteristic red coloration, which could be attributed to the degradation of anthocyanins in the chalcones [50].

The ΔE increased with time, reaching values greater than 5 units at 35 days. According to de Moura et al. [11], the human eye can distinguish the color change between samples with ΔE values greater than 3.5. The color difference was more noticeable in the CBJ stored at 25 °C (ΔE ≈18 units), whereas GG and GM exhibited lower values (ΔE ≈5 units). Comparable results were reported by Rivero et al. [9], de Moura et al. [11], and Yan et al. [51].

From a practical point of view, the present study can be useful in the food industry and molecular gastronomy, since the findings obtained are the use of the starting point for the cryoconcentrates used in gelling materials, which in turn allows multiple possibilities for food innovations.

3. Conclusions

The findings of this study suggest that it is possible to fortified gelatin-based confectionery with cryoconcentrated juice without detriment to their physicochemical characteristics and textural properties. CBJ improved nutritional value due to providing bioactive compounds with high antioxidant capacity. Specifically, on day 0, gelatin-based foods showed high TBC and AA. Regarding the stability study, a significant decrease in the TBC content and AA was observed in the CBJ and gelatin-based foods over storage (4 °C and 25 °C). Despite this, a protective effect of gelatin against loss of bioactive compounds and color changes was noticed. Furthermore, to preserve TBC, it can be estimated that GG and G could be stored for at least 21 days at 4 °C and 25 °C, respectively, while in aerated products (AGG and AGM), the storage time would be 15 days under storage at 4 °C and 25 °C, respectively. These findings support the idea that CBJ could be considered an innovative and functional ingredient. Furthermore, further studies must be carried out to evaluate the bioactivity of gelatin-based confectionery with cryoconcentrates to demonstrate their potential beneficial health effects.

4. Materials and Methods

4.1. Materials

Blueberries (*Vaccinium corymbosum* L.) were purchased from a local market in Chillán (36°36′24″ S 72°06′12″ W, XVI Región del Ñuble, Chile), and the fruits were stored at 4 °C until processing. Gelatin from bovine powder skin type B (G9382), sucrose, Folin–Ciocalteu reagent, gallic acid, cyanidin-3-glucoside, catequin, 1,1-diphenyl-2-picryl-hydrazyl (DPPH), 2,4,6-tris(2-pyridyl)-S-triazine (TPTZ), and 6-hydroxyl-2,5,7,8-tetramethyl-2-carboxylic acid (Trolox) were purchased from Sigma-Aldrich (St. Loius, MO, USA). Glucose syrup 42 DE was acquired locally from Furet Ltd. (Chillán, Chile). Standards and enzymes used to simulate in vitro digestion, such as human salivary α-amylase (1031), porcine pepsin (P6887), porcine pancreatin (H2625), and bovine bile (B3883), were purchased from Sigma-Aldrich (St. Loius, MO, USA). Calcium chloride ($CaCl_2$), sodium carbonate (Na_2CO_3), ferric chloride hexahydrate ($FeCl_3 \cdot 6H_2O$), sodium acetate (CH_3COONa), sodium hydroxide (NaOH), hydrochloric acid (HCl), sodium nitrite ($NaNO_2$), aluminum chloride ($AlCl_3$), and potassium chloride (KCl) were obtained from Merck (Darmstadt, Germany). Ultrapure and distilled water were used for the preparation of all aqueous solutions.

4.2. General Experimental Procedure

Figure 6 shows the experimental procedure of the hydrogels fortified with CBJ, the characterization of hydrogels samples, storage stability, determination of total bioactive compounds (TBC) and antioxidant activity (AA), color measurement, and half-life time analysis.

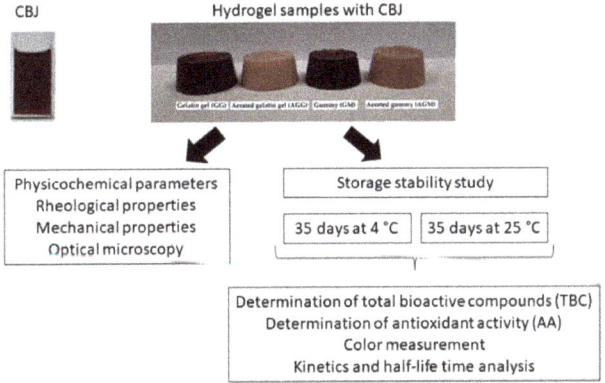

Figure 6. General experimental procedure.

4.3. CBJ Preparation

Blueberries were washed with tap water to remove dirt, and later, the juice was extracted in a tabletop juicer (JE2001, Nex, Barcelona, Spain). Immediately, the juice was filtered through a nylon cloth (0.8 mm mesh) to remove skin and seeds. Then, the juice was concentrated through three cryoconcentration cycles using the centrifugal method described previously by Casas-Forero et al. [16]. Hence, CBJ corresponds to the cryoconcentrated sample acquired in the final cycle. After the third cycle, the CBJ reached values of 45 °Brix and a 4.1 pH. Thus, the cryoconcentrated juice was stored until use as an ingredient in the elaboration of hydrogel samples.

4.4. Preparation of Hydrogel Products

For the commercial hydrogels products based on confectionery foods, four types of samples were prepared, gelatin gel (GG), aerated gelatin gel (AGG), gummy (GM), and aerated gummy (AGM). For GG and AGG samples, the hydrogels with CBJ were performed according to Casas-Forero et al. [15], with slight modifications. Separately, the gelatin powder (3 g for GG and 8 g for AGG) was hydrated in 70 mL distilled water at room temperature for 10 min. Then, the solutions were mechanically stirred (300 rpm) for 10 min at 60 °C. After, the solutions were cooled and maintained for 5 min at 40 °C, and later, 30 g of CBJ was added to each solution, and the solution (with CBJ) was stirred (300 rpm) for 5 min at 40 °C. Later, the gel solution with CBJ was poured into plastic vessels (approximately 34 mm inner diameter and 15 mm height). For AGG samples, air bubbles were incorporated using an Oster® mixer (Oster 2532, 250-watt, 6-speed, Rianxo, Spain) for 9 min, and then, the aerated gel solution with CBJ was deposited into plastic vessels. Finally, the GG and AGG samples were maintained overnight at 4 °C until the formation of a solid gel.

For GM and AGM samples, separately, 6 g of gelatin powder was hydrated in 25 mL distilled water for 10 min at room temperature. Then, the solution was heated at 60 °C and mechanically stirred at 300 rpm until complete dissolution ($\approx$10 min). The gelatin solution was cooled and maintained until at 40 °C. In parallel, a syrup solution was prepared with 350 mL of distilled water, 350 mL of sucrose, and 300 mL of glucose syrup. Thus, once the solution was homogenized with constant agitation, it was heated for 5 min at 110 °C, and then, it was cooled to 50 °C. After, 30 g of CBJ and 45 g of syrup were mixed with the gelatin solution, and the mixture was stirred (300 rpm) for 5 min at 40 °C. Thus, for GM, the gel solution with CBJ and syrup was placed into plastic vessels. For AGM, the aeration process was performed for 4 min using an Oster® mixer (Oster 2532, 250-watt, 6-speed, Rianxo, Spain). Finally, the aerated gel solution with CBJ and syrup was placed into plastic vessels. Thus, the GM and AGM were solidified overnight at room temperature.

4.5. Characterization of Hydrogels Samples

4.5.1. Physicochemical Parameters

Total soluble solids (TSS) expressed as °Brix were analyzed using a digital PAL-1 refractometer (PAL-3, $\approx$1 mL, range: 0–93 °Brix, precision: ±0.1 °Brix, Atago Inc., Tokyo, Japan). The pH was measured using a digital pH meter (HI 2210, Hanna Instruments, Woonsocket, RI, USA). The moisture content was obtained gravimetrically by vacuum drying at 60 °C in a vacuum oven (3618–1CE, Lab Line Instruments Inc., Melrose Park, IL, USA) for 24 h (AOAC, 20.013, 2000). The water activity (a_w) was determined using a dew-point hygrometer (AquaLab Model 4TE, Decagon Devices Inc., Pullman, WA, USA). The density was measured by the flotation method according to Zúñiga and Aguilera [31], and the results were expressed as kg/m^3. Gas hold-up (ε) of AGG and AGM was obtained by comparing the density of the aerated sample (ρ_{AGG} or ρ_{AGM}) with the respective gas-free sample (ρ_{GG} or ρ_{GM}) density. The ε was calculated according to Equation (1).

$$\varepsilon(\%) = \left(1 - \frac{\rho_{AGG} \text{ or } \rho_{AGM}}{\rho_{GG} \text{ or } \rho_{GM}}\right) * 100 \tag{1}$$

where ρ_{AGG}, ρ_{AGM}, ρ_{GG}, and ρ_{GM} are the density of aerated gelatin gel (AGG), aerated gummy (AGM), gelatin gel (GG), and gummy (GM), respectively.

4.5.2. Rheological Properties

The rheological properties were evaluated by a rotational-type rheometer (Physica MCR300, Anton Paar GmbH, Stuttgart, Germany) using a parallel plate geometry (50 mm diameter) with a 1 mm gap. The apparent viscosity of the samples was determined at 25 °C with a shear rate of 0.1 to 100 s^{-1} [27]. The frequency sweep measurement was performed from 0.1 to 100 Hz at a constant stress of 3.0 Pa within the linear viscoelastic region at 25 °C [52].

4.5.3. Mechanical Properties

The uniaxial compression test was carried out using a texture analyzer (TAXT plus100, Stable Micro Systems Ltd., Surrey, UK) with a load cell of 5.0 kg. The samples were compressed with a 50 mm diameter cylindrical aluminum probe (P50) at a constant speed of 1 mm/s up to a compression strain of 70% [53]. The true stress (σ_H) and true strain (ε_T) were calculated from force-time curves by Equations (2) and (3), respectively.

$$\sigma_H = F \left[\frac{h_o - h}{h_o\, A} \right] \qquad (2)$$

$$\varepsilon_T = -\ln \left[\frac{h_o}{h_o - h} \right] \qquad (3)$$

where F is the compression force (N), A is the cross-sectional area of the sample (m^2), and h_o and h are the initial and final height after compression (m), respectively.

4.5.4. Optical Microscopy

The samples were cut into thick slices of 3 mm with a razor blade and the slices were placed on a slide. Photomicrographs were acquired using an Olympus Trinocular Microscope (Olympus Co., Tokyo, Japan) coupled to a digital camera, Olympus LC 20 (Olympus Co., Munster, Germany) with an objective lens Nikon 10× [12].

4.6. Storage Stability Study (SSS)

Specifically, GG and AGG were maintained at 4 °C in the dark in a refrigerated incubator (FOC 215E, Velp Scientific Inc., Milano, Italy), and GM and AGM were stored at 25 °C in the dark in a thermostatic chamber (Memmert UF110, Memmert, Schwabach, Germany). For each sample and the control (CBJ), total bioactive compounds (TBC), antioxidant activity (AA), and color were analyzed at 0, 7, 14, 21, 28, and 35 days of storage.

4.7. Determination of Total Bioactive Compounds (TBC)

Total polyphenol content (TPC) was measured according to the Folin–Ciocalteu method described by Waterhouse [54], with minor modifications. Gallic acid (GA) was used as standard. An amount of 100 µL of the sample was mixed with 500 µL of 10-fold diluted Folin–Ciocalteu reagent. Then, the solution was vigorously mixed with 1500 µL of Na$_2$CO$_3$ (20% w/v). After 90 min in the dark at room temperature (incubation), the absorbance was recorded at 760 nm. TPC was calculated as milligrams of GA equivalents (GAE) per 100 g of sample (mg GAE/100 g).

Total monomeric anthocyanin content (TAC) was quantified by the differential pH method according to Lee et al. [55], with some modifications. Cyanidin-3-O-glucoside (C3G) was used as standard. An amount of 200 µL of the sample was mixed with 800 µL of KCl (pH 1.0, 0.025 M) and 800 µL of CH$_3$COONa (pH 4.5, 0.4 M) buffers. After 30 min in the dark at room temperature (incubation), the absorbance was measured at 510 and 700 nm. TAC was calculated as milligrams of C3G equivalents per 100 g of sample (mg C3G/100 g).

Total flavonoid content (TFC) was determined using the aluminum chloride colorimetric method described by Dewanto et al. [56], with modifications. Catequin (C) was used as standard. An amount of 250 µL of the sample was mixed with 1000 µL of distilled water and 75 µL of NaNO$_2$ (5% w/v). After 6 min in the dark at room temperature (incubation), 75 µL of AlCl$_3$ (10% w/v), 500 µL of NaOH (1 M), and 600 µL of distilled water were added to the solution, and later, the absorbance was measured at 510 nm. TFC was calculated as milligrams of C equivalents (CEQ) per 100 g of sample (mg CEQ/100 g).

4.8. Determination of Antioxidant Activity (AA)

The DPPH assay was assessed using the method reported by Brand-Williams et al. [57], with minor modifications. An amount of 150 µL of the sample was mixed with 2850 µL of DPPH methanolic solution. The mixture was kept in the dark at room temperature for 30 min (incubation), and the absorbance was measured at 515 nm.

The ferric reducing antioxidant power (FRAP) assay was performed according to Benzie and Strain [58], with some modifications. Briefly, FRAP reagent was prepared with

50 mL of CH_3COONa buffer (pH 3.6, 300 mM), 5.0 mL of TPTZ (10 mM in HCl (40 mM)), and 5.0 mL of $FeCl_3 \cdot 6H_2O$ (20 mM) (10:1:1 ratio), and then the solution was incubated at 37 °C. An amount of 150 µL of the sample was mixed with 2850 µL of FRAP reagent. The solution was kept in the dark at 37 °C for 30 min (incubation), and the absorbance was measured at 593 nm.

For DPPH and FRAP assays, Trolox (T) was used as the standard curve, and the results were expressed as µmol Trolox equivalents (TE) per 100 g of sample (µM TE/100 g).

TBC (TPC and TAC) and AA (DPPH and FRAP) values were measured by spectrophotometric analysis (T70 UV/Vis spectrophotometer, Oasis Scientific Inc., Greenville, SC, USA).

4.9. Color Measurement

Color analysis was performed through CIELab coordinates (L*: darkness–lightness, a*: green-red axis, b*: blue-yellow axis) using a colorimeter (CM-5, Konica Minolta, Osaka, Japan), with illuminant D65 and an observer angle of 10°, where the samples were filled in a glass cuvette, and thus, the CIELab values were measured [59].

4.10. Kinetics and Half-Life Time Analysis

First-order kinetics was used to describe the degradation TBC and AA content [22]. The reaction rate constant (k) and half-life time ($t_{1/2}$) were calculated according to Equations (4) and (5), respectively.

$$\frac{C_t}{C_o} = e^{-kt} \quad (4)$$

$$t_{1/2} = \frac{\ln 2}{k} \quad (5)$$

where C_o is the initial TBC and AA content, and C_t is the TBC and AA content at time t (days).

4.11. Statistical Analysis

All experiments were replicated three times and measurements were carried out in triplicate, and the data were expressed as mean ± standard deviation (SD). A one-way analysis of variance (ANOVA) was performed for each variable to identify differences between samples, and least significant difference (LSD) tests were performed for the comparison of means at a significance level of 5% ($p \leq 0.05$) using Statgraphics Centurion XVI software (v. 16.2.04, StatPoint Technologies Inc., Warrenton, VA, USA).

Author Contributions: Conceptualization, N.C.-F., R.N.Z. and G.P.; methodology, N.C.-F. and I.T.-M.; software, N.C.-F. and I.T.-M.; validation, N.C.-F., R.N.Z., G.P. and P.O.-P.; formal analysis, N.C.-F. and I.T.-M.; investigation, N.C.-F., R.N.Z., G.P. and P.O.-P.; resources, G.P. and P.O.-P.; data curation, N.C.-F. and I.T.-M.; writing—original draft preparation, N.C.-F. and I.T.-M.; writing—review and editing, N.C.-F., R.N.Z., G.P. and P.O.-P.; visualization, N.C.-F., I.T.-M., G.P. and P.O.-P.; supervision, G.P. and P.O.-P.; project administration, G.P. and P.O.-P.; funding acquisition, N.C-F., G.P. and P.O.-P. All authors have read and agreed to the published version of the manuscript.

Funding: Patricio Orellana-Palma acknowledges the financial support of ANID-Chile (Agencia Nacional de Investigación y Desarrollo de Chile) through the FONDECYT Postdoctoral Grant 2019 (Folio 3190420).

Institutional Review Board Statement: Not applicable.

Informed Consent Statement: Not applicable.

Data Availability Statement: The data is contained within the article.

Acknowledgments: Nidia Casas-Forero would like to thank the Vice-Chancellery for Research and Graduate at Universidad del Bío-Bío for the scholarship granted to complete her doctoral studies.

Conflicts of Interest: The authors declare no conflict of interest.

References

1. Yang, Z.; Chen, L.; McClements, D.J.; Qiu, C.; Li, C.; Zhang, Z.; Ming, M.; Tian, Y.; Zhu, K.; Jin, Z. Stimulus-responsive hydrogels in food science: A review. *Food Hydrocoll.* **2022**, *124*, 107218. [CrossRef]
2. Klein, M.; Poverenov, E. Natural biopolymer-based hydrogels for use in food and agriculture. *J. Sci. Food Agric.* **2020**, *100*, 2337–2347. [CrossRef] [PubMed]
3. Rubio-Arraez, S.; Capella, J.V.; Castelló, M.L.; Ortolá, M.D. Physicochemical characteristics of citrus jelly with non-cariogenic and functional sweeteners. *J. Food Sci. Technol.* **2016**, *53*, 3642–3650. [CrossRef] [PubMed]
4. Ge, H.; Wu, Y.; Woshnak, L.; Mitmesser, S. Effects of hydrocolloids, acids and nutrients on gelatin network in gummies. *Food Hydrocoll.* **2021**, *113*, 106549. [CrossRef]
5. Periche, Á.; Castelló, M.L.; Heredia, A.; Escriche, I. *Stevia rebaudiana*, oligofructose and isomaltulose as sugar replacers in marshmallows: Stability and antioxidant properties. *J. Food Process. Preserv.* **2016**, *40*, 724–732. [CrossRef]
6. Dai, H.; Li, X.; Du, J.; Ma, L.; Yu, Y.; Zhou, H.; Guo, T.; Zhang, Y. Effect of interaction between sorbitol and gelatin on gelatin properties and its mechanism under different citric acid concentrations. *Food Hydrocoll.* **2020**, *101*, 105557. [CrossRef]
7. Šeremet, D.; Mandura, A.; Cebin, A.; Martinić, A.; Galić, K.; Komes, D. Challenges in confectionery industry: Development and storage stability of innovative white tea-based candies. *J. Food Sci.* **2020**, *85*, 2060–2068. [CrossRef]
8. Cedeño-Pinos, C.; Martínez-Tomé, M.; Murcia, M.A.; Jordán, M.J.; Bañón, S. Assessment of rosemary (*Rosmarinus officinalis* L.) extract as antioxidant in jelly candies made with fructan fibres and stevia. *Antioxidants* **2020**, *9*, 1289. [CrossRef]
9. Rivero, R.; Archaina, D.; Sosa, N.; Schebor, C. Development and characterization of two gelatin candies with alternative sweeteners and fruit bioactive compounds. *LWT-Food Sci. Technol.* **2021**, *141*, 110894. [CrossRef]
10. Mandura, A.; Šeremet, D.; Ščetar, M.; Vojvodić Cebin, A.; Belščak-Cvitanović, A.; Komes, D. Physico-chemical, bioactive, and sensory assessment of white tea-based candies during 4-months storage. *J. Food Process. Preserv.* **2020**, *44*, e14628. [CrossRef]
11. de Moura, S.C.; Berling, C.L.; Garcia, A.O.; Queiroz, M.B.; Alvim, I.D.; Hubinger, M.D. Release of anthocyanins from the hibiscus extract encapsulated by ionic gelation and application of microparticles in jelly candy. *Food Res. Int.* **2019**, *121*, 542–552. [CrossRef] [PubMed]
12. Otálora, M.C.; de Jesús Barbosa, H.; Perilla, J.E.; Osorio, C.; Nazareno, M.A. Encapsulated betalains (*Opuntia ficus-indica*) as natural colorants. Case study: Gummy candies. *LWT-Food Sci. Technol.* **2019**, *103*, 222–227. [CrossRef]
13. Marinelli, V.; Lucera, A.; Incoronato, A.L.; Morcavallo, L.; Del Nobile, M.A.; Conte, A. Strategies for fortified sustainable food: The case of watermelon-based candy. *J. Food Sci. Technol.* **2021**, *58*, 894–901. [CrossRef] [PubMed]
14. Cano-Lamadrid, M.; Calín-Sánchez, Á.; Clemente-Villalba, J.; Hernández, F.; Carbonell-Barrachina, Á.A.; Sendra, E.; Wojdyło, A. Quality parameters and consumer acceptance of jelly candies based on pomegranate juice "Mollar de Elche". *Foods* **2020**, *9*, 516. [CrossRef] [PubMed]
15. Casas-Forero, N.; Orellana-Palma, P.; Petzold, G. Comparative study of the structural properties, color, bioactive compounds content and antioxidant capacity of aerated gelatin gels enriched with cryoconcentrated blueberry juice during storage. *Polymers* **2020**, *12*, 2769. [CrossRef] [PubMed]
16. Casas-Forero, N.; Moreno-Osorio, L.; Orellana-Palma, P.; Petzold, G. Effects of cryoconcentrate blueberry juice incorporation on gelatin gel: A rheological, textural and bioactive properties study. *LWT-Food Sci. Technol.* **2021**, *138*, 110674. [CrossRef]
17. Orellana-Palma, P.; Guerra-Valle, M.; Gianelli, M.P.; Petzold, G. Evaluation of freeze crystallization on pomegranate juice quality in comparison with conventional thermal processing. *Food Biosci.* **2021**, *41*, 101106. [CrossRef]
18. da Silva Haas, I.C.; de Espindola, J.S.; de Liz, G.R.; Luna, A.S.; Bordignon-Luiz, M.T.; Prudêncio, E.S.; de Gois, J.S.; Fedrigo, I.M.T. Gravitational assisted three-stage block freeze concentration process for producing enriched concentrated orange juice (*Citrus sinensis* L.): Multi-elemental profiling and polyphenolic bioactives. *J. Food Eng.* **2022**, *315*, 110802. [CrossRef]
19. Orellana-Palma, P.; Tobar-Bolaños, G.; Casas-Forero, N.; Zúñiga, R.N.; Petzold, G. Quality attributes of cryoconcentrated calafate (*Berberis microphylla*) juice during refrigerated storage. *Foods* **2020**, *9*, 1314. [CrossRef]
20. Qin, F.G.; Ding, Z.; Peng, K.; Yuan, J.; Huang, S.; Jiang, R.; Shao, Y. Freeze concentration of apple juice followed by centrifugation of ice packed bed. *J. Food Eng.* **2021**, *291*, 110270. [CrossRef]
21. Chen, J.Y.; Du, J.; Li, M.L.; Li, C.M. Degradation kinetics and pathways of red raspberry anthocyanins in model and juice systems and their correlation with color and antioxidant changes during storage. *LWT-Food Sci. Technol.* **2020**, *128*, 109448. [CrossRef]
22. Tutunchi, P.; Roufegarinejad, L.; Hamishehkar, H.; Alizadeh, A. Extraction of red beet extract with β-cyclodextrin-enhanced ultrasound assisted extraction: A strategy for enhancing the extraction efficacy of bioactive compounds and their stability in food models. *Food Chem.* **2019**, *297*, 124994. [CrossRef] [PubMed]
23. Rodríguez-Sánchez, J.A.; Cruz, M.T.C.; Barragán-Huerta, B.E. Betaxanthins and antioxidant capacity in *Stenocereus pruinosus*: Stability and use in food. *Food Res. Int.* **2017**, *91*, 63–71. [CrossRef]
24. Periche, Á.; Heredia, A.; Escriche, I.; Andrés, A.; Castelló, M.L. Optical, mechanical and sensory properties of based-isomaltulose gummy confections. *Food Biosci.* **2014**, *7*, 37–44. [CrossRef]
25. Mardani, M.; Yeganehzad, S.; Ptichkina, N.; Kodatsky, Y.; Kliukina, O.; Nepovinnykh, N.; Naji-Tabasi, S. Study on foaming, rheological and thermal properties of gelatin-free marshmallow. *Food Hydrocoll.* **2019**, *93*, 335–341. [CrossRef]
26. Mikulic-Petkovsek, M.; Schmitzer, V.; Slatnar, A.; Stampar, F.; Veberic, R. Composition of sugars, organic acids, and total phenolics in 25 wild or cultivated berry species. *J. Food Sci.* **2012**, *77*, C1064–C1070. [CrossRef]

27. Sadahira, M.S.; Rodrigues, M.I.; Akhtar, M.; Murray, B.S.; Netto, F.M. Influence of pH on foaming and rheological properties of aerated high sugar system with egg white protein and hydroxypropylmethylcellulose. *LWT-Food Sci. Technol.* **2018**, *89*, 350–357. [CrossRef]
28. Hartel, R.W.; Joachim, H.; Hofberger, R. (Eds.) Aerated Confections. In *Confectionery Science and Technology*, 1st ed.; Springer Nature: Cham, Switzerland, 2018; pp. 301–327.
29. Yang, F.; Zhang, M.; Bhandari, B.; Liu, Y. Investigation on lemon juice gel as food material for 3D printing and optimization of printing parameters. *LWT-Food Sci. Technol.* **2018**, *87*, 67–76. [CrossRef]
30. Laurindo, J.B.; Peleg, M. Mechanical measurements in puffed rice cakes. *J. Texture Stud.* **2007**, *38*, 619–634. [CrossRef]
31. Zúñiga, R.N.; Aguilera, J.M. Structure–fracture relationships in gas-filled gelatin gels. *Food Hydrocoll.* **2009**, *23*, 1351–1357. [CrossRef]
32. Zúñiga, R.N.; Kulozik, U.; Aguilera, J.M. Ultrasonic generation of aerated gelatin gels stabilized by whey protein β-lactoglobulin. *Food Hydrocoll.* **2011**, *25*, 958–967. [CrossRef]
33. Kaletunc, G.; Normand, M.D.; Johnson, E.A.; Peleg, M. "Degree of elasticity" determination in solid foods. *J. Food Sci.* **1991**, *56*, 950–953. [CrossRef]
34. Kumar, S.S.; Chauhan, A.S.; Giridhar, P. Nanoliposomal encapsulation mediated enhancement of betalain stability: Characterisation, storage stability and antioxidant activity of *Basella rubra* L. fruits for its applications in vegan gummy candies. *Food Chem.* **2020**, *333*, 127442. [CrossRef] [PubMed]
35. Oliveira, A.; Amaro, A.L.; Pintado, M. Impact of food matrix components on nutritional and functional properties of fruit-based products. *Curr. Opin. Food Sci.* **2018**, *22*, 153–159. [CrossRef]
36. Hani, N.; Romli, S.; Ahmad, M. Influences of red pitaya fruit puree and gelling agents on the physico-mechanical properties and quality changes of gummy confections. *Int. J. Food Sci.* **2015**, *50*, 331–339. [CrossRef]
37. Barba, F.; Jäger, H.; Meneses, N.; Esteve, M.; Frígola, A.; Knorr, D. Evaluation of the quality changes of blueberry juice during refrigerated storage after processing of high pressure and pulsed electric fields. *Innov. Food Sci. Emerg. Technol.* **2012**, *14*, 18–24. [CrossRef]
38. Cortellino, G.; Rizzolo, A. Storage stability of novel functional drinks based on ricotta cheese whey and fruit juices. *Beverages* **2018**, *4*, 67. [CrossRef]
39. Zhang, L.; Wu, G.; Wang, W.; Yue, J.; Yue, P.; Gao, X. Anthocyanin profile, color and antioxidant activity of blueberry (*Vaccinium ashei*) juice as affected by thermal pretreatment. *Int. J. Food Prop.* **2019**, *22*, 1035–1046. [CrossRef]
40. Mar, J.M.; Silva, L.S.; Rabelo, M.S.; Muniz, M.P.; Nunomura, S.M.; Correa, R.F.; Kinupp, V.F.; Campelo, P.H.; Bezerra, J.A.; Sanches, E. Encapsulation of amazonian blueberry juices: Evaluation of bioactive compounds and stability. *LWT-Food Sci. Technol.* **2020**, *124*, 109152. [CrossRef]
41. Maier, T.; Fromm, M.; Schieber, A.; Kammerer, D.R.; Carle, R. Process and storage stability of anthocyanins and non-anthocyanin phenolics in pectin and gelatin gels enriched with grape pomace extracts. *Eur. Food Res. Technol.* **2009**, *229*, 949–960. [CrossRef]
42. Tavares, I.M.; Sumere, B.R.; Gómez-Alonso, S.; Gomes, E.; Hermosín-Gutiérrez, I.; da Silva, R.; Lago-Vanzela, E.S. Storage stability of the phenolic compounds, color and antioxidant activity of jambolan juice powder obtained by foam mat drying. *Food Res. Int.* **2020**, *128*, 108750. [CrossRef] [PubMed]
43. Teribia, N.; Buve, C.; Bonerz, D.; Aschoff, J.; Hendrickx, M.; van Loey, A. Impact of processing and storage conditions on color stability of strawberry puree: The role of PPO reactions revisited. *J. Food Eng.* **2021**, *294*, 110402. [CrossRef]
44. Kia, E.M.; Langroodi, A.M.; Ghasempour, Z.; Ehsani, A. Red beet extract usage in gelatin/gellan based gummy candy formulation introducing *Salix aegyptiaca* distillate as a flavouring agent. *J. Food Sci. Tech.* **2020**, *57*, 3355–3362. [CrossRef]
45. Tarone, A.; Cazarin, C.; Junior, M. Anthocyanins: New techniques and challenges in microencapsulation. *Food Res. Int.* **2020**, *133*, 109092. [CrossRef]
46. Amjadi, S.; Ghorbani, M.; Hamishehkar, H.; Roufegarinejad, L. Improvement in the stability of betanin by liposomal nanocarriers: Its application in gummy candy as a food model. *Food Chem.* **2018**, *256*, 156–162. [CrossRef]
47. da Rosa, J.R.; Nunes, G.L.; Motta, M.H.; Fortes, J.P.; Weis, G.C.C.; Hecktheuer, L.H.R.; Muller, E.I.; de Menezes, C.R.; da Rosa, C.S. Microencapsulation of anthocyanin compounds extracted from blueberry (*Vaccinium* spp.) by spray drying: Characterization, stability and simulated gastrointestinal conditions. *Food Hydrocoll.* **2019**, *89*, 742–748. [CrossRef]
48. Muhamad, I.I.; Jusoh, Y.M.; Nawi, N.M.; Aziz, A.A.; Padzil, A.M.; Lian, H.L. Advanced natural food colorant encapsulation methods: Anthocyanin plant pigment. In *Natural and Artificial Flavoring Agents and Food Dyes*, 1st ed.; Grumezescu, A., Holban, A.M., Eds.; Elsevier: Amsterdam, The Netherlands, 2018; pp. 495–526.
49. Vukoja, J.; Pichler, A.; Kopjar, M. Stability of anthocyanins, phenolics and color of tart cherry jams. *Foods* **2019**, *8*, 255. [CrossRef]
50. Sinela, A.; Rawat, N.; Mertz, C.; Achir, N.; Fulcrand, H.; Dornier, M. Anthocyanins degradation during storage of *Hibiscus sabdariffa* extract and evolution of its degradation products. *Food Chem.* **2017**, *214*, 234–241. [CrossRef]
51. Yan, B.; Davachi, S.M.; Ravanfar, R.; Dadmohammadi, Y.; Deisenroth, T.W.; van Pho, T.; Odorisio, P.A.; Darji, R.H.; Abbaspourrad, A. Improvement of vitamin C stability in vitamin gummies by encapsulation in casein gel. *Food Hydrocoll.* **2021**, *113*, 106414. [CrossRef]
52. Huang, T.; Zhao, H.; Fang, Y.; Lu, J.; Yang, W.; Qiao, Z.; Lou, Q.; Xu, D.; Zhang, J. Comparison of gelling properties and flow behaviors of microbial transglutaminase (MTGase) and pectin modified fish gelatin. *J. Texture Stud.* **2019**, *50*, 400–409. [CrossRef]

53. Li, X.; Liu, X.; Lai, K.; Fan, Y.; Liu, Y.; Huang, Y. Effects of sucrose, glucose and fructose on the large deformation behaviors of fish skin gelatin gels. *Food Hydrocoll.* **2020**, *101*, 105537. [CrossRef]
54. Waterhouse, A.L. Determination of total phenolics. *Curr. Protoc. Food Anal. Chem.* **2002**, *6*, I1.1.1–I1.1.8. [CrossRef]
55. Lee, J.; Durst, R.; Wrolstad, R. Determination of total monomeric anthocyanin pigment content of fruit juices, beverages, natural colorants, and wines by the pH differential method: Collaborative study. *J. AOAC Int.* **2005**, *88*, 1269–1278. [CrossRef]
56. Dewanto, V.; Wu, X.; Adom, K.K.; Liu, R.H. Thermal processing enhances the nutritional value of tomatoes by increasing total antioxidant activity. *J. Agric. Food Chem.* **2002**, *50*, 3010–3014. [CrossRef] [PubMed]
57. Brand-Williams, W.; Cuvelier, M.E.; Berset, C.L.W.T. Use of a free radical method to evaluate antioxidant activity. *LWT-Food Sci. Technol.* **1995**, *28*, 25–30. [CrossRef]
58. Benzie, I.F.; Strain, J.J. The ferric reducing ability of plasma (FRAP) as a measure of "antioxidant power": The FRAP assay. *Anal. Biochem.* **1996**, *239*, 70–76. [CrossRef]
59. Orellana-Palma, P.; Guerra-Valle, M.; Zúñiga, R.N. Centrifugal filter-assisted block freeze crystallization applied to blueberry juice. *Processes* **2021**, *9*, 421. [CrossRef]

Article

Nano Matrix Soft Confectionary for Oral Supplementation of Vitamin D: Stability and Sensory Analysis

Mohammad Zubair Ahmed [1], Anshul Gupta [2], Musarrat Husain Warsi [3,*], Ahmed M. Abdelhaleem Ali [3], Nazeer Hasan [1], Farhan J. Ahmad [1], Ameeduzzafar Zafar [4] and Gaurav K. Jain [2,5,*]

[1] Department of Pharmaceutics, School of Pharmaceutical Education and Research, Jamia Hamdard, New Delhi 110062, India; zubair_amgaz@hotmail.com (M.Z.A.); nazeerhasan1994@gmail.com (N.H.); farhanja_2000@yahoo.com (F.J.A.)
[2] Department of Pharmaceutics, Delhi Pharmaceutical Sciences and Research University, New Delhi 110017, India; gupta.ansh.198@gmail.com
[3] Department of Pharmaceutics and Industrial Pharmacy, College of Pharmacy, Taif University, P.O. Box 11099, Taif 21944, Saudi Arabia; a.mali@tu.edu.sa
[4] Department of Pharmaceutics, College of Pharmacy, Jouf University, Sakaka 72341, Al-Jouf, Saudi Arabia; azafar@ju.edu.sa
[5] Center for Advanced Formulation Technology, Delhi Pharmaceutical Sciences and Research University, New Delhi 110017, India
* Correspondence: mvarsi@tu.edu.sa or mhwarsi@gmail.com (M.H.W.); gkjdpsru@gmail.com (G.K.J.)

Abstract: Vitamin D deficiency distresses nearly 50% of the population globally and multiple studies have highlighted the association of Vitamin D with a number of clinical manifestations, including musculoskeletal, cardiovascular, cerebrovascular, and neurological disorders. In the current study, vitamin D oil-in-water (O/W) nanoemulsions were developed and incorporated in edible gummies to enhance bioavailability, stability, and patient compliance. The spontaneous emulsification method was employed to produce a nano-emulsion using corn oil with tween 20 and lecithin as emulsifiers. Optimization was carried out using pseudo-ternary phase diagrams and the average particle size and polydispersity index (PDI) of the optimized nanoemulsion were found to be 118.6 ± 4.3 nm and 0.11 ± 0.30, respectively. HPLC stability analysis demonstrated that the nano-emulsion prevented the degradation and it retained more than 97% of active vitamin D over 15 days compared to 94.5% in oil solution. Similar results were obtained over further storage analysis. Vitamin D gummies based on emulsion-based gelled matrices were then developed using gelatin as hydrocolloid and varying quantities of corn oil. Texture analysis revealed that gummies formulated with 10% corn oil had the optimum hardness of 3095.6 ± 201.7 g on the first day which remained consistent on day 45 with similar values of 3594.4 ± 210.6 g. Sensory evaluation by 19 judges using the nine-point hedonic scale highlighted that the taste and overall acceptance of formulated gummies did not change significantly ($p > 0.05$) over 45 days storage. This study suggested that nanoemulsions consistently prevent the environmental degradation of vitamin D, already known to offer protection in GI by providing sustained intestinal release and enhancing overall bioavailability. Soft chewable matrices were easy to chew and swallow, and they provided greater patient compliance.

Keywords: vitamin D; nanoemulsion; gelled matrices; texture analysis; gelatin; sensory evaluation; gummy

1. Introduction

Vitamin D or calciferol is a lipid-soluble vitamin which is a combination of various steroidal derivatives. These derivatives include: cholecalciferol (vitamin D3), which is a derivative of cholesterol, calcidiol (partially active hydroxylated form of cholecalciferol), calcitriol (dihydroxylated active form), ergocalciferol (D2), and its mono and dihydroxylated derivatives [1]. Vitamin D is both endogenous and exogenous. The human skin epidermis, when exposed to ultraviolet-B (UVB) radiation present in sunlight, produces vitamin D3 from 7-dehydrocholesterol endogenously. Exogenous vitamin D is sourced from dietary food, such as dairy, oily fish, liver, egg yolk, etc., and supplements. Since the natural food options which can provide exogenous vitamin D are quite limited, supplementation is increasingly encouraged to counter its deficiency. Vitamin D in general, refers to vitamin D3. Its primary circulating form 25-hydroxyvitamin D is converted to active metabolite 1,25-dihydroxyvitamin D within the human body [2].

Individuals having serum vitamin D levels (i.e., 25-hydroxyvitamin D [25(OH)D]) below 20 ng per milliliter or 50 nmol per liter (deemed appropriate level by the Institute of Medicine in 2011) are considered vitamin D deficient. In 2011, after a comprehensive analysis of the literature, the Institute of Medicine (IOM) reached the conclusion that a 25(OH)D concentration of 20 ng/mL or above in blood was adequate for optimal bone health [3–5]. Vitamin D deficiency is associated with a number of clinical manifestations in people of different age groups. It is predominantly associated with bone health, and vitamin D deficiency is well understood to contribute to musculoskeletal disorders. Without vitamin D, just 10–15% of dietary calcium and about 60% of phosphorus are absorbed. [2]. Vitamin D deficiency results in musculoskeletal fatigue, and low vitamin D levels have been observed in conjunction with rheumatoid arthritis, multiple sclerosis, and arthritis, closely associated with many cardiovascular risk factors, e.g., myocardial heart infarction, congestive heart failure [6], diabetic cardiovascular disease, and peripheral arterial disease. A number of studies have suggested the role of vitamin D deficiency in increased cerebrovascular accident (CVA) risk and it has been linked to an elevated risk of depression and schizophrenia [7–9]. Vitamin D modulates the role of B and T-lymphocytes and regulates cell differentiation and proliferation [10,11]. Multiple studies have acknowledged that women represent a high risk population for vitamin D deficiency, with pregnant women being at higher risks, leading to multiple pregnancy complications, e.g., gestational diabetes and preeclampsia [12]. Vitamin D deficiency is prevalent across the world, with an estimated 1 billion people suffering from the condition [13]. According to several reports, 40–100% of elderly people in the U.S. and European countries are vitamin D deficient. Furthermore, various studies have shown low vitamin D status regardless of age, gender, and geography [2,13–15]. According to a meta-analysis report from 2015, vitamin D deficiency was 35% more frequent among obese people irrespective of age or demography [16]. Vitamin D deficiency has been associated with populations having higher levels of melanin in their skin and the populations that use extensive skin care, notably in Middle East nations [13,17].

The Institute of Medicine (IOM) has published guidelines for vitamin D supplementation, suggesting 600 IU/day vitamin D for most of the population aging between one and 70 years. The Recommended Dietary Allowance (RDA) of vitamin D for individuals above 70 years of age is 800 IU per day and for infants up to one year of age is 400 IU a day [18]. For the treatment and prevention of vitamin D deficiency, the Endocrine Society in the United States recommended more than 30 ng/mL of serum 25(OH)D concentrations to be achieved, more preferably in the range of 40–60 ng/mL [4]. To reach a serum 25(OH)D level of 30 ng/mL, all deficient adults should be provided with either 50,000 IU weekly of vitamin D3 or 6000 IU daily for 8 weeks. A daily maintenance dose of 1000 IU in patients up to 18 years and 1500–2000 IU daily in patients aging 18–50 years is recommended by The Endocrine Society [4,19].

Limitations with supplemental solid oral delivery: Oral delivery is perceived as most reliable route for the administration of active pharmaceutical ingredients (APIs). More than 60% of all medicines comprise of solid oral dosage forms due to ease of manufacture, availability of large number of excipients, cheaper development process, more accurate dosing compared to oral liquids, and ease of handling and dispensing [20]. Since it is non-invasive, patients are more comfortable compared to other routes. Major disadvantages with oral formulations include low gastric and intestinal solubility, which reduces the bioavailability of large number of pharmaceuticals and rapid stomach breakdown, possibly followed by degradation of the API by gastrointestinal (GI) tract enzymes [21]. Another critical concern is patient compliance among children and the elderly majorly due to dysphagia. A questionnaire research revealed that 26% of 6158 patients reported difficulties swallowing tablets [22]. Another study found that 37.4% of all participants (*n* = 1051) reported having trouble swallowing tablets and capsules [17]. When it comes to supplementation (either prophylactic or treatment), where a medication or dose is not usually regarded equally relevant or critical as other urgent needs or treatment medications, the graph of compliance further falls. Adults with dysphagia or those reluctant to take medications usually ignore the supplementary pills. Parents usually miss the supplements when given as solid orals due to reluctance of children to take the pills.

Limitations with vitamin D stability and bioavailability: Cholecalciferol is highly susceptible to environmental degradation (sunlight, heat, and oxygen), and readily undergoes isomerization or oxidation when exposed to open air in its raw form, resulting in a loss of its functioning and physiological implications. Several studies report that it is poorly water-soluble and solubilizing vitamin D in some carrier oils protects it from the effects of heat or oxidation [23–25]. Vitamin D has very limited bioavailability, which limits its effectiveness when given as an oral supplement, and higher doses are required to maintain effective levels. It has been observed that only around half (~50%) of the Vitamin D3 consumed orally is absorbed [26].

Lipid-based formulations have gained commercial recognition by improving the oral bioavailability and solubility of poor water-soluble drugs. They enhance the drug absorption by solubilizing the poorly water soluble or insoluble drugs and promoting the emulsification process in the GIT. The application of nano-emulsions (NEs) in the food sector has attracted growing interest owing to the changes in the physicochemical attributes and biological efficiency associated with particle size modifications. Particle size reduction in emulsion-based formulations can have a variety of effects that may benefit various food and supplemental applications: (i) enhanced stability against aggregation and separation of droplets under gravity; (ii) greater visual clarity; and (iii) improved oral bioavailability [27,28]. In one of the studies conducted by Salvia-Trujillo et al., lipid droplets of a relatively smaller size were digested considerably quicker in simulated GI fluid suggesting mixed micelles that solubilize lipophilic vitamins may develop more quickly in the small intestine [27]. Because of their potential to yield robust and transparent delivery systems with excellent oral bioavailability, NEs are particularly well suited for encapsulating lipophilic nutraceuticals [29]. NEs are dual phased thermodynamically stable systems comprising of at least two immiscible liquids having droplet sizes in the nanometer range usually below 200 nm. They have good kinetic stability upon storage and can be fabricated with minimal surfactant concentrations [30,31]. High-energy emulsification technologies, such as using ultrasonic generators or high-pressure homogenizers, can be used to synthesize NEs. Nano emulsion-based soft chewable matrices that are easy to chew and swallow and have the potential to entrap the medication, giving sustained, controlled release while avoiding quick disintegration in the stomach, have shown promise as delivery systems.

Vitamins in chewable gummy form have become increasingly common today among both the young and adults. Gummy supplements are reported to accommodate the 'pill fatigue' of consumers and to mask the 'vitamin taste' that is often very poor or has an odious smell [32]. Nowadays, children with complete dentition readily embrace gelled gummies and candies because of their good taste and chewability, since they are frequently flavored with various fruity and other lucrative flavors and extracts and are sweet in nature. Presently, two out of the top 10 vitamin D supplements are gummies. This is due to the widespread acceptability of gummy soft chew supplements among people of all age groups. It has also been evident in many studies that children's nutritional status has been improved with chewable vitamins [32,33]. In the light of above considerations, a nano-emulsion-based soft confectionary system was developed, having a soft chewable matrix, which was followed by an evaluation of its texture profile, dissolution, and consumer compliance.

2. Materials

Vitamin D (97–99% pure) was procured from HiMedia Laboratories Pvt. Ltd. Mumbai, India. Cold pressed pure corn oil was purchased from Deve herbes, Delhi, India. Gelatin

160 g bloom strength, Poloxamer 188, lecithin, and sucralose were purchased from Sigma-Aldrich, Inc., St. Louis, MO, USA. Transcutol and Poloxamer 407 were purchased from Gattefosse, Saint-Priest, France. Tween 20, Tween 80, Tween 60, and Span 20 were bought from Loba Chemie Pvt. Ltd., Mumbai, India. HPLC grade methanol and water were bought from Merck, Mumbai, India. Food Safety and Standards Authority of India (FSSAI) grade sorbitol/xylitol, sucralose, citric acid, and flavors were purchased from local vendors.

3. Experimental Design

3.1. Preparation and Optimization of Vitamin D Nanoemulsion

The oil phase of nanoemulsion was selected based on the reported solubility of vitamin D in various oils [34–39]. Surfactant and co-surfactant were selected based on their ability to spontaneously emulsify corn oil [40]. Briefly, 100 mg of corn oil was dissolved in 3 mL of methylene chloride and then surfactant/co-surfactant (10 mL) was added and stirred at 200 rpm (at 40 °C) using magnetic stirrer to remove methylene chloride. Thereafter, 1 mL of sample was withdrawn and in it 10 mL of distilled water was added gradually, and percentage transmittance was determined at 339 nm wavelength using a UV spectrophotometer [41]. All the measurements were performed in triplicate.

Nanoemulsion optimization was performed by aqueous titration method followed by pseudo-ternary phase diagram as described previously [42,43]. Corn oil containing a weighed quantity of vitamin D was admixed with the combination of surfactant and co-surfactant (Smix) in different ratios (1:1, 2:1, 1:2, 1:3 and 3:1). The mixtures were then titrated with water, vortexed, and analyzed visually for their appearance, clarity, or turbidity (Table 1) and plotted as ternary phase diagram. Clear dispersions (nanoemulsions) with a minimum concentration of Smix were selected for further evaluation [44].

Table 1. (**A**). Grading system of oil and Smix based on appearance and emulsification time. (**B**). Visual screening data of nano emulsions using different ratio of Smix and oil.

A. Grade	Parameter	Self-Emulsion Time (min)
1.	Rapid, clear nanoemulsion	<1
2.	Rapid, slight hazy nanoemulsion	<2
3.	Slow, turbid emulsion	>3
4.	No emulsification	>4

B. Grading based on appearance					
Ratio of Smix: Oil	Co-surfactant to surfactant ratio				
	1:1	2:1	1:2	1:3	3:1
10:3	1	1	1	1	1
09:3	1	1	1	1	1
08:3	2	1	1	1	1
07:3	2	1	1	2	1
06:3	2	1	2	2	1
05:3	3	1	2	2	2
04:3	3	1	2	3	2
03:3	4	1	2	4	2
02:3	4	2	3	4	3
01:3	4	4	4	4	4

3.2. Characterization of Vitamin D Nanoemulsion

The average particle size of the fabricated nanoemulsion was evaluated using Zetasizer Nano ZS90 (Malvern Instruments, Malvern, UK) via dynamic light scattering (DLS). An ambient temperature of 25 °C and scatter angle of 173° were set to carry out the measurements. To eliminate the overlapping effects of scattering, a dilution of 500 times was performed with buffer solution of equivalent pH before measurements. A disposable capillary cell of 1 mL was employed for zeta potential measurement.

To analyze morphology, a Morgagni 268D transmission electron microscope (TEM, Fei Electron Optics, Eindhoven, The Netherlands) functioning at 80 kV was employed. On a 400-mesh copper panel coated with carbon foil, one drop of each sample previously diluted 100 times was applied and allowed to dry for two minutes. A negative stain of 2% phosphotungstic acid was then applied to the sample slide and allowed to dry before examination. Observations were conducted at room temperature, using a smaller objective aperture, a lower accelerating voltage of 80 kV, and at increasing magnifications [44].

3.3. Stability Study of Vitamin D Nanoemulsion

The high-performance liquid chromatography (HPLC) analysis was employed for quantification of vitamin D. Waters e2695 separations Module HPLC system (Waters Co., Milford, MA, USA) equipped with autosampler, Waters 2489 dual wavelength UV detector and a column heater was used. The symmetry C8 column by Waters, having 100 Å pore size with 4.6 mm internal diameter, 5 µm particle size, and 250 mm length, was used for the stationary phase and Empower™ software (Version 3, feature release 4) from Waters was used to process and analyze the results.

A primary stock solution of 1 µg/mL was prepared by dissolving vitamin D in HPLC grade methanol. Hence, 50 mg vitamin D was serially diluted to obtain standard solution of desired concentration. The following process parameters were utilized for separation: temperature was maintained at 25 ± 2 °C, methanol and HPLC grade water in a ratio of 90:10 (v/v) were employed as mobile phase, 1.0 mL/min flow rate was set and maintained, sample cooler was set to 15 °C, and detection was carried out at 265 nm. The standard solution was analyzed 5 times ($n = 5$) for regression.

Vitamin D nanoemulsion was freshly prepared and stored in light resistant container. Since vitamin D degrades rapidly in the aqueous solution, a comparison was made with the oil solution of vitamin D prepared by mixing a weighed quantity of vitamin D in corn oil. One mL of both samples was diluted to 100 mL using HPLC grade methanol and were then analyzed on day 0, day 15 and day 30.

3.4. Soft Confectionary (Nano Gummy) Preparation

The nanoemulsion based gelled matrix system was developed using gelatin (hydrocolloids), as shown in Figure 1. The matrices were made from a blend of type A and type B gelatin with a bloom strength of 160 g. Different combinations of sweeteners (sorbitol, xylitol, sucralose), citric acid, and flavors (coffee, lime, vanilla, orange, candy) were tested.

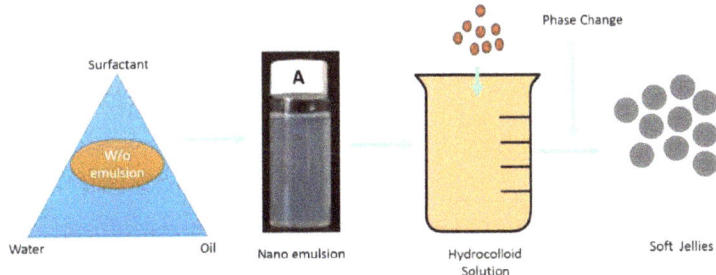

Figure 1. Formulation development process from nano emulsion optimization to gummy formation.

Briefly, gelatin types A and B in equal quantity were dissolved in a small quantity of water heated to 60 °C by stirring at 100 rpm. To this, the previously developed o/w nanoemulsion was added, and the mixture was stirred for 2 min. Sweeteners were then added followed by citric acid, flavoring agent, and corn oil (plasticizer). The influence of corn oil on the matrix properties and product acceptability was studied. The formula of the developed Nano gummy using 10% corn oil is shown in Table 2.

Table 2. Nano gummy formula with 10 wt. % corn oil.

Ingredients	Amount (wt. %)
O/W nanoemulsion	34.29
Gelatin	10.29
Sorbitol	13.20
Xylitol	30.79
Sucralose	0.38
Citric acid	0.38
Lime flavor	0.66
Corn Oil	10.00

3.5. Texture Profile Analysis of Nano Gummy

The textural properties of gummy confections were evaluated using two-bite compression tests [45]. Gummies were formulated with different corn oil concentrations (0%, 10%, 20% and 30%) and textural attributes, including hardness and stickiness, were measured. A TA.XT2i Texture Analyzer (Stable Micro Systems, Godalming, UK) was used for the evaluation. All measurements were made using a 5 kg load cell. A cylinder aluminum plate probe with a diameter of 7.5 cm, lubricated using the corn oil was lowered at a speed of 1 mm/s to compress 50% of confectionary jelly cube measuring 10 mm^3 placed on the fixed bottom surface. The equipment was adjusted to zero, trigger force was kept at 0.05 N, and the plate was automatically lowered until the bottom surface of the plate just contacted the table. All the measurements were made in triplicate. Data were obtained and analyzed using Stable Micro System's Exponent software, which evaluates the force–time curve formed during the compression of the sample [46,47].

3.6. Sensory Evaluation of Nano Gummy

Sensory assessment is accomplished by documenting the outcome of the assessor's replies based on the parameters of taste, texture/appearance, aroma, and overall liking after a comprehensive study. A questionnaire was prepared, and sensory assessments were performed using a qualitative 9-point hedonic scale, with 1 being least likely and 9 being most. Participant were given three different gummies, viz. Placebo gummy (nano gummy without vitamin D), Nano gummy, and Marketed gummy (Azveston Healthcare Pvt. Ltd., Bengaluru, India) on Day 1 and again on Day 45 after storage. Volunteers were screened based on inclusion and exclusion criteria and were informed about the study objectives and recruited after providing informed consent. Gender-neutral healthy volunteers who spoke and understood English, aged between 18 and 45 years, and had no medical history or known sensitivities to the materials used in gummy fabrication were eligible to participate. Participants younger than 18 years old, smokers, those having flu or similar manifestation, those who did not speak or comprehend the language of choice, and those who had known sensitivities to the materials used in gummy production were all excluded. Oral hygiene product, food and aerated drinks were restricted 2-h before and throughout the study period. Responses were collected based on the questionnaire, including: How do you like the taste? How do you like the aroma? How do you like the appearance and texture and overall product? One-way ANOVA (regression analysis) and *t*-test with Welch correction, with $p < 0.05$ as significantly different, were used for the evaluation of differences between the gummy and changes over 45 days of storage.

4. Results

4.1. Preparation and Optimization of Vitamin D Nanoemulsion

Based on the previously reported solubility of vitamin D, corn oil was selected as the oily phase. Results of the emulsification study (Figure 2) showed that tween 20 has maximum potential to emulsify corn oil followed by lecithin. The decreasing order of emulsification potential among was as follows: tween 20 (96%) > lecithin (87%) > tween 80 (84%) > tween 60 (84%) > transcutol (72%) > poloxamer 407 (61%) > poloxamer 188 (53%) > span 20 (49%).

From the results obtained, tween 20 and lecithin were selected as surfactant and co-surfactant, respectively, for the preparation of the vitamin D nanoemulsion.

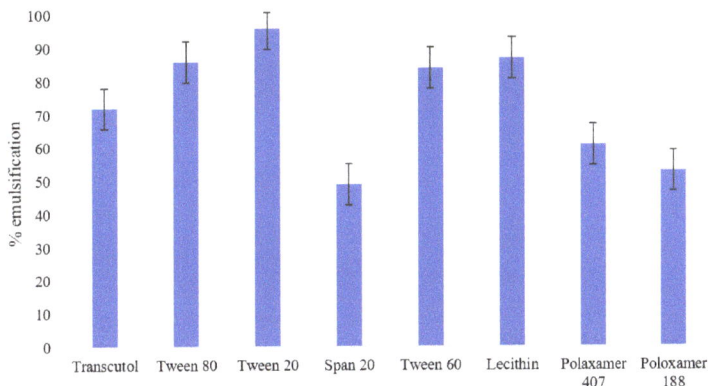

Figure 2. Comparative corn oil emulsification potential of various screened surfactants.

Data obtained from a grading system based on the visual screening, appearance, and emulsification capacity of formulations prepared using different ratios of Smix and Oil were obtained and are tabulated in Table 1. Data revealed that Smix ratio 2:1 resulted in a clear and stable nanoemulsion. Thus, Smix 2:1 was selected for pseudo ternary analysis to obtain the optimized formulation. The nanoemulsion region is shown by a colored region in the pseudo-ternary phase diagram (Figure 3). From the results, it was evident that 3% corn oil, 3% tween 20 and lecithin in a ratio of 2:1 and 94% water resulted in the optimized vitamin D nanoemulsion utilized for further evaluation.

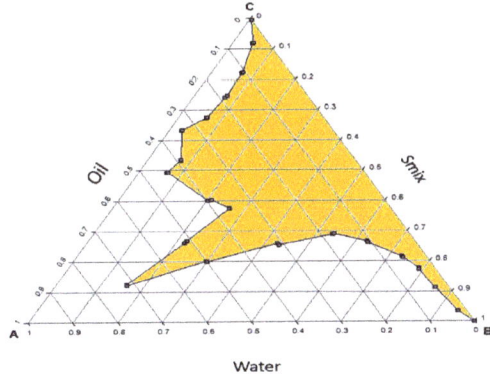

Figure 3. Pseudo-ternary Phase diagram of optimized nano-emulsion with corn oil, Smix (tween 20 + lecithin; 2:1) and water.

4.2. Characterization of Vitamin D Nanoemulsion

As illustrated in Figure 4B, the average particle size and particle size distribution of the optimized vitamin D nanoemulsion were found to be 118.6 ± 4.3 nm and 0.11 ± 0.30, respectively. The zeta-potential was found to be -27 ± 0.82 mV, indicating that the formed nanoemulsion was stable.

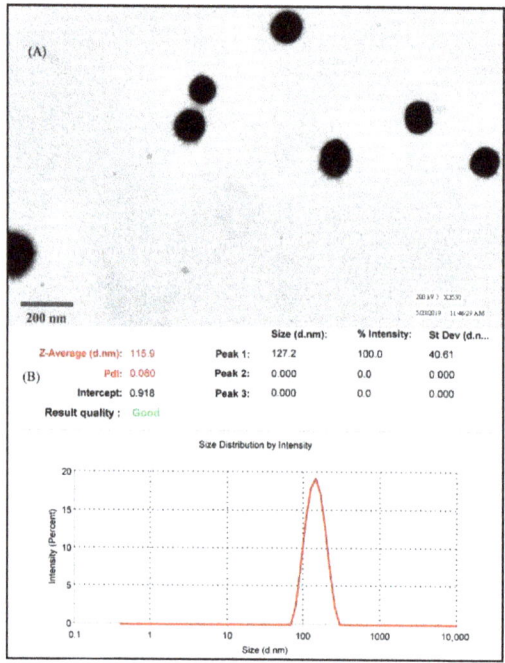

Figure 4. (**A**) HI-RES TEM micrograph of NE. (**B**) Particle size distribution of optimized NE.

TEM revealed that the vitamin D nanoemulsion was spherical (Figure 4A). The droplet size measured by TEM (127 ± 7.2 nm) was slightly larger, but not significantly different from that measured by dynamic light scattering (118.6 ± 4.3 nm). The slightly larger size observed in TEM could be due to the measurement of the hydration layer surrounding the globule, which was not measured using DLS due to difference in principle.

4.3. Stability Study of Vitamin D Nanoemulsion

A representative chromatogram of vitamin D is shown in Figure 5A,B. The standard solution of a known concentration of vitamin D (1 µg/mL) was analyzed five times for robustness. The data obtained for the regression plot of vitamin D by HPLC are shown in Table 3.

The linear equation for the calibration plot was used to quantify the concentration of vitamin D on days 0, 15, and 30. Vitamin D in corn oil solution degraded by 5.49% in 15 days and 8.97% in 30 days, whereas the nanoemulsion formulation degraded by 2.94% and 7.63% on day 15 and day 30, respectively (Table 4).

Table 3. Average retention area, retention time and peak height of stock solution.

Concentration (µg/mL)	Average Area	Retention Time	Height	Area %
1	4,132,567	7.643	284,993	100
1	4,134,052	7.620	285,379	100
1	4,142,904	7.660	286,712	100
1	4,130,736	7.663	284,132	100
1	4,140,783	7.677	285,363	100
Mean	4,136,208.4	7.652	285,315.8	100
Std. Dev.	5329.52	0.022	930.21	NA

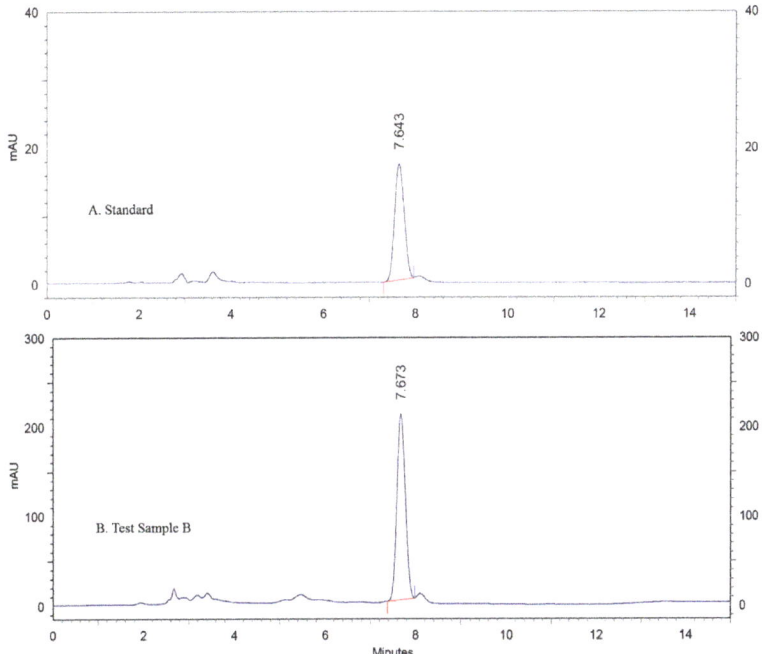

Figure 5. HPLC Chromatograph of (**A**) Standard solution (**B**) Test sample B.

Table 4. Concentration reduction and percent reduction of vitamin D on day 15 and 30.

Test Day	Sample A (IU/mL)	% Reduction	Sample B (IU/mL)	% Reduction
Day 0	289,302.01	0	38,282.13	0
Day 15	273,420.01	5.49	37,155.68	2.94
Day 30	263,360.20	8.97	35,360.46	7.63

4.4. Texture Analysis

The gummies prepared using different concentrations of corn oil as plasticizer were subjected to texture analysis. The maximum force during the initial compression was used to determine hardness (g) and the negative force with which the probe disconnects from the sample over its way up after compressing was termed as stickiness (g). The mean hardness values and mean stickiness values of each gummy type on day 1, 15, 30, and 45 are shown in Table 5. Gummies without corn oil were softest on day 1 (1540 ± 125 g) but they were extremely soft and difficult to handle and not in the range of an acceptable marketed gummy. Although gummies with 40% corn oil had the lowest stickiness value of the four (76 ± 8 gm on day 45) due the excess of oil present, they were the hardest. As shown in Figure 6, gummies with 10% corn oil had the optimum hardness value (3100 ± 200 g on day 1 and 3600 ± 210 g on day 45), and stickiness was also minimal. It was noted that the gummies prepared with 10% corn oil had excellent hardness and texture up to 30 days and this was retained up to 45 days on further storage.

Table 5. (**A**) Mean hardness value (Peak force (g) mean ± STDV) and (**B**) mean stickiness value (negative force (g) mean ± STDV) of each gummy on day 1, 15, 30 and 45.

Oil Content in Gummy	Day 1	Day 15	Day 30	Day 45
A. Mean Hardness Value (Peak Force (g) Mean ± STDV) of Each Gummy				
No corn oil	1540 ± 125	1620 ± 100	2030 ± 150	1820 ± 155
10% corn oil	3100 ± 200	3260 ± 155	3330 ± 165	3600 ± 210
20% corn oil	3410 ± 265	3630 ± 230	3930 ± 195	3970 ± 265
40% corn oil	4260 ± 165	4600 ± 185	5000 ± 205	4650 ± 195
B. Mean Stickiness Value (Negative Force (g) Mean ± STDV) of Each Gummy				
No corn oil	325 ± 45	400 ± 35	425 ± 60	550 ± 60
10% corn oil	160 ± 20	180 ± 30	125 ± 25	145 ± 33
20% corn oil	100 ± 15	85 ± 15	80 ± 7	85 ± 12
40% corn oil	90 ± 14	80 ± 14	75 ± 9	76 ± 8

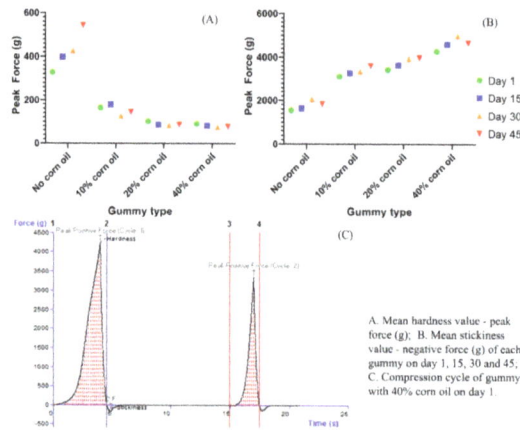

Figure 6. Graphical representation of (**A**) peak hardness cum (**B**) peak stickiness of each gummy on day 1, 15, 30 and 45 and (**C**) force-time curve of 40% corn oil gummy on day 1.

4.5. Sensory Analysis

The Placebo, Nano, and Marketed gummies were compared and the data obtained via sensory evaluation conducted at Day 1 and at Day 45 following storage are presented in Table 6. The highest score for taste was observed for the Marketed gummy. However, the difference between the score compared to placebo and nano gummy was insignificant ($p > 0.05$). Further, storage for a period of 45 days had no detrimental effect on the taste and the scores for all the gummies remained same. The aroma of Placebo and Nano gummies was significantly better than the Marketed gummy ($p = 0.01$; $p < 0.05$). Following 45 days storage, the loss of aroma was observed for all the gummies, but the loss was maximum for the Marketed gummy. Although the appearance and texture of the Marketed gummy was better than Nano gummy ($p = 0.029$; $p < 0.05$), the nano gummy retained both texture and appearance without any significant change ($p = 0.45$) over 45 days of storage. Overall liking of the Nano gummy was comparable to Marketed gummy and was similar on day 1 and day 45 without any significant difference ($p = 0.4$).

Table 6. Average scores from the sensory evaluation on Day 1 and day 45.

Day 1	Placebo Gummy	Nano Gummy	Marketed Gummy
Taste	7.1 ± 1.7	6.7 ± 1.6	7.3 ± 1.4
Aroma	7.4 ± 1.3	7.0 ± 0.8	5.4 ± 0.7
Appearance/texture	7.0 ± 0.6	6.8 ± 0.9	7.6 ± 1.2
Overall liking	6.9 ± 0.9	6.6 ± 0.8	7.2 ± 1.0
Day 45			
Taste	7.2 ± 1.4	6.6 ± 1.0	7.4 ± 0.7
Aroma	6.7 ± 0.7	6.4 ± 1.1	4.5 ± 0.6
Appearance/texture	7.1 ± 0.8	6.9 ± 1.2	7.4 ± 0.9
Overall liking	6.7 ± 0.6	6.5 ± 0.6	7.3 ± 1.1

5. Discussion

The present study explores the potential of a novel nano matrix-based gummy (Nano gummy) to deliver vitamin D. The novel matrix not only prevented the degradation of vitamin D, but also enhances the compliance. Our results suggested that vitamin D was well preserved in O/W nanoemulsion and remained stable over storage. Further, sensory analysis studies suggested that palatability and consumer acceptability were well achieved.

The Nano gummy described in the present work was prepared by a two-step procedure. Firstly, vitamin D loaded nanoemulsions were formed using spontaneous emulsification and aqueous titration procedure. Based on the solubility studies and published literature, corn oil was selected as oil phase. A previous study supported that the in vitro digestibility and bioaccessibility of vitamin D were maximum in long chain triglycerides such as corn oil compared to medium chain triglycerides [34,35]. Studies conducted by Yang et al. [36], Qian et al. [37], and Rao et al. [38] showed similar results where long chain triacylglycerols, viz. corn oil, displayed better bioavailability characteristics of fat-soluble vitamins in contrast to indigestible oils due to the improved solubilization ability of mixed micelles produced by long chain fatty acids [39]. From the results of the emulsification study, tween 20 and lecithin with maximum potential to emulsify corn oil (96% and 87%, respectively) were selected as surfactant and co-surfactant, respectively.

Pseudo ternary phase diagrams were drawn to locate the nanoemulsion zones, where oil, surfactant, co-surfactant, and water exist as clear and homogeneous dispersions. Ternary phase diagrams provide information on the best combinations to use for the preparation of nanoemulsion. The droplet size of the nanoemulsion is critical as it influences the solubility as well as bioavailability and it is well established that smaller size droplets result in enhanced solubility and bioavailability owing to the higher surface area [34,35].

The phase analysis demonstrated that when the surfactant: co-surfactant ratio was 2:1, the largest proportion of oil was integrated in the nanoemulsion system and smaller size nanoemulsions. The 3% corn oil, 2% tween 20, 1% lecithin, and 94% water resulted in an optimized vitamin D nanoemulsion with globule size 118.6 ± 4.3 nm and particle size distribution of 0.11 ± 0.30. The zeta potential of nanoemulsion is important because its value can be linked to colloidal dispersion stability. In a dispersion, the zeta potential demonstrates the extent of repulsion between contiguous, identically charged particles. A higher zeta potential indicates stability for particles and molecules of sufficiently small size. In cases of lower zeta potential, repulsive forces are weaker compared to attractive forces, leading to the breaking and flocculation of dispersions. Colloids having a large zeta potential (either negative or positive) are deemed electrically stable [48,49]. The optimized formulation displayed zeta-potential of -27 ± 0.82 mV, indicating that the developed nanoemulsions are stable. Using TEM, the structural and dimensional characterization of the optimized vitamin D nanoemulsion was accomplished. The nanoemulsion droplet structure was confirmed to be spherical, and each droplet had a discrete region surrounding it, which suggested vitamin D encapsulation. The size of the nanoemulsion droplets obtained by TEM was in agreement with that obtained by DLS. Blackness in the core of particles as observed in TEM image could be attributed to the solubilization of vitamin D in the oil.

During the stability study, the results showed that vitamin D nanoemulsion retained the vitamin D concentration quite well, although it was exposed to water in the emulsion. Jiang et al. [23] showcased similar behavior when encapsulating vitamin D in carrier oils and storing away from light, which prevents its otherwise easy degradation. This study confirmed the stability of vitamin D in the nanoemulsion upon storage for extended time periods. The vitamin D degradation was only 7.63% in nanoemulsion compared to 8.97% in oil following 30-day storage. The stability results do not appear to be distinctly different but are significant because in the nanoemulsion, the vitamin D, prone to hydrolysis, was in contact with water.

In the second step, the optimized nanoemulsion was mixed with hydrocolloid matrix to form the Nano gummy. It was hypothesized that the Nano gummy might improve the bioavailability of vitamin D, although this was not tested in the present research. The improved bioavailability of vitamin D from Nano gummy could be explained by two mechanisms. Firstly, gummies are meant for chewing in the mouth where they are mixed with saliva and begin to dissolve. Most likely, the released nanoemulsion from the Nano gummy in the buccal cavity will allow enhanced permeation of vitamin D and thus improved bioavailability. Such enhancement of buccal permeation by nanoemulsions has been previously reported [39,50,51]. Secondly, the released nanoemulsion from the Nano gummy will disperse in the gastric fluid. The nanoemulsion results in the enhanced solubilization of vitamin D and its subsequent absorption from GIT, resulting in improved bioavailability. Thus, the Nano gummy could significantly enhance the bioavailability of vitamin D encapsulated in nanosized emulsion droplets in hydrocolloid matrix [50].

While designing gummies, oil incorporation played an important in the texture properties of gummies. Gummies with corn oil 10% exhibit optimum hardness value throughout the study. The addition of oil led to a steep increase in hardness value, but for gummies the optimum hardness value ranges around 3100 g force. After 3100 g force, the gummies become harder and firmer. Texture profile analysis may further be used to track changes in texture during the drying process of gummy confections. The stickiness results showed that the gummies with the lowest content of oil, i.e., 10% oil, exhibited more adhesiveness throughout the study, as can be observed in Table 5. The stickiness value increases with a decrease in oil content. Stickiness for gummies with 40% oil content was even lower, but they became hard and deteriorated the chewy feel. The texture of the gummies is also affected by their age, because the water content changes over time and water plays an essential element in maintaining texture [52]. Concentrations of other additives were kept constant throughout the experiment. Sucralose is an artificial sweetener, while sorbitol and xylitol are sugar alcohols. These act as sweeteners and impart that likeliness among children to willingly consume the product. Various flavors improve the aroma and overall taste of the product since these factors play an important role in overall liking of the product. Homogenization speeds were limited to 1000 rpm as higher homogenization speeds may lead to instability or breaking of the incorporated emulsion.

Although texture can be measured using both instrumental and sensory techniques, it has been demonstrated that sensory analysis when coupled with instrumental analysis yields better attributes [53] for a consumer-oriented product. The sensory assessment technique is widely used in food product development, its quality control, monitoring of shelf life, consumer acceptance, and some experts suggest it to be one of the most sensitive and reliable approach in certain situations [54]. In the current study, the formulated gummy retained its texture and overall acceptance over 45 days of storage at room temperature. This signifies a stable product in terms of consumer perception. Unvarying taste over storage indicated that excipients were inter compatible and incorporated nano-emulsion jacketed vitamin D quite well. Although the formulated confectionary lost significant aroma on storage based on the sensory scores, this could be improved by modifying flavoring agents.

6. Conclusions

The nanoemulsion-based matrix system using edible oil, emulsifier and hydrocolloid proved to be an effective approach to develop stable, nano enhanced vitamin D gummies that are readily consumer acceptable. The developed Nano gummy could enhance the vitamin D bioavailability. However, prolonged stability studies and gummy evaluation are required before scale up.

Author Contributions: Conceptualization, G.K.J. and M.H.W.; methodology, A.G.; software, A.M.A.A.; validation, A.G. and A.Z.; formal analysis, M.Z.A.; investigation, M.Z.A.; resources, F.J.A.; data curation, N.H.; writing—original draft preparation, M.Z.A.; writing—review and editing, M.H.W. and G.K.J.; visualization, M.H.W. and A.M.A.A.; supervision, G.K.J. and F.J.A.; project administration, F.J.A.; funding acquisition, F.J.A. and A.M.A.A. All authors have read and agreed to the published version of the manuscript.

Funding: The authors of this research would like to acknowledge the financial support offered by Taif University Researchers Supporting Project number (TURSP-2020/50), Taif University, Taif, Saudi Arabia. The author: Mohammad Zubair Ahmed is indebted to the Jamia Hamdard Silver Jubilee Research Fellowship, Jamia Hamdard, New Delhi for providing the fellowship.

Institutional Review Board Statement: Ethical review and approval were waived for this study as its not required for supplemental food products made with standard established process. No biospecimen or sensitive information like demographics was collected from panelists in any form during the sensory evaluation.

Informed Consent Statement: Informed consent was obtained from all subjects involved in the study.

Data Availability Statement: This study did not report any data (All the data is included in the current manuscript).

Acknowledgments: The authors of this research would like to acknowledge the financial support offered by Taif University Researchers Supporting Project number (TURSP-2020/50), Taif University, Taif, Saudi Arabia. The author: Mohammad Zubair Ahmed is indebted to the Jamia Hamdard Silver Jubilee Research Fellowship, Jamia Hamdard, New Delhi for providing the fellowship. The authors also want to extend their gratitude to Department of Scientific and Industrial Research, Government of India for providing grant under CRTDH scheme.

Conflicts of Interest: The authors declare no conflict of interests.

References

1. Demer, L.L.; Hsu, J.J.; Tintut, Y. Steroid hormone vitamin D: Implications for cardiovascular disease. *Circ. Res.* **2018**, *122*, 1576–1585. [CrossRef] [PubMed]
2. Holick, M.F. Vitamin D deficiency. *N. Engl. J. Med.* **2007**, *357*, 266–281. [CrossRef] [PubMed]
3. Del Valle, H.B.; Yaktine, A.L.; Taylor, C.L.; Ross, A.C. *Dietary Reference Intakes for Calcium and Vitamin D*; National Academies Press: Washington, DC, USA, 2011.
4. Holick, M.F.; Binkley, N.C.; Bischoff-Ferrari, H.A.; Gordon, C.M.; Hanley, D.A.; Heaney, R.P.; Murad, M.H.; Weaver, C.M. Evaluation, Treatment, and Prevention of Vitamin D Deficiency: An Endocrine Society Clinical Practice Guideline. *J. Clin. Endocrinol. Metab.* **2011**, *96*, 1911–1930. [CrossRef] [PubMed]
5. American Geriatrics Society Workgroup on Vitamin D Supplementation for Older Adults. Recommendations abstracted from the American Geriatrics Society consensus statement on vitamin D for prevention of falls and their consequences. *J. Am. Geriatr. Soc.* **2014**, *62*, 147–152. [CrossRef]
6. Aparna, P.; Muthathal, S.; Nongkynrih, B.; Gupta, S.K. Vitamin D deficiency in India. *J. Fam. Med. Prim. Care* **2018**, *7*, 324–330. [CrossRef]
7. Pilz, S.; Tomaschitz, A.; Drechsler, C.; Zittermann, A.; Dekker, J.M.; Marz, W. Vitamin D Supplementation: A Promising Approach for the Prevention and Treatment of Strokes. *Curr. Drug Targets* **2011**, *12*, 88–96. [CrossRef]
8. Milaneschi, Y.; Hoogendijk, W.J.G.; Lips, P.; Heijboer, A.C.; Schoevers, R.; Van Hemert, A.M.; Beekman, A.T.F.; Smit, J.H.; Penninx, B.W.J.H. The association between low vitamin D and depressive disorders. *Mol. Psychiatry* **2014**, *19*, 444–451. [CrossRef]
9. Sarris, J.; Murphy, J.; Mischoulon, D.; Papakostas, G.I.; Fava, M.; Berk, M.; Ng, C.H. Adjunctive Nutraceuticals for Depression: A Systematic Review and Meta-Analyses. *Am. J. Psychiatry* **2016**, *173*, 575–587. [CrossRef]
10. Borkar, V.V.; Devidayal; Verma, S.; Bhalla, A.K. Low levels of vitamin D in North Indian children with newly diagnosed type 1 diabetes. *Pediatr. Diabetes* **2010**, *11*, 345–350. [CrossRef]

11. Zipitis, C.S.; Akobeng, A.K. Vitamin D supplementation in early childhood and risk of type 1 diabetes: A systematic review and meta-analysis. *Arch. Dis. Child.* **2008**, *93*, 512–517. [CrossRef]
12. Holick, M.F. A call to action: Pregnant women in-deed require vitamin D supplementation for better health outcomes. *J. Clin. Endocrinol. Metab.* **2019**, *104*, 13–15. [CrossRef] [PubMed]
13. Van Schoor, N.; Lips, P. Global overview of vitamin D status. *Endocrinol. Metab. Clin.* **2017**, *46*, 845–870. [CrossRef] [PubMed]
14. Lips, P.; Hosking, D.; Lippuner, K.; Norquist, J.; Wehren, L.; Maalouf, G.; Ragi-Eis, S.; Chandler, J. The prevalence of vitamin D inadequacy amongst women with osteoporosis: An international epidemiological investigation. *J. Intern. Med.* **2006**, *260*, 245–254. [CrossRef] [PubMed]
15. Manson, J.E.; Brannon, P.M.; Rosen, C.J.; Taylor, C.L. Vitamin D deficiency-is there really a pandemic. *N. Engl. J. Med.* **2016**, *375*, 1817–1820. [CrossRef] [PubMed]
16. Pereira-Santos, M.; Costa, P.D.F.; Assis, A.D.; Santos, C.D.S.; Santos, D.D. Obesity and vitamin D deficiency: A systematic review and meta-analysis. *Obes. Rev.* **2015**, *16*, 341–349. [CrossRef]
17. Holick, M.F. The vitamin D deficiency pandemic: Approaches for diagnosis, treatment and prevention. *Rev. Endocr. Metab. Disord.* **2017**, *18*, 153–165. [CrossRef]
18. Pludowski, P.; Holick, M.F.; Grant, W.B.; Konstantynowicz, J.; Mascarenhas, M.R.; Haq, A.; Povoroznyuk, V.; Balatska, N.; Barbosa, A.P.; Karonova, T.; et al. Vitamin D supplementation guidelines. *J. Steroid Biochem. Mol. Biol.* **2018**, *175*, 125–135. [CrossRef]
19. Chung, M.; Lee, J.; Terasawa, T.; Lau, J.; Trikalinos, T.A. Vitamin D With or Without Calcium Supplementation for Prevention of Cancer and Fractures: An Updated Meta-analysis for the U.S. Preventive Services Task Force. *Ann. Intern. Med.* **2011**, *155*, 827–838. [CrossRef]
20. Schiele, J.T.; Quinzler, R.; Klimm, H.-D.; Pruszydlo, M.G.; Haefeli, W.E. Difficulties swallowing solid oral dosage forms in a general practice population: Prevalence, causes, and relationship to dosage forms. *Eur. J. Clin. Pharmacol.* **2013**, *69*, 937–948. [CrossRef]
21. Fields, J.; Go, J.T.; Schulze, K.S. Pill Properties that Cause Dysphagia and Treatment Failure. *Curr. Ther. Res.* **2015**, *77*, 79–82. [CrossRef]
22. Andersen, O.; Zweidorff, O.K.; Hjelde, T.; Rødland, E.A. Problems when swallowing tablets. A questionnaire study from general practice. *Tidsskr. Nor. Laegeforening Tidsskr. Prakt. Med. Raekke* **1995**, *115*, 947–949.
23. Jiang, S.; Yildiz, G.; Ding, J.; Andrade, J.; Rababahb, T.M.; Almajwalc, A.; Feng, H. Pea protein nano-emulsion and nanocomplex as carriers for protection of cholecalciferol (vitamin D3). *Food Bioprocess Technol.* **2019**, *12*, 1031–1040. [CrossRef]
24. Zareie, M.; Abbasi, A.; Faghih, S. Thermal Stability and Kinetic Study on Thermal Degradation of Vitamin D_3 in Fortified Canola Oil. *J. Food Sci.* **2019**, *84*, 2475–2481. [CrossRef] [PubMed]
25. Donsì, F.; Annunziata, M.; Vincensi, M.; Ferrari, G. Design of nano-emulsion-based delivery systems of natural antimicrobials: Effect of the emulsifier. *J. Biotechnol.* **2012**, *159*, 342–350. [CrossRef] [PubMed]
26. Bothiraja, C.; Pawar, A.; Deshpande, G. Ex-Vivo absorption study of a nanoparticle based novel drug delivery system of vitamin D3 (Arachitol Nano™) using everted intestinal sac technique. *J. Pharm. Investig.* **2016**, *46*, 425–432. [CrossRef]
27. Salvia-Trujillo, L.; Fumiaki, B.; Park, Y.; McClements, D.J. The influence of lipid droplet size on the oral bioavailability of vitamin D_2 encapsulated in emulsions: An in vitro and in vivo study. *Food Funct.* **2017**, *8*, 767–777. [CrossRef]
28. McClements, D.J.; Rao, J. Food-grade nano-emulsions: Formulation, fabrication, properties, performance, biolog-ical fate, and potential toxicity. *Crit. Rev. Food Sci. Nutr.* **2011**, *51*, 285–330. [CrossRef]
29. Golfomitsou, I.; Mitsou, E.; Xenakis, A.; Papadimitriou, V. Development of food grade O/W nano-emulsions as carriers of vitamin D for the fortification of emulsion based food matrices: A structural and activity study. *J. Mol. Liq.* **2018**, *268*, 734–742. [CrossRef]
30. McClements, D.J. Nanoemulsions versus microemulsions: Terminology, differences, and similarities. *Soft Matter* **2012**, *8*, 1719–1729. [CrossRef]
31. Silva, H.D.; Cerqueira, M.Â.; Vicente, A.A. Nanoemulsions for food applications: Development and characteriza-tion. *Food Bioprocess Technol.* **2012**, *5*, 854–867. [CrossRef]
32. Kleiman-Weiner, M.; Luo, R.; Zhang, L.; Shi, Y.; Medina, A.; Rozelle, S. Eggs versus chewable vitamins: Which intervention can increase nutrition and test scores in rural China? *China Econ. Rev.* **2013**, *24*, 165–176. [CrossRef]
33. Stewart, R.; Askew, E.; Mcdonald, C.; Metos, J.; Jackson, W.; Balon, T.; Prior, R. Antioxidant Status of Young Children: Response to an Antioxidant Supplement. *J. Am. Diet. Assoc.* **2002**, *102*, 1652–1657. [CrossRef]
34. Ozturk, B.; Argin, S.; Ozilgen, M.; McDonald, D.J. Nanoemulsion delivery systems for oil-soluble vitamins: Influence of carrier oil type on lipid digestion and vitamin D3 bioaccessibility. *Food Chem.* **2015**, *187*, 499–506. [CrossRef]
35. Schoener, A.L.; Zhang, R.; Lv, S.; Weiss, J.; McClements, D.J. Fabrication of plant-based vitamin D 3-fortified nano-emulsions: Influence of carrier oil type on vitamin bioaccessibility. *Food Funct.* **2019**, *10*, 1826–1835. [CrossRef] [PubMed]
36. Yang, Y.; McClements, D.J. Vitamin E bioaccessibility: Influence of carrier oil type on digestion and release of emulsified α-tocopherol acetate. *Food Chem.* **2013**, *141*, 473–481. [CrossRef] [PubMed]
37. Qian, C.; Decker, E.A.; Xiao, H.; McClements, D.J. Nanoemulsion delivery systems: Influence of carrier oil on β-carotene bio-accessibility. *Food Chem.* **2012**, *135*, 1440–1447. [CrossRef] [PubMed]
38. Rao, J.; Decker, E.A.; Xiao, H.; McClements, D.J. Nutraceutical nano-emulsions: Influence of carrier oil compo-sition (digestible versus indigestible oil) on β-carotene bioavailability. *J. Sci. Food Agric.* **2013**, *93*, 3175–3183. [CrossRef]

39. Hsu, C.-Y.; Wang, P.-W.; Alalaiwe, A.; Lin, Z.-C.; Fang, J.-Y. Use of Lipid Nanocarriers to Improve Oral Delivery of Vitamins. *Nutrients* **2019**, *11*, 68. [CrossRef]
40. Imran, M.; Iqubal, M.K.; Imtiyaz, K.; Saleem, S.; Mittal, S.; Rizvi, M.M.A.; Ali, J.; Baboota, S. Topical nanostructured lipid carrier gel of quercetin and resveratrol: Formulation, optimization, in vitro and ex vivo study for the treatment of skin cancer. *Int. J. Pharm.* **2020**, *587*, 119705. [CrossRef]
41. Hasan, N.; Imran, M.; Kesharwani, P.; Khanna, K.; Karwasra, R.; Sharma, N.; Rawat, S.; Sharma, D.; Ahmad, F.J.; Jain, G.K.; et al. Intranasal delivery of Naloxone-loaded solid lipid nanoparticles as a promising simple and non-invasive approach for the management of opioid overdose. *Int. J. Pharm.* **2021**, *599*, 120428. [CrossRef]
42. McClements, D.J. Edible nano-emulsions: Fabrication, properties, and functional performance. *Soft Matter* **2011**, *7*, 2297–2316. [CrossRef]
43. Vandamme, T.F.; Anton, N. Low-energy nanoemulsification to design veterinary controlled drug delivery devices. *Int. J. Nanomed.* **2010**, *5*, 867. [CrossRef] [PubMed]
44. Zakkula, A.; Gabani, B.B.; Jairam, R.K.; Kiran, V.; Todmal, U.; Mullangi, R.; Babulal, B.G. Preparation and optimization of nilotinib self-micro-emulsifying drug delivery systems to enhance oral bioavailability. *Drug Dev. Ind. Pharm.* **2020**, *46*, 498–504. [CrossRef] [PubMed]
45. Phillips, G.O.; Williams, P.A. *Handbook of Hydrocolloids*; Elsevier: Amsterdam, The Netherlands, 2009.
46. Teixeira-Lemos, E.; Almeida, A.R.; Vouga, B.; Morais, C.; Correia, I.; Pereira, P.; Guiné, R.P.F. Development and characterization of healthy gummy jellies containing natural fruits. *Open Agric.* **2021**, *6*, 466–478. [CrossRef]
47. Marfil, P.H.; Anhê, A.C.; Telis, V.R. Texture and Microstructure of Gelatin/Corn Starch-Based Gummy Confections. *Food Biophys.* **2012**, *7*, 236–243. [CrossRef]
48. Sari, T.P.; Mann, B.; Kumar, R.; Singh, R.R.; Sharma, R.; Bhardwaj, M.; Athira, S. Preparation and characterization of nanoemulsion encapsulating curcumin. *Food Hydrocoll.* **2015**, *43*, 540–546. [CrossRef]
49. Gurpret, K.; Singh, S.K. Review of Nanoemulsion Formulation and Characterization Techniques. *Indian J. Pharm. Sci.* **2018**, *80*, 781–789. [CrossRef]
50. Grossmann, R.E.; Tangpricha, V. Evaluation of vehicle substances on vitamin D bioavailability: A systematic review. *Mol. Nutr. Food Res.* **2010**, *54*, 1055–1061. [CrossRef]
51. Kotta, S.; Khan, A.W.; Pramod, K.; Ansari, S.H.; Sharma, R.K.; Ali, J. Exploring oral nanoemulsions for bioavailability enhancement of poorly water-soluble drugs. *Expert Opin. Drug Deliv.* **2012**, *9*, 585–598. [CrossRef]
52. Ergun, R.; Lietha, R.; Hartel, R.W. Moisture and Shelf Life in Sugar Confections. *Crit. Rev. Food Sci. Nutr.* **2010**, *50*, 162–192. [CrossRef]
53. Rahman, M.S.; Al-Attabi, Z.H.; Al-Habsi, N.; Al-Khusaibi, M. Measurement of Instrumental Texture Profile Analysis (TPA) of Foods. In *Techniques to Measure Food Safety and Quality*; Springer: Cham, Switzerland, 2021; pp. 427–465. [CrossRef]
54. Kemp, S.E. IFST PFSG committee Application of sensory evaluation in food research. *Int. J. Food Sci. Technol.* **2008**, *43*, 1507–1511. [CrossRef]

Review

Biomimetic Hydrogels in the Study of Cancer Mechanobiology: Overview, Biomedical Applications, and Future Perspectives

Ayse Z. Sahan [1], Murat Baday [2,3,*] and Chirag B. Patel [4,5,6,*]

1. Biomedical Sciences Graduate Program, Department of Pharmacology, School of Medicine, University California at San Diego, 9500 Gilman Drive, San Diego, CA 92093, USA
2. Department of Neurology and Neurological Sciences, School of Medicine, Stanford University, Stanford, CA 94305, USA
3. Precision Health and Integrated Diagnostics Center, School of Medicine, Stanford University, Stanford, CA 94305, USA
4. Department of Neuro-Oncology, The University of Texas MD Anderson Cancer Center, Houston, TX 77030, USA
5. Neuroscience Graduate Program, The University of Texas MD Anderson Cancer Center UTHealth Graduate School of Biomedical Sciences (GSBS), Houston, TX 77030, USA
6. Cancer Biology Program, The University of Texas MD Anderson Cancer Center UTHealth Graduate School of Biomedical Sciences (GSBS), Houston, TX 77030, USA
* Correspondence: baday@stanford.edu (M.B.); cbpatel@mdanderson.org (C.B.P.)

Abstract: Hydrogels are biocompatible polymers that are tunable to the system under study, allowing them to be widely used in medicine, bioprinting, tissue engineering, and biomechanics. Hydrogels are used to mimic the three-dimensional microenvironment of tissues, which is essential to understanding cell–cell interactions and intracellular signaling pathways (e.g., proliferation, apoptosis, growth, and survival). Emerging evidence suggests that the malignant properties of cancer cells depend on mechanical cues that arise from changes in their microenvironment. These mechanobiological cues include stiffness, shear stress, and pressure, and have an impact on cancer proliferation and invasion. The hydrogels can be tuned to simulate these mechanobiological tissue properties. Although interest in and research on the biomedical applications of hydrogels has increased in the past 25 years, there is still much to learn about the development of biomimetic hydrogels and their potential applications in biomedical and clinical settings. This review highlights the application of hydrogels in developing pre-clinical cancer models and their potential for translation to human disease with a focus on reviewing the utility of such models in studying glioblastoma progression.

Keywords: cancer; glioblastoma; hydrogel; mechanobiology; mechanoreceptor; mechanotransduction

Citation: Sahan, A.Z.; Baday, M.; Patel, C.B. Biomimetic Hydrogels in the Study of Cancer Mechanobiology: Overview, Biomedical Applications, and Future Perspectives. *Gels* 2022, 8, 496. https://doi.org/10.3390/gels8080496

Academic Editors: Kiat Hwa Chan and Yang Liu

Received: 30 May 2022
Accepted: 2 July 2022
Published: 10 August 2022

Publisher's Note: MDPI stays neutral with regard to jurisdictional claims in published maps and institutional affiliations.

Copyright: © 2022 by the authors. Licensee MDPI, Basel, Switzerland. This article is an open access article distributed under the terms and conditions of the Creative Commons Attribution (CC BY) license (https://creativecommons.org/licenses/by/4.0/).

1. Cellular Microenvironment

The cellular microenvironment is characterized by a mixture of extracellular matrix proteins, soluble signaling factors, neighboring cells, and the physical properties of the niche that affect cell behavior through direct or indirect biomechanical and biochemical signals [1,2]. Properties of the microenvironment, such as stiffness and composition, have been shown to direct cell physiology and lineage [3,4]. Such findings have inspired research to define the cellular microenvironment and its links to cellular behaviors using in vitro tissue models that can mimic biomechanical conditions [5,6]. Here, we will discuss the cellular niche, namely, the biological and mechanical properties of the extracellular matrix (ECM), how cells sense these properties, and the dysregulation of cell-ECM interactions in various disease states. How these factors, which are involved in crosstalk with cells, contribute to cellular activities and overall health, will also be presented through a review of the various research publications on the topic. This review will provide an in-depth overview of what is known and what is unknown about the biomechanics involved in cell-microenvironment interactions, how the use of biomimetic hydrogel models can fill these gaps in the knowledge, and the utility of biomimetic hydrogels in biomedical applications.

1.1. Extracellular Matrix

Tissues may be described as having two components: cellular and non-cellular. The extracellular matrix is the non-cellular component and is composed of proteins, polysaccharides, growth factors, signaling molecules, proteases, and water [7] (Figure 1). These components are distributed heterogeneously rather than homogenously, resulting in unique niche microenvironments for each cell as well as tissue-specific mechanical, physical, and biochemical properties [7]. These play a large role in regulating and mediating cell behaviors. While the ECM is a mixture of many components, a large portion of it is composed of proteins. These include proteoglycans such as hyaluronan, and fibrous proteins such as collagens, elastins, fibronectins, and laminins [8]. These proteins function to anchor cells to the ECM via focal adhesions and aid propagating signals between cells [8]. ECM composition also affects physical properties such as elasticity, stiffness, porosity, static architecture, and dynamic deformations of the matrix [9]. For instance, based on their concentration, assembly, and crosslinking densities, the structural collagens and elastins of the ECM significantly alter its mechanical properties such as composite strength, elasticity, and mechanical resistance [10,11]. ECM properties are not static; they undergo dynamic changes as the ECM is continuously being remodeled through protein degradation, deposition, or modifications that can be self-contained or caused by cellular activity [10]. In cases such as tissue repair, the activity of growth factors and cytokines in the ECM cause matrix metalloproteinases to activate for ECM remodeling [12]. Wound healing and tissue remodeling processes activate growth factors through mechanical and biochemical stimuli to change ECM composition [13]. There are also specialized forms of ECMs that have proven to be important regulators of tissue and cell behavior. For instance, basement membrane is important structurally and functionally for blood vessels because of its involvement in angiogenesis [12]. The ECM has also been shown to mediate or regulate stem cell fate, cell proliferation, cell differentiation, cell migration, and tissue regeneration [11,14–16].

1.1.1. Cell-ECM Interactions

The cell and its ECM are involved in a dynamic reciprocity, through which cues from the ECM and cellular activities are in crosstalk to maintain a healthy state [17]. Signaling processes are one of the ways that the cell-ECM interactions are facilitated. Cell-matrix adhesion sites, or focal contacts, enable communication between cells and the ECM through physical connections of cellular integrins and cadherins to ligands in the ECM [15,16]. Focal contacts are important to cellular processes that require physical attachment to the ECM, such as migration and angiogenesis [18]. Engagement of integrins and cadherins to certain ligands can activate signaling pathways such as that of the Rho family of GTPases to stimulate structural changes in the cell or induce other processes, therefore serving as an important step in biochemical cell-ECM interactions [6,16].

Proteoglycans present in the ECM have functions such as inducing aggregation and participating in ECM structure by adhering to structural proteins [19]. Some proteoglycans reside on the surfaces of epithelial cells, where they act similar to cell-adhesion molecules and bind collagens and fibronectin to anchor cells to the ECM [20,21]. In addition to these vital functions, proteoglycans are active co-receptors that mediate cellular signaling by binding soluble ligands in the ECM and encouraging the formation of receptor complexes on cell surfaces [22,23]. Proteoglycan co-receptors are vital to various developmental processes, and the loss of co-receptor function or mutation has been implicated in diseases such as cancer and ischemic heart disease [23]. Therefore, the ECM also plays a crucial role in mediating cell–cell communication, which will be discussed more thoroughly in Section 1.2 of this review.

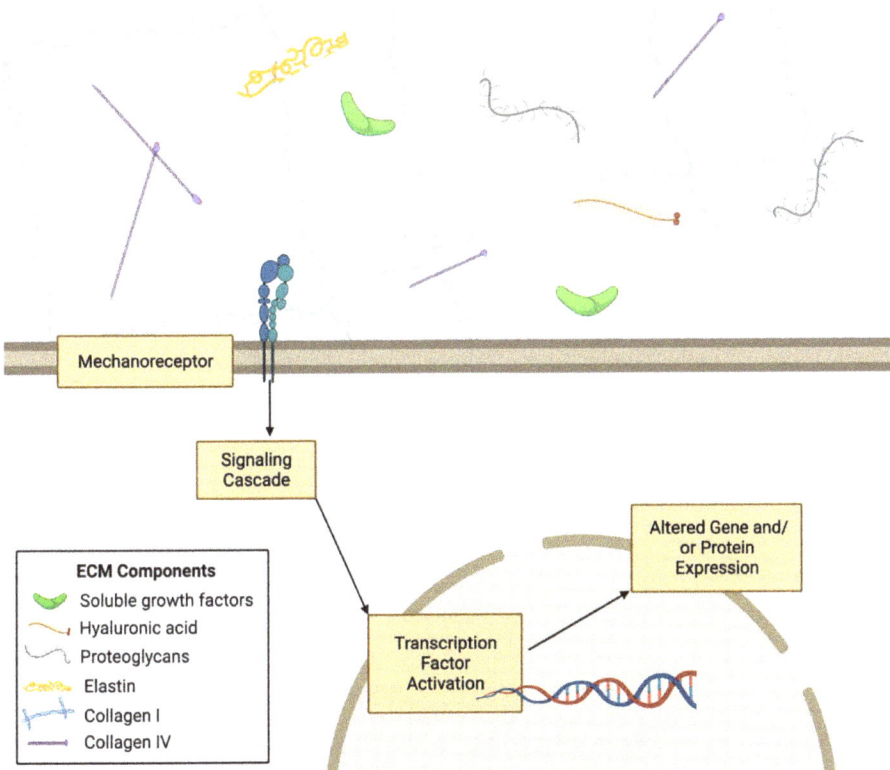

Figure 1. Extracellular matrix (ECM) and cellular mechanotransduction. ECM components such as collagens can alter mechanical properties to induce changes in cellular signaling and gene or protein expression via mechanoreceptors and mechanotransduction proteins. Legend on the bottom left shows which ECM component is represented by each symbol. (Created with BioRender.com, accessed on 29 May 2022).

Mechanobiology of the Cellular Microenvironment

The ECM is involved in mechanical crosstalk with cells [24], which relies on mechanotransduction proteins that help to regulate intracellular tensile response to mechanical forces from the ECM [25]. Mechanical stimuli that cells may receive include shear stress, membrane tension, force, strain, stiffness, and drag force [25]. These stimuli are listed in Table 1 with the mechanotransduction proteins identified to be involved in recognizing the stimuli and eliciting the response in cells. While the extracellular matrix of many cancers, including colon, breast, and prostate cancer, is stiffer than that of healthy tissues [26,27], cell deformability or reduced stiffness has been correlated to increased metastatic potential and invasiveness in cancer. In a study of ovarian cancer cells of varying invasiveness, Xu et al. found that the more-invasive ovarian cancer cell line was more deformable compared to the less-invasive cell line [28]. In another study, Hayashi and Iwata confirmed that cancer cells are softer (i.e., lower Young's modulus) than normal cells using atomic force microscopy [29]. Another study reported that cisplatin treatment caused decreased stiffness and invasiveness of prostate cancer cells [30], suggesting that various cell lines may have varied mechanical properties in the cancer state. Further controversy exists within the field of glioblastoma biomechanics. Gliomas are highly variable; therefore, measurements of tumor stiffness may vary depending on location of measurement [31]. Although some GBM tissues were stiffer than healthy reference tissues, GBM tissues, on average, were less stiff than healthy tissues [31–33]. The controversy regarding GBM tumor stiffness could by fueled by differences in methods of measurement, as there is currently no standard practice in the field [34–36].

Mechanical stimuli coming from the extracellular environment can be processed by cells through mechanotransduction pathways. The proteins of these pathways translate mechanical cues to induce biochemical and genetic responses. For instance, integrins and focal adhesion proteins are mechanotransduction proteins that communicate mechanical forces to the cell cytoskeleton, and there have been some studies that show force-dependent integrin activation [37–39]. Yes-associated protein (YAP) and other transcription factors have been shown to translocate to the nucleus in stiffer substrates [40]. Similarly, E-cadherin is a mechanotranducer of shear stress [41–43]. Many mechanotransduction proteins have been studied in the context of specific cell types, including platelet endothelial cell adhesion molecule-1 (PECAM-1) in skeletal muscle cells, G protein-coupled receptors in endothelial cells, and vascular endothelial growth factor receptor 2 (VEGFR2) in chondrocytes [44–47]. Cell surface receptors also transduce mechanical signals to cells upon recognition of a ligand that sustained force from the ECM, which can then cause conformational changes in the mechanoreceptor to activate a protein signaling pathway to alter cellular processes [48]. Such membrane proteins are not the sole propagators of mechanical force; studies have shown that there are mechanotransduction systems in cells that enable the progression of force through a long distance [49]. Src and Rac1 have been shown to be activated at distances from 30-60 μm from the original area the force was applied to, via cytoskeletal mediation of the force [49,50]. Mechanical signals can also be transmitted throughout the cytoskeleton to the nucleus via proteins such as linker of nucleo- and cyto-skeleton (LINC) complex 9 to change chromatin structure and cause nuclear stiffening [49,51]. Mechanotransduction systems translate the numerous mechanical cues from the ECM into biochemical signals interpreted by the cell that lead to signaling cascades that control transcription, proliferation, migration, and many other cellular processes [52–55] (Figure 2). For instance, smooth muscle cells can migrate along gradients of substrate stiffness through durotaxis [56]. ECM biomechanics is not only vital for cellular processes, but also regulates tissue and organ-level processes such as tissue differentiation, morphogenesis, and development [57–60]. Therefore, elucidating the interplay of ECM biomechanical and biochemical signals with mechanotransduction proteins and pathways is critical to understanding diverse aspects of cellular and tissue health.

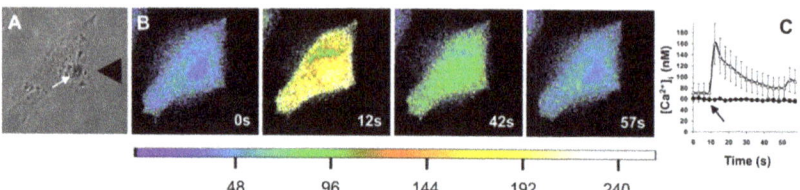

Figure 2. Application of extracellular stress leads to increased intracellular calcium concentrations. Matthews et al. applied high levels of stress to cells (**A**) and found that when imaged via Fura-2AM ratio imaging, it led to a transient increase in calcium concentrations, as shown in pseudocolor images ranging from blue to yellow (**B**) that is quantified (**C**) as a function of time for control and gadolinium chloride-treated cells. (Figure reprinted/adapted with permission from Ref. [61]. Copyright 2006, National Academy of Sciences.

Table 1. Mechanotransducers of various mechanical properties and human cellular responses.

Functional Category	Mechano-Transducers	Mechanical Signal	Examples of Cellular Responses
Cell Mechanical and Physical Properties	Integrins	Force	RhoA activation leading to increased cell stiffness [62,63]
	Focal Adhesions	Force	Actin polymerization [55]
	Yes-associated protein (YAP)	Force	Oligodendrocyte morphology and maturation [40]
	Titin	Force	Implicated in development of mechanical unloading-induced diaphragm weakness [64]
	Stress Fibers (actin filaments, myosin II, etc.)	Force	Transmit tension to other proteins, regulate assembly of filaments [65]
	Vinculin	Force	Transmit tensile force [66]
	Myosin II	Force	Increased cortical tension and cell membrane fusion promotion [67]
	Vasodilator stimulated phosphoprotein (VASP), zyxin, and Testin LIM domain protein (TES)	Force	Regulate junction dynamics [68]
	Neurogenic locus notch homolog protein 1 (NOTCH1)	Shear Stress	Altered cell morphology [69]
	Piezo1	Force	Vascular structure [70]
	Lamin A	Rigidity	Nuclear mechanics [71,72]
	Integrins	Force	Tyrosine Phosphorylation, MAPK signaling [15]
	Focal Adhesions	Force	Integrin convergence [73]
	Fibronectin	Force	Altered integrin binding [74]
	T-cell receptor (TCR)	Force	T-cell calcium and IL-2 secretion [75]
Alters Signaling Pathways	Talin	Force	Recruitment of vinculin to focal adhesion complexes [76]
	Piezo2	Force	Serotonin release [77]
	Vinculin	Force	Enhanced PI3K activation [78]
	p130Cas	Force	Activation of Cas signaling pathway [79]
	Syndecan-1	Force	Activation of pro-inflammatory and growth-stimulating pathways [80]

Table 1. *Cont.*

Functional Category	Mechano-Transducers	Mechanical Signal	Examples of Cellular Responses
Alters Signaling Pathways	Transient Receptor Potential Cation Channel Subfamily V Member 4 (TRPV4)	Force	Reorientation and flow-mediated nitric oxide production [81]
	Ion Channels	Force	Cell signaling [82]
	von Willebrand factor—glycoprotein Ib complex (VWF-GPIb)	Shear Stress	Enhanced calcium triggering in platelets and T cells [83]
	Platelet endothelial cell adhesion molecule-1 (PECAM-1)	Shear Stress	Tyrosine kinase Src and PI3K signaling activated [84]
	G-protein coupled receptor 68 (GPR68)	Shear Stress	Component in signaling for cardiovascular pathophysiology [85]
	β-catenin	Shear Stress	Activated expression of FOXC2 transcription factor [86]
	Caveolin-1 and β1 Integrin	Stiffness	FA assembly and turnover [62]
	rho-associated, coiled-coil-containing protein kinase (ROCK) 1 and 2	Stiffness	Regulation of RhoA signaling pathways [87]
	YAP	Stiffness	Altered translocation depending on surrounding stiffness [88]
	Piezo1	Force	Ion Permeation and selection [89]
	C-X-C motif chemokine receptor (CXCR1/2)	Shear Stress	Mediates laminar shear-stress-induced endothelial cell migration [90]
	Transforming growth factor beta 1 (TGFβ1)	Shear Stress	Collagenase-dependent fibroblast migration [91]
Migration	RhoA	Force	Collective cell migration [92]
	Vinculin and metavinculin	Force	Regulation of cell adhesion and motility [66]
	NOTCH1	Shear Stress	Decreased proliferation [69]
	Caveolin 1	Rigidity	Decreased proliferation [93]

Table 1. *Cont.*

Functional Category	Mechano-Transducers	Mechanical Signal	Examples of Cellular Responses
Cancer	YAP1	Shear Stress / Stiffness	Cancer cell motility [54] / Nuclear localization of YAP1 [94]
	TGFβ1	Shear Stress	Human melanoma cell tumor invasiveness [91]
	PI3K/Akt pathway	Stiffness	Overexpression of VEGF in hepatocarcinoma cells [63]
	TRPV4 ion channel	Stiffness	Tumor vascularization through down-regulation of Rho kinase activity [95]
	microRNAs	Stiffness	Altered expression in different stiffness conditions [96]
	Twist1	Stiffness	Induction of EMT and tumor metastasis [97]
	Myocardin related transcription factor A (MRTF-A)	Stiffness	Regulates miRNAs involved in myogenic differentiation [88]
Differentiation	Focal Adhesions	Force	Osteogenic differentiation [98] / Myofibroblastic differentiation [99]
	Transient Receptor Potential Cation Channel Subfamily M Member 7 (TRPM7)	Shear Stress	Osteogenic differentiation of mesenchymal stromal cells [100]

1.1.2. Cell-ECM Interactions in Cancer

Problems in mechanotransduction can result from changes in ECM mechanical properties and defects in proteins involved in mechano-sensitivity [101]. Since mechanotransduction is essential for modulating cellular homeostasis, its failure is linked to metastasis and cancer progression [96]. For instance, there are many proteins implicated in mechanotransduction in glioblastoma (GBM). Talin1 inhibition has been observed to decrease cell spreading and limit cell stiffness changes of glioma cells in response to ECM stiffness, proving its role as a mechanosensory [76]. Non-muscle myosin II depletion reduced the effect of matrix confinement on GBM cell motility [102]. Constitutive activation of RhoA GTPase caused lower sensitivity to matrix stiffness of GBM cells in toxicity assays [55]. Increased matrix stiffness was correlated to Hras, RhoA, and rho-associated, coiled-coil-containing *protein* kinase 1 (ROCK1) upregulation, which are mechanosensor proteins that are implicated in migration and proliferation in cancers in general [61,87]. Integrins are also particularly important in mechanotransduction by relaying signals from the ECM to the cell actin cytoskeleton and are essential to cell migration and cell-matrix adhesion in cancer [103].

Identifying proteins involved in mechanotransduction and their roles in cancer progression can be an essential part of developing therapeutic strategies to hinder cancer progression and malignancy. Certain studies have shown the potential effects of targeting mechanotransduction proteins on cancer cells. Knock-down of CD44 led to decreased structural microtubule, vimentin, and glial fibrillary acidic protein expression and decreased migration and cell stiffness [104]. Targeted inhibition of integrins in the tumor microenvironment has been shown to reduce angiogenesis and inhibit tumor growth [103]. While these studies show partial inhibition of mechanical sensitivity as decreasing invasive properties of tumor cells, other studies have shown that tumor initiating cells are mostly insensitive to mechanical cues from the ECM and that mechanically-insensitive cells have increased motility and invasiveness in vitro [28,105–109]. A complex approach is needed to target mechanical sensitivity in cells through mechanotransduction-based therapeutics for it to become a promising mode of cancer treatment. Therefore, extensive research in mechanotransduction and cell-ECM crosstalk is essential.

1.2. Neighboring Cells and Secreted Factors

In addition to the ECM components, cells are surrounded by heterogeneous populations of neighboring cells that are unique to the tissue and sub-location within the tissue [110]. How cells communicate and influence one another is crucial to maintaining homeostasis and coordinating processes that require several cells, such as tissue formation and regeneration.

Cells can interact through secreted signals that are recognized by membrane-bound receptors through either paracrine signaling, between cells, or autocrine signaling, which is from one cell to itself. In fact, cancer cells are often able to "override" signals from neighboring cells through autocrine pro-survival and proliferation signals. It is crucial to better understand the specific interactions between cells that promote healthy conditions or lead to disease states. For instance, Zervantonakis et al. found that fibroblasts in the tumor microenvironment of HER2 positive breast-cancer cells reduced drug sensitivity through paracrine signaling that activates mechanistic target of rapamycin (mTOR, anti-autophagic) and anti-apoptotic signals [111]. In addition to elucidating mechanisms of tumor resistance, cell–cell interaction dynamics can provide insights to developing self-assembled multicellular structures in vitro. Mueller et al. demonstrated that by utilizing pulsed light activation to control engineered photo-switchable cell-cell interactions, they were able to control the spatial organization of multicellular structures without a scaffold [112]. This highlights the importance of maintaining cell interactions in efforts to mimic physiological conditions in vitro.

Cells are also involved in communications through physical contacts with one another. This is another way that mechanical stimuli may play a role in influencing cellular processes.

Physical contacts are also involved in collective migration, which has been exhibited by metastatic and invasive cancer cells [113]. The cytoskeletal tension at cell-cell contacts can serve as a significant regulator of mechanotransduction pathways, and there is a wide field of study on the mechano-sensing implications of cell-cell contacts such as focal adhesions and adherens [114]. One method in particular, physically interacting cell sequencing (PIC-seq), is a novel sequencing approach that combines cell sorting, RNA-sequencing, and computational modeling to describe complex cellular interactions in different contexts [115]. However, the heterogeneity and complexity of the interactions between cells and with the ECM need to be better recapitulated in vitro to provide a more accurate understanding of cellular communication networks.

1.3. Hydrogels as In Vitro Models of the Cellular Microenvironment

Three-dimensional hydrogel models of the cellular microenvironment have gained interest in recent years for their improved mimicry of in vivo conditions as opposed to two-dimensional cell cultures (Table 2). These hydrogel models have been extensively studied and compared to 2D culture and in vivo conditions. Cells have been shown to have differing spatio-physical properties in 3D and conventional 2D culture conditions [116–119]. Hsieh and colleagues showed that cells cultured in 2D exhibit greater drug sensitivity than in vivo, and 3D cultures exhibit chemosensitivities comparable to solid tumors provided that they have similar cell density [120]. Expanding the cellular environment from 2D to 3D has also been shown to affect proliferation and metabolism [120–122]. Using 3D cultures has led to significant advancements in the understanding of cell migration strategies as well, since some of these could not be observed in 2D cultures [123–125]. For instance, fibroblasts have several different migration strategies that are utilized in different microenvironmental conditions, demonstrating cellular plasticity [125]. Such findings prompt questions about the mechano- and bio-sensing abilities of cells in relation to their environments. In this section, we will elaborate on the different methods utilized thus far to improve cell-microenvironment mimicry in the form of hydrogel models, the cellular response to such models in contrast with conventional cell-culture methods, and factors to be considered in crafting a hydrogel model.

Table 2. 2D versus 3D cultures [117–119,123,126–128].

2-Dimensional Culture		3-Dimensional Culture	
Advantages	Disadvantages	Advantages	Disadvantages
Simple	Does not mimic in vivo structure	More like in vivo structure	Expensive
Reproducible	Fewer interactions with environment	Niches are available	Time consuming
Inexpensive	Access to unlimited amount of nutrients from media	Access to nutrients is not unlimited, varies	Less reproducible
	Less diverse phenotype and polarity	Can form organs or spheroid clusters of cells	More complex and difficult to carry out
	Altered cell morphology	Allows study of cell-cell and cell-ECM interactions	Fewer interactions with environment

1.3.1. Mimicking Cellular Microenvironment Biomechanics

There have been many diverse approaches to capturing the complexity of the ECM and the biomechanical cues directed by it via in vitro hydrogel models. Tissue specificity is a crucial factor to consider in developing an ECM-mimetic hydrogel-based cell culture system, since the cells of each tissue produce and degrade matrix constituents, leading to variable ECM composition, bioactivity, and mechanics amongst tissues [129] (Figure 3). In the bone, the ECM is calcified to provide structural support to the tissue, while in tendons it is structured to provide tensile strength [130,131].

Most natural ECM mimics in vitro have been hydrogels composed of ECM components such as collagen, fibronectin, or hyaluronic acid. Such hydrogels have been significant in studying cellular behavior in 3D environments as well as cell response to controlled mechanical properties such as stiffness, elasticity, and rigidity [63,132–139]. Photo-crosslinking, chemical crosslinking, changing fiber density, and the development of 'smart' hydrogels that are responsive to external stimuli (e.g., changes in pH, temperature, and light) have all been used to fine-tune hydrogels to mimic the cellular microenvironment mechanically [140–151]. For example, photo-crosslinking of gelatin methacryloyl (GelMa) hydrogels alters hydrogel stiffness based on the light intensity, exposure time, and concentration of photo-initiator used in photo-crosslinking [143]. Nanocomposite smart hydrogels have been produced to change volume, Young's modulus, and breaking strength based on applied chemical and physical stimuli [148,150].

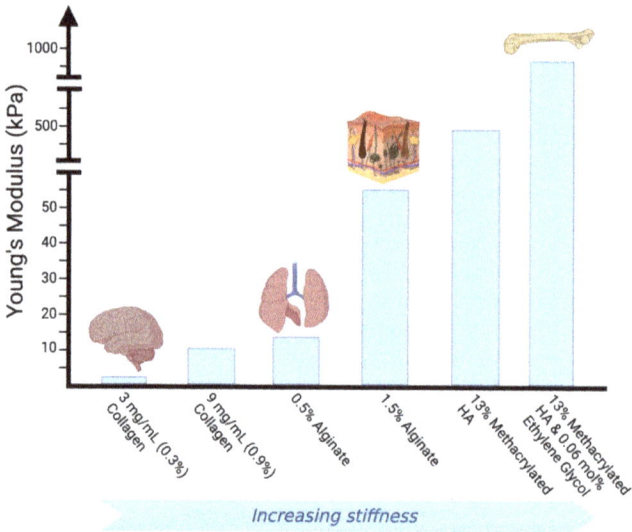

Figure 3. Young's modulus of hydrogels. The reported Young's modulus (kPa) values of various hydrogels are plotted with an image of the tissue that has similar Young's modulus. It is crucial to choose a hydrogel model that corresponds to the in vivo mechanical properties of tissue that is relevant to the work. (Created with BioRender.com, accessed on 14 July 2022).

1.3.2. Recapitulating Cellular Microenvironment Heterogeneity

While the above-discussed methods are useful to study the effects of semi-isolated mechanical conditions on cells, they often disregard the biological complexity of the ECM. In vivo ECMs can also serve as a medium through which cells communicate with one another via secreted growth and signaling factors in vitro. Several protocols have been developed to decellularize ECM isolated from animal or human tissue, or to harvest ECM secreted by fibroblasts in vitro [152–155]. To better capture the biological complexity of the ECM, hydrogel models have been developed to incorporate fibroblasts, growth factors, and more diverse components of the ECMs specific to the cell type being studied. Incorporating fibroblasts in a co-culturing system with the cell type of interest is a relatively newer approach, which allows for fine-tuning mechanical properties using a hydrogel model while adding biological complexity to the culture through the addition of the secreted matrix from fibroblast cells [156,157]. Lee et al. developed a 3D hydrogel system for co-culturing human liver-cancer cell spheroids with fibroblasts on a micropatterned fibrous scaffold, thereby modeling the three dimensional structure of tumors and the cross-talk between cancer cells and neighboring fibroblasts [156]. Several groups have also developed methods of controlled growth factor release into hydrogel-based cell cultures [158–160]. It is crucial to combine the biochemical components of the ECM with the mechanical properties to better mimic the cellular microenvironment in vitro.

2. Hydrogels and Their Applications

Hydrogels are defined as hydrophilic polymer networks that form a three-dimensional structure [161]. A critical property of hydrogels is their ability to swell with water without dissolving due to hydrophilic functional groups present on the polymers of the hydrogel. Their high water content makes them flexible and resemble soft tissue, implicating their use in biomedical studies [162]. There are many polymeric substances that can be classified as hydrogels, and they are generally composed of one or more natural or synthetic materials for use in a wide variety of applications. The three-dimensional structure of hydrogels is preserved despite swelling by chemical and/or physical crosslinks within the polymeric network. Changes in composition, protein concentration, and crosslinking density lead to changes in elasticity, polymer density, and biodegradation rate [163]. These and other tunable properties of hydrogels make them suitable for a wide range of studies on mechanical, chemical, and biological conditions in vitro as well as ideal candidates for use in in vivo clinical applications such as in drug delivery systems [161,162]. Here, we describe the major classifications of hydrogels based on their composition, the general properties of hydrogels, and current and potential experimental and clinical applications of hydrogels with a focus on pre-clinical cancer models.

2.1. Types of Hydrogels

There are several classification methods for hydrogels, such as those based on ionic charge, biodegradability, physical properties, crosslinking, and preparation. They are most commonly classified by their source polymer(s): naturally occurring biomaterial, synthetic bio-mimetic, or a hybrid of these two sources [163]. Synthetic materials are traditionally used extensively due to low biodegradation rate, ease of manipulation, and greater control over biochemical interactions [149,164,165]. Natural biomaterials, however, are preferred due to their ability to biologically mimic the structural and biochemical properties of the cellular niche in vitro, and are responsive to cellular activities in terms of biochemical reactivity and degradability [164]. Numerous studies have presented methods for forming hybrid hydrogels that possess properties of both natural and synthetic biomaterials that are mechanically and biochemically responsive and tunable by the surrounding environment [165,166].

2.1.1. Natural Hydrogels

Natural hydrogels are composed of naturally occurring polysaccharides and proteins such as collagen, Matrigel®, hyaluronan, gelatin, and their derivatives, including alginate and chitosan [167,168]. While most studies cite hydrogel formation from one ECM protein, there are several that utilized hydrogels composed of several proteins [167,168]. Either protein concentration or crosslinking density is altered in the hydrogels in order to change the Young's modulus of the gels [167,168]. However, these are not finely tunable or as well-understood as the mechanical properties of synthetic hydrogels [169]. In contrast, the biocompatibility of natural hydrogels and their responsiveness to cellular degradation add an important dimension to biological studies [170].

Many studies incorporate several proteins into biomaterial hydrogels to improve hydrogel stability and similarity to the in vivo cellular microenvironment. Comparing cellular properties in vitro in composite hydrogels versus single-material hydrogels can give more specific information on how cell mechano-sensitivity is influenced by specific cell-ECM protein interactions. A useful representative investigation of the insight obtained by studies undertaken with composite hydrogels is a study by Rao et al. [171] (Figure 4).

They mixed collagens I and III with hyaluronic acid (HA), collagen IV with HA, and used collagen (I and III) as a standalone in hydrogel formation to determine modulus values of the different hydrogels and observe changes in cell morphology, spreading, and migration. Composite hydrogels had much higher elastic modulus values than the collagen hydrogel, and increased HA content correlated to increased modulus values and greater cell spreading and migration [171]. Interestingly, cell morphology differed by the type of collagen used in the hydrogels: cancer cells cultured in collagen IV had a rounder cell shape, while those in collagen I and III were spindle shaped [171].

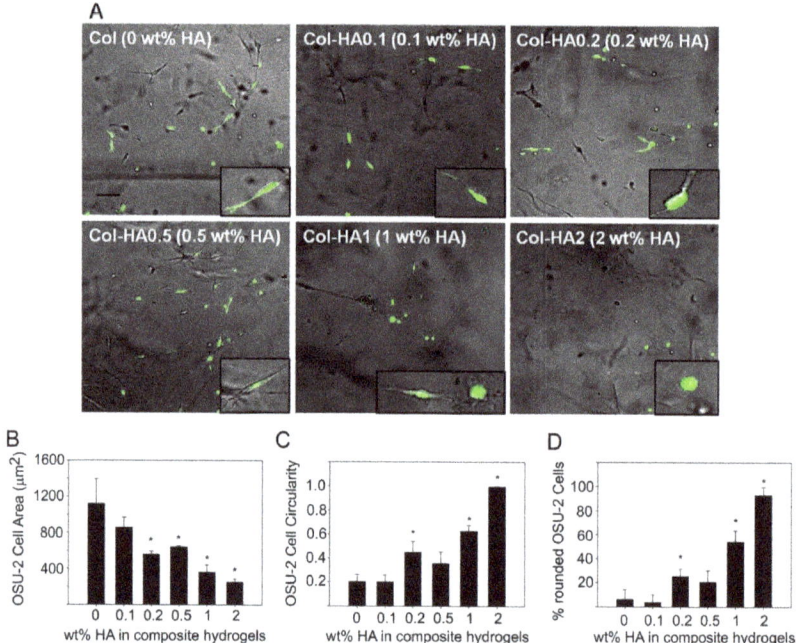

Figure 4. Glioblastoma cell morphologies and migration when cultured on different substrates. OSU-2 glioma cells were cultured on collagen I and III and HA composite hydrogels with different concentrations of HA. Cell morphologies shown in (**A**) were quantified via cell area (**B**), circularity (**C**), and roundness (**D**). Increased HA content led to lower cell area but increased cell circularity and roundness. The scale bar in (**A**) indicates 100 μm, * in (**B**–**D**) represents a p-value < 0.05 compared to 0% HA condition. Figure reprinted/adapted with permission from Ref. [171]. Copyright 2013, American Chemical Society.

Another common composite hydrogel used in several cell mechanics studies is Matrigel®, a mixture of ECM proteins secreted from mouse sarcoma cells, which is commonly used for spheroid formation in 3D cultures [172]. Several mechanics studies use Matrigel® for the formation of spheroids before seeding onto other types of hydrogels or for studies of cell invasion. For instance, Grundy et al. reported that primary GBM cells that were insensitive to rigidity were more invasive in spheroid cultures in Matrigel® as opposed to rigidity sensitive cells [173].

Hydrogels using natural biomaterials are also advantageous, since they allow for the design of an in vitro extracellular environment that can biologically mimic the in vivo ECM and various cellular conditions. Natural hydrogels have been used for tissue engineering and regenerative medicine because they form structures similar to tissue ECM due to the natural proteins and polysaccharides that they are composed of [168]. While the biological similarity is closer to in vivo when natural biomaterials are used in hydrogel preparation, the mechanical properties are not adjustable or robust compared to those of synthetic hydrogels. Additionally, natural hydrogels can lead to great variation from batch to batch due to the components being sourced naturally [168,174]. Recently, however, 'smart' hydrogels, which are robust, have localized mechanical properties, and are responsive to changes in pH, temperature, and light have been developed from natural materials [148]. Smart hydrogels composed of natural materials have many possible applications in tissue engineering, regenerative medicine, and stem cell and cancer research.

2.1.2. Synthetic Hydrogels

Synthetic hydrogels are composed of synthetic materials such as polyethylene glycol (PEG), poly(vinyl alcohol), and poly-2-hydroxy ethyl methacrylate [175,176]. They offer advantages to natural hydrogels in the sense that they offer greater control over gel mechanical properties, have higher capacity for water absorption, do not degrade as rapidly as biomaterial hydrogels, and have great reproducibility [177]. For instance, PEG-based hydrogels are useful in mechanics studies due to stiffness tunability and ability to support long culture periods (several weeks) [178]. Synthetic hydrogels have more utility in studies focused on the effect of isolated mechanical properties on cell behavior without the additional influences of interactions between cells and biochemically active biomaterials [179]. Biomimetic polymers have also been synthesized which may contain a similar component to a natural material, such as a specific amino-acid sequence, which can add biological activity such as degradability or biochemical signaling to the gel in a more controlled environment than if a natural biomaterial were used [180]. Therefore, synthetic hydrogels are extremely customizable and can be synthesized for specific medical or research applications.

Synthetic hydrogels have been used to study the mechanics of cancer, as reported in several previous publications. For instance, 2D "films" of poly-methylphenyl siloxane with increasing stiffness values resulted in increased cell spreading and migration of glioblastoma cells compared to more compliant films [181]. In developing an in vitro drug-screening platform for cancer, synthetic HA derivatives, HA-aldehyde and HA-hydrazide, were crosslinked and formed into a hydrogel where cells were able to form clustered structures similar to tumors, and had greater drug resistance than in 2D cultures [182]. PEG hydrogels are often used for hydrogel preparation due to their tunable properties and high biocompatibility compared to other synthetic materials. They were used in mechanical studies of lung adenocarcinoma where matrix stiffness alterations resulted in changes in cellular morphology [183].

While there are several advantages to using synthetic hydrogels, they are limiting in their ability to mimic the complexity of native ECM. This highlights the importance of efforts to add biochemical reactivity to synthetic hydrogels for in vitro studies that are aiming to mimic cellular microenvironments [184]. For instance, Lutolf et al. engineered synthetic hydrogels that were degradable by matrix metalloproteinases through crosslinking of synthetic substrates into the hydrogels [164]. Smart synthetic hydrogels have also been studied to engineer stimuli-responsive synthetic biomaterials that change when faced with altered temperature, pH, light, and other stimuli [185].

2.1.3. Hybrid Hydrogels

Many hydrogels are composed of synthetic materials that are mixed, conjugated, or coated with biomaterials to provide researchers with insight into the controlled mechanical response and specific cell-protein interactions while maintaining the ability to adjust mechanical properties, and keep a low rate of degradation [186]. In general, a limiting property of hydrogels is their low stiffness and rigidity, which is the opposite of in vivo tissue properties. Hybrid hydrogels have been of interest in studies to improve both the stiffness and rigidity of hydrogels [177,187].

Many types of hybrid hydrogels have been developed for cancer mechanics studies. Fibronectin-coated PA substrates were used to study the invasiveness of different human glioma cell lines with a focus on cell structure, migration, and proliferation [188]. Cells were rounder and had lower migration and proliferation rates in ECM substrates with lower rigidity. GBM migration patterns in the brain white matter tract were mimicked in vitro with electrospun alignment of nanofibers in a mixture of gelatin, poly-ethersulfone, poly-dimethylsiloxane, HA, and collagen in order to study migration patterns of GBM cells, and the addition of HA was seen to have a converse effect on migration [189]. The addition of HA to gelatin and PEG composite hydrogels resulted in dose-dependent glioma malignancy marker expression changes and cell clustering [190]. Another study combined PA hydrogels with HA and either laminin or collagen I and found that collagen and laminin presence was correlated to mechanical response to substrate stiffness [134].

2.2. General Properties of Hydrogels

Hydrogels are defined by their hydrophilicity, which allows them to store water and swell without dissolving [191]. The bio-responsive properties of hydrogels, such as biochemical activity and degradability, allow for culture conditions to be dynamic and receptive to cellular cues. Conversely, since hydrogels are tunable in mechanical properties such as elasticity, compliance, and stiffness, it is possible to study the responses of cells to microenvironmental mechanics. While synthetic hydrogels are easier to adjust mechanically, natural hydrogels are more bioresponsive. In crafting a study, these are important properties to keep in mind, as they introduce variables to whichever system is being studied.

Two important descriptors of hydrogels that determine many physical properties are ionization degree and crosslink density. Crosslinks in a hydrogel are either chemical or physical, and can be introduced to a gel by methods such as irradiation, sulfur vulcanization, or chemical reactions aided by temperature and pressure [144]. Swelling and elastic modulus values are determined by cross-linking degree and charge densities or ionic strength of the polymers in the hydrogel [192]. Greater concentration of cross-linked polymers and the number of ionic groups cause higher elastic modulus values and greater swelling capacity [193]. Similarly, the distribution of proteins or polymers and cross-links in hydrogels, which are generally non-homogenous, are affected by cross-link density and degree of ionization [194,195].

There are numerous properties of hydrogels that can be influenced by internal factors, e.g., composition, protein concentration, and polymer modifications, and by external factors, e.g., UV radiation and temperature [161]. Chemical and physical reproducibility of hydrogels depends on controlled conditions and utility in research can be greatly enhanced by proper knowledge of hydrogel properties and subsequent unique modifications of the scaffold to better mimic whichever system is being studied [196].

2.3. Research Applications of Hydrogels

Research applications of hydrogels are varied in vitro. They are often used as scaffolds in 3D cultures for the study of cellular physiology and characteristics, as they interact with tissue-mimetic hydrogels. While 2D culturing methods have provided valuable information on cellular characteristics, recent studies have undertaken a shift toward preference for 3D-culturing methods for more advanced and sensitive studies that imitate native tissues and cells more closely [126,197]. These 3D platforms provide more physiologically relevant information on cell-environment interactions biochemically and mechanically, including: morphology, cell and environmental stiffness, motility, and signaling [117]. Some 3D platforms that have been used for cell culture include microporous and nanofibrous scaffolds encapsulating cells. However, these either have pore sizes that are too large, which negate the 3D structure and act as a 2D scaffold, or are too weak for mechanical studies [198,199]. Hydrogels are useful as mechanically and biochemically tunable matrices for 3D cultures, which simulate soft tissue structure and have potential for translation and clinical applications [184]. For instance, Jiang et al. found that the formation of a hybrid hydrogel with ultralong hydroxyapatite nanowires and sodium alginate allows for improved mechanical properties of the hydrogels and enhanced biocompatibility for in vitro studies [200].

Drug screening or efficacy assays are prone to showing promising effects in vitro, only to fail or be substantially less effective in animal models and clinical trials [201]. In the search for in vitro cell-culture systems that can provide more accurate and relevant results, 3D scaffolds such as hydrogels have been gaining attention, in part due to their potential to mimic the ECM and inhibit drug delivery [119]. Huber et al. compared the response of 2D cultures of non-small cell lung cancer cultured to 3D microtissues of the same cell line to various drugs and found that drug efficacy was significantly different between the models [127]. Singh et al. developed a hydrogel microarray assay to generate uniform microtumors and subsequently study tumor response to epidermal growth factor (EGF) and cetuximab treatments [202]. Such systems increase the chances of success for translation because they enable studies of treatment response more closely aligned to clinical response.

2.4. Clinical Applications of Hydrogels

A clinical approach to hydrogel scaffolds is found tissue engineering applications, which have gained traction recently as potential solutions to donor shortage problems for tissue or organ transplantations [203,204]. These approaches generally combine cells from a donor with a hydrogel scaffold that is prepared to mimic the extracellular matrix of the tissue or organ being engineered [203]. Hydrogels are suitable tissue-engineering applications due to their ability to uptake water, to encapsulate cells, and to be bio-reactive [8]. Various hydrogels have been used as bio-ink for 3D organ or tissue printing applications in which tissues are built with direct deposition of cells with the bio-ink [205]. Studies have shown applicability of alginate, collagen, and various composite hydrogels for tissue printing methods [206–208]. For instance, natural protein and polysaccharide hydrogels have been used for articular cartilage-tissue engineering applications to promote cartilage regeneration [209]. Latifi and colleagues demonstrated the potential of an injectable hybrid hydrogel (collagen I and III) to be applied in soft-tissue engineering, specifically human vocal fold engineering [210]. A long-standing problem of the tissue engineering field is the need for in vitro tissue vascularization to be able to transplant or implant larger portions of tissue into patients. A couple of groups have recently made strides in promoting in vitro vascularization of hydrogel constructs of bone [211] and soft tissue [212].

Injectable hydrogels have been studied for drug delivery and wound healing or dressing applications [213]. Hydrogels have been applied in drug delivery applications due to their ability to give control over the time and/or site of drug delivery for enhanced treatment [214,215]. The biocompatibility, similarity to native tissues, and high-water content all contribute to great applicability of hydrogels for controlled drug release and delivery [216]. Naturally derived injectable hydrogels for the controlled delivery of small molecules to the central nervous system have also been extensively studied. Wang et al. have shown that an injectable hyaluronan-methylcellulose hydrogel enhanced delivery of growth factors [217]. Further, the great tunability of hydrogels has led to development of thermosensitive, pH-sensitive, and temperature-sensitive hydrogels that can be used for drug delivery in distinct biological environments [218–220]. Similarly, temperature-sensitive hydrogels have been developed for various wound-healing applications. A PEG-PLGA-PEG composite hydrogel has been developed for delivery of a growth factor linked to tissue repair for diabetic wound healing [221].

Tissue regeneration is another field in which hydrogels are being investigated and have shown to be promising for translation to the clinic due to their biocompatibility and ability to be fine-tuned or adapted for specific applications. For instance, Zheng et al. developed a polyacrylic, acid-alginate-demineralized, bone matrix hybrid double-network hydrogel, which was shown to promote vascular endothelial growth factor (VEGF) synthesis and basic fibroblast growth factor (bFGF) and alkaline phosphatase activity of MG63 osteosarcoma cells to enhance bone regeneration [211]. Stem-cell therapy is a promising solution for injuries that require tissue regeneration, but it is limiting since uncontrolled differentiation can lead to the presence of unnecessary cells at the site of injury and lead to stem cell metastasis and tumorigenesis [43,60,129]. Application of hydrogels can improve stem cell therapy by introducing stem cells and differentiation factors to the injury in a site-specific manner [5]. In a study of spinal cord injury, Mothe et al. investigated a hydrogel-integrated stem cell therapy by encapsulating neural stem cells and differentiation factors in a hyaluronan-based hybrid hydrogel and found that treatment enhanced graft survival, increased oligodendrocytic differentiation, and reduced cavitation in the injury site in rats [222].

Contact lenses are an example of the clinical application of hydrogels [223,224]. While hard contact lenses are composed of hydrophobic materials, soft contact lenses are hydrogel-based [225]. Owing to the wide variety in hydrogel-forming substances, attempts are continually being made to improve the physical and chemical properties of contact lenses [225]. Hydrogels have also been used clinically in dermatology applications such as wound healing and skin regeneration [226,227].

3. Cancer and the Tumor Microenvironment

Biomechanical properties of the tumor microenvironment have been shown to be altered compared to the healthy state in many types of cancer to promote processes crucial to tumorigenesis, including cellular proliferation and migration. Since changes in extracellular mechanical properties can induce structural reorganization, morphological changes, and altered signaling, they can cause cancer cells to exhibit mechanical properties differently than healthy cells, e.g., stiffness [228]. This, in turn, can further promote invasive or metastatic phenotypes [229]. For instance, cancer cells also usually have a lower Young's modulus compared to healthy cells of the same type, which can influence deformability and influence migratory ability [230]. Cancer cells also have a more robust ability to respond to ECM conditions, and can alter cytoplasm viscoelasticity in response to increased ECM stiffness and collagen I deposition [231].

Breast cancer is a context in which mechanical properties of tissue and cells have been well described, and many biomechanical contributions to carcinogenesis and metastasis have been identified [96]. Many of the studies involving cell response to mechanical stimuli have been conducted with aggressive breast-cancer cells, especially the MDA-MB-231 cell line, due to their robust response to changes in extracellular mechanics. Clinically, tissue stiffness has served as an indicator of breast tumors and risk of breast cancer [232,233], and the biomolecular consequences of this phenotype has been studied rigorously in many types of 3D hydrogel models in the laboratory, especially in the last decade. Various mechanical stimuli can cause cellular stress and lead to carcinogenesis or increased invasiveness of cancer cells. For instance, several mechanotransduction pathways are linked to carcinogenesis and invasiveness and upregulated in cancer [95,100,234,235]. Application of mechanical load was shown to regulate breast-cancer cell proliferation independent of matrix deformations or stiffness [234]. Mechanical stretch, ECM stiffness, and fluid shear-stress all led to more invasive phenotypes of breast-cancer cells. At the level of response to treatment, ECM stiffness has been characterized to contribute to chemoresistance of breast-cancer cells to doxorubicin [235]. Therefore, a potential therapeutic approach may be to introduce proteinases or drugs that reduce ECM stiffness by degrading certain components to reduce the number of treatment-resistant cells. Lastly, certain microenvironmental properties have been correlated to improved prognosis in breast cancer and can be used to identify diagnostic and prognostic signatures. For instance, the tumor-associated collagen signature consisting of aligned collagen fibers in biopsied tissues from breast-cancer patients has been identified as a prognostic signature for survival [236].

The positive contributions that studying biomechanics has made to our overall understanding of breast cancer highlight the importance of incorporating a biomechanics-based approach to cancer biology in vitro studies. Inspired by the benefits of biomechanics studies on our understanding of breast cancer, we will present the current research in glioblastoma (GBM), which is a disease that has not been well-defined biomechanically. We believe that a better understanding of the biomechanical properties of GBM and its microenvironment can produce translatable results that may contribute to the development of diagnostic and therapeutic approaches to improve the prognosis of this invasive cancer.

3.1. Glioblastoma and the Tumor Microenvironment

Biomechanical and biophysical studies can help to glean valuable insight into a wide variety of diseases, since biomechanics is an integral part of cell proliferative, migration, and survival signaling, all of which are crucial to carcinogenesis and tumorigenesis. GBM is the most common and lethal form of primary brain cancer in adults [237], but there are only two clinically approved chemotherapies targeting it. It is an example of a disease that has not been traditionally studied in terms of biomechanics until approximately the last ten years but is one that may greatly benefit from such studies. Here, we will present an overview of the biomechanical properties known about GBM, the methods and contributions of available 3D culture and mechanotransduction studies of GBM, and the potential translational impact of those studies on the clinic and patient survival.

3.1.1. The Blood-Brain Barrier

The blood-brain barrier (BBB) is also an essential part of the brain microenvironment that is altered in the cancer state. The BBB is formed by vascular endothelial cells lining microvessels in the brain and is essential in regulating brain extracellular conditions to ensure neuronal signaling [238]. The endothelial cells of the BBB limit transcellular and paracellular transit into, and thereby protect, the brain by regulating permeability through tight junctions, adherens junctions, charged moieties, pericytes, etc. [239,240]. In tumor microvessels, however, loss of claudin-1 and claudin-3 and down-regulation of claudin-5 was observed, which correlated to increased permeability [241,242].

3.1.2. Extracellular Matrix of the Brain

The ECM of the brain (Figure 5) is altered when tumorigenesis occurs. Several ECM components, including HA, tenascin-C, and vitronectin, are upregulated in the tumor microenvironment [243]. Studies show that proteins of the tumor niche also tend to be different than healthy brain tissue and that tumor invasion alters ECM composition [244]. Basement membrane components such as laminin, fibronectin, and collagen type IV are more highly secreted by glioma cells and, in turn, alter composition of the local ECM [245,246]. Tumor-associated mesenchymal stem-like cells induce HA synthase 2 activity and lead to greater HA abundance in the tumor niche [247]. In addition to a distinct microenvironment, the GBM tumor has a hypoxic and necrotic core that aids in the cancer cell-induced blood vessel formation by increasing expression of pro-angiogenic VEGF, VEGFR2, and angiopoietin 2, which results in the disorganized network of blood vessels observed in GBM [243,248–250].

Many cellular properties are altered in the cancer state. For instance, cell proliferation, migration, and deformability is increased in the GBM state when compared to healthy cells [67,138,251–254]. The impact of the extracellular niche on these cellular properties and the biology behind the changes have been elucidated by various in vitro and in vivo studies focusing on mechanical cell-ECM interactions [17,101,255–257].

3.1.3. Overview of Microenvironment and Biomechanics of Glioblastoma

A large body of research has been dedicated to studying the differences between the microenvironments, or niche, of tumor and healthy cells. These include studies on overall tissue stiffness, ECM composition, cellular signaling, and the presence/activation of mechanotransducers. There is a general consensus that while cancer cells are less stiff and more deformable than healthy cells [28,105,106,255,256], tumor tissue tends to be stiffer by variable magnitude compared to non-tumor tissues, a trend that has been shown in thyroid, breast, prostate, bladder, and kidney tissues [107–109]. The tumor niche is also characterized by altered ECM composition, which may lead to increased invasiveness and metastatic properties of cancer cells [257]. Characterization of the ECM, both biologically and mechanically, in GBM has not reached the depth of understanding that has been achieved for several other types of cancer, such as prostate and breast cancer. However, there are certain differences that have been noted between healthy and cancerous tissue microenvironments, which can lay the groundwork for future studies of GBM microenvironment and mechanics.

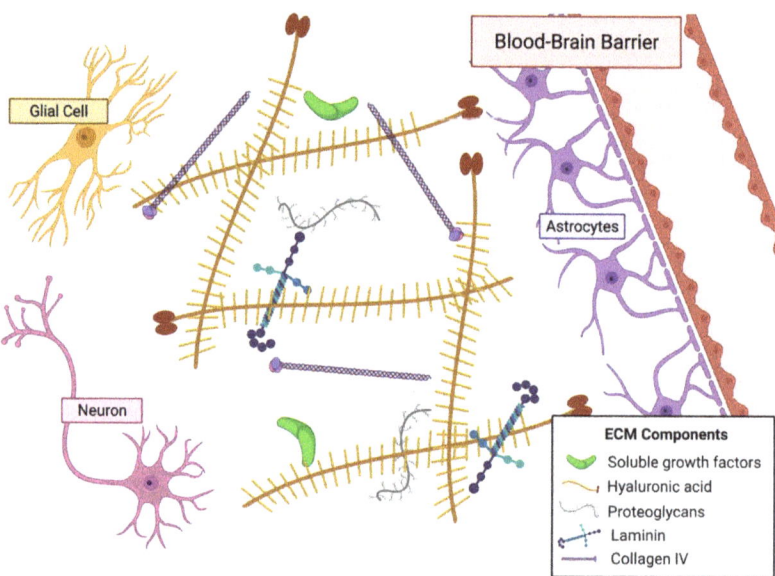

Figure 5. Extracellular matrix (ECM) of the brain. The ECM of the brain has a unique composition, including hyaluronic acid, collagen IV, and other ECM components along with glial cells, neurons, and astrocytes. The blood-brain barrier (BBB) is a neurovascular unit composed of vascular endothelial cells with surface charge modifications, tight junction proteins, pericytes, astrocytes, and other components. The BBB is selectively permeable and can block solutes in the systemic blood from entering the environment of the central nervous system (Created with BioRender.com, accessed on 14 July 2022).

Biomechanics of the Glioblastoma Extracellular Matrix

The mechanical properties of the tumor niche in GBM are different than those of the healthy brain ECM due to the altered composition, protein-protein, cell-protein, and cell-cell interactions. Even within the GBM tumor, there are distinct mechanical regions for necrotic and non-necrotic portions [250]. In general, GBM tissue has been found to be stiffer than healthy ECM, with increasing stiffness generally correlating to increased malignancy [109,147]. From a set of human brain biopsies, one study showed that increasing malignancy of tumors gave higher Young's modulus values, with primary GBMs exhibiting stiffness values varying from 70 to 13,500 Pa [258]. Stewart et al. showed that brain tumors had elastic moduli ranging from 170 to 16,060 Pa using a custom-built indenter [259]. Another study correlated increased ECM stiffness to decreased survival of human patients [260]. Altered stiffness results in changed mechanical cues that are relayed to the cell, which then impact cellular gene expression so significantly that the overall behavior of the cell can be drastically changed [9] (Table 1). Several studies have been published that note differences in cellular morphology, deformability, motility, proliferation, and signaling in response to changes in environmental stiffness [104,134].

Modification of the Extracellular Matrix in the Brain by Glioblastoma Cells

The tumor also modifies the local ECM through protein degradation. Some of the most studied cell-secreted proteins to be up-regulated in the cancer state are matrix metalloproteinases (MMPs), which are proteases that remodel the ECM by degrading certain component proteins [261]. The hypoxic core of GBMs has been shown to be a significant contributor to the increased MMP activity [262]. Various MMPs that are up-regulated in GBMs have also been shown to aid in glioma cell invasion [263]. Plasminogen activation by the urokinase pathway, which includes urokinase (uPA), urokinase receptor (uPAR), and plasminogen, is also prominent in GBMs. The urokinase pathway aids in ECM degradation and remodeling by converting plasminogen into active plasmin, which is a serine kinase that degrades certain ECM proteins, and activates MMP-2 and MMP-9 [264]. MMP-2 is a type IV collagenase that has been implicated in invasion and metastasis of GBMs [263]. Lastly, various cathepsins, which are lysosomal cysteine proteases that can be secreted into the ECM, are also up-regulated in GBMs and have been linked to tumorigenesis and invasion [265–267].

3.1.4. Glioblastoma Migration, Invasion, and Mechanotransduction

Some hallmarks of GBM are enhanced cellular migration and aggressive invasiveness [268]. These properties are achieved through complex mechano-chemical signaling mechanisms that enable crosstalk between the tumor cells and the tumor microenvironment. The proteins that sense and translate mechanical cues from the ECM or microenvironment and relay them to the cell are called mechanotransducers [54]. Interestingly, one study found that keeping the mechanotransducer RhoA GTPase constitutively active in vitro in U87 cells caused similar toxic responses in 3D environments with varying stiffness, indicating the importance of mechanotransduction in cell response to environmental conditions [133]. Another study found that by altering mechanotransducers in GBM tumor-initiating cells, they were able to alter cell motility and invasion in 3D cultures [67]. Knock-down of CD44, a transmembrane glycoprotein receptor for HA and other ECM components, resulted in decreased microtubule and vimentin expression, hampered migration, and decreased nuclear stiffness compared to control cells [104]. Integrins are well-studied mechanotransducers that are linked to malignancy and are primary communicators in cell-ECM adhesion and signal transduction [134,245,263]. Clinical nuclear medicine studies have evaluated the role of various positron emission tomography (PET) radiotracers (e.g., [^{18}F]Galacto-RGD [269], [^{68}Ga]PRGD2 [270], and [^{18}F]FPPRGD2 [271]) to enable molecular imaging of integrin $a_v b_3$, a member a class of adhesion molecules that mediate cell–cell and cell-ECM interactions, and which plays an important role in cancer metastasis and angiogenesis [272]. These studies reveal the importance of understanding how mechanotransducers process mechanical cues from the ECM to influence cancer-cell properties [93,94].

Cellular structure and cytoskeletal alterations are under the direct influence of ECM mechanical cues and cellular mechanotransducers [93]. Pathak and Kumar found that culturing cells in narrow versus wide channels of various extracellular stiffnesses led to altered cell morphology, migration, and myosin alignment (Figure 6), underlining the importance of extracellular culture conditions in determining cell behavior. Various integrins serve to form attachments to proteins of the ECM. Once an integrin is bound to an extracellular ligand, focal adhesion clusters form at the surface of the integrin receptor within the cell, which link to the cytoskeleton and function in cell motility by recruiting proteases for ECM degradation and activating signaling pathways that induce cytoskeletal rearrangement [55]. Cytoskeletal rearrangement as orchestrated by cell-ECM interactions are vital to increasing cell deformability and, in turn, enhancing migration.

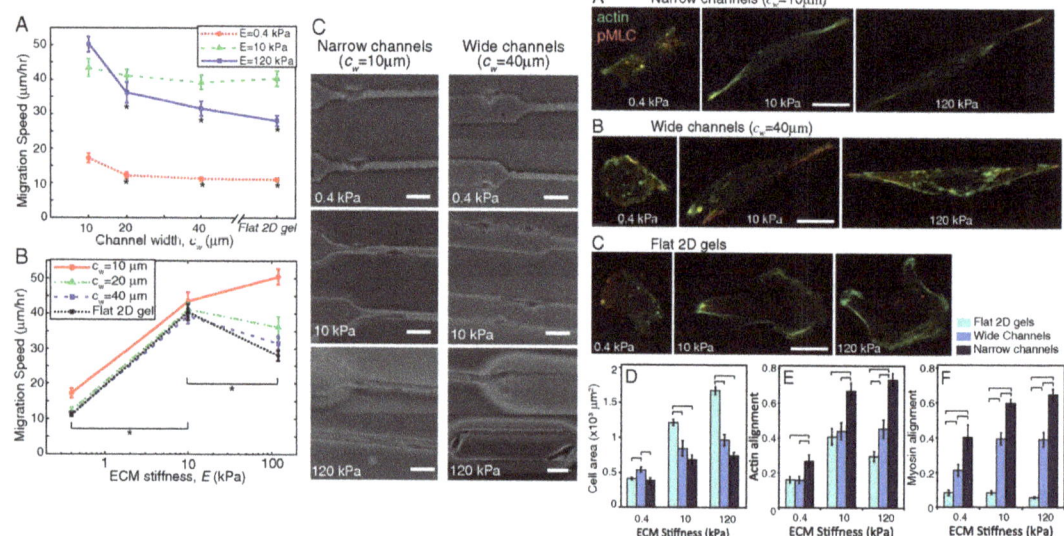

Figure 6. Cell migration, cell area, actin alignment, and myosin alignment when exposed to different channel widths and hydrogel stiffnesses. (**A–C, left panel**) Migration speed was quantified for cells cultured in narrow or wide channels with varying stiffness. With increased stiffness but decreased channel width, cell migration speed was higher. * $p < 0.05$. (**A–D, right panel**) Cell area, actin alignment, and myosin alignment were quantified from cells imaged after culture in wide or narrow channels with varying stiffness. Cells cultured in stiffer conditions with narrow channels exhibited increased actin and myosin alignment but lower cell spreading than those in narrow channels. Figure reprinted/adapted with permission from Ref. [102]. Copyright 2012, National Academy of Sciences.

3.1.5. In Vitro Studies of Mechanotransduction in Glioblastoma

Studies of GBM cell-ECM interactions and mechanics rely on in vitro matrix-mimetics that are composed of either biologically occurring proteins or synthetic materials. In 2D cultures, this is accomplished by coating glass or polystyrene with bio- or synthetic materials and culturing cells on top of the coating. In 3D cultures, hydrogels, defined as crosslinked polymer networks that can retain water, are composed of one or several synthetic or biomaterials to mimic the cell microenvironment. Changes in composition, component concentration, and crosslinking density lead to changes in rheological properties such as elasticity and stiffness [177]. Cell migration, morphology, cytoskeletal structure, invasiveness, and signaling are some of the properties characterized in gels of varying stiffness.

It is common practice in the study of GBM mechanics to use various naturally occurring proteins of the brain ECM to construct hydrogels in which cells are seeded. While most studies cite hydrogel formation from one ECM protein, there are several that utilize hydrogels mixed with several proteins, which are discussed in the next sections. Either protein concentration or crosslinking density is altered in the hydrogels to change Young's modulus of the gels and study the effect of elasticity or stiffness in the microenvironment on the cells. The use of biomaterials in hydrogels offers the additional variable of gel degradation by cells or over time in incubation, which adds an important dimension to understanding tumor-cell function in remodeling the ECM.

3.1.6. Hydrogel Culture Methods
Collagen

Despite collagen type I not being an abundant protein in the GBM and healthy brain ECM, it is often used in mechanics studies of these cells, due to its ability to easily form a gel.

Even though collagen IV is upregulated and more abundant in the GBM microenvironment, there is a limited amount of studies in which collagen IV is used for hydrogel formation to investigate GBM mechanics [171]. There are commercially available pre-coated glass slides that make mechanics studies using collagen I simpler. In studies where glioma cells were seeded onto collagen I-coated substrates (glass or polystyrene), there was a trend of cells spreading more extensively in stiffer ECMs [76,273]. One group observed a nearly linear 80 µm extension called an "invadopodia" that stretched from one cell to another, which they hypothesized was a significant factor in cell signaling [273]. The drawback of studies using collagen-coated glass or polystyrene is that it forms a 2D culture, which does not accurately mimic the 3D environment that cells are exposed to in vivo. It was shown that GBM cells exhibited chemoresistance to sunitinib, a kinase inhibitor, in vitro in a 3D collagen-based environment, but not on plastic or collagen-coated surfaces [136]. Additionally, 2D cell migration does not require MMP activity, while 3D culture does, so MMP expression can be studied extensively in relation to mechanics only in 3D culture [274]. For greater insight into in vivo characteristics, it is more useful to employ 3D rather than 2D cultures.

Collagen fiber density or concentration is an important factor to consider in experimental design. A study confirmed that varying the collagen I concentration resulted in varying elastic modulus of hydrogels, but that the difference is not significant with collagen concentrations from 1.5 to 2.5 mg/mL [275]. They also found that a higher concentration of collagen inhibited growth of the GBM in vitro. Decreasing collagen gel stiffness was shown to increase migration distance and velocity of GBM cells [124]. Another study investigated the effects of gelation temperature on collagen I gel pore size and how this impacted glioma invasion. Gelation temperature caused variations in pore size for hydrogels composed of 1 mg/mL collagen and was a significant factor in determining speed of invasion of glioma cells with smaller pore sizes (5-12 µm) hindering cell motility [276]. A study by Hwang et al. showed that actin filamentation of migrating cells in collagen gels is dynamic and undergoes rapid changes, and produces of stress fibers and lamellipodia [277]. This study also drew attention to the migratory patterns of glioma cells, which was shown to mostly be composed of double-nucleated cells that migrated in clusters and had extensive interactions, through actin filament extensions, to collagen in the surrounding gel and nearby cells [277]. Distinct cell shapes, in terms of elongation and roundness, have been observed in different collagen gels [278]. Collagen gels were also used to study the effect of the microenvironment on gene expression and the effects of depletion or overexpression of certain genes on cell ability to retain mechano-responsiveness [278–280].

Three-dimensional culturing methods are being modified with novel technologies to model more than just a homogenous tumor microenvironment; they are being used to form models of interacting systems. In a study by Chonan et al., the tumor niche including ECM and blood vessels was mimicked in a microfluidic device using collagen I and human umbilical vein endothelial cells (HUVECs) [281]. Further studies with collagens are required for a more thorough understanding of the effect of these proteins on GBM cells.

Hyaluronan

There is extensive research on the production of hyaluronan (HA)-based gels for studies of GBM mechanics. HA hydrogels have been used clinically as implants for neural regeneration and reduction of scar formation [282,283]. In in vitro studies, HA gels are generally functionalized with peptides or other ECM proteins, such as laminin, RGD (Arg-Gly-Asp peptides), poly-D-lysine, and poly-L-lysine, which allow for cell adhesion to the gel, since HA alone does not attach to the cells [283,284]. In a study with short Arg-Gly-Asp (RGD) peptides incorporated to HA gels, it was noted that glioma cells' actin stress fiber assembly and cell spreading was greater in stiffer gels [147]. Various HA crosslinkers can also be used to increase gel stiffness. Divinyl sulfone, for example, was used to crosslink HA carboxyl groups to varying degrees when different concentrations were used [285]. Chitosan-HA scaffolds, when compared to 2D surfaces, increased invasiveness and chemotherapeutic resistance and were proposed as possible in vitro mimics of the tumor microenvironment for pre-clinical drug effectiveness studies [286].

Other Proteins

Proteins such as fibronectin, laminin, and gelatin on their own are not commonly used for hydrogel formation. Rather, they are more frequently used to coat glass or polystyrene for two-dimensional cell-protein interaction studies. In the formation of hydrogels, they are generally used in combination with synthetic materials. These were also discussed previously in the synthetic hydrogels section of this review.

Methacrylated gelatin (GelMA) has been used on its own as a scaffold for glioma cells in a study of gel biophysical properties and their effect on cell morphology, proliferation, motility, and gene expression. Pedron et al. observed that GelMA biophysical properties could be varied by the methacrylation degree and bulk density. Cell morphology, motility and expression of hypoxia markers (VEGF, MMP-2, MMP-9, and HIF-1) and the ECM protein fibronectin were affected by the environmental properties [287]. Ramamoorthi et al. used an alginate hydrogel with varying stiffness values and observed greater cell sensitivity to toxins in less stiff gels [133].

Composite Biomaterial Hydrogels

There is a paucity of research involving composite hydrogels in the study of GBM mechanics. Researchers developed a composite matrix of HA and collagen oligomers with the addition of Matrigel-coated microfibers with tunable stiffness by varying component protein concentrations [124]. While low stiffness correlated to lower migration velocity and distance, collagen source and concentration was shown to affect these parameters variably. They also reported that in the HA hydrogel, cells exhibited collective migration while in collagen, they relied on single-cell migration [124]. These studies outline some of the different cellular responses to each of the ECM components and serve to emphasize the importance of studying specific cell-ECM interactions when attempting to describe GBM mechanics.

Synthetic Hydrogels

There are several types of synthetic polymers that have been used in the study of GBM mechanics. In 2D culture, "poly-methylphenyl-siloxane film" with greater stiffness values resulted in increased cell spreading and migration compared to more compliant films [181]. Wan et al. developed nanotextured polydimethylsiloxane (PDMS) surfaces with aptamers overexpressing epidermal growth factor receptor (EGFR) to isolate human GBM cells from a mix with fibroblast cells [288]. PA hydrogels are commonly used in GBM mechanics studies. In a novel study of cell mechano-sensitivity to matrix confinement, cells that were seeded onto PA hydrogels in more narrow channels exhibited greater migration speeds when compared to cells seeded on PA hydrogels in wider channels or directly onto 2D surfaces. This effect was abrogated by the inhibition of non-muscle myosin II, implicating this protein as a mechanosensory [102]. In a study investigating patient-derived primary GBM cell sensitivity to gel stiffness or rigidity, cells were seeded onto PA gels with different rigidity measures and the migration rate was correlated to rigidity-sensitivity [173]. Human glioma cell lines U373 and U118 cultured on polyacrylamide gels of normal brain stiffness (1 kPa) and GBM tumor stiffness (12 kPa), had greater proliferation rates in the stiffer substrate [289]. Umesh et al. studied the effect of PA gel stiffness on expression of proteins related to the cell cycle and dependency on EGFR signaling in human GBM cells, and saw that increased stiffness caused increased expression and phosphorylation of EGFR and Akt. Conversely, loss of EGFR, Akt, or phosphoinositide 3-kinase (PI3K) function resulted in decreased stiffness-sensitivity of the cells [252].

4. Challenges and Future Directions

We have endeavored to provide a thorough explanation of the various hydrogel-based methods that are employed for biomechanical studies and an overview of how studying cell-ECM interactions can lead to significant advancements in understanding the pathology of cancers, with a focus on GBM. It is abundantly clear from the state of the field that cellular response to biomechanical cues is a key player in maintaining health and homeostasis. To incorporate biomechanical effects into our current understanding of cell biology, hydrogel-based three-dimensional models of cell culture must become more widely used. The number of publications mentioning hydrogels has increased exponentially over the past 25 years (Figure 7). Furthermore, publications mentioning both hydrogels and cancer are making up a greater proportion of the hydrogel-related manuscripts published each year since 2012, indicating a growing interest in the development and use of hydrogels in the study of cancer. However, a major drawback of these approaches is that any one hydrogel model is not universally applicable due to the unique ECM composition and mechanical properties of the suite of cellular states that may be studied. Additionally, differences in hydrogel composition (and lot-to-lot variability of commercially available hydrogels), cell passage number, cell seeding density, hydrogel crosslinking density, time between cell seeding and microscopy, and other factors make it difficult to obtain consistent results [290,291]. It is worthy of note that ECM composition still has not been characterized based on the abundance of each component and elastic modulus for many tissues and conditions. Protocols unique to the disease model (cell type, microenvironment, perfusion, etc.) need to be established to enable wider adoption of hydrogel-based cell culture studies as an improved biomimetic replacement for conventional two-dimensional cell culture.

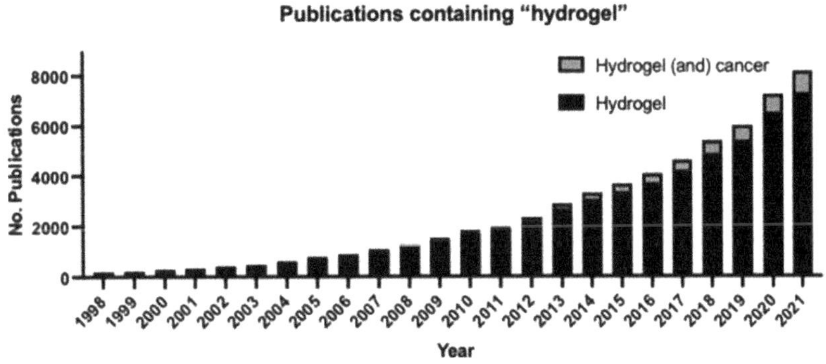

Figure 7. A bar chart showing the number of publications containing the key word "hydrogel" or ("hydrogel" AND "cancer") from the years 1998 to 2021 found in the PubMed database.

The use of hydrogels as model systems for studying cell response to biomechanical stimuli has been discussed in this review, with a focus on GBM as a case study for what has been achieved versus what has yet to be understood in the field. Breast cancer is an exemplar of the impact that biomechanical studies can have on diagnostic, prognostic, and therapeutic approaches to a disease. Therefore, we believe that future studies to understand how cells respond and adjust to mechanical stimuli, and how these responses may be dysregulated in various pathologies, are of utmost importance to craft a more systematic understanding of diseases that have been difficult to treat and cure, such as GBM. An improved understanding of tissue and cellular mechanics would facilitate the development of mechanotherapies for regenerative rehabilitation [292–295]. Not only will such studies further our understanding of the disease, but they may also provide clues for how to take advantage of mechanical stimuli to treat them. Hydrogels are used in contact lenses and as vehicles for drug delivery; they also have the potential to be used as therapeutic agents in cases where mechanical cues such as stiffness or stress can influence therapeutic resistance in certain cancers. For example, hydrogels with low stiffness may be implanted after tumor removal in patients with invasive cancers to reduce the ability of cancer cells to migrate. Such innovative uses for hydrogels and applications of knowledge that can be gained by biomechanical studies would improve our understanding and treatment of cancer.

Author Contributions: Conceptualization, A.Z.S., M.B. and C.B.P.; Methodology, A.Z.S. and C.B.P.; Validation, A.Z.S.; Investigation, A.Z.S.; Resources, A.Z.S. and C.B.P.; Data Curation, A.Z.S., M.B. and C.B.P.; Writing—Original Draft Preparation, A.Z.S. and C.B.P.; Writing—Review and Editing, A.Z.S., M.B. and C.B.P.; Visualization, A.Z.S.; Supervision, M.B. and C.B.P.; Project Administration, C.B.P.; Funding Acquisition, C.B.P. All authors have read and agreed to the published version of the manuscript.

Funding: This research received no external funding. The APC was funded by the McNair Medical Institute at The Robert and Janice McNair Foundation.

Institutional Review Board Statement: Not applicable.

Informed Consent Statement: Not applicable.

Data Availability Statement: Not applicable.

Acknowledgments: CBP is a McNair Scholar supported by the McNair Medical Institute at The Robert and Janice McNair Foundation.

Conflicts of Interest: The authors declare no conflict of interest.

Abbreviations

Arg-Gly-Asp peptides	RGD
basic fibroblast growth factor	bFGF
blood-brain barrier	BBB
C-X-C motif chemokine receptor	CXCR
epidermal growth factor	EGF
epidermal growth factor receptor	EGFR
extracellular matrix	ECM
G-protein coupled receptor 68	GPR68
gelatin methacryloyl	GelMa
glioblastoma	GBM
human umbilical vein endothelial cells	HUVECs
hyaluronic acid	HA
linker of nucleo- and cyto-skeleton	LINC
matrix metalloproteinase	MMP
Myocardin related transcription factor A	MRTF-A
Neurogenic locus notch homolog protein 1	NOTCH1
phosphoinositide 3-kinase	PI3K
physically interacting cell sequencing	PIC-seq
platelet endothelial cell adhesion molecule-1	PECAM-1
polydimethylsiloxane	PDMS
polyethylene glycol	PEG
positron emission tomography	PET
rho-associated, coiled-coil-containing protein kinase	ROCK
T-cell receptor	TCR
Testin LIM domain protein	TES
Three-dimensional	3D
Transforming growth factor beta 1	TGF b1
Transient Receptor Potential Cation Channel Subfamily M Member 7	TRPM7
Transient Receptor Potential Cation Channel Subfamily V Member 4	TRPV4

Two-dimensional	2D
urokinase	uPA
urokinase receptor	uPAR
vascular endothelial growth factor	VEGF
vascular endothelial growth factor receptor	VEGFR
Vasodilator stimulated phosphoprotein	VASP
von Willebrand factor—glycoprotein Ib complex	VWF-GPIb
Yes-associated protein	YAP

References

1. Shakiba, D.; Babaei, B.; Saadat, F.; Thomopoulos, S.; Genin, G.M. The Fibrous Cellular Microenvironment, and How Cells Make Sense of a Tangled Web. *Proc. Natl. Acad. Sci. USA* **2017**, *114*, 5772–5774. [CrossRef] [PubMed]
2. Lodish, H.; Berk, A.; Zipursky, S.L.; Matsudaira, P.; Baltimore, D.; Darnell, J. Noncollagen Components of the Extracellular Matrix. Molecular Cell Biology. 4th Edition. 2000. Available online: https://www.ncbi.nlm.nih.gov/books/NBK21706/ (accessed on 1 March 2022).
3. Frantz, C.; Stewart, K.M.; Weaver, V.M. The Extracellular Matrix at a Glance. *J. Cell Sci.* **2010**, *123*, 4195–4200. [CrossRef] [PubMed]
4. Chaffey, N. Alberts, B., Johnson, A., Lewis, J., Raff, M., Roberts, K. and Walter, P. Molecular Biology of the Cell. 4th Edn. *Ann. Bot.* **2003**, *91*, 401. [CrossRef]
5. Akhmanova, M.; Osidak, E.; Domogatsky, S.; Rodin, S.; Domogatskaya, A. Physical, Spatial, and Molecular Aspects of Extracellular Matrix of In Vivo Niches and Artificial Scaffolds Relevant to Stem Cells Research. Available online: https://www.hindawi.com/journals/sci/2015/167025/ (accessed on 25 May 2019).
6. Muiznieks, L.D.; Keeley, F.W. Molecular Assembly and Mechanical Properties of the Extracellular Matrix: A Fibrous Protein Perspective. *Biochim. Et Biophys. Acta (BBA)-Mol. Basis Dis.* **2013**, *1832*, 866–875. [CrossRef]
7. Yue, B. Biology of the Extracellular Matrix: An Overview. *J. Glaucoma* **2014**, S20–S23. [CrossRef] [PubMed]
8. Kular, J.K.; Basu, S.; Sharma, R.I. The Extracellular Matrix: Structure, Composition, Age-Related Differences, Tools for Analysis and Applications for Tissue Engineering. *J. Tissue Eng.* **2014**, *5*, 2041731414557112. [CrossRef] [PubMed]
9. Nelson, C.M.; Bissell, M.J. Of Extracellular Matrix, Scaffolds, and Signaling: Tissue Architecture Regulates Development, Homeostasis, and Cancer. *Annu. Rev. Cell Dev. Biol.* **2006**, *22*, 287–309. [CrossRef]
10. Gattazzo, F.; Urciuolo, A.; Bonaldo, P. Extracellular Matrix: A Dynamic Microenvironment for Stem Cell Niche. *Biochim. Biophys. Acta* **2014**, *1840*, 2506–2519. [CrossRef]
11. Ahmed, M.; ffrench-Constant, C. Extracellular Matrix Regulation of Stem Cell Behavior. *Curr. Stem Cell Rep.* **2016**, *2*, 197–206. [CrossRef]
12. Kalluri, R. Angiogenesis: Basement Membranes: Structure, Assembly and Role in Tumour Angiogenesis. *Nat. Rev. Cancer* **2003**, *3*, 422–433. [CrossRef]
13. Malik, R.; Lelkes, P.I.; Cukierman, E. Biomechanical and Biochemical Remodeling of Stromal Extracellular Matrix in Cancer. *Trends Biotechnol.* **2015**, *33*, 230–236. [CrossRef] [PubMed]
14. Katz, B.-Z.; Zamir, E.; Bershadsky, A.; Kam, Z.; Yamada, K.M.; Geiger, B.; Hynes, R. Physical State of the Extracellular Matrix Regulates the Structure and Molecular Composition of Cell-Matrix Adhesions. *Mol. Biol. Cell* **2000**, *11*, 1047–1060. [CrossRef] [PubMed]
15. Mui, K.L.; Chen, C.S.; Assoian, R.K. The Mechanical Regulation of Integrin–Cadherin Crosstalk Organizes Cells, Signaling and Forces. *J. Cell Sci.* **2016**, *129*, 1093–1100. [CrossRef] [PubMed]
16. Parsons, J.T.; Horwitz, A.R.; Schwartz, M.A. Cell Adhesion: Integrating Cytoskeletal Dynamics and Cellular Tension. *Nat. Rev. Mol. Cell Biol.* **2010**, *11*, 633–643. [CrossRef]
17. Latif, N.; Sarathchandra, P.; Taylor, P.M.; Antoniw, J.; Yacoub, M.H. Molecules Mediating Cell-ECM and Cell-Cell Communication in Human Heart Valves. *Cell Biochem. Biophys.* **2005**, *43*, 275–287. [CrossRef]
18. Seong, J.; Wang, N.; Wang, Y. Mechanotransduction at Focal Adhesions: From Physiology to Cancer Development. *J. Cell Mol. Med.* **2013**, *17*, 597–604. [CrossRef] [PubMed]
19. Bernfield, M.; Götte, M.; Park, P.W.; Reizes, O.; Fitzgerald, M.L.; Lincecum, J.; Zako, M. Functions of Cell Surface Heparan Sulfate Proteoglycans. *Annu. Rev. Biochem.* **1999**, *68*, 729–777. [CrossRef]
20. Salmivirta, M.; Jalkanen, M. Syndecan Family of Cell Surface Proteoglycans: Developmentally Regulated Receptors for Extracellular Effector Molecules. *Experientia* **1995**, *51*, 863–872. [CrossRef]
21. Mythreye, K.; Blobe, G.C. Proteoglycan Signaling Co–Receptors: Roles in Cell Adhesion, Migration and Invasion. *Cell. Signal.* **2009**, *21*, 1548–1558. [CrossRef]
22. Lin, X. Functions of Heparan Sulfate Proteoglycans in Cell Signaling during Development. *Development* **2004**, *131*, 6009–6021. [CrossRef]
23. Kirkbride, K.C.; Ray, B.N.; Blobe, G.C. Cell-Surface Co-Receptors: Emerging Roles in Signaling and Human Disease. *Trends Biochem. Sci.* **2005**, *30*, 611–621. [CrossRef] [PubMed]
24. Orr, A.W.; Helmke, B.P.; Blackman, B.R.; Schwartz, M.A. Mechanisms of Mechanotransduction. *Dev. Cell* **2006**, *10*, 11–20. [CrossRef]

25. Wang, N. Review of Cellular Mechanotransduction. *J. Phys. D Appl. Phys.* **2017**, *50*, 233002. [CrossRef] [PubMed]
26. Bauer, J.; Emon, M.A.B.; Staudacher, J.J.; Thomas, A.L.; Zessner-Spitzenberg, J.; Mancinelli, G.; Krett, N.; Saif, M.T.; Jung, B. Increased Stiffness of the Tumor Microenvironment in Colon Cancer Stimulates Cancer Associated Fibroblast-Mediated Prometastatic Activin A Signaling. *Sci. Rep.* **2020**, *10*, 50. [CrossRef] [PubMed]
27. Emon, B.; Bauer, J.; Jain, Y.; Jung, B.; Saif, T. Biophysics of Tumor Microenvironment and Cancer Metastasis-A Mini Review. *Comput. Struct. Biotechnol. J.* **2018**, *16*, 279–287. [CrossRef] [PubMed]
28. Xu, W.; Mezencev, R.; Kim, B.; Wang, L.; McDonald, J.; Sulchek, T. Cell Stiffness Is a Biomarker of the Metastatic Potential of Ovarian Cancer Cells. *PLoS ONE* **2012**, *7*, e46609. [CrossRef] [PubMed]
29. Hayashi, K.; Iwata, M. Stiffness of Cancer Cells Measured with an AFM Indentation Method. *J. Mech. Behav. Biomed. Mater.* **2015**, *49*, 105–111. [CrossRef]
30. Raudenska, M.; Kratochvilova, M.; Vicar, T.; Gumulec, J.; Balvan, J.; Polanska, H.; Pribyl, J.; Masarik, M. Cisplatin Enhances Cell Stiffness and Decreases Invasiveness Rate in Prostate Cancer Cells by Actin Accumulation. *Sci. Rep.* **2019**, *9*, 1660. [CrossRef]
31. Xiao, W.; Sohrabi, A.; Seidlits, S.K. Integrating the Glioblastoma Microenvironment into Engineered Experimental Models. *Future Sci. OA* **2017**, *3*, FSO189. [CrossRef]
32. Pepin, K.M.; McGee, K.P.; Arani, A.; Lake, D.S.; Glaser, K.J.; Manduca, A.; Parney, I.F.; Ehman, R.L.; Huston, J. MR Elastography Analysis of Glioma Stiffness and IDH1-Mutation Status. *Am. J. Neuroradiol.* **2017**. [CrossRef]
33. Streitberger, K.-J.; Reiss-Zimmermann, M.; Freimann, F.B.; Bayerl, S.; Guo, J.; Arlt, F.; Wuerfel, J.; Braun, J.; Hoffmann, K.-T.; Sack, I. High-Resolution Mechanical Imaging of Glioblastoma by Multifrequency Magnetic Resonance Elastography. *PLoS ONE* **2014**, *9*, e110588. [CrossRef] [PubMed]
34. Madsen, C.D.; Cox, T.R. Relative Stiffness Measurements of Tumour Tissues by Shear Rheology. *Bio-Protoc.* **2017**, *7*, e2265. [CrossRef] [PubMed]
35. Deng, X.; Xiong, F.; Li, X.; Xiang, B.; Li, Z.; Wu, X.; Guo, C.; Li, X.; Li, Y.; Li, G.; et al. Application of Atomic Force Microscopy in Cancer Research. *J. Nanobiotechnology* **2018**, *16*. [CrossRef] [PubMed]
36. Song, E.J.; Sohn, Y.-M.; Seo, M. Tumor Stiffness Measured by Quantitative and Qualitative Shear Wave Elastography of Breast Cancer. *Br. J. Radiol.* **2018**, *91*. [CrossRef]
37. Bondu, V.; Wu, C.; Cao, W.; Simons, P.C.; Gillette, J.; Zhu, J.; Erb, L.; Zhang, X.F.; Buranda, T. Low-Affinity Binding in Cis to P2Y2R Mediates Force-Dependent Integrin Activation during Hantavirus Infection. *Mol. Biol. Cell* **2017**, *28*, 2887–2903. [CrossRef]
38. Nordenfelt, P.; Elliott, H.L.; Springer, T.A. Coordinated Integrin Activation by Actin-Dependent Force during T-Cell Migration. *Nat. Commun.* **2016**, *7*, 13119. [CrossRef]
39. Li, J.; Springer, T.A. Integrin Extension Enables Ultrasensitive Regulation by Cytoskeletal Force. *Proc. Natl. Acad. Sci. USA* **2017**, *114*, 4685–4690. [CrossRef]
40. Shimizu, T.; Osanai, Y.; Tanaka, K.F.; Abe, M.; Natsume, R.; Sakimura, K.; Ikenaka, K. YAP Functions as a Mechanotransducer in Oligodendrocyte Morphogenesis and Maturation. *Glia* **2017**, *65*, 360–374. [CrossRef]
41. Conway, D.E.; Schwartz, M.A. Mechanotransduction of Shear Stress Occurs through Changes in VE-Cadherin and PECAM-1 Tension: Implications for Cell Migration. *Cell Adh. Migr.* **2014**, *9*, 335–339. [CrossRef]
42. Kale, G.R.; Yang, X.; Philippe, J.-M.; Mani, M.; Lenne, P.-F.; Lecuit, T. Distinct Contributions of Tensile and Shear Stress on E-Cadherin Levels during Morphogenesis. *Nat. Commun.* **2018**, *9*, 5021. [CrossRef]
43. Wang, C.-H.; Cherng, W.-J.; Verma, S. Drawbacks to Stem Cell Therapy in Cardiovascular Diseases. *Future Cardiol.* **2008**, *4*, 399–408. [CrossRef]
44. Snyder, J.L.; McBeath, E.; Thomas, T.N.; Chiu, Y.J.; Clark, R.L.; Fujiwara, K. Mechanotransduction Properties of the Cytoplasmic Tail of PECAM-1. *Biol. Cell* **2017**, *109*, 312–321. [CrossRef] [PubMed]
45. Gliemann, L.; Rytter, N.; Piil, P.; Nilton, J.; Lind, T.; Nyberg, M.; Cocks, M.; Hellsten, Y. The Endothelial Mechanotransduction Protein Platelet Endothelial Cell Adhesion Molecule-1 Is Influenced by Aging and Exercise Training in Human Skeletal Muscle. *Front. Physiol.* **2018**, *9*, 1807. [CrossRef] [PubMed]
46. Chachisvilis, M.; Zhang, Y.-L.; Frangos, J.A. G Protein-Coupled Receptors Sense Fluid Shear Stress in Endothelial Cells. *Proc. Natl. Acad. Sci. USA* **2006**, *103*, 15463–15468. [CrossRef] [PubMed]
47. Beckmann, R.; Houben, A.; Tohidnezhad, M.; Kweider, N.; Fragoulis, A.; Wruck, C.J.; Brandenburg, L.O.; Hermanns-Sachweh, B.; Goldring, M.B.; Pufe, T.; et al. Mechanical Forces Induce Changes in VEGF and VEGFR-1/SFlt-1 Expression in Human Chondrocytes. *Int. J. Mol. Sci.* **2014**, *15*, 15456–15474. [CrossRef] [PubMed]
48. Ju, L.; Chen, Y.; Xue, L.; Du, X.; Zhu, C. Cooperative Unfolding of Distinctive Mechanoreceptor Domains Transduces Force into Signals. *eLife* **2016**, *5*. [CrossRef]
49. Wang, N.; Tytell, J.D.; Ingber, D.E. Mechanotransduction at a Distance: Mechanically Coupling the Extracellular Matrix with the Nucleus. *Nat. Rev. Mol. Cell Biol.* **2009**, *10*, 75–82. [CrossRef]
50. Poh, Y.-C.; Na, S.; Chowdhury, F.; Ouyang, M.; Wang, Y.; Wang, N. Rapid Activation of Rac GTPase in Living Cells by Force Is Independent of Src. *PLoS ONE* **2009**, *4*, e7886. [CrossRef]
51. Wang, S.; Stoops, E.; CP, U.; Markus, B.; Reuveny, A.; Ordan, E.; Volk, T. Mechanotransduction via the LINC Complex Regulates DNA Replication in Myonuclei. *J. Cell Biol.* **2018**, *217*, 2005–2018. [CrossRef]
52. del Alamo, J.C.; Norwich, G.N.; Li, Y.J.; Lasheras, J.C.; Chien, S. Anisotropic Rheology and Directional Mechanotransduction in Vascular Endothelial Cells. *Proc. Natl. Acad. Sci. USA* **2008**, *105*, 15411–15416. [CrossRef]

53. Na, S.; Collin, O.; Chowdhury, F.; Tay, B.; Ouyang, M.; Wang, Y.; Wang, N. Rapid Signal Transduction in Living Cells Is a Unique Feature of Mechanotransduction. *Proc. Natl. Acad. Sci. USA* **2008**, *105*, 6626–6631. [CrossRef] [PubMed]
54. Dupont, S.; Morsut, L.; Aragona, M.; Enzo, E.; Giulitti, S.; Cordenonsi, M.; Zanconato, F.; Le Digabel, J.; Forcato, M.; Bicciato, S.; et al. Role of YAP/TAZ in Mechanotransduction. *Nature* **2011**, *474*, 179–183. [CrossRef] [PubMed]
55. Provenzano, P.P.; Keely, P.J. Mechanical Signaling through the Cytoskeleton Regulates Cell Proliferation by Coordinated Focal Adhesion and Rho GTPase Signaling. *J. Cell Sci.* **2011**, *124*, 1195–1205. [CrossRef] [PubMed]
56. Isenberg, B.C.; DiMilla, P.A.; Walker, M.; Kim, S.; Wong, J.Y. Vascular Smooth Muscle Cell Durotaxis Depends on Substrate Stiffness Gradient Strength. *Biophys. J.* **2009**, *97*, 1313–1322. [CrossRef] [PubMed]
57. Brouzés, E.; Farge, E. Interplay of Mechanical Deformation and Patterned Gene Expression in Developing Embryos. *Curr. Opin. Genet. Dev.* **2004**, *14*, 367–374. [CrossRef] [PubMed]
58. Wozniak, M.A.; Chen, C.S. Mechanotransduction in Development: A Growing Role for Contractility. *Nat. Rev. Mol. Cell Biol.* **2009**, *10*, 34–43. [CrossRef]
59. Mammoto, T.; Ingber, D.E. Mechanical Control of Tissue and Organ Development. *Development* **2010**, *137*, 1407–1420. [CrossRef]
60. Engler, A.J.; Sen, S.; Sweeney, H.L.; Discher, D.E. Matrix Elasticity Directs Stem Cell Lineage Specification. *Cell* **2006**, *126*, 677–689. [CrossRef]
61. Matthews, B.D.; Overby, D.R.; Mannix, R.; Ingber, D.E. Cellular Adaptation to Mechanical Stress: Role of Integrins, Rho, Cytoskeletal Tension and Mechanosensitive Ion Channels. *J. Cell Sci.* **2006**, *119*, 508–518. [CrossRef]
62. Yeh, Y.-C.; Ling, J.-Y.; Chen, W.-C.; Lin, H.-H.; Tang, M.-J. Mechanotransduction of Matrix Stiffness in Regulation of Focal Adhesion Size and Number: Reciprocal Regulation of Caveolin-1 and B1 Integrin. *Sci. Rep.* **2017**, *7*, 15008. [CrossRef]
63. Dong, Y.; Xie, X.; Wang, Z.; Hu, C.; Zheng, Q.; Wang, Y.; Chen, R.; Xue, T.; Chen, J.; Gao, D.; et al. Increasing Matrix Stiffness Upregulates Vascular Endothelial Growth Factor Expression in Hepatocellular Carcinoma Cells Mediated by Integrin B1. *Biochem. Biophys. Res. Commun.* **2014**, *444*, 427–432. [CrossRef] [PubMed]
64. van der Pijl, R.J.; Granzier, H.; Ottenheijm, C.A.C. Diaphragm Contractile Weakness Due to Altered Mechanical Loading: Role of Titin. *Am. J. Physiol. Cell Physiol.* **2019**. [CrossRef] [PubMed]
65. Burridge, K.; Wittchen, E.S. The Tension Mounts: Stress Fibers as Force-Generating Mechanotransducers. *J. Cell Biol.* **2013**, *200*, 9–19. [CrossRef] [PubMed]
66. Lee, H.T.; Sharek, L.; O'Brien, E.T.; Urbina, F.L.; Gupton, S.L.; Superfine, R.; Burridge, K.; Campbell, S.L. Vinculin and Metavinculin Exhibit Distinct Effects on Focal Adhesion Properties, Cell Migration, and Mechanotransduction. *PLoS ONE* **2019**, *14*, e0221962. [CrossRef]
67. Wong, S.Y.; Ulrich, T.A.; Deleyrolle, L.P.; MacKay, J.L.; Lin, J.-M.G.; Martuscello, R.T.; Jundi, M.A.; Reynolds, B.A.; Kumar, S. Constitutive Activation of Myosin-Dependent Contractility Sensitizes Glioma Tumor-Initiating Cells to Mechanical Inputs and Reduces Tissue Invasion. *Cancer Res.* **2015**, *75*, 1113–1122. [CrossRef]
68. Oldenburg, J.; van der Krogt, G.; Twiss, F.; Bongaarts, A.; Habani, Y.; Slotman, J.A.; Houtsmuller, A.; Huveneers, S.; de Rooij, J. VASP, Zyxin and TES Are Tension-Dependent Members of Focal Adherens Junctions Independent of the α-Catenin-Vinculin Module. *Sci. Rep.* **2015**, *5*, 17225. [CrossRef]
69. Mack, J.J.; Mosqueiro, T.S.; Archer, B.J.; Jones, W.M.; Sunshine, H.; Faas, G.C.; Briot, A.; Aragón, R.L.; Su, T.; Romay, M.C.; et al. NOTCH1 Is a Mechanosensor in Adult Arteries. *Nat. Commun.* **2017**, *8*, 1620. [CrossRef] [PubMed]
70. Li, J.; Hou, B.; Tumova, S.; Muraki, K.; Bruns, A.; Ludlow, M.J.; Sedo, A.; Hyman, A.J.; McKeown, L.; Young, R.S.; et al. Piezo1 Integration of Vascular Architecture with Physiological Force. *Nature* **2014**, *515*, 279–282. [CrossRef]
71. Swift, J.; Ivanovska, I.L.; Buxboim, A.; Harada, T.; Dingal, P.C.D.P.; Pinter, J.; Pajerowski, J.D.; Spinler, K.R.; Shin, J.-W.; Tewari, M.; et al. Nuclear Lamin-A Scales with Tissue Stiffness and Enhances Matrix-Directed Differentiation. *Science* **2013**, *341*. [CrossRef]
72. Lammerding, J.; Schulze, P.C.; Takahashi, T.; Kozlov, S.; Sullivan, T.; Kamm, R.D.; Stewart, C.L.; Lee, R.T. Lamin A/C Deficiency Causes Defective Nuclear Mechanics and Mechanotransduction. *J. Clin. Investig.* **2004**, *113*, 370–378. [CrossRef]
73. Wozniak, M.A.; Modzelewska, K.; Kwong, L.; Keely, P.J. Focal Adhesion Regulation of Cell Behavior. *Biochim. Biophys. Acta* **2004**, *1692*, 103–119. [CrossRef] [PubMed]
74. Kubow, K.E.; Vukmirovic, R.; Zhe, L.; Klotzsch, E.; Smith, M.L.; Gourdon, D.; Luna, S.; Vogel, V. Mechanical Forces Regulate the Interactions of Fibronectin and Collagen I in Extracellular Matrix. *Nat. Commun.* **2015**, *6*, 8026. [CrossRef] [PubMed]
75. Liu, B.; Chen, W.; Evavold, B.D.; Zhu, C. Accumulation of Dynamic Catch Bonds between TCR and Agonist Peptide-MHC Triggers T Cell Signaling. *Cell* **2014**, *157*, 357–368. [CrossRef] [PubMed]
76. Sen, S.; Ng, W.P.; Kumar, S. Contributions of Talin-1 to Glioma Cell–Matrix Tensional Homeostasis. *J. R. Soc. Interface* **2012**, *9*, 1311–1317. [CrossRef]
77. Alcaino, C.; Knutson, K.R.; Treichel, A.J.; Yildiz, G.; Strege, P.R.; Linden, D.R.; Li, J.H.; Leiter, A.B.; Szurszewski, J.H.; Farrugia, G.; et al. A Population of Gut Epithelial Enterochromaffin Cells Is Mechanosensitive and Requires Piezo2 to Convert Force into Serotonin Release. *Proc. Natl. Acad. Sci. USA* **2018**, *115*, E7632–E7641. [CrossRef]
78. Rubashkin, M.G.; Cassereau, L.; Bainer, R.; DuFort, C.C.; Yui, Y.; Ou, G.; Paszek, M.J.; Davidson, M.W.; Chen, Y.-Y.; Weaver, V.M. Force Engages Vinculin and Promotes Tumor Progression by Enhancing PI3K Activation of Phosphatidylinositol (3,4,5)-Triphosphate. *Cancer Res.* **2014**, *74*, 4597–4611. [CrossRef]
79. Sawada, Y.; Tamada, M.; Dubin-Thaler, B.J.; Cherniavskaya, O.; Sakai, R.; Tanaka, S.; Sheetz, M.P. Force Sensing by Mechanical Extension of the Src Family Kinase Substrate P130Cas. *Cell* **2006**, *127*, 1015–1026. [CrossRef]

80. Julien, M.A.; Haller, C.A.; Wang, P.; Wen, J.; Chaikof, E.L. Mechanical Strain Induces a Persistent Upregulation of Syndecan-1 Expression in Smooth Muscle Cells. *J. Cell. Physiol.* **2007**, *211*, 167–173. [CrossRef]
81. Cappelli, H.C.; Kanugula, A.K.; Adapala, R.K.; Amin, V.; Sharma, P.; Midha, P.; Paruchuri, S.; Thodeti, C.K. Mechanosensitive TRPV4 Channels Stabilize VE-Cadherin Junctions to Regulate Tumor Vascular Integrity and Metastasis. *Cancer Lett.* **2019**, *442*, 15–20. [CrossRef]
82. Martinac, B. Mechanosensitive Ion Channels: Molecules of Mechanotransduction. *J. Cell Sci.* **2004**, *117*, 2449–2460. [CrossRef]
83. Zhang, W.; Deng, W.; Zhou, L.; Xu, Y.; Yang, W.; Liang, X.; Wang, Y.; Kulman, J.D.; Zhang, X.F.; Li, R. Identification of a Juxtamembrane Mechanosensitive Domain in the Platelet Mechanosensor Glycoprotein Ib-IX Complex. *Blood* **2015**, *125*, 562–569. [CrossRef] [PubMed]
84. Science, A.A. for the A. of PECAM-1 as Fluid Shear Stress Sensor. *Sci. Signal.* **2005**, *2005*, tw333. [CrossRef]
85. Xu, J.; Mathur, J.; Vessières, E.; Hammack, S.; Nonomura, K.; Favre, J.; Grimaud, L.; Petrus, M.; Francisco, A.; Li, J.; et al. GPR68 Senses Flow and Is Essential for Vascular Physiology. *Cell* **2018**, *173*, 762–775.e16. [CrossRef] [PubMed]
86. Cha, B.; Srinivasan, R.S. Mechanosensitive β-Catenin Signaling Regulates Lymphatic Vascular Development. *BMB Rep.* **2016**, *49*, 403–404. [CrossRef]
87. Peng, Y.; Chen, Y.; Qin, X.; Li, S.; Liu, Y. Unveiling the Mechanotransduction Mechanism of Substrate Stiffness-Modulated Cancer Cell Motility via ROCK1 and ROCK2 Differentially Regulated Manner. *FASEB J.* **2019**, *33*, 644.4. [CrossRef]
88. Hadden, W.J.; Young, J.L.; Holle, A.W.; McFetridge, M.L.; Kim, D.Y.; Wijesinghe, P.; Taylor-Weiner, H.; Wen, J.H.; Lee, A.R.; Bieback, K.; et al. Stem Cell Migration and Mechanotransduction on Linear Stiffness Gradient Hydrogels. *Proc. Natl. Acad. Sci. USA* **2017**, *114*, 5647–5652. [CrossRef]
89. Wang, Y.; Xiao, B. The Mechanosensitive Piezo1 Channel: Structural Features and Molecular Bases Underlying Its Ion Permeation and Mechanotransduction. *J. Physiol.* **2018**, *596*, 969–978. [CrossRef]
90. Zeng, Y.; Shen, Y.; Huang, X.-L.; Liu, X.-J.; Liu, X.-H. Roles of Mechanical Force and CXCR1/CXCR2 in Shear-Stress-Induced Endothelial Cell Migration. *Eur. Biophys. J.* **2012**, *41*, 13–25. [CrossRef]
91. Shieh, A.C.; Rozansky, H.A.; Hinz, B.; Swartz, M.A. Tumor Cell Invasion Is Promoted by Interstitial Flow-Induced Matrix Priming by Stromal Fibroblasts. *Cancer Res.* **2011**, *71*, 790–800. [CrossRef]
92. Burridge, K.; Monaghan-Benson, E.; Graham, D.M. Mechanotransduction: From the Cell Surface to the Nucleus via RhoA. *Philos. Trans. R. Soc. Lond. B Biol. Sci.* **2019**, *374*. [CrossRef]
93. Chin, L.; Xia, Y.; Discher, D.E.; Janmey, P.A. Mechanotransduction in Cancer. *Curr. Opin. Chem. Eng.* **2016**, *11*, 77–84. [CrossRef] [PubMed]
94. McKenzie, A.J.; Hicks, S.R.; Svec, K.V.; Naughton, H.; Edmunds, Z.L.; Howe, A.K. The Mechanical Microenvironment Regulates Ovarian Cancer Cell Morphology, Migration, and Spheroid Disaggregation. *Sci. Rep.* **2018**, *8*, 7228. [CrossRef] [PubMed]
95. Thoppil, R.J.; Cappelli, H.C.; Adapala, R.K.; Kanugula, A.K.; Paruchuri, S.; Thodeti, C.K. TRPV4 Channels Regulate Tumor Angiogenesis via Modulation of Rho/Rho Kinase Pathway. *Oncotarget* **2016**, *7*, 25849–25861. [CrossRef] [PubMed]
96. Broders-Bondon, F.; Ho-Bouldoires, T.H.N.; Fernandez-Sanchez, M.-E.; Farge, E. Mechanotransduction in Tumor Progression: The Dark Side of the Force. *J. Cell Biol.* **2018**, *217*, 1571–1587. [CrossRef] [PubMed]
97. Wei, S.C.; Fattet, L.; Tsai, J.H.; Guo, Y.; Pai, V.H.; Majeski, H.E.; Chen, A.C.; Sah, R.L.; Taylor, S.S.; Engler, A.J.; et al. Matrix Stiffness Drives Epithelial-Mesenchymal Transition and Tumour Metastasis through a TWIST1-G3BP2 Mechanotransduction Pathway. *Nat. Cell Biol.* **2015**, *17*, 678–688. [CrossRef]
98. Biggs, M.J.P.; Dalby, M.J. Focal Adhesions in Osteoneogenesis. *Proc. Inst. Mech Eng. H* **2010**, *224*, 1441–1453. [CrossRef]
99. Dugina, V.; Fontao, L.; Chaponnier, C.; Vasiliev, J.; Gabbiani, G. Focal Adhesion Features during Myofibroblastic Differentiation Are Controlled by Intracellular and Extracellular Factors. *J. Cell Sci.* **2001**, *114*, 3285–3296. [CrossRef]
100. Liu, Y.-S.; Liu, Y.-A.; Huang, C.-J.; Yen, M.-H.; Tseng, C.-T.; Chien, S.; Lee, O.K. Mechanosensitive TRPM7 Mediates Shear Stress and Modulates Osteogenic Differentiation of Mesenchymal Stromal Cells through Osterix Pathway. *Sci. Rep.* **2015**, *5*, 16522. [CrossRef]
101. Jaalouk, D.E.; Lammerding, J. Mechanotransduction Gone Awry. *Nat. Rev. Mol. Cell Biol.* **2009**, *10*, 63–73. [CrossRef]
102. Pathak, A.; Kumar, S. Independent Regulation of Tumor Cell Migration by Matrix Stiffness and Confinement. *Proc. Natl. Acad. Sci. USA* **2012**, *109*, 10334–10339. [CrossRef]
103. Desgrosellier, J.S.; Cheresh, D.A. Integrins in Cancer: Biological Implications and Therapeutic Opportunities. *Nat. Rev. Cancer* **2010**, *10*, 9–22. [CrossRef] [PubMed]
104. Maherally, Z.; Smith, J.R.; Ghoneim, M.K.; Dickson, L.; An, Q.; Fillmore, H.L.; Pilkington, G.J. Silencing of CD44 in Glioma Leads to Changes in Cytoskeletal Protein Expression and Cellular Biomechanical Deformation Properties as Measured by AFM Nanoindentation. *BioNanoScience* **2016**, *6*, 54–64. [CrossRef]
105. Prabhune, M.; Belge, G.; Dotzauer, A.; Bullerdiek, J.; Radmacher, M. Comparison of Mechanical Properties of Normal and Malignant Thyroid Cells. *Micron* **2012**, *43*, 1267–1272. [CrossRef] [PubMed]
106. Ramos, J.R.; Pabijan, J.; Garcia, R.; Lekka, M. The Softening of Human Bladder Cancer Cells Happens at an Early Stage of the Malignancy Process. *Beilstein J. Nanotechnol.* **2014**, *5*, 447–457. [CrossRef]
107. Lekka, M.; Pabijan, J. Measuring Elastic Properties of Single Cancer Cells by AFM. *Methods Mol. Biol.* **2019**, *1886*, 315–324. [CrossRef]

108. Plodinec, M.; Loparic, M.; Monnier, C.A.; Obermann, E.C.; Zanetti-Dallenbach, R.; Oertle, P.; Hyotyla, J.T.; Aebi, U.; Bentires-Alj, M.; Lim, R.Y.H.; et al. The Nanomechanical Signature of Breast Cancer. *Nat. Nanotechnol.* **2012**, *7*, 757–765. [CrossRef]
109. Paszek, M.J.; Zahir, N.; Johnson, K.R.; Lakins, J.N.; Rozenberg, G.I.; Gefen, A.; Reinhart-King, C.A.; Margulies, S.S.; Dembo, M.; Boettiger, D.; et al. Tensional Homeostasis and the Malignant Phenotype. *Cancer Cell* **2005**, *8*, 241–254. [CrossRef]
110. Nishida-Aoki, N.; Gujral, T.S. Emerging Approaches to Study Cell–Cell Interactions in Tumor Microenvironment. *Oncotarget* **2019**, *10*, 785–797. [CrossRef]
111. Zervantonakis, I.K.; Poskus, M.D.; Scott, A.L.; Selfors, L.M.; Lin, J.-R.; Dillon, D.A.; Pathania, S.; Sorger, P.K.; Mills, G.B.; Brugge, J.S. Fibroblast–Tumor Cell Signaling Limits HER2 Kinase Therapy Response via Activation of MTOR and Antiapoptotic Pathways. *Proc. Natl. Acad. Sci. USA* **2020**, *117*, 16500–16508. [CrossRef]
112. Mueller, M.; Rasoulinejad, S.; Garg, S.; Wegner, S.V. The Importance of Cell–Cell Interaction Dynamics in Bottom-Up Tissue Engineering: Concepts of Colloidal Self-Assembly in the Fabrication of Multicellular Architectures. *Nano Lett.* **2020**, *20*, 2257–2263. [CrossRef]
113. Mayor, R.; Etienne-Manneville, S. The Front and Rear of Collective Cell Migration. *Nat. Rev. Mol. Cell Biol.* **2016**, *17*, 97–109. [CrossRef] [PubMed]
114. Chen, C.S.; Tan, J.; Tien, J. Mechanotransduction at Cell-Matrix and Cell-Cell Contacts. *Annu. Rev. Biomed. Eng.* **2004**, *6*, 275–302. [CrossRef] [PubMed]
115. Giladi, A.; Cohen, M.; Medaglia, C.; Baran, Y.; Li, B.; Zada, M.; Bost, P.; Blecher-Gonen, R.; Salame, T.-M.; Mayer, J.U.; et al. Dissecting Cellular Crosstalk by Sequencing Physically Interacting Cells. *Nat. Biotechnol.* **2020**, *38*, 629–637. [CrossRef] [PubMed]
116. Melissaridou, S.; Wiechec, E.; Magan, M.; Jain, M.V.; Chung, M.K.; Farnebo, L.; Roberg, K. The Effect of 2D and 3D Cell Cultures on Treatment Response, EMT Profile and Stem Cell Features in Head and Neck Cancer. *Cancer Cell Int.* **2019**, *19*, 16. [CrossRef]
117. Pampaloni, F.; Reynaud, E.G.; Stelzer, E.H.K. The Third Dimension Bridges the Gap between Cell Culture and Live Tissue. *Nat. Rev. Mol. Cell Biol.* **2007**, *8*, 839–845. [CrossRef]
118. Lee, J.; Cuddihy, M.J.; Kotov, N.A. Three-Dimensional Cell Culture Matrices: State of the Art. *Tissue Eng. Part. B Rev.* **2008**, *14*, 61–86. [CrossRef] [PubMed]
119. Edmondson, R.; Broglie, J.J.; Adcock, A.F.; Yang, L. Three-Dimensional Cell Culture Systems and Their Applications in Drug Discovery and Cell-Based Biosensors. *Assay Drug Dev. Technol.* **2014**, *12*, 207–218. [CrossRef] [PubMed]
120. Hsieh, C.-H.; Chen, Y.-D.; Huang, S.-F.; Wang, H.-M.; Wu, M.-H. The Effect of Primary Cancer Cell Culture Models on the Results of Drug Chemosensitivity Assays: The Application of Perfusion Microbioreactor System as Cell Culture Vessel. *Biomed. Res. Int.* **2015**, *2015*, 470283. [CrossRef]
121. Bonnier, F.; Keating, M.E.; Wróbel, T.P.; Majzner, K.; Baranska, M.; Garcia-Munoz, A.; Blanco, A.; Byrne, H.J. Cell Viability Assessment Using the Alamar Blue Assay: A Comparison of 2D and 3D Cell Culture Models. *Toxicol. Vitr.* **2015**, *29*, 124–131. [CrossRef]
122. Duval, K.; Grover, H.; Han, L.-H.; Mou, Y.; Pegoraro, A.F.; Fredberg, J.; Chen, Z. Modeling Physiological Events in 2D vs. 3D Cell Culture. *Physiology* **2017**, *32*, 266–277. [CrossRef]
123. Petrie, R.J.; Yamada, K.M. At the Leading Edge of Three-Dimensional Cell Migration. *J. Cell Sci.* **2012**, *125*, 5917–5926. [CrossRef] [PubMed]
124. Herrera-Perez, M.; Voytik-Harbin, S.L.; Rickus, J.L. Extracellular Matrix Properties Regulate the Migratory Response of Glioblastoma Stem Cells in Three-Dimensional Culture. *Tissue Eng. Part A* **2015**, *21*, 2572–2582. [CrossRef] [PubMed]
125. Petrie, R.J.; Gavara, N.; Chadwick, R.S.; Yamada, K.M. Nonpolarized Signaling Reveals Two Distinct Modes of 3D Cell Migration. *J. Cell Biol.* **2012**, *197*, 439–455. [CrossRef] [PubMed]
126. Antoni, D.; Burckel, H.; Josset, E.; Noel, G. Three-Dimensional Cell Culture: A Breakthrough in Vivo. *Int. J. Mol. Sci.* **2015**, *16*, 5517–5527. [CrossRef] [PubMed]
127. Huber, J.M.; Amann, A.; Koeck, S.; Lorenz, E.; Kelm, J.M.; Obexer, P.; Zwierzina, H.; Gamerith, G. Evaluation of Assays for Drug Efficacy in a Three-Dimensional Model of the Lung. *J. Cancer Res. Clin. Oncol.* **2016**, *142*, 1955–1966. [CrossRef]
128. Das, S.; Sahan, A.Z.; Lim, E.; Suarez, K. The Potential Use of a Three Dimensional Cell Culture of Spheroids in the Study of Various Therapeutic Approaches. *MOJ Immunol.* **2016**, *3*, 11–12. [CrossRef]
129. Barthes, J.; Özçelik, H.; Hindié, M.; Ndreu-Halili, A.; Hasan, A.; Vrana, N.E. Cell Microenvironment Engineering and Monitoring for Tissue Engineering and Regenerative Medicine: The Recent Advances. *Biomed. Res. Int.* **2014**, *2014*. [CrossRef]
130. Gentili, C.; Cancedda, R. Cartilage and Bone Extracellular Matrix. *Curr. Pharm. Des.* **2009**, *15*, 1334–1348. [CrossRef]
131. Liu, M.; Zeng, X.; Ma, C.; Yi, H.; Ali, Z.; Mou, X.; Li, S.; Deng, Y.; He, N. Injectable Hydrogels for Cartilage and Bone Tissue Engineering. *Bone Res.* **2017**, *5*, 17014. [CrossRef]
132. Charrier, E.E.; Pogoda, K.; Wells, R.G.; Janmey, P.A. Control of Cell Morphology and Differentiation by Substrates with Independently Tunable Elasticity and Viscous Dissipation. *Nat. Commun.* **2018**, *9*, 449. [CrossRef]
133. Ramamoorthi, K.; Hara, J.; Ito, C.; Asuri, P. Role of Three-Dimensional Matrix Stiffness in Regulating the Response of Human Neural Cells to Toxins. *Cell. Mol. Bioeng.* **2014**, *7*, 278–284. [CrossRef]
134. Pogoda, K.; Bucki, R.; Byfield, F.J.; Cruz, K.; Lee, T.; Marcinkiewicz, C.; Janmey, P.A. Soft Substrates Containing Hyaluronan Mimic the Effects of Increased Stiffness on Morphology, Motility, and Proliferation of Glioma Cells. *Biomacromolecules* **2017**, *18*, 3040–3051. [CrossRef] [PubMed]

135. Koch, D.; Rosoff, W.J.; Jiang, J.; Geller, H.M.; Urbach, J.S. Strength in the Periphery: Growth Cone Biomechanics and Substrate Rigidity Response in Peripheral and Central Nervous System Neurons. *Biophys. J.* **2012**, *102*, 452–460. [CrossRef] [PubMed]
136. Fernandez-Fuente, G.; Mollinedo, P.; Grande, L.; Vazquez-Barquero, A.; Fernandez-Luna, J.L. Culture Dimensionality Influences the Resistance of Glioblastoma Stem-like Cells to Multikinase Inhibitors. *Mol. Cancer* **2014**, *13*, 1664–1672. [CrossRef]
137. Wong, S.; Guo, W.-H.; Wang, Y.-L. Fibroblasts Probe Substrate Rigidity with Filopodia Extensions before Occupying an Area. *Proc. Natl. Acad. Sci. USA* **2014**, *111*, 17176–17181. [CrossRef]
138. Rao, S.S.; Bentil, S.; DeJesus, J.; Larison, J.; Hissong, A.; Dupaix, R.; Sarkar, A.; Winter, J.O. Inherent Interfacial Mechanical Gradients in 3D Hydrogels Influence Tumor Cell Behaviors. *PLoS ONE* **2012**, *7*, e35852. [CrossRef]
139. Stabenfeldt, S.E.; LaPlaca, M.C. Variations in Rigidity and Ligand Density Influence Neuronal Response in Methylcellulose–Laminin Hydrogels. *Acta Biomater.* **2011**, *7*, 4102–4108. [CrossRef]
140. Pereira, R.F.; Bártolo, P.J. 3D Bioprinting of Photocrosslinkable Hydrogel Constructs. *J. Appl. Polym. Sci.* **2015**, *132*. [CrossRef]
141. Bessonov, I.V.; Rochev, Y.A.; Arkhipova, A.Y.; Kopitsyna, M.N.; Bagrov, D.V.; Karpushkin, E.A.; Bibikova, T.N.; Moysenovich, A.M.; Soldatenko, A.S.; Nikishin, I.I.; et al. Fabrication of Hydrogel Scaffolds via Photocrosslinking of Methacrylated Silk Fibroin. *Biomed. Mater.* **2019**, *14*, 034102. [CrossRef]
142. Seiffert, S.; Oppermann, W.; Saalwächter, K. Hydrogel Formation by Photocrosslinking of Dimethylmaleimide Functionalized Polyacrylamide. *Polymer* **2007**, *48*, 5599–5611. [CrossRef]
143. O'Connell, C.D.; Zhang, B.; Onofrillo, C.; Duchi, S.; Blanchard, R.; Quigley, A.; Bourke, J.; Gambhir, S.; Kapsa, R.; Bella, C.D.; et al. Tailoring the Mechanical Properties of Gelatin Methacryloyl Hydrogels through Manipulation of the Photocrosslinking Conditions. *Soft Matter* **2018**, *14*, 2142–2151. [CrossRef] [PubMed]
144. Maitra, J.; Shukla, V.K. Cross-Linking in Hydrogels—A Review. *Am. J. Polym. Sci.* **2014**, *4*, 25–31.
145. Baday, M.; Ercal, O.; Sahan, A.Z.; Sahan, A.; Ercal, B.; Inan, H.; Demirci, U. Density Based Characterization of Mechanical Cues on Cancer Cells Using Magnetic Levitation. *Adv. Healthc. Mater.* **2019**, *8*, 1801517. [CrossRef] [PubMed]
146. Akalin, O.B.; Bayraktar, H. Alteration of Cell Motility Dynamics through Collagen Fiber Density in Photopolymerized Polyethylene Glycol Hydrogels. *Int. J. Biol. Macromol.* **2020**, *157*, 414–423. [CrossRef]
147. Ananthanarayanan, B.; Kim, Y.; Kumar, S. Elucidating the Mechanobiology of Malignant Brain Tumors Using a Brain Matrix-Mimetic Hyaluronic Acid Hydrogel Platform. *Biomaterials* **2011**, *32*, 7913–7923. [CrossRef] [PubMed]
148. Samal, S.K.; Dash, M.; Dubruel, P.; Van Vlierberghe, S. Chapter 8—Smart Polymer Hydrogels: Properties, Synthesis and Applications. In *Smart Polymers and their Applications*; Aguilar, M.R., San Román, J., Eds.; Woodhead Publishing: Sawston, UK, 2014; pp. 237–270, ISBN 978-0-85709-695-1.
149. Kopecek, J. Hydrogel Biomaterials: A Smart Future? *Biomaterials* **2007**, *28*, 5185–5192. [CrossRef] [PubMed]
150. Liu, Z.; Faraj, Y.; Ju, X.-J.; Wang, W.; Xie, R.; Chu, L.-Y. Nanocomposite Smart Hydrogels with Improved Responsiveness and Mechanical Properties: A Mini Review. *J. Polym. Sci. Part B Polym. Phys.* **2018**, *56*, 1306–1313. [CrossRef]
151. Lim, H.L.; Hwang, Y.; Kar, M.; Varghese, S. Smart Hydrogels as Functional Biomimetic Systems. *Biomater. Sci.* **2014**, *2*, 603–618. [CrossRef]
152. Dzobo, K.; Motaung, K.S.C.M.; Adesida, A. Recent Trends in Decellularized Extracellular Matrix Bioinks for 3D Printing: An Updated Review. *Int. J. Mol. Sci.* **2019**, *20*, 4628. [CrossRef]
153. Tam, R.Y.; Smith, L.J.; Shoichet, M.S. Engineering Cellular Microenvironments with Photo- and Enzymatically Responsive Hydrogels: Toward Biomimetic 3D Cell Culture Models. *Acc. Chem. Res.* **2017**, *50*, 703–713. [CrossRef]
154. DeQuach, J.A.; Mezzano, V.; Miglani, A.; Lange, S.; Keller, G.M.; Sheikh, F.; Christman, K.L. Simple and High Yielding Method for Preparing Tissue Specific Extracellular Matrix Coatings for Cell Culture. *PLoS ONE* **2010**, *5*, e13039. [CrossRef]
155. Scherzer, M.T.; Waigel, S.; Donninger, H.; Arumugam, V.; Zacharias, W.; Clark, G.; Siskind, L.J.; Soucy, P.; Beverly, L. Fibroblast-Derived Extracellular Matrices: An Alternative Cell Culture System That Increases Metastatic Cellular Properties. *PLoS ONE* **2015**, *10*, e0138065. [CrossRef] [PubMed]
156. Lee, H.W.; Kook, Y.-M.; Lee, H.J.; Park, H.; Koh, W.-G. A Three-Dimensional Co-Culture of HepG2 Spheroids and Fibroblasts Using Double-Layered Fibrous Scaffolds Incorporated with Hydrogel Micropatterns. *RSC Adv.* **2014**, *4*, 61005–61011. [CrossRef]
157. Hunt, N.C.; Shelton, R.M.; Grover, L. An Alginate Hydrogel Matrix for the Localised Delivery of a Fibroblast/Keratinocyte Co-Culture. *Biotechnol. J.* **2009**, *4*, 730–737. [CrossRef] [PubMed]
158. Yamamoto, M.; Ikada, Y.; Tabata, Y. Controlled Release of Growth Factors Based on Biodegradation of Gelatin Hydrogel. *J. Biomater. Sci. Polym. Ed.* **2001**, *12*, 77–88. [CrossRef]
159. Tabata, Y. Tissue Regeneration Based on Growth Factor Release. *Tissue Eng.* **2003**, *9* (Suppl. S1), S5–S15. [CrossRef]
160. Wang, Z.; Wang, Z.; Lu, W.W.; Zhen, W.; Yang, D.; Peng, S. Novel Biomaterial Strategies for Controlled Growth Factor Delivery for Biomedical Applications. *NPG Asia Mater.* **2017**, *9*, e435. [CrossRef]
161. Zhu, J.; Marchant, R.E. Design Properties of Hydrogel Tissue-Engineering Scaffolds. *Expert Rev. Med. Devices* **2011**, *8*, 607–626. [CrossRef]
162. Saha, N.; Saarai, A.; Roy, N.; Kitano, T.; Saha, P. Polymeric Biomaterial Based Hydrogels for Biomedical Applications. *J. Biomater. Nanobiotechnology* **2011**, *2*, 85. [CrossRef]
163. Patel, A.; Mequanint, K. Hydrogel Biomaterials. *Biomed. Eng.-Front. Chall.* **2011**. [CrossRef]

164. Lutolf, M.P.; Lauer-Fields, J.L.; Schmoekel, H.G.; Metters, A.T.; Weber, F.E.; Fields, G.B.; Hubbell, J.A. Synthetic Matrix Metalloproteinase-Sensitive Hydrogels for the Conduction of Tissue Regeneration: Engineering Cell-Invasion Characteristics. *Proc. Natl. Acad. Sci. USA* **2003**, *100*, 5413–5418. [CrossRef] [PubMed]
165. Li, J.; Illeperuma, W.R.K.; Suo, Z.; Vlassak, J.J. Hybrid Hydrogels with Extremely High Stiffness and Toughness. *ACS Macro Lett.* **2014**, *3*, 520–523. [CrossRef]
166. Dehne, H.; Hecht, F.M.; Bausch, A.R. The Mechanical Properties of Polymer–Colloid Hybrid Hydrogels. *Soft Matter* **2017**, *13*, 4786–4790. [CrossRef] [PubMed]
167. Singh, M.R.; Patel, S.; Singh, D. Chapter 9—Natural Polymer-Based Hydrogels as Scaffolds for Tissue Engineering. In *Nanobiomaterials in Soft Tissue Engineering*; Grumezescu, A.M., Ed.; William Andrew Publishing: Norwich, NY, USA, 2016; pp. 231–260, ISBN 978-0-323-42865-1.
168. Vieira, S.; da Silva Morais, A.; Silva-Correia, J.; Oliveira, J.M.; Reis, R.L. Natural-Based Hydrogels: From Processing to Applications. In *Encyclopedia of Polymer Science and Technology*; American Cancer Society: Atlanta, GA, USA, 2017; pp. 1–27, ISBN 978-0-471-44026-0.
169. Caliari, S.R.; Burdick, J.A. A Practical Guide to Hydrogels for Cell Culture. *Nat. Methods* **2016**, *13*, 405–414. [CrossRef]
170. Tan, H.; Marra, K.G. Injectable, Biodegradable Hydrogels for Tissue Engineering Applications. *Materials* **2010**, *3*, 1746. [CrossRef]
171. Rao, S.S.; DeJesus, J.; Short, A.R.; Otero, J.J.; Sarkar, A.; Winter, J.O. Glioblastoma Behaviors in Three-Dimensional Collagen-Hyaluronan Composite Hydrogels. *ACS Appl. Mater. Interfaces* **2013**, *5*, 9276–9284. [CrossRef] [PubMed]
172. Hughes, C.S.; Postovit, L.M.; Lajoie, G.A. Matrigel: A Complex Protein Mixture Required for Optimal Growth of Cell Culture. *Proteomics* **2010**, *10*, 1886–1890. [CrossRef] [PubMed]
173. Grundy, T.J.; De Leon, E.; Griffin, K.R.; Stringer, B.W.; Day, B.W.; Fabry, B.; Cooper-White, J.; O'Neill, G.M. Differential Response of Patient-Derived Primary Glioblastoma Cells to Environmental Stiffness. *Sci. Rep.* **2016**, *6*, 23353. [CrossRef]
174. Cruz-Acuña, R.; Quirós, M.; Farkas, A.E.; Dedhia, P.H.; Huang, S.; Siuda, D.; García-Hernández, V.; Miller, A.J.; Spence, J.R.; Nusrat, A.; et al. Synthetic Hydrogels for Human Intestinal Organoid Generation and Colonic Wound Repair. *Nat. Cell Biol.* **2017**, *19*, 1326–1335. [CrossRef]
175. Sawhney, A.S.; Pathak, C.P.; Hubbell, J.A. Bioerodible Hydrogels Based on Photopolymerized Poly(Ethylene Glycol)-Co-Poly(.Alpha.-Hydroxy Acid) Diacrylate Macromers. *Macromolecules* **1993**, *26*, 581–587. [CrossRef]
176. Chirila, T.V.; Constable, I.J.; Crawford, G.J.; Vijayasekaran, S.; Thompson, D.E.; Chen, Y.C.; Fletcher, W.A.; Griffin, B.J. Poly(2-Hydroxyethyl Methacrylate) Sponges as Implant Materials: In Vivo and in Vitro Evaluation of Cellular Invasion. *Biomaterials* **1993**, *14*, 26–38. [CrossRef]
177. Ahmed, E.M. Hydrogel: Preparation, Characterization, and Applications: A Review. *J. Adv. Res.* **2015**, *6*, 105–121. [CrossRef] [PubMed]
178. Wang, C.; Tong, X.; Yang, F. Bioengineered 3D Brain Tumor Model to Elucidate the Effects of Matrix Stiffness on Glioblastoma Cell Behavior Using PEG-Based Hydrogels. *Mol. Pharm.* **2014**, *11*, 2115–2125. [CrossRef] [PubMed]
179. Wu, J.; Li, P.; Dong, C.; Jiang, H.; Xue, B.; Gao, X.; Qin, M.; Wang, W.; Chen, B.; Cao, Y. Rationally Designed Synthetic Protein Hydrogels with Predictable Mechanical Properties. *Nat. Commun.* **2018**, *9*, 620. [CrossRef]
180. Langer, R.; Tirrell, D.A. Designing Materials for Biology and Medicine. *Nature* **2004**, *428*, 487. [CrossRef]
181. Thomas, T.W.; DiMilla, P.A. Spreading and Motility of Human Glioblastoma Cells on Sheets of Silicone Rubber Depend on Substratum Compliance. *Med. Biol. Eng. Comput.* **2000**, *38*, 360–370. [CrossRef]
182. Gurski, L.A.; Jha, A.K.; Zhang, C.; Jia, X.; Farach-Carson, M.C. Hyaluronic Acid-Based Hydrogels as 3D Matrices for in Vitro Evaluation of Chemotherapeutic Drugs Using Poorly Adherent Prostate Cancer Cells. *Biomaterials* **2009**, *30*, 6076–6085. [CrossRef]
183. Gill, B.J.; Gibbons, D.L.; Roudsari, L.C.; Saik, J.E.; Rizvi, Z.H.; Roybal, J.D.; Kurie, J.M.; West, J.L. A Synthetic Matrix with Independently Tunable Biochemistry and Mechanical Properties to Study Epithelial Morphogenesis and EMT in a Lung Adenocarcinoma Model. *Cancer Res.* **2012**, *72*, 6013–6023. [CrossRef]
184. Tibbitt, M.W.; Anseth, K.S. Hydrogels as Extracellular Matrix Mimics for 3D Cell Culture. *Biotechnol. Bioeng.* **2009**, *103*, 655–663. [CrossRef]
185. Ratner, B.D.; Bryant, S.J. Biomaterials: Where We Have Been and Where We Are Going. *Annu. Rev. Biomed. Eng.* **2004**, *6*, 41–75. [CrossRef]
186. Ratner, B.D.; Hoffman, A.S. Synthetic Hydrogels for Biomedical Applications. In *Hydrogels for Medical and Related Applications*; ACS Symposium Series; American Chemical Society: Washington, NY, USA, 1976; Volume 31, pp. 1–36. ISBN 978-0-8412-0338-9.
187. Zheng, S.Y.; Ding, H.; Qian, J.; Yin, J.; Wu, Z.L.; Song, Y.; Zheng, Q. Metal-Coordination Complexes Mediated Physical Hydrogels with High Toughness, Stick–Slip Tearing Behavior, and Good Processability. *Macromolecules* **2016**, *49*, 9637–9646. [CrossRef]
188. Ulrich, T.A.; de Juan Pardo, E.M.; Kumar, S. The Mechanical Rigidity of the Extracellular Matrix Regulates the Structure, Motility, and Proliferation of Glioma Cells. *Cancer Res.* **2009**, *69*, 4167–4174. [CrossRef] [PubMed]
189. Rao, S.S.; Nelson, M.T.; Xue, R.; DeJesus, J.K.; Viapiano, M.S.; Lannutti, J.J.; Sarkar, A.; Winter, J.O. Mimicking White Matter Tract Topography Using Core-Shell Electrospun Nanofibers to Examine Migration of Malignant Brain Tumors. *Biomaterials* **2013**, *34*, 5181–5190. [CrossRef] [PubMed]
190. Pedron, S.; Becka, E.; Harley, B.A.C. Regulation of Glioma Cell Phenotype in 3D Matrices by Hyaluronic Acid. *Biomaterials* **2013**, *34*, 7408–7417. [CrossRef]

191. Shibayama, M.; Tanaka, T. Volume Phase Transition and Related Phenomena of Polymer Gels. In *Responsive Gels: Volume Transitions I*; Dušek, K., Ed.; Advances in Polymer Science; Springer: Berlin/Heidelberg, Germany, 1993; pp. 1–62, ISBN 978-3-540-47737-2.
192. Taraban, M.B.; Feng, Y.; Hammouda, B.; Hyland, L.L.; Yu, Y.B. Chirality-Mediated Mechanical and Structural Properties of Oligopeptide Hydrogels. *Chem. Mater.* **2012**, *24*, 2299–2310. [CrossRef]
193. *Hydrogel Sensors and Actuators: Engineering and Technology*; Gerlach, G.; Arndt, K.-F. (Eds.) Springer Series on Chemical Sensors and Biosensors; Springer: Berlin/Heidelberg, Germany, 2010; ISBN 978-3-540-75644-6.
194. Kizilay, M.Y.; Okay, O. Effect of Initial Monomer Concentration on Spatial Inhomogeneity in Poly(Acrylamide) Gels. *Macromolecules* **2003**, *36*, 6856–6862. [CrossRef]
195. Lindemann, B.; Schröder, U.P.; Oppermann, W. Influence of the Cross-Linker Reactivity on the Formation of Inhomogeneities in Hydrogels. *Macromolecules* **1997**, *30*, 4073–4077. [CrossRef]
196. Oldenziel, W.H.; Beukema, W.; Westerink, B.H.C. Improving the Reproducibility of Hydrogel-Coated Glutamate Microsensors by Using an Automated Dipcoater. *J. Neurosci. Methods* **2004**, *140*, 117–126. [CrossRef]
197. Petersen, O.W.; Rønnov-Jessen, L.; Howlett, A.R.; Bissell, M.J. Interaction with Basement Membrane Serves to Rapidly Distinguish Growth and Differentiation Pattern of Normal and Malignant Human Breast Epithelial Cells. *Proc. Natl. Acad. Sci. USA* **1992**, *89*, 9064–9068. [CrossRef]
198. Baino, F.; Novajra, G.; Vitale-Brovarone, C. Bioceramics and Scaffolds: A Winning Combination for Tissue Engineering. *Front. Bioeng. Biotechnol.* **2015**, *3*. [CrossRef]
199. Viswanathan, P.; Chirasatitsin, S.; Ngamkham, K.; Engler, A.J.; Battaglia, G. Cell Instructive Microporous Scaffolds through Interface Engineering. *J. Am. Chem. Soc.* **2012**, *134*, 20103–20109. [CrossRef] [PubMed]
200. Jiang, Y.-Y.; Zhu, Y.-J.; Li, H.; Zhang, Y.-G.; Shen, Y.-Q.; Sun, T.-W.; Chen, F. Preparation and Enhanced Mechanical Properties of Hybrid Hydrogels Comprising Ultralong Hydroxyapatite Nanowires and Sodium Alginate. *J. Colloid Interface Sci.* **2017**, *497*, 266–275. [CrossRef] [PubMed]
201. Suggitt, M.; Bibby, M.C. 50 Years of Preclinical Anticancer Drug Screening: Empirical to Target-Driven Approaches. *Clin. Cancer Res.* **2005**, *11*, 971–981. [CrossRef]
202. Singh, M.; Close, D.A.; Mukundan, S.; Johnston, P.A.; Sant, S. Production of Uniform 3D Microtumors in Hydrogel Microwell Arrays for Measurement of Viability, Morphology, and Signaling Pathway Activation. *Assay Drug Dev. Technol.* **2015**, *13*, 570–583. [CrossRef] [PubMed]
203. Lee, K.Y.; Mooney, D.J. Hydrogels for Tissue Engineering. *Chem. Rev.* **2001**, *101*, 1869–1879. [CrossRef]
204. Putnam, A.J.; Mooney, D.J. Tissue Engineering Using Synthetic Extracellular Matrices. *Nat. Med.* **1996**, *2*, 824–826. [CrossRef]
205. Wu, Z.; Su, X.; Xu, Y.; Kong, B.; Sun, W.; Mi, S. Bioprinting Three-Dimensional Cell-Laden Tissue Constructs with Controllable Degradation. *Sci. Rep.* **2016**, *6*, 24474. [CrossRef]
206. Lee, K.Y.; Mooney, D.J. Alginate: Properties and Biomedical Applications. *Prog. Polym. Sci.* **2012**, *37*, 106–126. [CrossRef]
207. Kim, G.; Ahn, S.; Kim, Y.; Cho, Y.; Chun, W. Coaxial Structured Collagen–Alginate Scaffolds: Fabrication, Physical Properties, and Biomedical Application for Skin Tissue Regeneration. *J. Mater. Chem.* **2011**, *21*, 6165–6172. [CrossRef]
208. Tan, Z.; Parisi, C.; Silvio, L.D.; Dini, D.; Forte, A.E. Cryogenic 3D Printing of Super Soft Hydrogels. *Sci. Rep.* **2017**, *7*, 16293. [CrossRef]
209. Zhao, W.; Jin, X.; Cong, Y.; Liu, Y.; Fu, J. Degradable Natural Polymer Hydrogels for Articular Cartilage Tissue Engineering. *J. Chem. Technol. Biotechnol.* **2013**, *88*, 327–339. [CrossRef]
210. Yu, L.; Ding, J. Injectable Hydrogels as Unique Biomedical Materials. *Chem. Soc. Rev.* **2008**, *37*, 1473–1481. [CrossRef] [PubMed]
211. Zheng, Y.; Huang, K.; You, X.; Huang, B.; Wu, J.; Gu, Z. Hybrid Hydrogels with High Strength and Biocompatibility for Bone Regeneration. *Int. J. Biol. Macromol.* **2017**, *104*, 1143–1149. [CrossRef]
212. Wu, D.-Q.; Chu, C.-C. Biodegradable Hydrophobic-Hydrophilic Hybrid Hydrogels: Swelling Behavior and Controlled Drug Release. *J. Biomater. Sci. Polym. Ed.* **2008**, *19*, 411–429. [CrossRef] [PubMed]
213. Thambi, T.; Li, Y.; Lee, D.S. Injectable Hydrogels for Sustained Release of Therapeutic Agents. *J. Control. Release* **2017**, *267*, 57–66. [CrossRef] [PubMed]
214. Gil, M.S.; Cho, J.; Thambi, T.; Giang Phan, V.H.; Kwon, I.; Lee, D.S. Bioengineered Robust Hybrid Hydrogels Enrich the Stability and Efficacy of Biological Drugs. *J. Control. Release* **2017**, *267*, 119–132. [CrossRef]
215. Kang, C.E.; Poon, P.C.; Tator, C.H.; Shoichet, M.S. A New Paradigm for Local and Sustained Release of Therapeutic Molecules to the Injured Spinal Cord for Neuroprotection and Tissue Repair. *Tissue Eng. Part A* **2008**, *15*, 595–604. [CrossRef]
216. Nguyen, M.K.; Lee, D.S. Injectable Biodegradable Hydrogels. *Macromol. Biosci.* **2010**, *10*, 563–579. [CrossRef]
217. Wang, Y.; Lapitsky, Y.; Kang, C.E.; Shoichet, M.S. Accelerated Release of a Sparingly Soluble Drug from an Injectable Hyaluronan-Methylcellulose Hydrogel. *J. Control. Release* **2009**, *140*, 218–223. [CrossRef]
218. Jeong, B.; Bae, Y.H.; Lee, D.S.; Kim, S.W. Biodegradable Block Copolymers as Injectable Drug-Delivery Systems. *Nature* **1997**, *388*, 860–862. [CrossRef]
219. Qiu, Y.; Park, K. Environment-Sensitive Hydrogels for Drug Delivery. *Adv. Drug Deliv. Rev.* **2001**, *53*, 321–339. [CrossRef]
220. Yubao, G.; Hecheng, M.; Jianguo, L. Controlled WISP-1 ShRNA Delivery Using Thermosensitive Biodegradable Hydrogel in the Treatment of Osteoarthritis. *J. Bionic Eng.* **2015**, *12*, 285–293. [CrossRef]

221. Lee, P.Y.; Cobain, E.; Huard, J.; Huang, L. Thermosensitive Hydrogel PEG–PLGA–PEG Enhances Engraftment of Muscle-Derived Stem Cells and Promotes Healing in Diabetic Wound. *Mol. Ther.* **2007**, *15*, 1189–1194. [CrossRef] [PubMed]
222. Mothe, A.J.; Tam, R.Y.; Zahir, T.; Tator, C.H.; Shoichet, M.S. Repair of the Injured Spinal Cord by Transplantation of Neural Stem Cells in a Hyaluronan-Based Hydrogel. *Biomaterials* **2013**, *34*, 3775–3783. [CrossRef]
223. Caló, E.; Khutoryanskiy, V.V. Biomedical Applications of Hydrogels: A Review of Patents and Commercial Products. *Eur. Polym. J.* **2015**, *65*, 252–267. [CrossRef]
224. Agaoglu, S.; Diep, P.; Martini, M.; Kt, S.; Baday, M.; Emre Araci, I. Ultra-Sensitive Microfluidic Wearable Strain Sensor for Intraocular Pressure Monitoring. *Lab A Chip* **2018**, *18*, 3471–3483. [CrossRef]
225. Lloyd, A.W.; Faragher, R.G.; Denyer, S.P. Ocular Biomaterials and Implants. *Biomaterials* **2001**, *22*, 769–785. [CrossRef]
226. Salmon, D.; Roussel, L.; Gilbert, E.; Kirilov, P.; Pirot, F. *Percutaneous Penetration Enhancers Chemical Methods in Penetration Enhancement*; Springer: Berlin, Germany, 2015.
227. Qi, C.; Xu, L.; Deng, Y.; Wang, G.; Wang, Z.; Wang, L. Sericin Hydrogels Promote Skin Wound Healing with Effective Regeneration of Hair Follicles and Sebaceous Glands after Complete Loss of Epidermis and Dermis. *Biomater. Sci.* **2018**, *6*, 2859–2870. [CrossRef]
228. Baker, E.L.; Lu, J.; Yu, D.; Bonnecaze, R.T.; Zaman, M.H. Cancer Cell Stiffness: Integrated Roles of Three-Dimensional Matrix Stiffness and Transforming Potential. *Biophys. J.* **2010**, *99*, 2048–2057. [CrossRef]
229. Gkretsi, V.; Stylianopoulos, T. Cell Adhesion and Matrix Stiffness: Coordinating Cancer Cell Invasion and Metastasis. *Front. Oncol.* **2018**, *8*, 145. [CrossRef]
230. Quan, F.-S.; Kim, K.S. Medical Applications of the Intrinsic Mechanical Properties of Single Cells. *Acta Biochim. Biophys. Sin.* **2016**, *48*, 865–871. [CrossRef] [PubMed]
231. Wullkopf, L.; West, A.-K.V.; Leijnse, N.; Cox, T.R.; Madsen, C.D.; Oddershede, L.B.; Erler, J.T. Cancer Cells' Ability to Mechanically Adjust to Extracellular Matrix Stiffness Correlates with Their Invasive Potential. *Mol. Biol. Cell* **2018**, *29*, 2378–2385. [CrossRef]
232. Boyd, N.F.; Li, Q.; Melnichouk, O.; Huszti, E.; Martin, L.J.; Gunasekara, A.; Mawdsley, G.; Yaffe, M.J.; Minkin, S. Evidence That Breast Tissue Stiffness Is Associated with Risk of Breast Cancer. *PLoS ONE* **2014**, *9*, e100937. [CrossRef] [PubMed]
233. Fenner, J.; Stacer, A.C.; Winterroth, F.; Johnson, T.D.; Luker, K.E.; Luker, G.D. Macroscopic Stiffness of Breast Tumors Predicts Metastasis. *Sci. Rep.* **2014**, *4*, 5512. [CrossRef] [PubMed]
234. Olcum, M.; Ozcivici, E. Daily Application of Low Magnitude Mechanical Stimulus Inhibits the Growth of MDA-MB-231 Breast Cancer Cells in Vitro. *Cancer Cell Int.* **2014**, *14*, 102. [CrossRef]
235. Joyce, M.H.; Lu, C.; James, E.R.; Hegab, R.; Allen, S.C.; Suggs, L.J.; Brock, A. Phenotypic Basis for Matrix Stiffness-Dependent Chemoresistance of Breast Cancer Cells to Doxorubicin. *Front. Oncol.* **2018**, *8*, 337. [CrossRef]
236. Conklin, M.W.; Eickhoff, J.C.; Riching, K.M.; Pehlke, C.A.; Eliceiri, K.W.; Provenzano, P.P.; Friedl, A.; Keely, P.J. Aligned Collagen Is a Prognostic Signature for Survival in Human Breast Carcinoma. *Am. J. Pathol* **2011**, *178*, 1221–1232. [CrossRef]
237. Wen, P.Y.; Weller, M.; Lee, E.Q.; Alexander, B.M.; Barnholtz-Sloan, J.S.; Barthel, F.P.; Batchelor, T.T.; Bindra, R.S.; Chang, S.M.; Chiocca, E.A.; et al. Glioblastoma in Adults: A Society for Neuro-Oncology (SNO) and European Society of Neuro-Oncology (EANO) Consensus Review on Current Management and Future Directions. *Neuro-Oncol.* **2020**, *22*, 1073–1113. [CrossRef]
238. Abbott, N.J.; Rönnbäck, L.; Hansson, E. Astrocyte-Endothelial Interactions at the Blood-Brain Barrier. *Nat. Rev. Neurosci.* **2006**, *7*, 41–53. [CrossRef]
239. Blanchette, M.; Daneman, R. Formation and Maintenance of the BBB. *Mech. Dev.* **2015**, *138*, 8–16. [CrossRef]
240. Tajes, M.; Ramos-Fernández, E.; Weng-Jiang, X.; Bosch-Morató, M.; Guivernau, B.; Eraso-Pichot, A.; Salvador, B.; Fernàndez-Busquets, X.; Roquer, J.; Muñoz, F.J. The Blood-Brain Barrier: Structure, Function and Therapeutic Approaches to Cross It. *Mol. Membr. Biol.* **2014**, *31*, 152–167. [CrossRef] [PubMed]
241. Wolburg, H.; Wolburg-Buchholz, K.; Kraus, J.; Rascher-Eggstein, G.; Liebner, S.; Hamm, S.; Duffner, F.; Grote, E.-H.; Risau, W.; Engelhardt, B. Localization of Claudin-3 in Tight Junctions of the Blood-Brain Barrier Is Selectively Lost during Experimental Autoimmune Encephalomyelitis and Human Glioblastoma Multiforme. *Acta Neuropathol.* **2003**, *105*, 586–592. [CrossRef] [PubMed]
242. Liebner, S.; Fischmann, A.; Rascher, G.; Duffner, F.; Grote, E.H.; Kalbacher, H.; Wolburg, H. Claudin-1 and Claudin-5 Expression and Tight Junction Morphology Are Altered in Blood Vessels of Human Glioblastoma Multiforme. *Acta Neuropathol.* **2000**, *100*, 323–331. [CrossRef]
243. Bellail, A.C.; Hunter, S.B.; Brat, D.J.; Tan, C.; Van Meir, E.G. Microregional Extracellular Matrix Heterogeneity in Brain Modulates Glioma Cell Invasion. *Int. J. Biochem. Cell Biol.* **2004**, *36*, 1046–1069. [CrossRef]
244. Demuth, T.; Berens, M.E. Molecular Mechanisms of Glioma Cell Migration and Invasion. *J. Neurooncol.* **2004**, *70*, 217–228. [CrossRef]
245. Mahesparan, R.; Read, T.-A.; Lund-Johansen, M.; Skaftnesmo, K.O.; Bjerkvig, R.; Engebraaten, O. Expression of Extracellular Matrix Components in a Highly Infiltrative in Vivo Glioma Model. *Acta Neuropathol.* **2003**, *105*, 49–57. [CrossRef]
246. Payne, L.S.; Huang, P.H. The Pathobiology of Collagens in Glioma. *Mol. Cancer Res.* **2013**, *11*, 1129–1140. [CrossRef]
247. Lim, E.-J.; Suh, Y.; Yoo, K.-C.; Lee, J.-H.; Kim, I.-G.; Kim, M.-J.; Chang, J.H.; Kang, S.-G.; Lee, S.-J. Tumor-Associated Mesenchymal Stem-like Cells Provide Extracellular Signaling Cue for Invasiveness of Glioblastoma Cells. *Oncotarget* **2017**, *8*, 1438–1448. [CrossRef]
248. Gilbertson, R.J.; Rich, J.N. Making a Tumour's Bed: Glioblastoma Stem Cells and the Vascular Niche. *Nat. Rev. Cancer* **2007**, *7*, 733–736. [CrossRef]

249. Heddleston, J.M.; Li, Z.; McLendon, R.E.; Hjelmeland, A.B.; Rich, J.N. The Hypoxic Microenvironment Maintains Glioblastoma Stem Cells and Promotes Reprogramming towards a Cancer Stem Cell Phenotype. *Cell Cycle* **2009**, *8*, 3274–3284. [CrossRef]
250. Ciasca, G.; Sassun, T.E.; Minelli, E.; Antonelli, M.; Papi, M.; Santoro, A.; Giangaspero, F.; Delfini, R.; Spirito, M.D. Nano-Mechanical Signature of Brain Tumours. *Nanoscale* **2016**, *8*, 19629–19643. [CrossRef]
251. Diao, W.; Tong, X.; Yang, C.; Zhang, F.; Bao, C.; Chen, H.; Liu, L.; Li, M.; Ye, F.; Fan, Q.; et al. Behaviors of Glioblastoma Cells in in Vitro Microenvironments. *Sci. Rep.* **2019**, *9*, 85. [CrossRef]
252. Umesh, V.; Rape, A.D.; Ulrich, T.A.; Kumar, S. Microenvironmental Stiffness Enhances Glioma Cell Proliferation by Stimulating Epidermal Growth Factor Receptor Signaling. *PLoS ONE* **2014**, *9*, e101771. [CrossRef]
253. Memmel, S.; Sisario, D.; Zöller, C.; Fiedler, V.; Katzer, A.; Heiden, R.; Becker, N.; Eing, L.; Ferreira, F.L.R.; Zimmermann, H.; et al. Migration Pattern, Actin Cytoskeleton Organization and Response to PI3K-, MTOR-, and Hsp90-Inhibition of Glioblastoma Cells with Different Invasive Capacities. *Oncotarget* **2017**, *8*, 45298–45310. [CrossRef]
254. Masoumi, S.; Harisankar, A.; Gracias, A.; Bachinger, F.; Fufa, T.; Chandrasekar, G.; Gaunitz, F.; Walfridsson, J.; Kitambi, S.S. Understanding Cytoskeleton Regulators in Glioblastoma Multiforme for Therapy Design. *Drug Des. Devel.* **2016**, *10*, 2881–2897. [CrossRef]
255. Efremov, Y.M.; Dokrunova, A.A.; Efremenko, A.V.; Kirpichnikov, M.P.; Shaitan, K.V.; Sokolova, O.S. Distinct Impact of Targeted Actin Cytoskeleton Reorganization on Mechanical Properties of Normal and Malignant Cells. *Biochim. Biophys. Acta* **2015**, *1853*, 3117–3125. [CrossRef]
256. Li, Q.S.; Lee, G.Y.H.; Ong, C.N.; Lim, C.T. AFM Indentation Study of Breast Cancer Cells. *Biochem. Biophys. Res. Commun.* **2008**, *374*, 609–613. [CrossRef]
257. Lu, P.; Weaver, V.M.; Werb, Z. The Extracellular Matrix: A Dynamic Niche in Cancer Progression. *J. Cell Biol.* **2012**, *196*, 395–406. [CrossRef]
258. Miroshnikova, Y.A.; Mouw, J.K.; Barnes, J.M.; Pickup, M.W.; Lakins, J.N.; Kim, Y.; Lobo, K.; Persson, A.I.; Reis, G.F.; McKnight, T.R.; et al. Tissue Mechanics Promote IDH1-Dependent HIF1α-Tenascin C Feedback to Regulate Glioblastoma Aggression. *Nat. Cell Biol.* **2016**, *18*, 1336–1345. [CrossRef]
259. Stewart, D.C.; Rubiano, A.; Dyson, K.; Simmons, C.S. Mechanical Characterization of Human Brain Tumors from Patients and Comparison to Potential Surgical Phantoms. *PLoS ONE* **2017**, *12*, e0177561. [CrossRef]
260. Barnes, J.M.; Przybyla, L.; Weaver, V.M. Tissue Mechanics Regulate Brain Development, Homeostasis and Disease. *J. Cell Sci.* **2017**, *130*, 71–82. [CrossRef]
261. Wiranowska, M.; Rojiani, M.V. Extracellular Matrix Microenvironment in Glioma Progression. *Glioma-Explor. Its Biol. Pract. Relev.* **2011**. [CrossRef]
262. Brat, D.J.; Castellano-Sanchez, A.A.; Hunter, S.B.; Pecot, M.; Cohen, C.; Hammond, E.H.; Devi, S.N.; Kaur, B.; Van Meir, E.G. Pseudopalisades in Glioblastoma Are Hypoxic, Express Extracellular Matrix Proteases, and Are Formed by an Actively Migrating Cell Population. *Cancer Res.* **2004**, *64*, 920–927. [CrossRef]
263. Nakada, M.; Okada, Y.; Yamashita, J. The Role of Matrix Metalloproteinases in Glioma Invasion. *Front. Biosci.* **2003**, *8*, e261–e269. [CrossRef]
264. Schuler, P.J.; Bendszus, M.; Kuehnel, S.; Wagner, S.; Hoffmann, T.K.; Goldbrunner, R.; Vince, G.H. Urokinase Plasminogen Activator, UPAR, MMP-2, and MMP-9 in the C6-Glioblastoma Rat Model. *In Vivo* **2012**, *26*, 571–576.
265. Mikkelsen, T.; Yan, P.-S.; Ho, K.-L.; Sameni, M.; Sloane, B.F.; Rosenblum, M.L. Immunolocalization of Cathepsin B in Human Glioma: Implications for Tumor Invasion and Angiogenesis. *J. Neurosurg.* **1995**, *83*, 285–290. [CrossRef]
266. Kobayashi, T.; Honke, K.; Gasa, S.; Fujii, T.; Maguchi, S.; Miyazaki, T.; Makita, A. Proteolytic Processing Sites Producing the Mature Form of Human Cathepsin D. *Int. J. Biochem.* **1992**, *24*, 1487–1491. [CrossRef]
267. Rempel, S.A.; Rosenblum, M.L.; Mikkelsen, T.; Yan, P.S.; Ellis, K.D.; Golembieski, W.A.; Sameni, M.; Rozhin, J.; Ziegler, G.; Sloane, B.F. Cathepsin B Expression and Localization in Glioma Progression and Invasion. *Cancer Res.* **1994**, *54*, 6027–6031.
268. Armento, A.; Ehlers, J.; Schötterl, S.; Naumann, U. Molecular Mechanisms of Glioma Cell Motility. In *Glioblastoma*; De Vleeschouwer, S., Ed.; Codon Publications: Brisbane, Australia, 2017; ISBN 978-0-9944381-2-6.
269. Schnell, O.; Krebs, B.; Carlsen, J.; Miederer, I.; Goetz, C.; Goldbrunner, R.H.; Wester, H.-J.; Haubner, R.; Pöpperl, G.; Holtmannspötter, M.; et al. Imaging of Integrin Alpha(v)Beta(3) Expression in Patients with Malignant Glioma by [18F] Galacto-RGD Positron Emission Tomography. *Neuro-Oncol.* **2009**, *11*, 861–870. [CrossRef]
270. Li, D.; Zhao, X.; Zhang, L.; Li, F.; Ji, N.; Gao, Z.; Wang, J.; Kang, P.; Liu, Z.; Shi, J.; et al. ^{68}Ga-PRGD2 PET/CT in the Evaluation of Glioma: A Prospective Study. *Mol. Pharm.* **2014**, *11*, 3923–3929. [CrossRef]
271. Iagaru, A.; Mosci, C.; Mittra, E.; Zaharchuk, G.; Fischbein, N.; Harsh, G.; Li, G.; Nagpal, S.; Recht, L.; Gambhir, S.S. Glioblastoma Multiforme Recurrence: An Exploratory Study of (18)F FPPRGD2 PET/CT. *Radiology* **2015**, *277*, 497–506. [CrossRef]
272. Ellert-Miklaszewska, A.; Poleszak, K.; Pasierbinska, M.; Kaminska, B. Integrin Signaling in Glioma Pathogenesis: From Biology to Therapy. *Int. J. Mol. Sci.* **2020**, *21*, 888. [CrossRef]
273. Fillmore, H.L.; Chasiotis, I.; Cho, S.W.; Gillies, G.T. Atomic Force Microscopy Observations of Tumour Cell Invadopodia: Novel Cellular Nanomorphologies on Collagen Substrates. *Nanotechnology* **2002**, *14*, 73–76. [CrossRef]
274. Kim, H.-D.; Guo, T.W.; Wu, A.P.; Wells, A.; Gertler, F.B.; Lauffenburger, D.A. Epidermal Growth Factor–Induced Enhancement of Glioblastoma Cell Migration in 3D Arises from an Intrinsic Increase in Speed But an Extrinsic Matrix- and Proteolysis-Dependent Increase in Persistence. *Mol. Biol. Cell* **2008**, *19*, 4249–4259. [CrossRef]

275. Kaufman, L.J.; Brangwynne, C.P.; Kasza, K.E.; Filippidi, E.; Gordon, V.D.; Deisboeck, T.S.; Weitz, D.A. Glioma Expansion in Collagen I Matrices: Analyzing Collagen Concentration-Dependent Growth and Motility Patterns. *Biophys. J.* **2005**, *89*, 635–650. [CrossRef]
276. Yang, Y.; Motte, S.; Kaufman, L.J. Pore Size Variable Type I Collagen Gels and Their Interaction with Glioma Cells. *Biomaterials* **2010**, *31*, 5678–5688. [CrossRef]
277. Hwang, Y.-J.; Kolettis, N.; Yang, M.; Gillard, E.R.; Sanchez, E.; Sun, C.-H.; Tromberg, B.J.; Krasieva, T.B.; Lyubovitsky, J.G. Multiphoton Imaging of Actin Filament Formation and Mitochondrial Energetics of Human ACBT Gliomas. *Photochem. Photobiol.* **2011**, *87*, 408–417. [CrossRef]
278. Zhong, J.; Bach, C.T.; Shum, M.S.Y.; O'Neill, G.M. NEDD9 Regulates 3D Migratory Activity Independent of the Rac1 Morphology Switch in Glioma and Neuroblastoma. *Mol. Cancer Res.* **2014**, *12*, 264–273. [CrossRef]
279. Ardebili, S.Y.; Zajc, I.; Gole, B.; Campos, B.; Herold-Mende, C.; Drmota, S.; Lah, T.T. CD133/Prominin1 Is Prognostic for GBM Patient's Survival, but Inversely Correlated with Cysteine Cathepsins' Expression in Glioblastoma Derived Spheroids. *Radiol. Oncol.* **2011**, *45*, 102–115. [CrossRef]
280. Chigurupati, S.; Venkataraman, R.; Barrera, D.; Naganathan, A.; Madan, M.; Paul, L.; Pattisapu, J.V.; Kyriazis, G.A.; Sugaya, K.; Bushnev, S.; et al. Receptor Channel TRPC6 Is a Key Mediator of Notch-Driven Glioblastoma Growth and Invasiveness. *Cancer Res.* **2010**, *70*, 418–427. [CrossRef]
281. Chonan, Y.; Taki, S.; Sampetrean, O.; Saya, H.; Sudo, R. Endothelium-Induced Three-Dimensional Invasion of Heterogeneous Glioma Initiating Cells in a Microfluidic Coculture Platform. *Int. Biol.* **2017**, *9*, 762–773. [CrossRef] [PubMed]
282. Burdick, J.A.; Prestwich, G.D. Hyaluronic Acid Hydrogels for Biomedical Applications. *Adv. Mater. Weinh.* **2011**, *23*, H41–H56. [CrossRef] [PubMed]
283. Ouasti, S.; Donno, R.; Cellesi, F.; Sherratt, M.J.; Terenghi, G.; Tirelli, N. Network Connectivity, Mechanical Properties and Cell Adhesion for Hyaluronic Acid/PEG Hydrogels. *Biomaterials* **2011**, *32*, 6456–6470. [CrossRef] [PubMed]
284. Wang, X.; He, J.; Wang, Y.; Cui, F.-Z. Hyaluronic Acid-Based Scaffold for Central Neural Tissue Engineering. *Interface Focus* **2012**, *2*, 278–291. [CrossRef]
285. Borzacchiello, A.; Russo, L.; Malle, B.M.; Schwach-Abdellaoui, K.; Ambrosio, L. Hyaluronic Acid Based Hydrogels for Regenerative Medicine Applications. Available online: https://www.hindawi.com/journals/bmri/2015/871218/ (accessed on 17 June 2020).
286. Florczyk, S.J.; Wang, K.; Jana, S.; Wood, D.L.; Sytsma, S.K.; Sham, J.; Kievit, F.M.; Zhang, M. Porous Chitosan-Hyaluronic Acid Scaffolds as a Mimic of Glioblastoma Microenvironment ECM. *Biomaterials* **2013**, *34*, 10143–10150. [CrossRef]
287. Pedron, S.; Harley, B.A.C. Impact of the Biophysical Features of a 3D Gelatin Microenvironment on Glioblastoma Malignancy. *J. Biomed. Mater. Res. Part A* **2013**, *101*, 3404–3415. [CrossRef]
288. Wan, Y.; Mahmood, M.A.I.; Li, N.; Allen, P.B.; Kim, Y.; Bachoo, R.; Ellington, A.D.; Iqbal, S.M. Nano-Textured Substrates with Immobilized Aptamers for Cancer Cell Isolation and Cytology. *Cancer* **2012**, *118*, 1145–1154. [CrossRef]
289. Niu, C.J.; Fisher, C.; Scheffler, K.; Wan, R.; Maleki, H.; Liu, H.; Sun, Y.; A Simmons, C.; Birngruber, R.; Lilge, L. Polyacrylamide Gel Substrates That Simulate the Mechanical Stiffness of Normal and Malignant Neuronal Tissues Increase Protoporphyin IX Synthesis in Glioma Cells. *J. Biomed. Opt.* **2015**, *20*, 098002. [CrossRef]
290. Cruz-Acuña, R.; García, A.J. Synthetic Hydrogels Mimicking Basement Membrane Matrices to Promote Cell-Matrix Interactions. *Matrix Biol.* **2017**, *57–58*, 324–333. [CrossRef]
291. Trujillo, S.; Gonzalez-Garcia, C.; Rico, P.; Reid, A.; Windmill, J.; Dalby, M.J.; Salmeron-Sanchez, M. Engineered 3D Hydrogels with Full-Length Fibronectin That Sequester and Present Growth Factors. *Biomaterials* **2020**, *252*, 120104. [CrossRef]
292. Huang, C.; Holfeld, J.; Schaden, W.; Orgill, D.; Ogawa, R. Mechanotherapy: Revisiting Physical Therapy and Recruiting Mechanobiology for a New Era in Medicine. *Trends Mol. Med.* **2013**, *19*, 555–564. [CrossRef] [PubMed]
293. Tadeo, I.; Berbegall, A.P.; Escudero, L.M.; Alvaro, T.; Noguera, R. Biotensegrity of the Extracellular Matrix: Physiology, Dynamic Mechanical Balance, and Implications in Oncology and Mechanotherapy. *Front. Oncol.* **2014**, *4*, 39. [CrossRef] [PubMed]
294. Vining, K.H.; Mooney, D.J. Mechanical Forces Direct Stem Cell Behaviour in Development and Regeneration. *Nat. Rev. Mol. Cell Biol.* **2017**, *18*, 728–742. [CrossRef] [PubMed]
295. Sheridan, C. Pancreatic Cancer Provides Testbed for First Mechanotherapeutics. *Nat. Biotechnol.* **2019**, *37*, 829–831. [CrossRef]

Review

Novel Hydrogels for Topical Applications: An Updated Comprehensive Review Based on Source

Yosif Almoshari

Department of Pharmaceutics, College of Pharmacy, Jazan University, Jazan 45142, Saudi Arabia; yalmoshari@jazanu.edu.sa

Abstract: Active pharmaceutical ingredients (API) or drugs are normally not delivered as pure chemical substances (for the prevention or the treatment of any diseases). APIs are still generally administered in prepared formulations, also known as dosage forms. Topical administration is widely used to deliver therapeutic agents locally because it is convenient and cost-effective. Since earlier civilizations, several types of topical semi-solid dosage forms have been commonly used in healthcare society to treat various skin diseases. A topical drug delivery system is designed primarily to treat local diseases by applying therapeutic agents to surface level parts of the body such as the skin, eyes, nose, and vaginal cavity. Nowadays, novel semi-solids can be used safely in pediatrics, geriatrics, and pregnant women without the possibility of causing any allergy reactions. The novel hydrogels are being used in a wide range of applications. At first, numerous hydrogel research studies were carried out by simply adding various APIs in pure form or dissolved in various solvents to the prepared hydrogel base. However, numerous research articles on novel hydrogels have been published in the last five to ten years. It is expected that novel hydrogels will be capable of controlling the APIs release pattern. Novel hydrogels are made up of novel formulations such as nanoparticles, nanoemulsions, microemulsions, liposomes, self-nano emulsifying drug delivery systems, cubosomes, and so on. This review focus on some novel formulations incorporated in the hydrogel prepared with natural and synthetic polymers.

Keywords: hydrogel; novel formulations; natural polymer; synthetic polymer; topical application

1. Introduction

Generally, active pharmaceutical ingredients (API) or drugs are not supplied as pure chemical compounds (for the prevention or the treatment of any diseases). Still, APIs are commonly delivered in pharmaceutical formulations, commonly known as dosage forms [1]. It is a multistep process that involves the combination of API with other components widely known as excipients or pharmaceutical inactive ingredients and converting into a final valuable medicinal compound [2]. They have been converted into a suitable dosage formulation and delivered in various administration routes [1]. The dosage forms are commonly administered in solid, semi-solid, and liquid forms. The various solid dosage forms include tablets, capsules, and powders. The various semi-solid dosage forms include creams, gel, emulgel, ointments, paste, etc. The various types of liquid dosage forms include emulsion, injections, lotion, ocular formulations, suspension, syrup, etc. [3].

For delivering therapeutic agents locally, topical administration is mainly preferred due to its convenience and affordability. A topical drug delivery system (TDDS) is mainly meant for treating local diseases by applying the therapeutic agents to superficial body parts, including the skin, eyes, nose, and vaginal cavity. In contrast to oral administration, topical administration avoids first-pass metabolism in the liver, gastric pH variations in the stomach, and fluctuations in plasma concentrations. The other benefits of a topical drug delivery system consist of patient fulfilment and acceptance, easy and convenient application, less pain and noninvasive system, increase in API bioavailability, improved physiological and pharmacological action, as well as minimal systemic toxicity and exposure of API to non-infectious tissue/sites [4].

The production of semi-solid products has been going on for decades, and they are often used as pharmaceuticals, cosmetics, and health supplements. Typically, semi-solid dosage form (SSDF) products are administered topically or inserted into a body orifice [5]. This dosage form does have some benefits, such as applying the medications directly to the affected area and the ease of administration to patients of any age. A challenge in SSDFs is the need to deliver the API across the skin or other physical membranes of the patient to reach the desired system [6]. Since ancient times, there have been several kinds of topical semi-solid dosage forms widely used in human society for treating various skin diseases [7]. The semi-solid dosage forms used topically usually come in creams, gels, ointments, or pastes [8]. Dermatology products are available in various formulations and consistencies, but semi-solid dosage forms are the most popular kind [9].

Skin is the most important and largest organ in the body. It is responsible for a variety of necessary functions, mainly protection (from an entry of microorganisms, from the external environment, to prevent excessive water loss, etc.,) as well as to regulate body temperature [10,11]. In TDDS, skin is one of the primary and accessible organs [9]. Most topical semi-solid formulations are designed to target the skin or underlying tissues [7]. Among the three layers (stratum corneum (SC), epidermis, and dermis) of skin, the SC is primarily responsible for protecting the tissues underneath the skin. Most of the API's cannot easily pass through the skin due to the effective barrier of the SC [10]. The SC restricts the penetration of nearly all large and hydrophilic API molecules, including proteins, peptides, nucleotides, and oligonucleotides [12].

Despite their washable water bases, novel semi-solids are not greasy. Therefore, they are less irritating to the skin and more effective than conventional semi-solid dosage forms. Novel semi-solids can be used safely in pediatrics, geriatrics, and pregnant women without the possibility of causing any allergy reactions. It is expected that novel semi-solids will control the release pattern [13]. Topical dosage forms that are intended to give local or systemic effects. Conventional TDDS has significant drawbacks. Acne, alopecia, and psoriasis are all skin diseases with deep roots inside the skin. Traditional TDDS seems inadequate in treating the above skin disease due to the poor absorption of APIs in the skin.

Due to skin's poor retention, conventional topical dosage forms appear ineffective in treating these conditions. Based on patient fulfillment, security, effectiveness, feasibility, and shelf life, novel TDDS has gained popularity in recent decades. Based on the research results performed in novel TDDS, it can be concluded that transferring from conventional TDDS to novel TDDS through the use of carriers necessitates extensive research and will offer new hope for the treatment of various diseases [9].

According to the USP, gels (also known as jellies) are semi-solid systems made up of either a suspension containing small inorganic particles or organic macromolecules (primarily polymers) dissolved in a large quantity of liquid to form an infinite rigid network structure [14]. According to Rathod and Mehta [15] and also by Jeganath and Jeevitha [16], there are several ways to classify gels, including colloidal phases (inorganic and organic), solvents type (emulgel, hydrogels, organogels, and xerogels), composition (flexible and inflexible gels), and rheological property (plastic, pseudoplastic, and thixotropic gels). According to Paul et al. [17], the novel formulation can include into the following gels: (i) hydrogels, (ii) organogels, (iii) in situ gels, (iv) emulgels, (v) microgels, (vi) nanogels, and (vii) vesicular gels (liposomal gel, niosomal gel, and transferosomal gel).

A semi-solid state became advantageous for actual usage of the prescribed dose for applying in the skin. Pharmaceutical companies are gaining rapid attention or are increasingly interested in hydrogels due to their modified drug release [18]. Initially, many research studies were performed in hydrogels by simply incorporating various drugs in pure or dissolved in suitable solvents to the prepared hydrogel base. However, for the past five to ten years, numerous research articles on novel hydrogels have been published. Novel hydrogels are made up of novel formulations such as nanoparticles, nanoemulsions, microemulsions, liposomes, self-nano emulsifying drug delivery systems, cubosomes, and so on. The novel hydrogels are used for different applications.

Few review articles related to novel hydrogels have been previously published in various journals. Torres et al. [19] reviewed classification of hydrogels based on physical or chemical interactions as well as stimuli-sensitive substances. Taxol (a Paclitaxel loaded formulation) loaded with biocompatible nanocarriers (nanoparticles, microparticles, micelles, liposomes, etc.) incorporated into hydrogels prepared with natural polymers (hyaluronic acid, gellan gum, alginate, and chitosan) was reviewed by Voci et al. [20]. A chitosan and cellulose-based hydrogel for wound treatment was reviewed by Alven and Aderibigbe [21]. Michalik and Wandzik [22] reviewed chitosan-based hydrogel in the agriculture field for sustained action. Cai et al. [23] reviewed and discussed the latest findings in advanced hybrid-based hydrogel technologies combining nanostructures and microstructures, their formulation, and potential uses mainly in tissue engineering and antitumor delivery systems. The current review aims to highlight the most important developments that have come about from the increase of a variety of novel formulations containing the API embedded in different sources of polymer-based hydrogels. This review article discusses some previous research on topical novel hydrogels prepared with varying sources like natural and synthetic gelling agents.

2. Methodology

To obtain the appropriate literature, relevant keywords like hydrogels, topical hydrogels, novel topical hydrogels, etc. have been used in search engines such as Google Scholar, PubMed, etc. The information was mainly collected from research and review articles published between 2010 to 2022.

3. Hydrogels

Hydrogel refers to a three-dimensional arrangement made of hydrophilic polymers with a high capacity to interact with and retain a vast amount of water and biological fluids due to several functional grouping [like amino ($-NH_2$), the carboxylic acid ($-COOH$), a hydroxyl group ($-OH$), amide ($-CONH$), sulfo groups ($-SO_3H$)] in the polymer chains [24]. As per Lee, Kwon, and Park, the term "hydrogel" comes from an article published in 1894 and the first cross-linked network substance that showed up in publications [25]. In 1960, Wichterle O from Prague and Lim D from the Czech Republic were the first to discover hydrogels. They used synthetic polymer poly-2-hydroxyethyl methacrylate for preparing contact lenses [4,26]. Due to the high-water holding capacity, elastic in nature, compatibility with living tissue, and adaptability made hydrogels as a broad choice of use in various fields [27]. As part of the second stage, commencement in the 1970s, scientists began developing a more complex hydrogel capable of responding to pH and temperature and producing a precise response. Third-stage hydrogels are made from supramolecular inclusion complexes that are biocompatible and versatile. Hydrogels developed in the third stage of development led to creating "smart hydrogels" [28]. Since the 1980s, hydrogels have been used for various biomedical disciplines like contact lenses, absorbent cotton, suture, and cell engineering [29], and biosensor, drug delivery, cell therapy, and 3D cell culture [27] to incorporate various conventional and novel formulations [18,29].

4. Classifications of Hydrogels

Hydrogels can be classified into multiple ways with various viewpoints based on the literature. Based on phase transition like gel-sol reactions due to physical or chemical, or biochemical stimulation. The physical stimulants comprise temperature, electric fields, magnetic fields, solvent composition, light strength, and stress. In contrast, chemical stimulation includes pH, ionic strength, specific chemical compositions, and biochemical stimulation by enzymes and amino acids [24]. Based on their pore size (nanogels and microgels), based on the polymer structure (homopolymers and copolymers), based on cross-linking (physically cross-linked and chemically cross-linked), based on degradation (biodegradable, non-degradable) and their source (natural, synthetic, or hybrid), and based on physical properties (conventional, and smart) [20], the classification of hydrogel is shown in Figure 1.

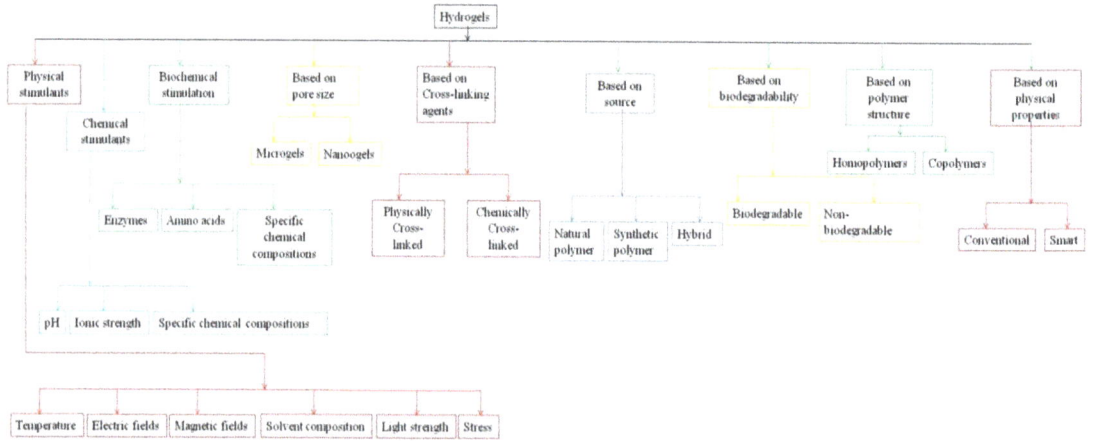

Figure 1. Classification of hydrogel.

Hydrogels provide comfortable drug delivery methods because of their tunable properties, rapid expandable degradation, and attainable preparation. [30]. A hydrogel is an excellent carrier for delivering APIs, especially to the skin. Ex vivo porcine ear skin is a widely used model due to its similarity to the human skin in terms of behavior, composition, and permeability [31].

5. Hydrogels Prepared with Natural Polymers

Natural hydrogels were prepared with natural constituents recommended for maximum biocompatibility as they are obtained from natural sources [28]. Novel formulation formulated with natural gelling agents, preparation methods, mixing with the hydrogels, and its use are shown in Table 1. Different novel formulation evaluation studies before incorporating into hydrogel are shown in Table 2. Physical evaluation studies for novel natural hydrogel formulation prepared with natural gelling agent are shown in Table 3. Novel natural hydrogel formulation in vitro, in vivo evaluation studies details are shown in Table 4.

Table 1. Novel formulation formulated with natural gelling agents, preparation methods, mixing with the hydrogels, and its use.

API	Novel Formulation	Novel Formulation Method	Gelling Agent for Making Hydrogels	Novel Formulation and Hydrogel Mixing Method	Use	References
Resveratrol	Liposomes	Film hydration method	Chitosan	Hand stirring method	Vaginal chlamydia infection	[32]
Vitamin C	Self-double-emulsifying drug delivery system	Two-step emulsification method	Xanthan gum	Mixing with a mechanical stirrer	Penetration enhance the skin	[33]
Rosmarinic acid	Nanoemulsions	Spontaneous emulsification method	Hydroxyethyl cellulose	Rosmarinic acid- nanoemulsion added to hydroxyethyl cellulose and stirred for 15 min	New anti-ageing skin products	[34]
Pentyl Gallate	Nanoemulsions	Spontaneous emulsification method	Chitosan	Chitosan added to pentyl gallate nanoemulsions	Increase skin penetration for herpis labialis	[35]
Phenytoin	Nanocapsule Nanoemulsion	Interfacial deposition Spontaneous emulsification method	Chitosan Chitosan	Chitosan was dispersed in the nanoemulsion and nanocapsule	Skin permeation and wound healing activity	[36]
Ibuprofen	Microemulsion	NM	Xanthan gum	Mixing	To enhance percutaneous delivery	[37]
Terbinafine hydrochloride	Microemulsion	NM	Chitosan, and Natrosol 250	Gelling agent added to Terbinafine hydrochloride microemulsion	Anti fungal activity	[38]
Curcumin	Microemulsion	NM	Xanthan and galactomannan	Curcumin microemulsion was added to hydrogel preparation	To increase skin penetration and anti-inflammatory activity	[39]
Baicalin	Nanocrystals	Coupling homogenization technology followed by spray-drying technology	Hyaluronic acid	Baicalin nanocrystals was added into hyaluronic acid hydrogel and mixed	To improve skin permeation	[40]
Silver sulfadiazine	Cubosome	Emulsification method	Chitosan	Cubosome incorporated into chitosan	Increasing skin permeation and for treating topical burn	[41]

NM—Not mentioned by the researcher.

Table 2. Different novel formulation evaluation studies before incorporating into hydrogel.

API-Novel Formulation	Droplet/Particle Range	Drug Content/ Entrapment Efficiency (%)	PDI	Zeta Potential (mV)	pH	Viscosity	Shape	Time	DR (%)	References
RVT-LS	100 to 200 nm (Average 158 ± 22 nm)	85 ± 2	0.077	−6.72 ± 2.47	NP	appropriate viscosity (NM)	Spherical	8 h	61	[32]
Vit C-SDEDDS	0.06 to 60.26 μm (Average 17.13 ± 2.50 μm)	NP	NP	NP	NP	NP	NP	NP	NP	[33]
ROS-NE	180.57 ± 1.82 to 224.67 ± 2.31 nm	98.59 ± 2.12 to 107.69 ± 6.28	0.123 ± 0.021 to 0.230 ± 0.036	−39.65 ± 1.53 to 46.17 ± 3.90	3.85 ± 0.07 to 4.73 ± 0.07	1.1 to 1.3 cps	NP	8 h	71.8 ± 1.98	[34]
PG-NE	164.3 ± 7.4 nm	96.2 ± 3.4	0.12 ± 0.03	−48.9 ± 2.1	5.5 ± 0.2	Near to 1 cps	NM	NP	NP	[35]
PENY-NC	161 ± 4 nm	95.2 ± 1.4	0.14 ± 0.01	−15.7 ± 0.3	5.6 ± 0.1	NP	Spherical	3 h 24 h	27 ± 1 67 ± 1	[36]
PENY-NE	125 ± 6 nm	88.7 ± 1.1	0.12 ± 0.01	−10.8 ± 0.4	5.0 ± 0.7	NP	Spherical	3 h 24 h	36 ± 2 77 ± 2	[36]
IBU-ME	14.34 ± 0.98 nm	NP	0.220 ± 0.075	NP	5.23	0.2025 ± 0.003 Pas	NP	12 h	46.78 ± 4.59	[37]
TER-ME	44.98 ± 27.34 nm	NP	NP	NP	NP	77.98 ± 0.75 cp	Spherical	NP	NP	[38]
CUR-ME	231.8 ± 7.6 nm	99.50	NP	NP	NP	NP	NP	NP	NP	[39]
BCN-NCY	189.21 ± 0.36 nm	NP	NP	NP	NP	NP	Spherical Gel TEM showed network structure	4 h	65.3 ± 3.2	[40]
SSD-CUBO	152.3 to 389.6 nm	86.05 ± 3.86 to 94.56 ± 1.40	0.25 ± 0.004 to 0.65 ± 0.45	NP	NP	NP	Cubic	NP	NP	[41]

NP: Not performed by the researchers; NM: Not mentioned by the researchers.

Table 3. Physical evaluation studies for novel hydrogel formulation prepared with natural gelling agent.

API-Novel Formulation	Hydrogel Made of	Hydrogel Concentration	Formulation Concentration	Loading Efficient (%)	Ph	Droplet SIZE (nm)	Zeta Potential	Viscosity	SEM	References
RVT-LS	CHI	2.5% w/w	20% w/w	NP	NP	NP	NP	NP	NP	[32]
Vit-C-SEDDS	XG	2%	5%	NP	5.5 ± 0.1	NP	NP	4.62 ± 0.50	The structure of Vit-C was completely destroyed in freeze-dried hydrogel	[33]
ROS-NE	HEC	1% w/v	0.1% w/v	98.50 ± 3.59 to 100.79 ± 1.98	3.83 ± 0.05 to 4.73 ± 0.07	NP	NP	NP	NP	[34]
PG-NE	CHI	2.5% w/w	0.5% w/w	94.4 ± 4.8	5.0 ± 0.3	297.0 ± 8.6	52.6 ± 0.1	NP	NP	[35]
PENY-NCY	CHI	2.75% w/v	0.025% w/v	0.24 ± 0.01 mg/gm	4.8 ± 0.1	NP	NP	24.23 ± 2.70 pasn	NP	[36]
PENY-NE	CHI	2.75% w/v	0.025% w/v	0.25 ± 0.01 mg/gm	4.7 ± 0.2	NP	NP	24.53 ± 3.71 pasn	NP	[36]
IBU-ME	XG	0.25–1%	5%	NP	NP	5.17 ± 0.01	NP	1.12 ± 0.15 to 6.80 ± 0.02	NP	[37]
TER-ME	CHI, NAT, and CAR	CHI-1%, NAT-4%, CAR-1%	1%	NP	3.04 ± 0.02	NP	NP	5044.03 ± 22.43	NP	[38]
CUR-ME	X-GAL	1.25%	NP	103.90	5.3	NP	NP	NP	Network structure	[39]
BCN-NCY	HA	0.5%, 1%, 1.5% and 2%, w/v	1%	NP	NP	NP	NP	NP	Porous structure	[40]
SSD-CUBO	CHI	NP	NP	NP	4	NP	NP	NP	NP	[41]

NP: Not performed by the researchers.

Table 4. Novel natural hydrogel formulation in vitro, in vivo evaluation studies details.

API-Formulation-Hydrogel	In Vitro Release Study		In Vitro Kinetics		In Vivo Skin Studies				Animal Used	References
	Time	Drug Released (%)	Model	Mechanism	Model Skin	Time	DR			
RVT-LS-CHI	8 h	38	NP	NP	NP	NP	NP		NP	[32]
Vit C-SEDDS-XG	6 h	72.33	Weibull model	Fickian diffusion and Case-II transport	Porcine abdominal skin	12 h	12%		NP	[33]
ROS-NE-HEC	8 h	57 ± 0.36	NP	NP	Pig ear skin	8 h	0.65 ± 0.08 µg/cm²		NP	[34]
PG-NE-CHI	24 h	Not shown	NP	NP	Porcine ear skin	NP	NP		NP	[35]
PENY-NC-CHI	12 h	43 ± 1	NP	NP	Porcine ear skin	12 h	NP		Male Wistar rats	[36]
PENY-NE-CHI	12 h	53 ± 1	NP	NP	Porcine ear skin	12 h	NP		Male Wistar rats	[36]
IBU-ME-XG	12	ME-XG-H1-23%, ME-XG-H2-16%, ME-XG-H3-14%, and ME-XG-4-11%	Zero order	NP	NP	NP	NP		Male Wistar rats	[37]
TER-ME-CHI TER-ME-NAT	7 h	8.70	Zero order	NP	NP	NP	NP		NP	[38]
CUR-ME-X-GAL	10 h	<60	Higuchi	Diffusion controlled	Porcine ear skin	NP	NP		NP	[39]
BCN-NCY-HA	6 h	0.5 and 1% CAR->95% 1.5%CAR-85.4% 2%CAR-72.3%	NP	NP	Mouse abdominal skin	12 h	NM		NP	[40]
SSD-CUBO-CHI	NP	NP	Zero order	Diffusion controlled	NP	NP	NP		NP	[41]

NP: Not performed by the researchers; NM: Not mentioned by the researchers.

5.1. Hydrogels Loaded with Liposomes

Liposomes (LS) are nanocarriers primarily composed of phospholipids and cholesterol [42]. *Chlamydia trachomatis* is the causative agent of sexually transmitted infections. The current treatment option is oral azithromycin or doxycycline, both of which have potential side effects. To reduce the side effect and effectively treat *C. trachomatis*, Jraholmen et al. [32] used natural polyphenol Resveratrol (RVT) LS incorporated into a natural CHI hydrogel. To maximize RVT's potential therapeutic activity, they used LS as the primary release medium and CHI hydrogel as a supplemental medium. Since RVT inhibits biofilm formation, LS are preferred nanocarriers for topical therapy since it does not interact with vaginal flora, and chitosan hydrogel effectively obstructs vaginal biofilms. RVT in LS preparation was found to increase RVT solubility, deliver sustained action, and enhance chemical stability, allowing for medical uses. In addition, RVT in LS formulation improves RVT's ability to bind to microbes, producing a more potent antimicrobial action even at a low dosage. The study showed that RVT-LS in CHI hydrogel could inhibit nitric oxide, the chief free radical that causes inflammation. The RVT-LS incorporated into the CHI hydrogel delivery system improved the anti-chlamydial effect of RVT at lower concentrations and highlighted the existence of a delivery system to ensure effective treatment.

5.2. Hydrogels Loaded with Self-Double-Emulsifying Drug Delivery System

Self-double-emulsifying drug delivery system (SDEDDS) is a mixture of water-soluble surfactants and water-in-oil (w/o) emulsions, which can instantaneously emulsify to water-in-oil-in-water (w/o/w) double emulsions, with water-soluble drugs present in the internal aqueous phase [43]. Vitamin C is one of the most effective antioxidants for preventing skin damage and a whitening agent. Once exposed to oxygen, basic pH, and high temperatures, vitamin C is highly unsteady, rapidly degraded, and discolored. As a result, a promising strategy for resolving these weaknesses is required to enable its clinical implementation. To overcome the above and improve the vitamin C penetration in the skin, Wang Q et al. [33] prepared a vitamin C-loaded SDEDDS, followed by blending it in a xanthan gum (XG) hydrogel. Vitamin C-loaded SDEDDS showed improved physical potency upon incorporation into hydrogels, implying that the shell is more secure to protect vitamin C from degradation, mainly from ionization solution and oxygen exposure. Incorporating vitamin C in SEDDS-based hydrogels increases the vitamin C permeation in the skin due to the bioadhesive property of XG. The oil vesicle of the SDEDDS could even act as a protective coating for vitamin C, allowing for improved vitamin C-controlled release from the SDEDDS formulation. Vitamin C encapsulation in SDEDDS, hydrogels, or even both might greatly increase vitamin C permeability and distribution inside the skin. Overall, vitamin C-incorporated SDEDDS-mixed with XG hydrogels will greatly enhance skin penetration.

5.3. Hydrogels Loaded with Microparticles

Polymeric microparticles (MP) have been extensively researched as a beneficial and novel carrier for the sustained and controlled release of vast drugs. Chronic wound healing treatment usually requires the administration of drugs at regular intervals for a more extended period. Long-term sustained release treatment might decrease the frequency of administration while maintaining drug concentration at the site of a wound. Based on this knowledge, Yasasvini S et al. [44] created simvastatin (SIM) CHI CHI-MP and incorporated them into polyvinyl alcohol (PVA) hydrogels to improve wound healing activity. After seven days, 92% of SIM was released from 5% PVA hydrogel for a dose of 2.5 mg. This SIM release from 5% PVA was correlated with the swelling index. In low dose (2.5 mg), the swelling index value was more when compared to 5 and 10 mg SIM concentration. The in vivo wound healing activity showed SIM released in controlled manner results in continuous wound healing. The combination of APIs in MP formulation incorporated into hydrogels could be best for releasing the APIs in a sustained manner and a successful topical wound healing activity from the above results.

5.4. Hydrogels Loaded with Nanoemulsion

Nanoemulsions (NE) are thermodynamically stable and spherical structures where a thin layer of emulsifying agent stabilizes the oil droplets that contain the drug. NE is getting popular for enhancing the skin permeation of lipophilic drugs [45]. Rosmarinic acid (ROS) is a natural polyphenol. Due to its antioxidant property, it can be used as anti-aging compound. But its use is limited due to poor water solubility, high instability, and poor permeability through biological barriers. To overcome the above drawbacks Marafon et al. [34] created ROS loaded NE for topical use. The pH of the ROS-NE was ranged from 3.80 to 5.80, making the formulation physically stable and suitable for topical application. The permeability studies in porcine ear skin showed that nearly 1.5 fold higher ROS was penetrated when compared to pure ROS. The absence of necrosis in keratinocyte cells indicates the formulation's safety. Their findings showed that ROS as a NE prepared with tween-80 and incorporated in HEC-based gelling agents could be used as an appropriate carrier for skin application in novel anti-aging skincare preparation.

Another example for NE encapsulated hydrogel formulation was prepared with Pentyl Gallate (PG) by Kelmann RG and colleagues [35]. PG-NE was prepared and encapsulated into a hydrogel formulation containing natural polymer CHI and a synthetic polymer Aristoflex AVC (ART). PG-NE-ART's viscosity is less than PG-NE-CHI; this increase in viscosity is due to the degree of deacetylation of CHI. The pH of both the PG-NE-CHI and PG-NE-ART was reduced after 90 days due to the hydrolysis of the triglyceride and phospholipid moieties, which resulted in the release of free fatty acids. The presence of a CHI or an ART did not affect the skin retention of PG from the NE. ART compared to the two gelling agents tested, ART appeared to be more intriguing since it initially did not affect PG-NE's droplet size and zeta potential. The PG-NE-ART also demonstrated improved intrinsic stability, viscosity, and spreadability. According to the authors, PG can be an attractive topical antiherpetic agent.

Phenytoin (PENY) is an antiepileptic drug on prolonged oral administration that results in gingival hyperplasia. Previous studies confirmed that PENY could be used successfully as a wound-healing agent, but none of the studies clearly states the permeation of PENY across the skin. Cardoso et al. [36] prepared PENY-NE and nanocapsule (NC), to which CHI was dispersed and converted into novel hydrogels, and then evaluated PENY skin penetration and wound healing activity. The in vitro release study results showed that PENY diffuses through the NE and NC and then through the hydrogel network to arrive at the release medium in a controlled manner. The higher PENY accumulation in the dermis of injured skin for PENY-NC-CHI and PENY-NE-CHI hydrogels compared to other skin layers suggested that such hydrogels might be helpful to formulations for wound treatment. When PENY-loaded NC and NE hydrogels were compared to the non-encapsulated form, the amount of PENY that reached the dermis and the receptor medium was significantly lower. It indicates that PENY as a wound-healing agent poses a low risk of systemic absorption.

5.5. Hydrogels Loaded with Microemulsion

The formation of microemulsion (ME) is straightforward, comprising an oil phase, a surfactant, a cosurfactant, and aqueous phases, and can easily be prepared through gentle stirring. ME is an encouraging nanocarrier for the topical administration of poorly water-soluble drugs [46]. Ibuprofen (IBU) is a non-steroidal anti-inflammatory poor water-soluble drug. The skin restricted its percutaneous penetration from conventional preparations. Nanocarriers play a vital role in enhancing the IBU permeation in the skin. To increase IBU percutaneous delivery, Djekic L et al. [37] created an o/w ME with IBU and incorporated it into hydrogel made up of XG. ME-XG-hydrogel prepared with the least XG (0.25%) had the most excellent drug release rate and spreadability. Artificially stimulated inflammation in male Wistar rats via intraplantar injection of carrageenan dispersed in saline. The ME-XG-H1 formulation produced approximately 65% antihyperalgesic effect and 74% antiedematous after 3 h, showing a considerable lowering in paw inflammation caused by carrageenan. The hydrogel-thickened ME could offer promise carriers with improved percutaneous delivery to prevent pain and, to a lesser extent, inflammation.

Invasive fungal infections pose a significant and growing risk to human healthiness. Terbinafine hydrochloride (TER) is an anti-fungal agent used in topical and oral forms. Celebi et al. [38] prepared TER-ME in which natural gelling agents such as CHI (1%) and natrosol 250 (4%), as well as synthetic gelling agents such as CAR-974 (1%), are dispersed. The anti-fungal activities of the TER-ME-loaded hydrogels were evaluated against a variety of strains. The anti-fungal activity was determined against three yeast species (*Candida albicans*, *Candida krusei*, and *Candida parapsilosis*), two nondermatophytic fungi (*Aspergillus niger* and *Penicillium*), and two dermatophytes (*Microsporum spp.* and *Trichophyton rubrum*). TER loaded in CHI hydrogel was the most effective against *Candida albicans* and *Candida krusei*. TER loaded in natrosol hydrogel was most effective against *Trichophyton rubrum*. Based on the findings of their study, they concluded that TER in CHI, CAR-974, natrosol hydrogel, and microemulsion formulations had similar anti-fungal activity as the marketable product (Lamisil).

Curcumin (CUR) administration via percutaneous route could be appropriate for topical and systemic therapeutic applications. Koop et al. [39] created an ME containing the hydrophobic drug CUR and incorporated it into a hydrogel combining xanthan and galactomannan (X-GAL) for topical use. About <60% of CUR was released from CUR-ME-X-GAL. The skin permeation study was carried out on porcine ear skin, and the results revealed that similar amounts of CUR (2.17 to 2.47 µg/mL) were found in the SC, epidermis, and dermis of the skin. In vivo anti-inflammatory study in male Swiss mice showed the inhibition of inflammation of 63.2% for CUR-ME-X-GAL. The researchers hypothesized that combining CUR's bioactivity with X-GAL hydrogels aided tissue restoration.

5.6. Hydrogel Loaded with Nanocrystals

Nanocrystals (NCY) are colloidal carriers of nano-sized particles stabilized in a dispersion medium by least polymeric and/or amphiphilic stabilizing agents. NCY was designed to enhance the absorption and permeation of lipophilic drugs [47]. Baicalin (BCN) cannot be used in topical applications due to its lipophilicity and poor permeation. Wei S et al. [40] attempted to prepare BCN-NCY and incorporate them into a hyaluronic acid (HA) hydrogel to improve BCN topical permeation. The 6-h in vitro release study revealed that more than 95% of BCN was released from BCN-NCY incorporated into 0.5 and 1% HA, while 85.4 and 72.3% of BCN was released from BCN-NCY incorporated into 1.5 and 2% HA, respectively. Over 12 h, the cumulative amount of BCN that permeated rat skin from BCN-NCY-hydrogel with 0.5 and 1% HA was significantly more significant than that of BCN-NCY hydrogel with 1.5 and 2% HA. This is because the viscosity of 0.5 and 1% HA gels was lower than that of 1.5 and 2% HA hydrogels. The 1% HA is relatively advantageous for preparing excellent BCN-NCY hydrogel. According to their study, HA-based NCY-hydrogel can deliver poorly water-soluble drugs topically, safely, and effectively.

5.7. Hydrogel Loaded with Cubosomes

Cubosomes (CUBO) are thermodynamically stable; self-assembled nano-sized liquid crystalline particles prepared by combining a specific ratio of lipids with suitable surfactants, water, and temperature conditions [48–50]. Silver sulfadiazine (SSD) helps treat burns, but it has drawbacks such as poor permeation into the burn wound and cytotoxicity. However, permeation can be increased and can reduce SSD cytotoxicity with the help of nanocarriers. Morsi et al. [41] prepared a CUBO based on silver sulfadiazine (SSD). They incorporated it into a CHI (1.5%), CAR 940 (1%), or a combination of CHI-CAR-based hydrogels to create cubosomal hydrogels or cubogel. The release of SSD from SSD-CUBO formulation is controlled; thereby, the cytotoxic effect of silver is avoided. Formulating SSD-CUBO into cubogels using CHI and CAR-934 showed advantages over the marketed product Dermazin. From the first day of treatment, there was no interference in the healing process as well as it being compatible with a biological fluid. Healing tissues started earlier (day 9) for SSD-cubogel than Dermazin (day 15). SSD cubogel could thus be used very effectively in administering deep second-degree burns, resulting in improved healing outcomes with few adverse effects compared to most formulations on the market.

6. Hydrogels Prepared with Synthetic Polymers

Synthetic hydrogels are prepared with synthetic gelling agents that seem to be more reproducible, have greater flexibility for adjusting their chemical or mechanical qualities, and have more closely controlled structures. However, synthetic hydrogels cannot offer similar biocompatibility to natural hydrogels [28]. Novel formulation formulated with synthetic gelling agents, preparation methods, mixing with the hydrogels, and its use are shown in Table 5. Different novel formulation evaluation studies before incorporating into hydrogel are shown in Table 6. Physical evaluation studies for novel hydrogel formulation prepared with synthetic gelling agent are shown in Table 7. Novel synthetic hydrogel formulation in vitro, in vivo evaluation studies details are shown in Table 8.

6.1. Hydrogels Loaded with Liposomes

Photothermal therapy transmitted by near-infrared light (NIR) has emerged as an attractive cancer cell treatment method and an alternative to conventional tumor therapy. Chen G et al. [51] prepared IR780 iodide (tumor-targeting photosensitizer) and IR792 perchlorate (tumor non-target photosensitizer) LS loaded in hydrogels for systematically targeted cancer photothermal therapy. Images of photosensitizers against CT-26 tumor behavior in Balb/c mice showed that compared to IR792, IR780 resulted in the most significant tumor accumulation and a strong fluorescence in tumor tissue parts for the patches applied to mice's backs. The IR780-LS in hydrogel could systematically target tumors after topical administration. IR780-LSs applied topically in hydrogel increased lung fluorescence signals, confirming IR780 to be targeted effectively against deep metastases. The anti-cancer effect in CT-26 colon tumor mice showed that IR780-LSs in hydrogel-treated mice showed outstanding anti-cancer activity (which could be attributed to a higher concentration of photosensitizers at the tumor site) are nontoxic under laser irradiation. The hydrogel's biosafety of the IR780-LS was tested in a mouse skin model. The absence of any skin reaction or toxicity during the seven-day course of therapy indicated that IR780/LP in the hydrogel applied topically were secure for the skin. IR780-LS in hydrogel forms a potent combination formulation for topical administration against tumor cells.

6.2. Hydrogels Loaded with Self-Nanoemulsifying Drug Delivery Systems

Self-nanoemulsifying drug delivery systems (SNEDDS) are homogeneous mixtures of API combined with lipids, emulsifiers, and hydrophilic co-solvents/solubilizers, which create nanosize emulsions (generally <50 nm) upon constant mixing in the aqueous phase. SNEDDS have a high capacity to solubilize lipophilic APIs, and they also improve the skin permeation [61]. Cutaneous leishmaniasis (CL) is a significantly ignored tropical skin disease that is the most widespread form of leishmaniasis, evidenced by skin lesions that can lead to ulcers, scars, impairment, and stigma [62]. Liposomal amphotericin B intravenously showed good efficacy, but treating with a topical application is preferred due to it being self-administration and cheaper. For the effective and safe treatment option, CL Lalatsa et al. [52] prepared topical buparvaquone (BQ) SNEDDS-enabled carbopol hydrogels. Treatment with BQ-SNEDDS gel indicates no skin modifications (like inflammatory conditions or skin redness); the epidermis and dermis layers were free of inflammatory cells, with no acanthosis or hyperkeratosis. After seven days, the topical application of BQ-SNEDDS gels reduced the parasite load by 99%, very similar to the intralesional administration of Glucantime. Histology studies confirmed that the Balb/c mice treated with BQ-SNEDDS hydrogels showed a reduction in parasitism and evidence of healing. In conclusion, their findings suggest that nano-enabled BQ hydrogels may offer a safe, non-invasive therapy for CL.

Table 5. Novel formulation formulated with synthetic gelling agents, preparation methods, mixing with the hydrogels, and its use.

API	Novel Formulation	Novel Formulation Method	Gelling Agent for Making Hydrogels	Formulation and Hydrogel Mixing Method	Use	References
IR780 iodide and IR792 perchlorate	LP	Thin-film hydration method	Poloxamer 407, and 188	Gelling agents added to novel formulation	Targeted tumor photothermal therapy	[51]
Buparvaquone	SNDDS	NM	Carbopol 940	Carbopol was mixed with novel formulation	Cutaneous leishmaniasis	[52]
Escin and escin β-sitosterol phytosome	PHY	NM	Carbopol 934	Hydrogel added dropwise to the novel formulation	Antihyperalgesic activity	[53]
Pentyl Gallate	NE	Spontaneous emulsification method	Aristoflex AVC	Gelling agent added to novel formulation	Increase skin penetration for herpis labialis	[35]
Simvastatin	MP	Ionic gelation method	Poly vinyl alcohol	Chemical cross linking method	Sustained SIM release and wound healing activity	[44]
5,10,15,20-tetrakis(1-methylpyridinium-4-yl)-porphyrin tetra-iodide	NP	Solvent evaporation method	Carbopol-940	Novel formulation added to CAR-940	Photodynamic applications	[31]
Phenytoin	NLC	Hot homogenization followed by ultrasonication method	Carbomer 934	CAR dispersed in the NLC suspension	Increasing the entrapment efficacy and to sustained release.	[54]
Tenoxicam	ME	NM	Carbopol 940	TEN-ME gelled with CAR-940	Arthritis	[55]
Genistein	NE	Spontaneous emulsification process	Carbopol 940	Hand stirring method	To enhance skin permeation	[56]
Terbinafine hydrochloride	ME	NM	Carbopol 974	Mixing in magnetic stirrer	Anti fungal activity	[38]
Silver sulfadiazine	CUBO	Emulsification method	Carbomer 934	Cubosome incorporated into CAR-934	For improving skin permeation and to treat topical burn	[41]
Resveratrol 3,5,4′-trihydroxy-trans-stilbene	ME	NM	Carbopol 940	CAR dispersed in novel formulation	Treatment of osteoarthritis	[57]
Articaine	NC and SLNP	NM	Aristoflex AVC	NC and SLNP was incorporated into ART hydrogel	In-vitro release studies	[58]
ketoconazole	CUBO	Hot emulsification method	Carbopol 971P	CUBO added to CAR-971P hydrogel and stirred (350 rpm)	In-vitro release and ex vivo penetration studies	[59]
Clobetasol propionate	NS	NM	Carbopol 934	CP-NS incorporated into CAR-934 hydrogel	Anti-psoriatic studies	[60]

NM: Not mentioned by the researchers.

Table 6. Different novel formulation evaluation studies before incorporating into hydrogel.

API-Novel Formulation	Droplet/Particle Range (nm)	Entrapment Efficiency (%)	PDI	Zeta Potential (mV)	pH	Viscosity	Shape	SEM	DR Time	DR %	Kinetics	References
IR 780-LS	Around 130	NP	0.185	NP	NP	NP	Spherical	Freeze dried formulation-porous sponge-like structures	NP	NP	NP	[51]
IR 792-LS	122	NP	NP	NP	NP	NP	Spherical	NP	NP	NP	NP	[52]
BQ-SEDDS	255 ± 37	NP	0.685 ± 0.085	−13.5 ± 0.2	NP	NP	Spherical	NP	NP	NP	NP	[44]
SIM-MP	between 0.5 μm <10 μm	51 ± 0.7 to 82 ± 0.3	NP	NP	NP	NP	Spherical	Rough	8 h	3.8 ± 1.1 to 9.9 ± 0.4	NP	[31]
TMPTI-NPT	118 ± 5 and 133 ± 2	55.8 ± 1.1 to 92.5 ± 3.5	0.17 ± 0.01 to 0.18 ± 0.03	−21.6 ± 1.0 to 26.7 ± 3.0	NP	NP	Spherical	NP	NP	NP	NP	[54]
PENV-NLC	121.45 ± 2.65 to 258.24 ± 6.59	55.24 ± 1.60 to 88.80 ± 4.13	0.18 ± 0.01 to 0.41 ± 0.02	−15.44 ± 0.87 to −32.26 ± 1.68	5.67 ± 0.02 to 6.49 ± 0.23	NP	Spherical	Smooth surface	48 h	73.47 ± 2.45	Higuchi model	[54]
TEN-ME	106 to 122	99	NP	Near zero	5.5 to 5.7	11,100 to 12,000 cps	Spherical	NP	NP	NP	NP	[55]
GEN-NE	GEN-NE-MCT-240 ± 28	93.00 ± 2.00	<0.25	−37 ± 4	5.8 ± 0.3	1.50 ± 0.10	NP	NP	NP	NP	NP	[56]
	GEN-NE-ODD-247 ± 23	96.00 ± 1.00	<0.25	−36 ± 4	5.9 ± 0.2	1.80 ± 0.07	NP	NP	NP	NP	NP	
SSD-CUBO	150 to 400	74.93 ± 0.903 to 92.10 ± 0.250	0.25 ± 0.004 to 0.65 ± 0.45	−17.61 to 7.41	NP	NP	Cubic	NP	NP	NP	Zero order	[41]
RTTS-ME	17.5 ± 1.4	NP	0068 ± 0.016	−11.8 ± 0.5	NP	14.2 ± 0.1 mPa s	Spherical	NP	NP	NP	NP	[57]
ART-NC	4455 ± 21	78.10	0068 ± 0005	NP	8.1 ± 1.2	NP	Spherical	Smooth surface	400 min	50	Higuchi model	[58]
ART-SLNP	2499 ± 22	65.70	0113 ± 0008	NP	7.9 ± 0.9	NP	Spherical	Smooth surface	300 min	50	NP	
KETO-CUBO	188.6 ± 5.992 to 381 ± 2.082	15.779 ± 1.23 to 72.22 ± 1.08	0.437 ± 0.032 to 0.918 ± 0.06	NP	NP	NP	Cubic	NP	24 h	67	Korsmeyer-Peppas model	[59]
CP-NS	194.27 ± 49.24 nm	56.33 ± 0.94%	0.498 ± 0.095	−21.83 ± 0.95	NP	NP	Porus and crystalline nature	Freeze dried formulation-porous sponge-like structures.	1st h 6th h 24th h	32.39 ± 0.10 55.81 ± 0.60 86.25 ± 0.28	Higuchi model	[60]

NP: Not performed by the researchers; NM: Not mentioned by the researchers.

Table 7. Physical evaluation studies for novel hydrogel formulation prepared with synthetic gelling agent.

API-Novel Formulation	Hydrogel Made of	Hydrogel Concentration	Formulation Concentration	Particle Size	PDI	ZP	Loading Efficient (%)	pH	Viscosity	References
IR 780-LS	Pol-407 and 188	NM	NM	NP	NP	NP	NP	NP	NP	[51]
Escin and escin β-sitosterol PHY	CAR-934	1%	1–5%	NP	NP	NP	NP	4.95 to 6.3	1.0 ± 0.4 to 31.7 ± 0.5 Pas	[53]
BQ-SNDDS	CAR-940	1%	2%	266 ± 59	0.609 ± 0.046	−28.7 ± 1.1	NP	NP	Appropriate for skin application (NM)	[52]
BQ-SNDDS	CAR-940	2%	2%	260 ± 35	0.758 ± 0.072	−34.5 ± 1.2	NP	NP	Appropriate for skin application (NM)	
SIM-MP	PVA	5, 7 and 9% w/v	2.5, 5, and 10 mg	NP	NP	NP	NP	NP	NP	[44]
TMPTI-NPT	CAR-940	NM	NM	NP	NP	NP	NP	5.7 to 6.6	NP	[31]
PENY-NLC	CAR-934	1% w/v	0.05%	NP	NP	NP	90 to 100	6.88 ± 0.30 and 7.27 ± 0.16	16 to 18 ps	[54]
TEN-ME	CAR-940	NM	NM	NP	NP	NP	NP	NP	NP	[55]
GEN-NE-MCT	CAR-940	0.5%	0.1 (1 mg/gm)	NP	NP	NP	92.00 ± 3.00	7	25–33 cP	[56]
GEN-NE-ODD	CAR-940	0.5%	0.1 (1 mg/gm)	NP	NP	NP	91.00 ± 6.00	7	58–64 cP	
TER-ME	CAR-940	1%	NP	NP	NP	NP	NP	3.04 ± 0.02	5044.03 ± 22.43	[38]
SSD-CUBO	CAR-934	0.5, 1, 1.5 and 2%	NP	NP	NP		76 to 91	8	925 to 982 cps at 10 rpm	[41]
RTTS-ME	CAR-940	1.50%	2%	NF	NP	NP	NP	6.7	171.1 ± 0.3 mPa·s	[57]
ATC-NC	ART	2%	20 mg/gm	463.2 ± 24.7 nm	0.190 ± 0.013	NP	NP	NP	19,554.99 Pa·s^{-1}	[58]
ATC-SLNP	ART	2%	20 mg/gm	315.3 ± 20.1 nm	0.206 ± 0.009	NP	NP	NP	22,090.23 Pa·s^{-1}	
PG-NE	ART	2% w/w	0.5% w/w	97.3 ± 2.7	5.1 ± 0.2	162.1 ± 1.1	−46.5 ± 1.3	NP	NP	[35]
KETO-CUBO	CAR-971P	1% w/w	0.2% w/w	NP	NP	NP	96.81 ± 4.50	NP	25,586.67 ± 743.32 at 1.0 rpm	[59]
CP-NS	CAR-934	NP	NP	NP	NP	NP	NP	NP	NP	[60]

NP: Not performed by the researchers; NM: Not mentioned by the researchers.

Table 8. Novel synthetic hydrogel formulation in vitro, in vivo evaluation studies details.

API-Novel Formulation-Hydrogel	In Vitro Release Study				Ex Vivo Skin Studies					Animal Used	References
	Time	Drug Released (%)	Kinetics	Mechanism	Model Skin	Time	DR (%)	Kinetics	Mechanism		
IR 780-LS-POL	12 h	>90	NP	NP	NP	NP	NP	NP	NP	CT-26 cancer bearing mice	[51]
BQ-SNDDS-CAR-940	NP	NP	NP	NP	BALB/c mouse skin	NP	NP	Zero order	Case II drug transport	BALB/c mouse	[52]
Escin and ES-PHY-CAR-934	NP	NP	NP	NP	NP	NP	NP	NP	NP	Wistar rat	[53]
SIM-MP-PVA	7 days	2.5 mg SIM-92% 5 mg SIM-60% 10 mg SIM-36%	NP	NP	NP	NP	NP	NP	NP	Wistar rat	[44]
TMPTI-NPT-CAR-940	4.5 h 24 h	20 40	Korsmeyer-Peppas model	Non-fickian diffusion	Porcine skin	24 h	Not detected	NP	NP	NP	[31]
PENY-NLC-CAR-934	48 h	51.13 ± 1.69	Korsmeyer-Peppas model	Non-fickian diffusion	NP	NP	NP	NP	NP	NP	[54]
TEN-ME-CAR-940	NP	NP	NP	NP	Laca mouse skin	24 h	64-71	NP	NP	Rat Sprague-Dawley	[55]
GEN-NE-MCT-CAR-940	8 h	NP	NP	NP	Porcine skin	In 8 h	100 µg/cm^2	NP	NP	NP	[56]
GEN-NE-ODD-CAR-940	8 h	NP	NP	NP	Porcine skin	In 8 h	150 µg/cm^2	NP	NP	NP	
SSD-CUBO-CAR-934	12 h	76 to 98	NP	NP	NP	NP	NP	NP	NP	Male adult Wister rats	[41]
RTTS-ME-CAR-940	NP	NP	NP	NP	Porcine abdominal skin	NP	NP	NP	NP	Rabbit	[57]
ATC-NC-ART	8 h	NP	Higuchi	Diffusion	NP	NP	NP	NP	NP	NP	[58]
ATC-SLNP-ART	NP	NP	NP	NP	NP	NP	NP	NP	NP	NP	
PG-NE	24	Not shown	NP	NP	Porcine ear skin	NP	NP	NP	NP	NP	[35]
KETO-CUBO-CAR-97TP	NP	NP	NP	NP	Goat skin	24 h	92.73	NP	NP	NP	[59]
CP-NS-CAR-934	NP	NP	NP	NP	NP	NP	NP	NP	NP	Male Swiss albino mice	[60]

NP: Not performed by the researchers.

6.3. Hydrogels Loaded with Phytosomes

Phytosomes (PHY) are typically made by combining APIs with phospholipids in a precise molar ratio under controlled conditions. PHY are an advanced lipid-based delivery method with a LS-like structure that can entrap various polyphenolic-based phytoconstituents to improve absorption when administered [63]. Topical escin-containing components have conventionally been used to relieve leg discomfort and heaviness caused by mild vascular circulatory disturbances and reduce bruises. However, it was unclear if escin itself could decrease pain hypersensitivity in the inflammatory process. To confirm the anti-inflammatory effect of escin, Djekic, et al. [53] and colleagues created topical hydrogels containing escin-sitosterol phytosomes (ES) and escin for antihyperalgesic activity. ES and escin-incorporated hydrogels had significant, concentration-dependent antihyperalgesic effects in a rat inflammatory pain model. CAR-934 hydrogels containing 1–2% ES can be recognized as potential topical formulations that are highly significant to hydrogels containing the same concentrations of escin. The skin irritation test on male Wistar rat skin demonstrated that it could use it safely on human skin. The above research results might imply that topical monocomponent hydrogels containing ES and escin could effectively treat inflammation-related pain. ES-incorporated hydrogels were considerably more efficient than those incorporated with escin.

6.4. Hydrogels Loaded with Nanoparticles

Alginate (ALG) hydrogel holds active compounds such as various drugs, signaling molecules, or stem cells with soft flexible gels in ALG rich in M-blocks and firm gels in ALG in rich G-blocks. A few examples of novel approaches based on alginate hydrogels are: Porous 3D hydrogel calcium alginate (Ca-ALG) has great swelling capacity in wounds, providing slow drug release. It is used to entrap cells for tissue regeneration and engineering, as physical support for cells or tissue, or as a hurdle between two media. It protects the cells from the host's immune system until it reaches the targeted area. The encapsulated fibroblasts represent an excellent example of a dual-layered structure made from alginate hydrogel with apical keratinocytes. Hydrogel film based on poly (N-vinyl caprolactam)-calcium alginate (PVCL/PV-Ca-ALG) loaded with thrombin receptor agonist peptide has shown a beneficial effect on wound healing and tissue regeneration. A relatively recent study compared a sodium alginate-acacia gum-based hydrogel loaded with zinc oxide nanoparticles (ZnO-NPs) to only ZnO-NPs by their healing effects and activity against *B. cereus* and *P. aeruginosa* [64].

Polymeric nanoparticles (NPT) are colloidal particles that are sub-micron (1 to 1000 nm) in size and contain APIs entrapped inside or adsorbed to the polymer [65]. The 5,10,15, 20-tetrakis(1-methylpyridinium-4-yl)-porphyrin tetra-iodide (TMPTI) is a well-known porphyrin that is broadly used for the inactivation of different types of microorganisms. TMPTI, on the other hand, is not widely used due to accumulation in healthy cells. Nanoparticles can rectify it; by keeping this concept, Gonzalez et al. [31], with his research team, synthesized TMPTI-NPT by a solvent evaporation method encapsulated into CAR hydrogel. A skin permeability study in domestic porcine skin showed a better TMPTI permeated deeper skin levels with no surrounding damages. It shows that the TMPTI-NPT encapsulated in hydrogel formulation does not affect the normal cells. These findings indicate that encapsulating topical pharmaceutical carriers like hydrogels may successfully treat topical skin diseases, including skin cancer.

Solid lipid nanoparticles (SLNP) are made up of solid lipid compounds, incorporated with an API in them, and coated with a surfactant to stabilize their structure [66]. Articaine (ATC) is a local anesthetic drug, but continuous usage leads to prolonged and permanent paresthesia. Melo et al. [58] aimed to create NC and SLNP containing ART, which was used topically by incorporating ART hydrogels attained a sustained release of ATC from the NC and SLNP formulation. The encapsulation of ATC in NC and SLNP results in higher cellular viability values. The formulation containing ATC-NC showed the best permeation characteristics compared to SLNP. This study opens up possibilities for the future use of nanostructured delivery systems for ATC local anesthetics.

6.5. Hydrogels Loaded with Nanostructured Lipid Carrier

Nanostructured lipid carrier (NLC) are second-generation lipid nanomaterials [67] composed of a lipid component with solid and liquid lipids distributed in an aqueous emulsifier solution [68]. PENY is an antiepileptic medication; its noticeable stimulatory effect on connective tissue development suggests that it could be used topically as a wound-healing agent. However, it is not widely used due to its low solubility, bioavailability, and ineffective distribution during topical application. Motawea et al. [54] made an effort to sustain the release of PENY by making it into nanostructured lipid carriers (NLCs) and incorporating it into a hydrogel made up of CAR-934. The intermolecular hydrogen bond interaction between the PENY-NLC components evidenced the PENY entrapped inside the lipid matrix. After 48 h, the release of PENY from the hydrogel preparations was slower than their NLC. The obtained reduction in drug release results could be believed to be due to the polymer network of CAR-934. According to their results PENY-NLCs loaded in CAR-934 hydrogels could be novel, cost-efficient, and economically feasible carriers with significant potential in topical administration.

6.6. Hydrogels Loaded with Microemulsion

Tenoxicam (TEN) is an NSAID that is used to treat arthritis. However, long-term oral administration results in peptic ulcers. To overcome this adverse effect, Goindi S et al. [55] formulated a novel TEN-ME and incorporated it into a CAR-940 hydrogel for treating arthritis. The TEN-ME formulations produce higher levels of anti-inflammatory activity better than conventional topical dosage forms with effectiveness comparable to an oral formulation. Their research concluded that TEN-ME formulations could be a viable alternatives to the TEN oral formulations.

Res, 3,5,4'-trihydroxy-trans-stilbene (RTTS) is an effective treatment for osteoarthritis. Long-term oral administration of RTTS causes kidney damage. Because osteoarthritis is a topical disease, topical therapeutic delivery is much preferred. Hu et al. [57] developed RTTS-ME and incorporated it into CAR-940 hydrogel to treat osteoarthritis in a rabbit. Due to the decrease in particle size, the penetration of RTTS was more from ME and hydrogel formulations. It appears that papain-induced osteoarthritis in animals mimicked human osteoarthritis, as papain affects cartilage, but it did not interfere with the cartilage's repair mechanism. Topical delivery of RTTS-ME incorporated in hydrogel noticeably relieved osteoarthritis signs by decreasing pro-inflammatory cytokines and improved macroscopic cartilage restoration. As a result, the hydrogel preparation could be an effective carrier for the topical administration of RTTS-ME to treat osteoarthritis.

6.7. Hydrogels Loaded with Nanoemulsion

Genistein (GEN) is an isoflavone that has recently received a lot of attention because of its effects on avoiding skin carcinoma and dermal aging on exposure to ultraviolet light. Due to its poor aqueous solubility, it cannot incorporate it into a topically applied form. To improve the solubility and permeability, Vargas et al. [56] created topical hydrogels encapsulated in GEN nanoemulsions. CAR-940-based hydrogel containing GEN-NE penetrates the skin in a sustained manner. They concluded that compared to octyldodecanol, there was detected a higher amount of GEN in the skin from the formulation composed of medium chain triglycerides (MCT) as the oily core. It is a promising delivery system for the skin.

6.8. Hydrogel Loaded with Cubosomes

The higher concentration of surfactant affects the size of the particle, entrapment efficiency of API, release from the formulation, and causes an adverse or toxic effect to the body. To overcome the above-mentioned drawbacks, Rapalli et al. [59] used the 'Quality by Design (QbD) method to create a topical hydrogel comprising ketoconazole (KETO)-entrapped cubosomes with lesser surfactant concentrations. They were successful in their research, using the QbD method, by formulating the KETO-CUBO with a lower amount of surfactant than those reported in previous works of literature.

6.9. Hydrogel Loaded with Nanosponge

Nanosponge (NS) is a novel preparation that is a nanoporous carrier with a sponge-like network shaped by hyper cross-linking polymeric materials to form three-dimensional covalent structures used to incorporate nanoparticles with a non-collapsible and porous formation [69,70]. Clobetasol propionate (CP), a potent topical corticosteroid, possesses a high therapeutic potential for psoriasis. Meanwhile, common adverse effects, such as skin degeneration, steroidal acne, skin discoloration, and allergic skin reactions, limit its utility for topical administration. Kumar et al. [60] created a CP-NS based on these observations. They incorporated it into a CAR-934 hydrogel to reduce the adverse effects and control the CP release. They successfully prepared cyclodextrin-based CP-NS, with approximately 86% of CP released after 24 h. Incorporating CP-NS with CAR-934 hydrogel enhanced its suitability for topical administration. The anti-psoriatic potential of fabricated nanoformulation was further substantiated in vivo using a mouse tail model. Histological and biochemical findings showed appreciable anti-psoriatic activity of the prepared nanogel.

7. Conclusions

The formulations' main goal is to deliver the APIs effectively, and to be toxicity-free and long-lasting. The novel formulations have benefits over traditional formulations, such as improved solubility, bioavailability, toxicity protection, improved pharmacological action, stability improvement, better tissue macrophage dispersion, sustained delivery, and protection from physiochemical deterioration. Incorporation of different APIs into novel formulations like nanoparticles, nanocapsules, liposomes, phytosomes, nanoemulsions, microemulsions, cubosomes, SLNP, NLC, nanosponge, and other novel formulations fulfill the above benefits.

Common skin diseases normally necessitate topical preparations to ensure patient compliance, while causing negligible systemic adverse effects. Treating skin diseases with a simple semi solid dosage form is an appealing goal for dermatologists, patients, and pharmaceutical companies alike. The most significant challenge in achieving this goal is attaining adequate drug penetration, while reducing side effects. Numerous obstacles exist in the treatment of the skin on a topical basis. As an outcome, drug delivery through the skin is extremely complicated. Detailed physical and chemical properties and delivery systems are needed to estimate and analyze topical preparation and improve particle properties. In several cases, in vitro, ex vivo, and in vivo animal studies are the best method to study drug penetration and analyze the clinical efficiency and effectiveness of novel topical drugs delivery systems.

Hydrogels have been extensively used in topical applications because of their excellent biocompatibility, solubility in water, and three-dimensional pore structure that fits the extracellular matrix. Hydrogel research based on sources (mainly natural and synthetic polymers) has recently gained attraction that showed outstanding physicochemical properties that could be useful in treating a wide range of skin diseases. This innovation could allow for the localized delivery of APIs via topical applications to increase the action. There has been an increase in research on novel hydrogels over the last decade. Novel hydrogels were able to enhance the penetration of APIs, minimizing the risks of percutaneous absorption. These novel hydrogels have a strong pharmacological effect against different skin diseases. The combination of novel formulations with hydrogels offers a great opening to treat several currently available skin diseases. In this review, we learned that many researchers formulated novel formulations and a few did not conduct many characteristic studies (particle size, shape, entrapment efficacy, zeta potential, viscosity, pH, in-vitro release studies, etc.) for the prepared novel formulations. The novel formulation was then primarily mixed with hydrogel preparations. All of the researchers then carried out the evaluation studies for the novel hydrogel formulation by the studies. As evidenced by animal studies, many novel hydrogels demonstrated significant activity against various topical diseases and found this one to be stable, as evidenced by stability studies. There is a strong possibility that novel hydrogels based on natural and synthetic polymers will soon enter clinical trials and the market.

Funding: No funding to declare.

Institutional Review Board Statement: Not applicable.

Informed Consent Statement: Not applicable.

Conflicts of Interest: The authors declare no conflict of interest.

Abbreviations

ALG	Alginate
API	Active pharmaceutical ingredients
ATC	Articaine
BCN	Baicalin
BQ	Buparvaquone
Ca	Calcium
CAR	Carbopol
CHI	Chitosan
CL	Cutaneous leishmaniasis
CP	Clobetasol propionate
CUBO	Cubosomes
CUR	Curcumin
ES	Escin-sitosterol
GAL	Galactomannan
GEN	Genistein
HA	Hyaluronic acid
IBU	Ibuprofen
KETO	Ketoconazole
LS	Liposomes
MP	Microparticle
NC	Nanocapsule
NCY	Nanocrystals
NE	Nanoemulsions
NLC	Nanostructured lipid carrier
NM	Not mentioned
NP	Not performed
NPT	Nanoparticle
NS	Nanosponge
PENY	Phenytoin
PG	Pentyl Gallate
PHY	Phytosomes
POL	Poloxamer
PVA	Polyvinyl alcohol
QbD	Quality by Design
ROS	Rosmarinic acid
RTTS	Res, 3,5,4'-trihydroxy-trans-stilbene
RVT	Resveratrol
SC	Stratum corneum
SDEDDS	Self-double-emulsifying drug delivery system
SIM	Simvastatin
SLNP	Solid lipid nanoparticles
SNEDDS	Self-nanoemulsifying drug delivery systems
SSD	Silver sulfadiazine
SSDF	Semi-solid dosage form
TDDS	Topical drug delivery system
TEN	Tenoxicam
TER	Terbinafine hydrochloride
TMPTI	5,10,15,20-tetrakis(1-methylpyridinium-4-yl)-porphyrin tetra-iodide
X	Xanthan
XG	Xanthan gum
ZnO	Zinc oxide

References

1. Fakhree, M.; Ahmadian, S. Pharmaceutical dosage forms: Past, present, future. In *Drug Delivery*; Nova Science Publishers, Inc.: Hauppauge, NY, USA, 2011; pp. 51–89.
2. Afrin, S.; Gupta, V. *Pharmaceutical Formulation*; StatPearls Publishing: Treasure Island, FL, USA, 2020.

3. Noordin, M.I. Advance delivery system dosage form for analgesic, their rationale, and specialty. In *Pain Relief-From Analgesics to Alternative Therapies*; IntechOpen: London, UK, 2017.
4. Singh Malik, D.; Mital, N.; Kaur, G. Topical drug delivery systems: A patent review. *Expert Opin. Ther. Pat.* **2016**, *26*, 213–228. [CrossRef] [PubMed]
5. Sharma, S.; Singh, S. Dermatological preparations, formulation and evaluation of various semi-solid dosage form. *Asian J. Pharm. Res. Dev.* **2014**, *2*, 10–25.
6. Panwar, A.; Upadhyay, N.; Bairagi, M.; Gujar, S.; Darwhekar, G.; Jain, D. Emulgel: A review. *Asian J. Pharm. Life Sci.* **2011**, *2231*, 4423.
7. Ilić, T.; Pantelić, I.; Savić, S. The implications of regulatory framework for topical semisolid drug products: From critical quality and performance attributes towards establishing bioequivalence. *Pharmaceutics* **2021**, *13*, 710. [CrossRef]
8. Bora, A.; Deshmukh, S.; Swain, K. Recent advances in semisolid dosage form. *Int. J. Pharm. Sci. Res.* **2014**, *5*, 3594–3608.
9. Sharadha, M.; Gowda, D.; Gupta, V.; Akhila, A. An overview on topical drug delivery system–updated review. *Int. J. Res. Pharm. Sci* **2020**, *11*, 368–385. [CrossRef]
10. Yu, Y.-Q.; Yang, X.; Wu, X.-F.; Fan, Y.-B. Enhancing Permeation of Drug Molecules Across the Skin via Delivery in Nanocarriers: Novel Strategies for Effective Transdermal Applications. *Front. Bioeng. Biotechnol.* **2021**, *9*, 646554. [CrossRef] [PubMed]
11. Nastiti, C.M.; Ponto, T.; Abd, E.; Grice, J.E.; Benson, H.A.; Roberts, M.S. Topical nano and microemulsions for skin delivery. *Pharmaceutics* **2017**, *9*, 37. [CrossRef]
12. Choi, H.; Kwon, M.; Choi, H.E.; Hahn, S.K.; Kim, K.S. Non-Invasive Topical Drug-Delivery System Using Hyaluronate Nanogels Crosslinked via Click Chemistry. *Materials* **2021**, *14*, 1504. [CrossRef]
13. Bharat, P.; Paresh, M.; Sharma, R.; Tekade, B.; Thakre, V.; Patil, V. A review: Novel advances in semisolid dosage forms & patented technology in semisolid dosage forms. *Int. J. PharmTech Res.* **2011**, *3*, 420–430.
14. Badola, A.; Goyal, M.; Baluni, S. Gels And Jellies A Recent Technology In Semisolids: A. *World J. Pharm. Res.* **2021**, *10*, 461–475.
15. Rathod, H.J.; Mehta, D.P. A review on pharmaceutical gel. *Int. J. Pharm. Sci.* **2015**, *1*, 33–47.
16. Jeganath, S.; Jeevitha, E. Pharmaceutical Gels and Recent Trends-A Review. *Res. J. Pharm. Technol.* **2019**, *12*, 6181–6186.
17. Paul, S.D.; Sharma, H.; Jeswani, G.; Jha, A.K. Novel gels: Implications for drug delivery. In *Nanostructures for Drug Delivery*; Elsevier: Amsterdam, The Netherlands, 2017; pp. 379–412.
18. Zielińska, A.; Eder, P.; Rannier, L.; Cardoso, J.C.; Severino, P.; Silva, A.M.; Souto, E.B. Hydrogels for modified-release drug delivery systems. *Curr. Pharm. Des.* **2021**. [CrossRef] [PubMed]
19. Bustamante-Torres, M.; Romero-Fierro, D.; Arcentales-Vera, B.; Palomino, K.; Magaña, H.; Bucio, E. Hydrogels Classification According to the Physical or Chemical Interactions and as Stimuli-Sensitive Materials. *Gels* **2021**, *7*, 182. [CrossRef]
20. Voci, S.; Gagliardi, A.; Molinaro, R.; Fresta, M.; Cosco, D. Recent Advances of Taxol-Loaded Biocompatible Nanocarriers Embedded in Natural Polymer-Based Hydrogels. *Gels* **2021**, *7*, 33. [CrossRef]
21. Alven, S.; Aderibigbe, B.A. Chitosan and Cellulose-Based Hydrogels for Wound Management. *Int. J. Mol. Sci.* **2020**, *21*, 9656. [CrossRef]
22. Michalik, R.; Wandzik, I. A mini-review on chitosan-based hydrogels with potential for sustainable agricultural applications. *Polymers* **2020**, *12*, 2425. [CrossRef]
23. Cai, M.-H.; Chen, X.-Y.; Fu, L.-Q.; Du, W.-L.; Yang, X.; Mou, X.-Z.; Hu, P.-Y. Design and development of hybrid hydrogels for biomedical applications: Recent trends in anticancer drug delivery and tissue engineering. *Front. Bioeng. Biotechnol.* **2021**, *9*. [CrossRef]
24. Bahram, M.; Mohseni, N.; Moghtader, M. An introduction to hydrogels and some recent applications. In *Emerging Concepts in Analysis and Applications of Hydrogels*; IntechOpen: London, UK, 2016.
25. Russo, E.; Villa, C. Poloxamer hydrogels for biomedical applications. *Pharmaceutics* **2019**, *11*, 671. [CrossRef]
26. Patel, K.D.; Silva, L.B.; Park, Y.; Shakouri, T.; Keskin-Erdogan, Z.; Sawadkar, P.; Cho, K.J.; Knowles, J.C.; Chau, D.Y.; Kim, H.-W. Recent advances in drug delivery systems for glaucoma treatment. *Mater. Today Nano* **2022**, 100178. [CrossRef]
27. Veloso, S.R.; Andrade, R.G.; Castanheira, E.M. Review on the advancements of magnetic gels: Towards multifunctional magnetic liposome-hydrogel composites for biomedical applications. *Adv. Colloid Interface Sci.* **2021**, *288*, 102351. [CrossRef] [PubMed]
28. Harrison, I.P.; Spada, F. Hydrogels for atopic dermatitis and wound management: A superior drug delivery vehicle. *Pharmaceutics* **2018**, *10*, 71. [CrossRef] [PubMed]
29. Singh, G.; Lohani, A.; Bhattacharya, S.S. Hydrogel as a novel drug delivery system: A review. *J. Fundam. Pharm. Res* **2014**, *2*, 35–48.
30. Gallo, E.; Diaferia, C.; Rosa, E.; Smaldone, G.; Morelli, G.; Accardo, A. Peptide-Based Hydrogels and Nanogels for Delivery of Doxorubicin. *Int. J. Nanomed.* **2021**, *16*, 1617. [CrossRef]
31. Gonzalez-Delgado, J.A.; Castro, P.M.; Machado, A.; Araujo, F.; Rodrigues, F.; Korsak, B.; Ferreira, M.; Tome, J.P.; Sarmento, B. Hydrogels containing porphyrin-loaded nanoparticles for topical photodynamic applications. *Int. J. Pharm.* **2016**, *510*, 221–231. [CrossRef]
32. Jøraholmen, M.W.; Johannessen, M.; Gravningen, K.; Puolakkainen, M.; Acharya, G.; Basnet, P.; Škalko-Basnet, N. Liposomes-In-Hydrogel Delivery System Enhances the Potential of Resveratrol in Combating Vaginal Chlamydia Infection. *Pharmaceutics* **2020**, *12*, 1203. [CrossRef]

33. Wang, Q.; Zhang, H.; Huang, J.; Xia, N.; Li, T.; Xia, Q. Self-double-emulsifying drug delivery system incorporated in natural hydrogels: A new way for topical application of vitamin C. *J. Microencapsul.* **2018**, *35*, 90–101. [CrossRef]
34. Marafon, P.; Fachel, F.N.S.; Dal Prá, M.; Bassani, V.L.; Koester, L.S.; Henriques, A.T.; Braganhol, E.; Teixeira, H.F. Development, physico-chemical characterization and in-vitro studies of hydrogels containing rosmarinic acid-loaded nanoemulsion for topical application. *J. Pharm. Pharmacol.* **2019**, *71*, 1199–1208. [CrossRef]
35. Kelmann, R.G.; Colombo, M.; Nunes, R.J.; Simões, C.M.; Koester, L.S. Nanoemulsion-loaded hydrogels for topical administration of pentyl gallate. *Aaps Pharmscitech* **2018**, *19*, 2672–2678. [CrossRef]
36. Cardoso, A.M.; de Oliveira, E.G.; Coradini, K.; Bruinsmann, F.A.; Aguirre, T.; Lorenzoni, R.; Barcelos, R.C.S.; Roversi, K.; Rossato, D.R.; Pohlmann, A.R. Chitosan hydrogels containing nanoencapsulated phenytoin for cutaneous use: Skin permeation/penetration and efficacy in wound healing. *Mater. Sci. Eng. C* **2019**, *96*, 205–217. [CrossRef] [PubMed]
37. Djekic, L.; Martinovic, M.; Stepanović-Petrović, R.; Micov, A.; Tomić, M.; Primorac, M. Formulation of hydrogel-thickened nonionic microemulsions with enhanced percutaneous delivery of ibuprofen assessed in vivo in rats. *Eur. J. Pharm. Sci.* **2016**, *92*, 255–265. [CrossRef] [PubMed]
38. Çelebi, N.; Ermiş, S.; Özkan, S. Development of topical hydrogels of terbinafine hydrochloride and evaluation of their antifungal activity. *Drug Dev. Ind. Pharm.* **2015**, *41*, 631–639. [CrossRef] [PubMed]
39. Koop, H.S.; de Freitas, R.A.; de Souza, M.M.; Savi-Jr, R.; Silveira, J.L.M. Topical curcumin-loaded hydrogels obtained using galactomannan from Schizolobium parahybae and xanthan. *Carbohydr. Polym.* **2015**, *116*, 229–236. [CrossRef] [PubMed]
40. Wei, S.; Xie, J.; Luo, Y.; Ma, Y.; Tang, S.; Yue, P.; Yang, M. Hyaluronic acid based nanocrystals hydrogels for enhanced topical delivery of drug: A case study. *Carbohydr. Polym.* **2018**, *202*, 64–71. [CrossRef] [PubMed]
41. Morsi, N.M.; Abdelbary, G.A.; Ahmed, M.A. Silver sulfadiazine based cubosome hydrogels for topical treatment of burns: Development and in vitro/in vivo characterization. *Eur. J. Pharm. Biopharm.* **2014**, *86*, 178–189. [CrossRef]
42. Furlani, F.; Rossi, A.; Grimaudo, M.A.; Bassi, G.; Giusto, E.; Molinari, F.; Lista, F.; Montesi, M.; Panseri, S. Controlled Liposome Delivery from Chitosan-Based Thermosensitive Hydrogel for Regenerative Medicine. *Int. J. Mol. Sci.* **2022**, *23*, 894. [CrossRef]
43. Bhattacharjee, A.; Verma, S.; Verma, P.R.P.; Singh, S.K.; Chakraborty, A. Fabrication of liquid and solid self-double emulsifying drug delivery system of atenolol by response surface methodology. *J. Drug Deliv. Sci. Technol.* **2017**, *41*, 45–57. [CrossRef]
44. Yasasvini, S.; Anusa, R.; VedhaHari, B.; Prabhu, P.; RamyaDevi, D. Topical hydrogel matrix loaded with Simvastatin microparticles for enhanced wound healing activity. *Mater. Sci. Eng. C* **2017**, *72*, 160–167. [CrossRef]
45. Rashid, S.A.; Bashir, S.; Naseem, F.; Farid, A.; Rather, I.A.; Hakeem, K.R. Olive Oil Based Methotrexate Loaded Topical Nanoemulsion Gel for the Treatment of Imiquimod Induced Psoriasis-like Skin Inflammation in an Animal Model. *Biology* **2021**, *10*, 1121. [CrossRef]
46. Hung, W.-H.; Chen, P.-K.; Fang, C.-W.; Lin, Y.-C.; Wu, P.-C. Preparation and Evaluation of Azelaic Acid Topical Microemulsion Formulation: In Vitro and In Vivo Study. *Pharmaceutics* **2021**, *13*, 410. [CrossRef] [PubMed]
47. Im, S.H.; Jung, H.T.; Ho, M.J.; Lee, J.E.; Kim, H.T.; Kim, D.Y.; Lee, H.C.; Choi, Y.S.; Kang, M.J. Montelukast nanocrystals for transdermal delivery with improved chemical stability. *Pharmaceutics* **2020**, *12*, 18. [CrossRef] [PubMed]
48. Khan, S.; Jain, P.; Jain, S.; Jain, R.; Bhargava, S.; Jain, A. Topical delivery of erythromycin through cubosomes for acne. *Pharm. Nanotechnol.* **2018**, *6*, 38–47. [CrossRef] [PubMed]
49. Sharma, P.; Dhawan, S.; Nanda, S. Cubosome: A Potential Liquid Crystalline Carrier System. *Curr. Pharm. Des.* **2020**, *26*, 3300–3316. [CrossRef]
50. Kaur, S.D.; Singh, G.; Singh, G.; Singhal, K.; Kant, S.; Bedi, N. Cubosomes as Potential Nanocarrier for Drug Delivery: A Comprehensive Review. *J. Pharm. Res. Int.* **2021**, *33*, 118–135. [CrossRef]
51. Chen, G.; Ullah, A.; Xu, G.; Xu, Z.; Wang, F.; Liu, T.; Su, Y.; Zhang, T.; Wang, K. Topically applied liposome-in-hydrogels for systematically targeted tumor photothermal therapy. *Drug Deliv.* **2021**, *28*, 1923–1931. [CrossRef]
52. Lalatsa, A.; Statts, L.; de Jesus, J.A.; Adewusi, O.; Dea-Ayuela, M.A.; Bolas-Fernandez, F.; Laurenti, M.D.; Passero, L.F.D.; Serrano, D.R. Topical buparvaquone nano-enabled hydrogels for cutaneous leishmaniasis. *Int. J. Pharm.* **2020**, *588*, 119734. [CrossRef]
53. Djekic, L.; Čalija, B.; Micov, A.; Tomić, M.; Stepanović-Petrović, R. Topical hydrogels with escin β-sitosterol phytosome and escin: Formulation development and in vivo assessment of antihyperalgesic activity. *Drug Dev. Res.* **2019**, *80*, 921–932. [CrossRef]
54. Motawea, A.; Borg, T.; Abd El-Gawad, A.E.-G.H. Topical phenytoin nanostructured lipid carriers: Design and development. *Drug Dev. Ind. Pharm.* **2018**, *44*, 144–157. [CrossRef]
55. Goindi, S.; Narula, M.; Kalra, A. Microemulsion-based topical hydrogels of tenoxicam for treatment of arthritis. *Aaps Pharmscitech* **2016**, *17*, 597–606. [CrossRef]
56. De Vargas, B.A.; Bidone, J.; Oliveira, L.K.; Koester, L.S.; Bassani, V.L.; Teixeira, H.F. Development of topical hydrogels containing genistein-loaded nanoemulsions. *J. Biomed. Nanotechnol.* **2012**, *8*, 330–336. [CrossRef] [PubMed]
57. Hu, X.-B.; Kang, R.-R.; Tang, T.-T.; Li, Y.-J.; Wu, J.-Y.; Wang, J.-M.; Liu, X.-Y.; Xiang, D.-X. Topical delivery of 3, 5, 4'-trimethoxy-trans-stilbene-loaded microemulsion-based hydrogel for the treatment of osteoarthritis in a rabbit model. *Drug Deliv. Transl. Res.* **2019**, *9*, 357–365. [CrossRef] [PubMed]
58. Melo, N.F.S.d.; Campos, E.V.R.; Franz-Montan, M.; Paula, E.D.; Silva, C.M.G.d.; Maruyama, C.R.; Stigliani, T.P.; Lima, R.D.; Araújo, D.R.d.; Fraceto, L.F. Characterization of articaine-loaded poly (ε-caprolactone) nanocapsules and solid lipid nanoparticles in hydrogels for topical formulations. *J. Nanosci. Nanotechnol.* **2018**, *18*, 4428–4438. [CrossRef] [PubMed]

59. Rapalli, V.K.; Banerjee, S.; Khan, S.; Jha, P.N.; Gupta, G.; Dua, K.; Hasnain, M.S.; Nayak, A.K.; Dubey, S.K.; Singhvi, G. QbD-driven formulation development and evaluation of topical hydrogel containing ketoconazole loaded cubosomes. *Mater. Sci. Eng. C* **2021**, *119*, 111548. [CrossRef]
60. Kumar, S.; Prasad, M.; Rao, R. Topical delivery of clobetasol propionate loaded nanosponge hydrogel for effective treatment of psoriasis: Formulation, physicochemical characterization, antipsoriatic potential and biochemical estimation. *Mater. Sci. Eng. C* **2021**, *119*, 111605. [CrossRef]
61. Ponto, T.; Latter, G.; Luna, G.; Leite-Silva, V.R.; Wright, A.; Benson, H.A. Novel Self-Nano-Emulsifying Drug Delivery Systems Containing Astaxanthin for Topical Skin Delivery. *Pharmaceutics* **2021**, *13*, 649. [CrossRef]
62. Akuffo, R.; Sanchez, C.; Chicharro, C.; Carrillo, E.; Attram, N.; Mosore, M.-T.; Yeboah, C.; Kotey, N.K.; Boakye, D.; Ruiz-Postigo, J.-A. Detection of cutaneous leishmaniasis in three communities of Oti Region, Ghana. *PLoS Negl. Trop. Dis.* **2021**, *15*, e0009416. [CrossRef]
63. Alharbi, W.S.; Almughem, F.A.; Almehmady, A.M.; Jarallah, S.J.; Alsharif, W.K.; Alzahrani, N.M.; Alshehri, A.A. Phytosomes as an emerging nanotechnology platform for the topical delivery of bioactive phytochemicals. *Pharmaceutics* **2021**, *13*, 1475. [CrossRef]
64. Barbu, A.; Neamtu, B.; Zăhan, M.; Iancu, G.M.; Bacila, C.; Mireșan, V. Current trends in advanced alginate-based wound dressings for chronic wounds. *J. Pers. Med.* **2021**, *11*, 890. [CrossRef]
65. Prabha, A.S.; Dorothy, R.; Jancirani, S.; Rajendran, S.; Singh, G.; Kumaran, S.S. Recent advances in the study of toxicity of polymer-based nanomaterials. In *Nanotoxicity*; Elsevier: Amsterdam, The Netherlands, 2020; pp. 143–165.
66. Musielak, E.; Feliczak-Guzik, A.; Nowak, I. Synthesis and Potential Applications of Lipid Nanoparticles in Medicine. *Materials* **2022**, *15*, 682. [CrossRef]
67. Gundogdu, E.; Demir, E.-S.; Ekinci, M.; Ozgenc, E.; Ilem-Ozdemir, D.; Senyigit, Z.; Gonzalez-Alvarez, I.; Bermejo, M. An Innovative Formulation Based on Nanostructured Lipid Carriers for Imatinib Delivery: Pre-Formulation, Cellular Uptake and Cytotoxicity Studies. *Nanomaterials* **2022**, *12*, 250. [CrossRef] [PubMed]
68. Varrica, C.; Carvalheiro, M.; Faria-Silva, C.; Eleutério, C.; Sandri, G.; Simões, S. Topical Allopurinol-Loaded Nanostructured Lipid Carriers: A Novel Approach for Wound Healing Management. *Bioengineering* **2021**, *8*, 192. [CrossRef] [PubMed]
69. Ghose, A.; Nabi, B.; Rehman, S.; Md, S.; Alhakamy, N.A.; Ahmad, O.A.; Baboota, S.; Ali, J. Development and Evaluation of Polymeric Nanosponge Hydrogel for Terbinafine Hydrochloride: Statistical Optimization, In Vitro and In Vivo Studies. *Polymers* **2020**, *12*, 2903. [CrossRef] [PubMed]
70. Mashaqbeh, H.; Obaidat, R.; Al-Shar'i, N. Evaluation and Characterization of Curcumin-β-Cyclodextrin and Cyclodextrin-Based Nanosponge Inclusion Complexation. *Polymers* **2021**, *13*, 4073. [CrossRef] [PubMed]

MDPI
St. Alban-Anlage 66
4052 Basel
Switzerland
Tel. +41 61 683 77 34
Fax +41 61 302 89 18
www.mdpi.com

Gels Editorial Office
E-mail: gels@mdpi.com
www.mdpi.com/journal/gels

www.ingramcontent.com/pod-product-compliance
Lightning Source LLC
LaVergne TN
LVHW070149100526
838202LV00015B/1922